ROUTLEDGE HANDBOOK ON TOURISM AND SMALL ISLAND STATES IN THE PACIFIC

This timely handbook critically examines the development and role of tourism in small Pacific Island states located across Melanesia, Micronesia and Polynesia. The volume presents an expansive evaluation of current issues, challenges and potentialities for the 13 self-governing states.

Interdisciplinary in coverage and borne of a varied and international authorship, this handbook incorporates 27 specifically commissioned and original contributions. Structured into four thematic sections and embellished with insightful tables and illustrations throughout, the overarching ethos of this volume is to contribute to framing the role of tourism, tourism development and the tourism industry within the context of self-governing Pacific Island states faced with the challenge of pursuing an independent path of development. In doing so, the work highlights and deciphers various tourism development perplexities in the Pacific, examining closely the intersecting sociocultural, geopolitical, environmental, organizational, operational and strategic challenges. This volume, thus, discusses a range of issues: facilitators and inhibitors of tourism growth and development; climate change, ecological concerns, and ecotourism; non-tourism and undertourism; crisis management and the COVID-19 virus; transportation and tourism infrastructural concerns; tourism policy and planning (including tourism governance); sectoral links between tourism, food and agriculture; gender and micro-entrepreneurship; community management and participation; cultural and natural heritage sites; and the handicraft industry. The work pays critical attention to the various trajectories of sustainable tourism and the United Nations Sustainable Development Goals. Despite the many challenges and concerns raised, the book implicates the importance of good governance, progressive post-COVID-19 recovery strategies and directives, and creative and imaginative options in the successful development, re-development and advancement of tourism.

As a definitive reference resource for this subject area, this handbook will be of great interest to students, researchers and academics within tourism, development studies, geography, Pacific studies, sustainability and environmental studies.

Marcus L. Stephenson is Professor of Tourism and Hospitality Management and Dean of the School of Hospitality at Service Management at Sunway University (Malaysia). Prior to this appointment in October 2017, he was Professor and Head of the School of Tourism and Hospitality Management at the University of the South Pacific (Fiji). He is the co-author of *Tourism and Citizenship: Rights, Freedoms and Responsibilities in the Global Order* (2014, Routledge) and co-editor of *International Tourism Development and the Gulf Cooperation Council States: Challenges and Opportunities* (2017, Routledge). He has conducted tourism research in the Caribbean, Middle East, Pacific Island states, and the United Kingdom and has published extensively on the sociology of tourism.

ROUTLEDGE HANDBOOK ON TOURISM AND SMALL ISLAND STATES IN THE PACIFIC

Edited by Marcus L. Stephenson

Routledge
Taylor & Francis Group
LONDON AND NEW YORK

Cover image: Getty Images

First published 2023

by Routledge
4 Park Square, Milton Park, Abingdon, Oxon OX14 4RN

and by Routledge
605 Third Avenue, New York, NY 10158

Routledge is an imprint of the Taylor & Francis Group, an informa business

British Library Cataloguing-in-Publication Data
A catalogue record for this book is available from the British Library

Library of Congress Cataloging-in-Publication Data
Names: Stephenson, Marcus L., editor.
Title: Routledge handbook on tourism and small island states in the Pacific/edited by Marcus L. Stephenson.
Description: First Edition. | New York: Routledge, 2023. | Includes bibliographical references and index.
Identifiers: LCCN 2022029224 (print) | LCCN 2022029225 (ebook) | ISBN 9780367030322 (Hardback) | ISBN 9781032323909 (Paperback) | ISBN 9780429019968 (eBook)
Subjects: LCSH: Tourism–Pacific Area. | Tourism–Melanesia. | Tourism–Micronesia (Federated States) | Tourism–Polynesia.
Classification: LCC G155.P25 R68 2023 (print) | LCC G155.P25 (ebook) | DDC 338.4/79196–dc23/eng20221129
LC record available at https://lccn.loc.gov/2022029224
LC ebook record available at https://lccn.loc.gov/2022029225

ISBN: 978-0-367-03032-2 (hbk)
ISBN: 978-1-032-32390-9 (pbk)
ISBN: 978-0-429-01996-8 (ebk)

DOI: 10.4324/9780429019968

Typeset in Bembo
by Deanta Global Publishing Services, Chennai, India

In Memory of Professor David Harrison

CONTENTS

FIGURES

TABLES

CONTRIBUTORS

Nazia Ali is Senior Lecturer in Event and Leisure Management and Course Leader for MSc Sports Management at the Royal Docks School of Business and Law, University of East London (UK). She is a graduate of Sociology from Middlesex University (UK) and received her PhD from the University of Bedfordshire (UK) in Tourism Studies. Nazia is a sociologist, and her main subject expertise concerns matters of identity as they relate to the business of events, sports and tourism spaces. She has published on diaspora tourism mobilities, issues of race and racism in the experience economies of tourism and sport, and advanced qualitative approaches in research.

Alberto Amore is an Assistant Professor in Geography at the University of Oulu (Finland). Alberto was a former Senior Lecturer in Tourism and Event Management at Solent University, Southampton (UK) and Sustainability Champion for the Solent Business School. He holds a PhD in Management from the University of Canterbury (New Zealand) and has an MA in Tourism Territory and Local Development (Universita degli Studi di Milano-Bicocca, Italy). His research interests include urban planning, urban tourism and urban regeneration, post-disaster governance and planning for tourism, destination planning, destination resilience and tourism policymaking. He is Fellow of the Royal Geographical Society and the Regional Studies Association and Teaching Champion for the Tourism Education Future Initiative.

Vijaya Malar Arumugam is Senior Teaching Fellow at the School of Hospitality and Service Management, Sunway University (Malaysia). Vijaya obtained her MA in Human Resource Management and Industrial Relations from the University of Newcastle (Australia). She has extensive experience in teaching at tertiary levels and was the former Chair for the School Teaching and Learning Committee. She is actively involved as an internal auditor for Sunway University and has participated in attaining internal and external research grants. Her research interests are in the culture of health, human resource management, organisational behaviour and Islamic tourism.

Chris Ballard teaches and participates in research on history, archaeology and anthropology at the Australian National University in Canberra (Australia). He has more than 35 years of experience working with Indigenous communities, mostly in Indonesia and the Pacific (Papua New Guinea and Vanuatu), on a range of issues, including agricultural systems and colonial encounters, land rights and human rights, oral traditions and cultural heritage, and disasters and epidemics. Since 2016, he has been working with the Living Heritage Unit of the United Nations Educational, Scientific and Cultural Organization (UNESCO) on the relationship between culture and emergencies, including climate change, conflict and disaster.

Madelene Blaer (née McWha) is the Visitor Economy Programs and Pathways Manager at Victoria University (Australia), and previously a Lecturer in the Graduate Tourism Program at Monash University (Australia). Her research expertise is in tourism and media and digital technology applications. She is interested in how these can support and intersect with sustainable and responsible practices in travel and tourism. Madelene's PhD provided insights into how digital innovations in publishing and online interactions dramatically changed the contemporary travel writing profession and experience. She is Vice President on the board of the Travel and Tourism Research Association (TTRA) Asia Pacific Chapter. Madelene has published extensively in the field of tourism.

Anna Carr is Associate Professor at the University of Otago (New Zealand). Anna has Ngapuhi, Ngaruahine, Ngati Ruanui and Scottish ancestry. Prior to academia, she was a tourism owner-operator of two adventure tourism businesses. She specializes in research that explores cultural landscapes, protected area management and Indigenous tourism development. Outside of academia, Anna has served on various boards, including Te Ana Māori Rock Art Centre in Timaru (Aotearoa New Zealand) and as a government ministerial appointee to a conservation board. She has worked with local communities and Indigenous researchers in Samoa, Papua New Guinea, Fiji, North America, Canada and Scandinavia.

Jenny Cave is an Associate at the Tourism CoLab, catalyzing regenerative tourism and community engagement through education, consultancy and research. She holds adjunct roles in tourism development at Massey University (New Zealand) and Swansea University (Wales). Jenny was awarded a PhD in the field of tourism in 2009 from the University of Waikato (New Zealand). She has long-standing practice experience and research interest in community empowerment and co-design, born of long-term research partnerships with island communities in the Pacific and a global career in cultural industries, attractions design and evaluation.

Kamelia Chaichi is a Senior Lecturer in the field of Hospitality Management at the School of Hospitality and Service Management at Sunway University (Malaysia). She is a PhD holder and started her academic career in 2013, teaching undergraduate and postgraduate programmes. Prior to joining academia, she had five years of experience in the industry. As a quantitative and qualitative researcher, Kamelia has published in international journals and her research concerns the fields of hospitality, tourism and human resource management, including service development in the hospitality industry.

Joseph M. Cheer (PhD) is a Professor at the Center for Tourism Research, Wakayama University (Japan), board member of PATA (Pacific Asia Travel Association) and Co-Editor-in-Chief of the journal *Tourism Geographies*. He is Adjunct Professor, Auckland University of Technology (New Zealand) and Visiting Professor, UCSI University (Malaysia). His recent collaborative book projects include: *Global Tourism & COVID-19: Implications for Theory and Practice* (2022) and *Recentering Tourism Geographies in the Asian Century* (2022). Joseph is an *Australian Research Council* (ARC) Linkage Project grant recipient with colleagues at the University of Melbourne (Australia) examining Aboriginal futures through tourism. He is a recent *Australia-Japan Foundation* grant recipient, benchmarking rural tourism for community resilience in Australia and Japan. He is a board member of the International

Geographical Union Tourism Commission, Critical Tourism Studies Asia Pacific and American Association of Geographers, Recreation, Tourism and Sport.

Daniel Ka Leong Chong is Associate Professor and Associate Dean (International) at the School of Hospitality and Service Management, Sunway University (Malaysia). He has more than 16 years of experience in hospitality education and training. He has a doctorate in the field of business administration with a major in branding, awarded by Asia e University (Malaysia). He is a certified hotelier, sommelier, and barista who is actively involved in Malaysia's food and beverage industry through industrial projects and continuing education. He has significant experience in the field of management and training and consultancy in the food and beverage industry, and represented multinational hotel groups and multinational franchised restaurants. Daniel has published work concerning sustainable practices, challenges and opportunities confronting Asia's tourism, hospitality and food industries.

Rosemarie Fili Grover gained a Master of Management Studies in Tourism from the University of Waikato (New Zealand) in 2016. She was crowned Miss Heilala ('Miss Tonga') in 2013–2014 and became the Brand Ambassador for the Kingdom of Tonga. Rosemarie's research interests include cultural identity, comparative beauty standards between Tonga and the West, and the role of culture in Tongan business culture.

C. Michael Hall is a Professor at the University of Canterbury, Christchurch (New Zealand); Visiting Professor and Docent in Geography, University of Oulu (Finland); Guest Professor, Department of Service Management and Service Studies, Lund University, Helsingborg (Sweden); and a Visiting Professor in the Department of Organisation and Entrepreneurship, School of Business and Economics, Linnaeus University, Kalmar (Sweden) and the Centre for Research and Innovation in Tourism, Taylor's University, Subang Jaya (Malaysia).

Nicholas Halter is a Lecturer in history at the University of the South Pacific. He is an Australian historian who has lived and worked in Micronesia and Fiji. Born in Sydney, he studied history at the University of Wollongong and the Australian National University. Since 2016, he has lectured in Pacific history and historiography at the University of the South Pacific (Fiji). His latest book, entitled *Australian Travellers in the South Seas* (ANU Press 2021), offers a wide-ranging survey of Australian engagement with the Pacific Islands in the late nineteenth and early twentieth centuries.

David Harrison was a sociologist and an anthropologist. Professor Harrison held positions at the University of Sussex (UK), London Metropolitan University (UK) and the University of the South Pacific in Fiji. David was a Fellow of the International Academy for the Study of Tourism. David's work primarily examines society and its interaction with modernity and tourism development. In collaboration with Professor Richard Sharpley, David recently published the edited book, *A Research Agenda for Tourism and Development*, 2019. In semi-retirement, he was still very active, and he held appointments as Professor of Tourism, Middlesex University (UK), Visiting Senior Research Fellow, Kings College London (UK), Adjunct Professor of Tourism, University of the South Pacific (Fiji) and Visiting Professor, University of Surrey (UK). Sadly, David passed in April 2021. He leaves a huge legacy of excellent tourism scholarship.

Jeremy M. Hills is a Professor in the Office of the Deputy Vice-Chancellor (Research, Innovation & International) at the University of the South Pacific (Fiji). He was previously Director of the Institute of Marine Resources at the same university. He is a Chartered Environmentalist with more than 25 years of coastal and ocean-based research and consultancy, working in more than 50 countries across the world. Jeremy has worked for many actors including international institutions such as the European Union, UNESCO, United Nations Development Programme (UNDP), United Nations Environment Programme (UNEP), United Nations Economic and Social Commission for Asia and the Pacific (UNESCAP), World Bank, Asian Development Bank, bilateral aid organisations and a range of other non-governmental and commercial organizations. His research is on ocean policy and governance in developing countries, with a particular focus on development assistance, financing and the blue economy.

Emma Hughes is a Research Associate in Development Studies at Massey University and currently works at Allen + Clarke in New Zealand. She completed her PhD at the Institute of Development Studies at Massey University (New Zealand) examining community development and tourism from community perspectives. This focused on the impacts of corporate-led community development on local development outcomes and community wellbeing. She has collaborated with Roskilde University in Denmark in examining the commodification of tourist compassion to 'do good' for communities. Subsequent research has focused on how the tourism industry is engaging with the Sustainable Development Goals (SDGs). Emma has co-authored several collaborative articles on tourism and the SDGs, including the challenge of tourism in addressing poverty, Indigenous tourism and the SDGs, and how tourism businesses in Fiji are localizing the SDGs.

Marika Kuilamu is an Assistant Lecturer at the University of the South Pacific (Fiji). In 2012, he gained a Master of Arts degree in Tourism Studies from the University of the South Pacific, and his research interests include sustainable tourism development, ecotourism, community and Indigenous tourism. Prior to joining USP in 2006, he was the Principal Tourism Officer with the Ministry of Tourism and was responsible for sustainable tourism development policies, resource owners, stakeholder participation and best practice initiatives.

Navneet Nimesh Kumar is currently working as a Guest Experience Maker at Six Senses Fiji. He completed a Master of Commerce in Tourism and Hospitality Management from the University of the South Pacific (Fiji). He is an early career researcher in the field of tourism and hospitality management, and his research interests are sustainable tourism, tourism micro-entrepreneurship, community tourism and tourism resilience. He has published work on tourists and micro-entrepreneurs in the souvenir industry in Fiji and the tourism-led growth hypothesis.

Mei Kei Leong is a Senior Lecturer at Taylor's University (Malaysia). She was a Senior Lecturer at the School of Hospitality and Service Management at Sunway University (Malaysia) from 2018 to 2022. She completed her PhD in the field of consumer science from Universiti Putra Malaysia in 2018 and has a Bachelor of Consumer Studies from the same institution. Mei Kei teaches in the field of branding, marketing, strategic management and event planning. Her current research focus is on marketing management, consumer behaviour in tourism and hospitality, and quantitative research in hospitality and marketing.

Evelyn G.Y. Loh oversees the marketing, branding and PR initiatives within the School of Hospitality and Service Management at Sunway University. She is an avid writer and has contributed to numerous editorial writings on behalf of the school. She has also published several papers in peer-reviewed academic journals. Evelyn attained a Bachelor of Management (Major in Finance) from the Universiti of Sains Malaysia (USM) and has recently completed her MBA at the Management of Science University (MSU) (Malaysia). Evelyn's research interests are in the field of hospitality marketing, consumer behaviour and purchase intention in the service industry.

Brent Lovelock is Professor and Head of the Department of Tourism at the Otago Business School, University of Otago (New Zealand). His research and teaching focuses on sustainable tourism – in the broadest sense – socially, environmentally and politically. One strand of this work considers destination development and the challenges faced by communities around the world in trying to bring about positive change through tourism, especially peripheral and Indigenous communities. Brent's research also addresses the ethics of tourism in which he examines how niche tourism products, such as medical tourism and consumptive wildlife tourism, can be conducted in ways that are respectful of their participants and their environments. He is currently leading a project studying the relationship between tourists and the natural environment, with a focus on tourism and invasive species.

Marica Mafi-Stephens teaches in the field of tourism and hospitality management at the University of the South Pacific (Fiji). She holds a Master of Commerce in Tourism and Hospitality Management from this university, which she completed in 2018. At the same institution, she has also obtained a Postgraduate Certificate in Tertiary Teaching, Postgraduate Certificate in Human Resources Management and a Postgraduate Diploma in Tourism and Hospitality Management. Prior to joining the university, Marica gained extensive experience in the service industries working for more than ten years in the hospitality and financial sectors. Her research interests concern ecotourism, sustainable tourism, human resources management, and culture and heritage tourism.

Evanthie Michalena holds a PhD in Energy Policy and Sustainable Tourism Development from Sorbonne University (France). She is a researcher at Espaces, Nature et Culture (ENEC) research laboratory of Sorbonne University (France), the laboratory of Tourism Research and Studies (ETEM) of the Aegean University (Greece) and an adjunct Associate Professor at the University of the Sunshine Coast (Australia). Eva is a Senior Expert collaborating as an evaluator with the European Commission and works full time as a Special Adviser in the Regulatory Authority for Energy (Greece), which also assigns her as an expert to the Council of European Energy of Regulators for lecturing missions in countries such as Armenia and Azerbaijan. She has published extensively, with research interests in sustainable energy policy, sustainable tourism development, sustainable entrepreneurship, innovation and local communities, and local administration.

Farah Atiqah Mohamad Noor is a Lecturer at the School of Hospitality and Service Management at Sunway University, specializing in Event Management. She has been a lecturer at the tertiary level since 2013, having worked at two other institutions prior to joining Sunway University in 2017. She completed her MSc in Tourism at Taylor's University (Malaysia), where she conducted research from a sociocultural perspective, namely on slow tourism. Her research work for her MSc was funded by Malaysia's Higher Education Ministry through the Long-Term Research Grant Scheme, awarded to her principal supervisor at the time. She is currently a PhD candidate at Universiti Putra Malaysia, focusing her research on Muslim-friendly tourism.

Apisalome Movono is a Senior Lecturer at Massey University, where he continues his passion for the Pacific through researching and promoting development that is fair, resilient and sustainable for future generations. In 2018, he was awarded a PhD from Griffith University, Brisbane (Australia), concerning the adaptivity and resilience of tourism-dependent communities in Fiji. He is an active conservationist and community development advocate for the people of Fiji and the Pacific region. Apisalome has published papers and articles focusing on resilience, adaptivity and development in the Pacific. He is a current recipient of a Marsden Fast-Start Fellowship from the Te Āpārangi/Royal Society of New Zealand, which examines resilience building in the tourism sector in Fiji and Vanuatu.

Nicole Orsua is currently working in the public sector as an Economic and Tourism Development Officer in regional Tasmania (Australia), focusing on product development, marketing and promotion, and community development. She has a Master of International Sustainable Tourism Management, which she completed at Monash University in 2020. Prior to postgraduate study, Nicole had a long career as a sales professional in the hotel and resort industry in the United States. Her research interests include sustainable tourism development, female entrepreneurship and Indigenous tourism.

Jefferson Patovaki is the Chief Tourism Officer for the Honiara City Council (Solomon Islands). He previously served in the Western Provincial Government as the Chief Tourism Officer for five years and has also worked in the hotel sector as a cook and as a duty manager. He obtained a Master of Commerce in Tourism and Hospitality Management at the University of the South Pacific in Fiji in 2020. At the same institution, Jefferson acquired a Bachelor of Hotel Management in 2013. His research interests include sustainable tourism, heritage tourism and UNESCO World Heritage Sites, niche tourism and tourism product development.

Stephen Pratt is Professor and Department Chair of Tourism, Events and Attractions at Rosen College of Hospitality Management at University of Central Florida (United States) from August 2022. He held the previous position of Professor of Tourism at the School of Business and Management at the University of the South Pacific. He served as Deputy Head of School (Research, Innovation and Postgraduate Affairs) at the same institution. He worked at USP from 2009 to 2012 and from 2018 to 2022. In the interim, he worked at the School of Hotel and Tourism Management at the Hong Kong Polytechnic University. He completed his PhD at the University of Nottingham (UK). He obtained a Bachelor of Economics and Masters of Economics from the University of Sydney (Australia). Stephen is an Executive Council Member of the International Association for Tourism Economics and serves on the Board of Directors of the International Centre of Excellence in Tourism and Hospitality Education (THE-ICE) and Country Representation for the Republic of Fiji of the Asia Pacific Council on Hotel, Restaurant, and Institutional Education (APacCHRIE) Board. He is a Visiting Fellow at Wakayama University (Japan).

Bruce Prideaux is Professor of Sustainable Tourism Management at Central Queensland University, Australia. He has a wide range of research interests, including crisis management, transport, sustainability, small islands, climate change and rural tourism. Other active areas of research include mobilities research and implications of the transition to net-zero carbon. He has authored more than 220 journal articles and book chapters and authored or co-authored 13 books. Bruce serves on a number of editorial boards, including the *International Journal of Tourism Research*, *Journal of Travel and Tourism Marketing*, *Sustainability*, *Journal of Vacation Marketing*, *Journal of Destination Marketing & Management*, *Tourism Geographies* and *Asia Pacific Journal of Tourism Research*.

Jale Samuwai has a PhD in Climate Change from the University of the South Pacific in climate finance. He is an independent Pacific-based researcher with work interests in the Pacific and small islands states in general. Jale was the interim Economic Justice Lead for Oxfam in the Pacific. He is presently the Resilience Finance Analyst who supports the Framework for Resilient Development in the Pacific based at the Pacific Islands Forum Secretariat. His research areas include climate change, climate and ocean finance, equity and justice.

Chetan Shah is a Doctoral student at the University of Auckland (New Zealand). He has more than a decade of work experience in managing operations, teaching, training and consulting for the tourism and hospitality industry. He has previously worked in Fiji with the University of the South Pacific and Fiji National University. In India, he was associated with the World Class Skill Centre, New Delhi, and the Institute of Hotel Management, Mumbai. Chetan is a doctoral research scholarship recipient at the University of Auckland and a silver medallist from the National Council for Hotel Management and Catering Technology, New Delhi. As an Honorary Consultant for the Central Himalayan Environmental Association (India), he is responsible for tourism and rural development projects. His current research focuses on consumer behaviour, crowding, alternative tourism (rural development) and destination marketing.

Karishma Sharma teaches tourism and hospitality management at the University of the South Pacific in Fiji, where she has worked for the past six years. In addition to her Master of Commerce in Tourism and Hospitality Management and a Postgraduate Certificate in Tertiary Teaching from the University of South Pacific, she has more than a decade of experience in the tourism and hospitality industry in both Fiji and Australia. Karishma has used this experience not only to contribute to a relevant teaching space with real-world examples but also to further pursue her academic interests, including digital marketing and social media in tourism.

Marcus L. Stephenson is a Professor of Tourism and Hospitality Management and Dean of the School of Hospitality at Service Management at Sunway University (Malaysia). Prior to this appointment in October 2017, he was Professor and Head of the School of Tourism and Hospitality Management at the University of the South Pacific (Fiji). He is the co-author of *Tourism and Citizenship: Rights, Freedoms and Responsibilities in the Global Order* (2014) and co-editor of *International Tourism Development and the Gulf Cooperation Council States: Challenges and Opportunities* (2017). Both books are Routledge publications. Marcus has also published extensively on the sociology of tourism, especially concerning nationality, ethnicity, culture and religion. His current research focus is on tourism development in the South Pacific and the Middle East. He has been an external examiner and academic adviser for institutions in Australia, the Middle East, East Asia and Europe.

Ai Ling Tan is a Senior Lecturer at the Surrey International Institute, Dongbei University of Finance and Economics (SII-DUFE) (China). She was a Senior Lecturer at School of Hospitality and Service Management at Sunway University (Malaysia) from 2013 to 2022. She attained a PhD in the field of hospitality and tourism at Taylor's University (Malaysia) in 2015 and has worked as an academician for more than ten years. She is an experienced internal auditor at Sunway University and was recently appointed as a panel auditor by the Malaysian Qualifications Agency. Before joining higher education, Ai Ling was an operations manager as well as a quality assurance executive in the food and healthcare industry. Her research interests include food studies, service quality and recovery, and Islamic tourism. She has successfully secured several international and national research grants in the past few years.

Dallen J. Timothy is Professor of Community Resources and Development at Arizona State University (United States) and Senior Sustainability Scientist at the Julie Anne Wrigley Global Institute of Sustainability. He is also a Senior Research Associate at the University of Johannesburg (South Africa); Visiting Professor at Hunan Normal University and Guangxi University (China); and Guest Professor in the Erasmus Mundus European Master in Tourism Management programme based at the University of Girona (Spain). Dallen is the founding editor of the *Journal of Heritage Tourism* and currently serves on the editorial boards of 24 international journals.

He is commissioning or co-commissioning editor over four book series with Routledge and Channel View Publications. He has ongoing research projects in North America, Asia, Europe, the Middle East and Africa on topics related to borders and tourism, religious tourism, heritage and community empowerment.

Adam M. Trau is an Independent Consultant and Senior Technical Advisor, with expertise in community-based livelihoods and tourism within Pacific Island countries. He was awarded a PhD from Western Sydney University (Australia) in 2013. He has more than 15 years of experience in research and technical leadership, project and grant management, capacity building and monitoring and evaluation within the Pacific, Timor-Leste and Australia. Adam has worked as a technical adviser, evaluator, consultant, senior government officer, community development practitioner, university researcher and lecturer. His research has been published extensively in international academic peer-reviewed journals and edited volumes on international tourism and development.

Alexander Trupp is currently Associate Dean for Research and Postgraduate Studies cum Associate Professor at the School of Hospitality and Service Management, Sunway University (Malaysia). He is also the Acting Head of the Asia Pacific Centre for Hospitality Research (APCHR) and Editor-in-Chief of the *Austrian Journal of South-East Asian Studies*. Alexander obtained his PhD at the University of Vienna (Austria), where he was also a lecturer and assistant professor researching and teaching in the fields of tourism geographies, anthropology of tourism and Asian studies. Prior to joining Sunway University, Alexander worked at Mahidol University (Thailand) and the University of the South Pacific (Fiji). His research is nested in the fields of tourism geography, mobilities and sustainable tourism, with a regional focus on Southeast Asia and the Pacific Islands.

Lenara Tuipoloa-Utuva is currently a Principal Training and Development Consultant at Samoa's Public Service Commission and previously was a Lecturer in Tourism at the National University of Samoa. Lenara successfully completed her Master's degree in Tourism with Distinction at the University of Otago (New Zealand) and is a member of the first cohort of the East-West Centre Pacific Islands Tourism Professional Fellows Program, where she developed an interest in the contribution of farm-to-table tourism. Lenara is an avid tourism researcher, and her interests include tourism in Samoa in the areas of training and education, farm-to-table tourism and sustainable tourism relating to local communities, their interaction with tourism activities and resilience to impacts.

Anuradha Vyas is a Lecturer in Foundation Studies at Middlesex University Dubai, United Arab Emirates. She previously worked for ten years at Skyline University College, Sharjah (UAE). She holds two Master's degrees – the first in Mathematics from Kurukshetra University (India) and the second in Business Administration from Middlesex University, London (UK). Anuradha has worked in the field of higher education for the past 22 years. Her research interests include sustainable tourism and development, tourism marketing and strategic management.

ACKNOWLEDGEMENTS

This book, from start to finish, has been a fascinating journey and would not have been possible without the support of my colleagues from the School of Hospitality and Service Management at Sunway University (Malaysia). Therefore, I would like to express appreciation, particularly to June Quay, Evelyn Loh and Alexander Trupp, for their help and kindness during the intense periods of editing and writing. A special thank you also for support from Sunway University's President, Professor Sibrandes Poppema, and to my former employer, the University of the South Pacific (USP) (Fiji) for providing me with the opportunity to engage in various research projects in several Pacific Island countries. Thank you also to my former students at the various USP campuses in the Pacific for inspiring me to learn more about the region. I would also like to acknowledge the wonderful assistance from my Routledge colleagues, notably Emma Travis and Harriet Cunningham, especially for their patience and professionalism. Importantly, I would like to express warm appreciation to my partner Golnoosh Najafi-Sohi, for her constant patience and unwavering support. This book is dedicated to the memory of Professor David Harrison, a former colleague at London Metropolitan University (UK), where I had the pleasure of working with David from 1999 until 2005. David's love of the Pacific and his research insights had a significant and lasting impression on me.

PART I

Tourism and small island states in the Pacific

Conceptual overview and regional context

1

INTRODUCTION

Understanding small island states and tourism in the Pacific

Marcus L. Stephenson

Introduction

This chapter addresses the selection of 13 self-governing islands in the Pacific as a unit of investigation for the current Routledge *Handbook of Tourism and Small Island States in the Pacific*. The initial discussion acknowledges the complexities of several of these states being classified as fully sovereign states. The discussion then provides a historical, political and geographical overview of the Pacific Island states. The regional groupings of 'Melanesia', 'Micronesia' and 'Polynesia' are deconstructed, and a succinct outline of the economic and tourism development issues of such states is provided, including the presentation of secondary data concerning particular economic indicators and international tourist arrivals. Subsequently, this chapter reviews the scholarly contributions of edited volumes on tourism and the Pacific Islands helping to contextualize the current volume's purpose. The latter part of the chapter presents a synopsis of the volume's chapter contributions, structured under four main sections: Part I, 'Tourism and small island states in the Pacific: Conceptual overview and regional context'; Part II, 'Tourism and island states in Melanesia'; Part III, 'Tourism and island states in Micronesia' and Part IV, 'Tourism and island states in Polynesia'. Finally, the concluding chapter identifies future research trajectories, especially to advance a critical conception of tourism within the context of the wider study of Pacific Island tourism.

Conceptualizing Pacific Island states: the historical, political and geographical context

This volume represents all those self-governing small island states that are located in the Pacific Ocean of which there are 13 states in total: the Cook Islands, the Federated States of Micronesia (FSM), Fiji, Kiribati, the Marshall Islands, Nauru, Niue, Palau, Samoa, the Solomon Islands, Tonga, Tuvalu and Vanuatu. The classification includes those states which have free association with New Zealand (the Cook Islands and Niue) and with the US (the FSM, the Marshall Islands and Palau). The work does not include such politically dependent states as French Polynesia and New Caledonia (France), Hawaii and Guam (United States (US)), Easter Island (Chile) and Pitcairn Islands (United Kingdom (UK)). The conceptualization of small island states concerns those islands that have small quantities of accessible lands, making them rather unique in

DOI: 10.4324/9780429019968-2

terms of their social, cultural, environmental and political composition. It is in this context that tourism and its development will be examined. Papua New Guinea, too, is excluded from this volume because it is not considered to be a small island state.

Although the Cook Islands and Niue are self-governing, New Zealand is officially responsible for their defence and foreign affairs, which can be exercised on request from the two states. Citizens of both states have the right to live and work in New Zealand, as they possess New Zealand citizenship and, along with New Zealand, share the same head of state (Smith 2010; Townend 2003). Also, the Micronesian states of the Marshall Islands, Palau and the FSM have established Compacts of Free Association with the US in which the US has authority over their security and defence. Therefore, these three states, together with the Cook Islands and Niue, technically, can be termed 'associated states' (Smith 2010). Unlike the Cook Islands and Niue, the Micronesian states are member states of the United Nations. As all the associated states still have self-government, they have been included in the wider definitional context of a Pacific Island state. The current volume adopts both forms of terminology – 'Pacific Island states' and 'Pacific Island countries' (PICs) – and other similar related terminology, but in all cases, the reference point concerns the 13 self-governing states/countries.

Most of the Pacific Island states share colonial histories. The British Empire started to acquire islands largely in the Melanesia and Polynesia regions in the latter part of the nineteenth century. However, Vanuatu (formerly New Hebrides) was jointly managed by Britain and France, and Samoa was managed by Britain, Germany, and the US from 1889 to 1899, until the British withdrew its interests, whereupon the western part of Samoa became German Samoa, and the eastern part became American Samoa (Droessler 2017). Britain started to handover its colonies from the late 1960s, though their dominance in the region had been reduced due to the geopolitical roles played by Australia and New Zealand (Table 1.1). The former New Zealand and British colonies largely pursued a governance process that was similar to the former colonies in which a parliamentary government with a Westminster model was adopted. For the US colonies and dependencies, a presidential form of government was recognized, though the Republic of the Marshall Islands installed a parliamentary system, and for Kiribati, the Marshall Islands and Nauru, the president serves both as prime minister and head of state (Meller 1987). One atypical governance structure concerns the Kingdom of Tonga, a British protectorate that was not fully administered by the British government but by a monarchical government. Until recently, Tonga maintained an absolute monarchy in which its governance was based on the constitution enacted by King George Tupou 1 in 1875, emphasizing that the country's executive power resided with the monarchy. Due to constitutional reforms taking place in 2010, partly as a response to expressions of anti-monarchy sentiments in some sections of Tongan society, royal authority and power were partially devolved. However, the monarch still holds certain rights, such as the vetoing of bills and judiciary appointments (Corbett et al. 2017).

With anti-royalty agitation less profound in Tonga and national elections eventually taking place in Fiji from 2014, it would be reasonable to claim that Pacific Island states are largely politically stable, which is certainly conducive to tourism growth and development (Sönmez 1998). However, in November 2021 the region witnessed rioting in the Chinatown district of Honiara, the capital city of the Solomon Islands, indicating the re-emergence of ethnic tensions in a country that has experienced overt ethnic conflict since the late 1990s. Overall, and compared to sub-Saharan Africa, the democratic record of the Pacific Island states is described as 'impressive' (Firth 2018: 11), but, as this author notes, the more that countries, such as the Solomon Islands, Vanuatu and Fiji, experience increased urbanization, the more prone they are to public agitation and conflict, as linguistic and group diversities are very apparent in the Solomon Islands and Vanuatu.

Table 1.1 Past and present political administration of the 13 Pacific Island states by region

Regions and Pacific Islands states (formal title)	*Acquired*	*Original Administrator*	*Last Administrator*	*Date of Independence*
Melanesia				
Republic of Fiji	1874	Britain	Britain	1970
Solomon Islands	1892	Britain	Britain	1978
Republic Vanuatu [formerly New Hebrides]	1887	Britain and France	Britain and France	1980
Micronesia				
Federated States of Micronesia	1885	Spain	US	1986
Republic of Kiribati [formerly Gilbert Islands	1892	Britain	Britain	1979
Republic of Marshall Islands	1885	Germany	US	1986
Republic of Nauru	1888	Germany	Australia, Britain and New Zealand	1968
Republic of Palau	1885	Spain	US	1981*
Polynesia				
Cook Islands	1888	Britain	New Zealand	1965*
Niue	1900	Britain	New Zealand	1974*
Independent State of Samoa	1900	Germany	New Zealand	1962
Kingdom of Tonga	1900	Britain	Britain	1970
Tuvalu [formerly Ellice Islands]	1892	Britain	Britain	1978

Source: Adapted from Meller (1987: 115).
Notes: * This refers to the date that self-government was established.

The 13 states stretch across a vast Pacific Ocean (Figure 1.1), where the distance from Palau (northwest) to the Cook Islands (south) is approximately 7,850 kilometres. Therefore, the physical geography of the 13 states is diverse. The Pacific Islands comprise of three ethno-geographic groupings: Melanesia, Micronesia and Polynesia, which typically exclude the neighbouring island continent of Australia (West and Foster 2020). Polynesia represents a large triangle that encompasses the east-central Pacific Ocean, with the Hawaiian islands at the north, Easter Island (Rapa Nui) in the east and New Zealand (Aotearoa) in the west (Kiste et al. 2022). The island region of Polynesia encompasses the self-governing states of the Cook Islands, Samoa (formerly Western Samoa), Tonga, Tuvalu and Niue, with the rest being dependent states. Polynesia technically refers to 'many islands', though it consists of three types of islands: high islands, uplifted coral platforms and atolls – with coral atolls being the most prevalent in Polynesia and having a tendency to be narrow and low-lying (Berno and Douglas 1998: 65). The largest independent state of Polynesia is Samoa, representing 3,046 square kilometres. The combined size of the total area of these five self-governing states is 4,532 square kilometres, and there are 157 islands in total (Table 1.2). Given the large expanse of Polynesia, islanders have a prolific history as seafarers with sophisticated wayfinding skills (Martins 2020). The ability to navigate the Pacific Ocean without scientific instruments aided scientific thought on voyaging, disputing the colonial narrative which typified the region as being undeveloped (Finney 1993).

The island region of Melanesia incorporates the self-governing states of Fiji, the Solomon Islands and Vanuatu, including the much larger country of Papua New Guinea and other depend-

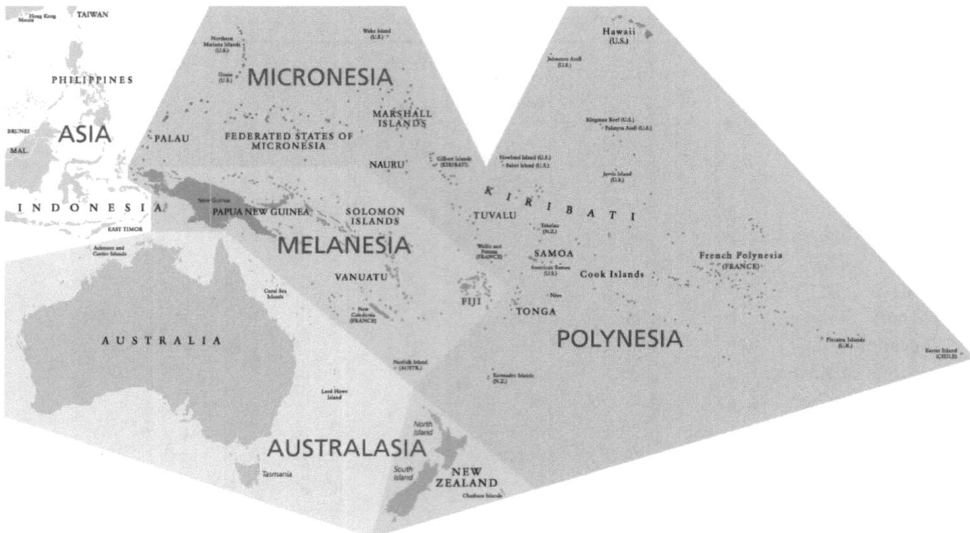

Figure 1.1 Map of the Pacific Island states, indicating the regions of Melanesia, Micronesia and Polynesia. Source: Peter Hermes Furian (https://www.shutterstock.com/image-vector/regions-oceania-political-map-colored-geographic-1724734174).

ent states. The region represents many high islands and forms parts of the 'Ring of Fire', expanding around the rim of the Pacific. It is, thus, one of the most turbulent expanses in the world in which volcanic eruptions and earthquakes have 'high explosion indices' (Shoji et al. 1993: 3). However, such turbulence is not confined to Melanesia, as was witnessed in early 2022 with the eruption of Tonga's Hunga Tonga-Hunga Ha'apai volcano (Pacha 2022). The three independent island states of Melanesia (Fiji, the Solomon Islands and Vanuatu) have a combined total area size of 64,058 square kilometres and collectively represent a total of 707 islands. The smallest of these three island countries is Vanuatu, whose total size is bigger than the total combined size of both the Micronesian and Polynesian self-governing states. Micronesia encompasses the small independent states of the FSM, Kiribati, the Marshall Islands, Nauru and Palau, including other dependent states. The five independent states collectively represent 228 islands and a combined total area size of 2,598 square kilometres (Table 1.2). Micronesia tends to be composed of 'reef islands' which are largely submerged volcanic islands where a reef has then formed, as in the case of the FSM, Kiribati and the Marshall Islands (Nunn et al. 2016).

The state with the average island maximum elevation is Samoa (504 meters) and the state with the lowest elevation is the Marshall Islands (3 meters), followed by Tuvalu (4 meters) and Kiribati (6 meters) (Nunn et al. 2016). The sea level rise for the Pacific Islands has increased 2–3 times more than the global average, where there has been an almost 0.3 meters net increase since 1990 (Storlazzi et al. 2021). It is ironic that, although the Pacific Island states largely avoided industrialization, they actually became victims of this process, producing greenhouse gas emissions that directly affect the fragile states (Opeskin and MacDermott 2009). The impact that climate change has on tourism development in the Pacific Islands cannot be underestimated, as well as the impact of tourism on climate change (Wolf et al. 2021).

From the 13 Pacific Islands countries, the smallest state is Nauru, representing just one island and is 23 square kilometres, whereas the largest state is the Solomon Islands represent-

Table 1.2 Number of islands for each Pacific Island state, total area and average maximum elevation

Pacific Island states	*Number of Islands**	*Total Area of Islands*** *(square kilometres)*	*Average Island Maximum Elevation (metres)*
Melanesia			
Fiji	211	20,857	134
Solomon Islands	415	29,675	88
Vanuatu	81	13,526	330
Micronesia			
Federated States of Micronesia	127	799	45
Kiribati	33	995	6
Marshall Islands	34	286	3
Nauru	1	23	71
Palau	33	495	58
Polynesia			
Cook Islands	15	297	73
Niue	1	298	60
Samoa	7	3,046	504
Tonga	124	847	56
Tuvalu	10	44	4

Source: Adapted from Nunn et al. (2016: 3).

Notes:

*The number of islands, in some cases, may also appear less than some other sources of information, but the data does not calculate strings of islands ('motus') (Nunn et al. 2016: 11–12).

**This calculation acknowledges that land size may differ from other sources of information, as the 'islands were calculated directly from the polygon shapefile layer in a GIS' (geographic information system) (Nunn et al. 2016: 6).

ing 29,675 square kilometres with a collection of 415 islands (Table 1.2). Nauru is ranked the third smallest state in the world, followed by Tuvalu (4) and the Marshall Islands (7) (World Population Review 2022). As countries such as Fiji and the Solomon Islands are much larger in size, there are more opportunities for human livelihoods to thrive (including tourism-related livelihoods), and the higher the physical elevation of the land the more opportunity to collect rainfall to satisfy a variety of human needs (Nunn et al. 2016). In fact, both the climate and ecology of Melanesian countries have ensured that the countries have, historically, maintained considerable forestry assets (Opeskin and MacDermott 2009), though countries such as the Solomon Islands have experienced extensive logging by international companies (Kabutaulaka 2000).

The study of small states attracts substantial attention from various disciplinary fields (Baldacchino and Wivel 2020; Lockhart et al. 1993; Ratter 2018). It is fundamental, however, not to intellectually prey on the 'smallness' of states as a source of inferiority or weakness or the main reason to justify being dependent on larger states. Being small does not necessary equate to being disadvantaged (Mehmet and Tahiroglu 2002). Stephenson and Al-Hamarneh's (2017) edited volume on *International Tourism Development and the Gulf Cooperation Council States* includes contributions from small states in the Middle East, where rapid modernization and economic growth have been significant and attempts to leverage tourism largely enable the diversification of the economies of these states (Stephenson 2017). Nearly three decades ago, the Tongan and Fijian writer Epeli Hau'ofa (1994), wrote a seminal paper entitled *Our Sea of Islands,* where he dismantled normative conceptualizations of Pacific geographies, economies and cultures, stating:

> The world of Oceania is not small; it is huge and growing bigger every day. The idea that the countries of Polynesia and Micronesia are too small, too poor, and too isolated to develop any meaningful degree of autonomy is an economistic and geographic deterministic view of a very narrow kind that overlooks culture history and the contemporary process of what may be called world enlargement that is carried out by tens of thousands of ordinary Pacific Islanders right across the ocean—from east to west and north to south, under the very noses of academic and consultancy experts, regional and international development agencies, bureaucratic planners and their advisers, and customs and immigration officials—making non sense of all national and economic boundaries, borders that have been defined only recently, crisscrossing an ocean that had been boundless for ages before Captain Cook's apotheosis (151)... Together with our exclusive economic zones, the areas of the earth's surface that most of our countries occupy can no longer be called small.
>
> *(1994: 158)*

Hau'ofa's vision recognizes the vastness of Oceania and its richness in terms of natural and cultural resources in which these states can be reconceptualized as large ocean states (see Chapter 3). However, Teaiwa (1996: 215) observes that some writers have had to 'gently remind Hau'ofa of the sad or ugly visions of rural poverty and urban deprivation in the Pacific'. Hau'ofa's work implies that conceptualizations of the Pacific Island region ought to develop organically. The conceptual applications of such terminology as 'Melanesia' and 'Polynesia' are regarded in some academic and political circles as 'discredited ethnological typifications' of Pacific Island peoples and societies (Thomas 1989: 27), especially as these concepts are rooted in colonial politics. In 1832, the French naval commander Jules Dumont d'Urville branded the wider Pacific into four regions: 'Malaysia', 'Melanesia', 'Micronesia' and 'Polynesia' – particularly on the basis of phenotypic traits. His descriptions are crude, generalized and stereotypical in both substance and form. He categorized Melanesia on the basis of the skin pigmentation of its inhabitants, describing them as 'the black race from Africa', and for Polynesians, he described them as the 'offshoot of the yellow race of Asia' (quoted in Burley 2013: 436). According to Douglas (2010: 209): 'd'Urville's racial geography was ultimately normalised in global geopolitics and modern indigenous identities'.

McGavin (2014: 133) believes that the conceptual application of 'Melanesia', 'Micronesia' and 'Polynesia' has logical use value, as 'the way these terms are used today by many Islanders suggests they have attained a place of emergent authenticity within Islander worldviews'. He further explains that the terms Melanesia, Micronesia and Polynesia can offer a 'quick and easily understood way to establish external boundaries of the Pacific and identify who is a Pacific Islander' (2014: 133). Moreover, the 'tourist gaze' (Urry 2002) is enchanted by the socio-cultural dimensions of such regional connotations as Melanesia and Polynesia (see Caneen 2014; Moulin 2017; White 2007). McGavin (2014: 134) further infers that even popular terminology can be contestable: Oceania (a derivative of the French word *Océanie*) has colonial inferences, and the localized term 'Moana' (a Māori term referring to 'wide expanse of water') is arguably too narrow, as it is a Polynesian term. Nonetheless, there are certain kinds of generalized observations, attributes and comparisons that can be made across the three regions. For instance, in terms of approaches to social organizations, 'hierarchical systems of chieftainships' traditionally exist in Polynesia but in Melanesian societies more 'egalitarian' principles apply (Berno and Douglas 1998: 70; Linnekin and Poyer 1990). However, Micronesian societies traditionally represent a complexity of hierarchy and egalitarianism (Petersen 2009).

Pacific Island economies and tourism development

Hau'ofa's (1994) vision can be conceived as one of hope and a rebuttal to the dependency thinking in the study of less-developed and developing states, which basically emphasizes that the Pacific Islands are locked into economically exploitative relationships for the benefit of developed nations. Britton's (1980, 1982, 1983) influential work in the early 1980s on tourism, neo-colonialism and dependency concludes that the international structure of tourism and the tourism industry in Pacific Island states represents the dependency scenario. Based on data from the Cook Islands, Fiji and Tonga, Britton (1982) contends that there is an inequitable relationship between the 'periphery' (Pacific Islands) and the 'metropole' (developed/post-colonial) country, with tourism principally benefiting organizations of the metropole. For Britton, the tourism industries of Fiji and the Cook Islands embody distinct colonial traits and socio-economic dependent relationships in which 'the greatest commercial gains therefore go to foreign and local elite interests' (1982: 355).

In describing the socio-economic context of the Pacific Islands, there appears to be a 'development paradox' at play on two fronts. One typical illogicality concerns the contention that, despite Pacific Island states becoming politically independent or having self-governing powers, these countries continue to replicate colonial economic relations (Bertram and Watters 1985). The other economic paradox is that, irrespective of high levels of aid and overseas investments reaching the Pacific Island states, economic growth has been slow-moving (Toatu 2021). Pacific Island economies have traditionally been based on primary industries which have been typified by narrow productive sectors (Duncan and Nakagawa 2006), though, as we have seen in colonial and post-colonial economies, primary resources have been targeted by foreign companies for commodity extraction, whether for phosphate in Nauru or Banaba (an island of Kiribati), or bauxite or nickel in the Solomon Islands (see Allen 2018). It is ironic that Fiji's Ministry of Forests has been promoting the development of sandalwood farming when the country was a major world supplier of sandalwood more than 200 years ago, until 1814 when commercial logging was no longer profitable due to over-exploitation by foreign companies (Fontenoy 1997). Resource concerns can be very restricting for Pacific Island states, and this can be a challenge to entrepreneurship activities and private enterprises, as the relatively small size of island populations reduce economies of scale, together with challenges concerning the high costs for raw materials and transportation (Juswanto and Ali 2016).

Dependence on overseas aid, remittances from migrant workers, residents overseas and tourism revenues have arguably led to what Bertram and Watters (1985: 497) term a 'MIRAB economy', representing the 'migration' (MI), 'remittances' (R), 'aid' (A) and 'bureaucracy' (B) framework. These authors emphasize that the aid dimension demonstrates 'the realities of the decolonisation process and subsequent geopolitical calculations by major regional powers' (1985: 513). In examining five Pacific Island states that display MIRAB characteristics (the Cook Islands, Kiribati, Niue, Tonga and Vanuatu), Milne (1992: 195) claims:

> All suffer from the economic handicaps associated with small size, including small internal markets that limit the ability to adopt import substitution development strategies, remoteness and high shipping costs that raise the price of imports and concomitantly place exports at a competitive disadvantage, a reliance on a limited range

of primary commodity exports, problems of underemployment, and a labor market characterized by a limited skills base.

Nonetheless, recent economic data indicates the ongoing economic concerns experienced by most of the Pacific Island states. As Table 1.3 indicates, from the latest figures available at the time of writing, the gross domestic product (GDP) per capita is the lowest for Kiribati, only representing US$1,636, followed closely by the Solomon Islands (US$2,295) and Vanuatu (US$3,223). The two associated states in Polynesia that have a close relationship with New Zealand demonstrate a better GDP per capita: the Cook Islands (US$24,913) and Niue (US$18,757). This was followed closely by the other associated state of Micronesia: Palau (US$15,673). However, the remaining two associated states are significantly low: the FSM (US$3,830) and the Marshall Islands (US$4,337).

In terms of public (national) debt, that which is borrowed by the government in order to meet its development budget, Fiji has the highest debt status of the 10 states where data is recorded, reaching US$3,732 million in 2021. The debt burden is also extensive for countries such as Vanuatu (US$491 million), Samoa (US$390 million) and Tonga (US$214 million) (Table 1.3). For Samoa and Vanuatu, China accounts for around 40% of total external public debt (ESCAP 2022: 3). Compared to the pre-COVID-19 pandemic year of 2019, public debt in 2020 increased by US$74 million for the Solomon Islands and US$62 million for Vanuatu, and from 2019 to 2021, debt increased by US$1,077 million for Fiji (Countryeconomy.com 2022). Debt as a percentage of GDP is the highest in Fiji (79.18%), followed by Samoa (49.56%), Vanuatu (49.39%) and Tonga (43.25%), and the country with the lowest debt as a percentage of GDP is Tuvalu (11.53%) followed by the Solomon Islands (13.12%) (Table 1.3).

As Table 1.3 also illustrates, of the 11 states where data is available, nine states recorded a negative trade balance in which the value of imports exceeds exports. This is highly noticeable

Table 1.3 Economic data for the Pacific Island states: GDP per capita and select indicators

Pacific Island States	*GDP Per Capita★ (US$ thousands)*	*Debt★ (US$ million)*	*Debt as a % of GDP★*	*Trade Balance for 2021 (Exports Minus Imports) (US$ million)*
Melanesia				
Fiji	6,152 (2021)	3,732 (2021)	79.18 (2021)	−1,306.0
Solomon Islands	2,295 (2021)	203 (2020)	13.12 (2020)	−190.8
Vanuatu	3,223 (2020)	491 (2020)	49.39 (2020)	−268.2
Micronesia				
FSM	3,830 (2021)	76 (2018)	18.83 (2018)	−92.0
Kiribati	1,636 (2021)	35 (2020)	19.02 (2020)	−147.3
Marshall Islands	4,337 (2021)	46 (2020)	19.01 (2020)	0
Nauru	11,666 (2021)	36 (2021)	27.06 (2021)	75.0
Palau	15,673 (2020)	–	–	−197.5
Polynesia				
Cook Is	24,913 (2021)	–	–	–
Niue	18,757 (2021)	–	–	–
Samoa	4,284 (2021)	390 (2021)	49.56 (2021)	−339.1
Tonga	5,081 (2021)	214 (2020)	43.25 (2020)	−270.4
Tuvalu	4,223 (2021)	6 (2019)	11.53 (2019)	−41.6

Source: Adapted from Countryeconomy.com (2022) and SPC (2022).
Notes: ★ These figures are based on the latest data available.

for Fiji (–US$1.306.0 million), Samoa (–US$339.1 million), Tonga (–US$270.4 million) and Vanuatu (–US$268.2 million). All countries have a negative balance of trade, with the exception of Nauru (US$75 million). Bertram (2018) informs that, although the Cook Islands has had a healthy balance of trade (as data in Table 1.3 does not show for the Cook Islands), overseas aid increased more than sixfold from 2004 to 2018, especially to meet vital infrastructure requirements and maintain pubic services. It would be fair to say that all the self-governing Pacific Island states variously experience economic vulnerabilities, and this is despite receiving different levels of economic support from donor states.

The prevailing pro-tourism lobby, particularly international tourism organizations, often perceive tourism as 'valuable aid to national, regional and local development', with a perceived ability to produce 'various kinds of Keynesian multipliers' (Harrison 1994: 249). Harrison and Prasad (2013) critically acknowledge the main problems of tourism development in the Pacific Islands, including limited natural resources, communal ownership of land which makes it a challenge to attract foreign investment, poor accessibility and reliance on international carriers to transport tourists, and the 'lack of financial, human and (sometimes) cultural capital to meet tourist demand and compete with other island destinations' (2013: 755).

The complexity, however, is that tourism trends indicate the popularity of some Pacific Island states more than others, which represent different economic impacts in different states, thus, making it difficult to make crude generalizations concerning tourism development's relationship to island economies – as acknowledged in this volume (see Part I, especially Chapters 5 and 7) and inherent in the content of various case examples (see Parts II, III and IV). From Table 1.4, it is clear that, prior to the 2020 global pandemic, the most popular Pacific Island is Fiji, with tourism arrivals reaching over 894,000 in 2019. Samoa was the second most popular destination (over 172,000), closely followed by the Cook Islands (just under 172, 000) – with the less visited island being Tuvalu (just under 4,000), followed by the Marshall Islands (around 6,000) and Kiribati (around 8,000). Collectively, the Melanesia countries have been the most popular,

Table 1.4 Breakdown of total visitor arrivals for 2010, 2019 and 2020 for the Pacific Island states

Pacific Island States	*2010 Annual (1000s)*	*2019 Annual (1000s)*	*2020 Annual (1000s)*
Melanesia			
Fiji	632	894	147
Solomon Islands	21	29	4
Vanuatu	97	121	22
Micronesia			
FSM	45	18	–
Kiribati	5	8	1
Marshall Islands	5	6	1
Nauru	–	–	–
Palau	85	94	18
Polynesia			
Cook Islands	104	172	25
Niue	6	10	1
Samoa	122	172	23
Tonga	47	67	9
Tuvalu	2	4	1

Source: Adapted from UNWTO (2021: 19) and UNWTO (2022: Annex 13).
Notes: As noted by UNWTO (2022), the data for 2020 is still deemed provisional.

which is not too surprising given the size of these territories, followed by the Polynesian countries (with Samoa and the Cook Islands taking the bulk of the tourists). With around 94,000 arrivals in 2019, Palau has been the most visited destination state in Micronesia.

There were fairly strong increases in tourism numbers over a 9-year period from 2010 to 2019, indicating, in some way, that tourism was targeted in the region as a growth industry until the COVID-19 pandemic struck. Only the FSM experienced a significant decrease in tourists during that period, with notable increases for Cook Islands (by around 68,000), Fiji (262,000) and Samoa (50,000) (Table 1.4). Countries such as Nauru (Chapter 20) and Tuvalu (Chapter 26) have been associated with 'non-tourism' or undertourism, even prior to the pandemic, and other countries, such as the Solomon Islands, are seen to have substantial potential for tourism growth, though the range of social, organizational and economic challenges have hindered development (Chapter 10). The tourism arrival numbers, however, are quite deceptive in the sense that, although countries like Niue may appear to have low tourism numbers, around 10,000 prior to the pandemic in 2019, annual arrivals are six times higher than the country's population (see Chapter 7). Therefore, the level of impact due to high levels of tourism intensity in some countries can be considerable despite the appearance of low visitation numbers. Nonetheless, what is unequivocal is the very significant impact that the COVID-19 pandemic has on visitor numbers, particularly since international borders were closed from mid-March 2020. International tourist arrivals to the Pacific Islands decreased from 2019 to 2020 by at least 80%, despite the fact that Fiji still managed to record around 147,000 tourist arrivals in 2020; which stands in stark contrast to Kiribati, which received just less than 1,500 arrivals (Table 1.4).

The COVID-19 pandemic has had a dramatic impact on both tourism and island economies as well as on society in general (McGarry et al. 2021; Remeikis 2021). However, as different Pacific Island states have had different experiences, there are various approaches to consider in the context of tourism development and the post-pandemic recovery. Therefore, approaches and ways forward are meaningfully exposed in this current volume. Nonetheless, Pacific Island states have had many other contentions to deal with over the years and other crises, too. Before an outline of the book's structure and an expose of each chapter are presented, attention will, initially, turn to understanding the evolution of edited volume contributions concerning the investigation of Pacific Island tourism. This approach should help to comprehend, albeit implicitly, the purpose of the current volume.

A review of the scholarly contributions of edited volumes on tourism and the Pacific Islands

There have been a number of edited volumes embracing scholarly discussions on Pacific Island tourism. One of the early contributions came more than 40 years ago from Rajotte and Crocombe (1980) who edited the volume *Pacific Tourism as Islanders See It*. This contribution of 24 chapters claims to represent the local and regional perspectives concerning the role and impact of tourism in the Pacific Islands, focusing on a range of issues such as: the social, cultural and economic ramifications of tourism development; working in the informal and formal tourism sectors; and the advancing relationship between tourism and urbanism (Akauola et al. 1980; Bolabola 1980; Fong, 1980; Sikivou 1980; Skinner 1980; Tong and Tenentoa 1980). The volume's overall approach to tourism was not necessarily unanticipated, especially as the 1970s witnessed a range of publications highlighting concerns with tourism development (de Kadt 1979; Greenwood 1972; Nash 1977; Turner and Ash 1975), including some critical treatment of Pacific Island countries (Farrell 1979; Finney and Watson 1975).

Finney and Watson's (1975) edited volume, *A New Kind of Sugar: Tourism in the Pacific,* had already set the scene in the mid-1970s to critically and collectively review tourism development in the Pacific Islands. This volume, consisting of 21 chapters, partly addresses some of the colonial and neo-colonial undertones of the tourism industry (see Kent 1975; Samy 1975), especially as it was published during a period when not all the Pacific Island states had become politically independent and when international tourism mobility was still at a developmental stage. Nonetheless, as both volumes from Rajotte and Crocombe (1980) and Finney and Watson (1975) exemplify, there were regional signs at that time that tourism was displaying overt socio-economic and cultural concerns. The criticality of tourism that is partly inherent in both volumes is largely supported by case illustrations from the region of Polynesia and Melanesia, with the occasional (almost tokenistic) representation from Micronesia (Ashman 1975; Tong and Tanentoa 1980). Accordingly, more than one-third of the chapters in the Finney and Watson (1975) volume focused on the Polynesian island of Hawaii, and almost one-third of the chapters in the Rajotte and Crocombe (1980) volume focused on the Melanesian island of Fiji. This is not too surprising given the level of tourism development that was taking place in both destinations and, thus, a range of issues and concerns to duly discuss

Around 16 years following the publication of Rajotte and Crocombe's (1980) work, Hall and Page's (1996a) edited volume was published, *Tourism in the Pacific Issues and Cases.* This volume, consisting of 18 chapters, significantly revisits tourism and its development in the Pacific region. With this publication, there was more awareness of the importance of tourism planning and marketing as well as an emerging interest in tourism and sustainable development (Fagence 1996; Milne 1996; Rudkin and Hall 1996), no doubt reflecting the evolutionary advancement of tourism in the region. This volume also drew attention to the political and economic presence of Australia and New Zealand in the region, predominantly in terms of travel, trade and aid (Hall and Page 1996b). As the volume largely conceptualizes the 'South Pacific' (Hall and Page 1996c), the work mainly utilizes case examples of politically independent and dependent countries from Melanesia and Polynesia (Buck and Hall 1996; de Burlo 1996; Minerbi 1996; Plange 1996).

Seven years following the Hall and Page (1996) publication, Harrison's (2003) edited volume, *Pacific Island Tourism,* was published. This book, consisting of 11 chapters, is arguably more concerned with 'micro-level ecologically aware tourism management' rather than 'macroeconomic implications of tourism in the Pacific, or about the economic benefits and costs of tourism development' (Throsby 2005: 212). Although the book explores a range of insightful case examples from the east to west Pacific (e.g., Russell and Stabile 2003 Stanton 2003), once again there is little attention to Micronesia. Subsequently, the 'thematic and geographic coverage' of this book is described as being 'fairly serendipitous' (King 2005: 632). Shortly after this volume, another edited text was produced by Cooper and Hall (2005), *Oceania: A Tourism Handbook,* containing 20 chapters. The work focuses on key issues around tourism demand and supply and the structure of the tourism industry. This volume covers a range of contemporary themes, though more in terms of providing an overview rather than a deeper academic treatment. Although important topics concerning risk management (Wilks 2005) and tourism's relationship with global warming (Craig-Smith 2005) are included in the volume, insights are rather brief, though the chapter concerning tourism and customary practices in the French Pacific is certainly insightful (d'Hauteserre 2005). The Pacific Islands were one of three geographical regions of Oceania presented, along with Australia and New Zealand, and, thus, the volume contributes more to a southwestern Oceania perspective. This volume is student friendly, illustrative and arguably more of a textbook type enquiry. At a similar time to this publication, a special edition on *Tourism and the Pacific* appeared in *The Journal of Pacific Studies* (2004), representing nine articles that were edited by Harrison (2004a). The volume includes several perceptive contributions, in particular the history of tourism in the Solomon Islands (Douglas 2004), the role of the UNESCO World Heritage Site of Levuka (Fiji) (Harrison 2004b: 199–232) and sustainable development in Samoa (Twining-Ward and Tuailemafua 2004: 77–103). Predictably, there was little treatment to islands in Micronesia, except for a solitary but informative contribution from Ringer (2004) concerning the qualitative effects of ecotourism on the island of Kosrae (FSM).

Over a decade later, Pratt and Harrison's (2015) edited work *Tourism in Pacific Islands: Current Issues and Future Challenges* was produced. This contribution of 15 chapters rightly acknowledges that (despite past contributions), in contrast to other regions of academic study, Pacific Island tourism is not significantly advanced. The book interestingly focuses on three conceptual areas: 'images of the South Pacific', 'socio-economic impacts of tourism' and 'Pacific Island countries and the outside world'. However, more than one-third of the volume's contributions focus on Fiji, which dominates the section on socio-economic impacts. Therefore, the volume is not entirely representative of the current issues and future challenges affecting tourism in the Pacific Islands. The work is cautioned for not dealing with pertinent contemporary issues that are considered to be important to the region, notably 'disaster risk management, or human and natural health related threats to the tourism sector in Pacific Islands' (Hess 2016: n.p.). Nonetheless, Pratt and Harrison's (2015) volume supports some inimitable insights concerning Pacific Island tourism, such as the role and potential of 'corporate social responsibility' strategies by businesses in the region (Scheyvens and Hughes 2015), tourism's intersections with community development (Gibson 2015), and emerging and non-normative tourism markets (Koth 2015).

Shortly following the Pratt and Harrison (2015) volume, Alexeyeff and Taylor (2016) published their edited volume, *Touring Pacific Cultures*. This rather ambitious book, consisting of 31 chapters, represents a mixture of scholastic, creative and personalized writings concerning tourism in the region, loosely aiming to merge contributions from tourism studies, tourism research, Pacific studies, anthropology and geography to study Pacific tourism (and culture). The main scholarly themes concern tourism's relationship to the 'commodification of culture', 'image creation and construction', and 'promotional imagery'. In their introductory chapter, Taylor and Alexeyeff express grave concern with tourism development in the Pacific Islands, emphasizing that it can represent 'destination inequality' (2016: 15) and 'leisure imperialism' (2016:17). This type of critique is familiar territory in critical tourism studies and an argument that is almost a mimicry of such classic works as Turner and Ash's (1975) *Golden Hordes*, which was written almost half a century ago. Nonetheless, there are some intriguing contributions in *Touring Pacific Cultures* with some valuable and historical insights concerning missionary imagery from the late nineteen to early twentieth century (Lindstrom 2016) and travellers and cruise ship tourists in the late nineteenth century (Steel 2016). Other academic and thought-provoking contributions come from an enquiry from Phipps (2016) concerning the role of festivals in relation to cultural expression and resistance, and an examination of constructions of paradise through tourism imagery by Alexeyeff (2016). On the whole, the volume manifests an overly critical perspective, provoking Campbell (2018: 212) to comment: 'we are all too familiar with the trope that casts tourism as an exploitative operation, but there are experiences that suggest other considerations as well. It is to these possibilities that I hope future enquiry turns'. Finally, like many of the enquiries on Pacific Island tourism, Polynesia and Melanesia dominate the contributions and discussions.

Hanlon (2009) observes that there is a tendency to not deeply engage with the Micronesian region when focusing on Pacific-based enquiries, emphasizing that the dominant trend is to focus more on 'Cook's Pacific'. Therefore, in a sardonic manner, Hanlon postulates that, if the

British explorer James Cook had the opportunity to seriously venture into the Micronesia region during his explorations in the eighteenth century, this region would have been far more represented in academic studies of the Pacific, but 'scholarship tends to follow the colonial flag' (2009: 92).

Consequently, this current volume, *Tourism and Small Island States in the Pacific*, will systematically attempt to bring together the study of self-governing Pacific Island states in an inclusive manner by ensuring representation across Melanesia, Micronesia and Polynesia. However, as indicated earlier, the unit of analysis concerns self-governing states which face socio-cultural, political, environmental/ecological, organizational, operational and strategic challenges. The volume aims to look at ways in which Pacific Island states can determine their own future and articulate their own tourism development pathways, and this is despite the many challenges the region faces and the ramifications of the COVID-19 pandemic to the tourism industry, island states and Pacific Island communities.

Structure of the book and chapter summaries

Part I (Chapters 1–8) provides deep academic treatment concerning the role of tourism within the 13 Pacific Island states, including an overview of the region covered and a focus on relevant topics of tourism enquiry that intersect with these states, though mainly in their collective form. The volume then goes on to deal with the island countries within their three regional clusters, Melanesia (Part II, Chapters 9–16), Micronesia (Part III, Chapters 17–21) and Polynesia (Part IV, Chapters 22–26), thus, ensuring an inclusive investigation of the wider region. The final chapter, the Conclusion (Chapter 27), largely deals with identifying and justifying future research trajectories that are valuable in extending the current field of enquiry.

Part I: tourism and small island states in the Pacific: conceptual overview and regional context

In Chapter 2, Hughes examines tourism's interrelationships with the United Nations' (UN) Sustainable Development Goals (SDGs) in the context of small Pacific Island states. The SDGs were developed as a framework to measure progress globally toward achieving long-term sustainable and inclusive development. As a major economic sector in many self-governing states, tourism arguably has a key role in achieving the SDGs. This chapter, thus, examines the relevance of the SDGs for tourism in the Pacific and the potential for the goals to respond to local priorities. It illustrates how specific tourism-focused goals and targets align with Pacific Island development plans and strategies at a regional and national level but also demonstrates where Pacific Island priorities are not as well-reflected in the SDGs. Findings show that strategies are founded on national values and principles with policies formulated from a position of states as stewards rather than managers of natural and cultural resources, reflecting the responsibility to protect cultural and environmental heritage for future generations. Hughes emphasizes that attention to local perspectives and traditional knowledge presents opportunities to understand how Pacific tourism can address the SDGs in which culture plays a central role within sustainable tourism.

In Chapter 3, Hall and Amore conceptualize tourism's association with the new regional governance of 'Pacific large ocean states' in which international tourism represents a key economic driver for development. The chapter addresses both global and inherent characteristics of these states to highlight governance issues in the current policy environment. The discussion looks at tourism-relevant perspectives such as global trade, hyper-neoliberalism, climate change, biodiversity loss and the increased frequency and magnitude of natural disasters. Significantly, the repositioning of the new regionalism that is emerging from the Pacific itself and the reframing of Pacific Island states as large ocean states rather than as small islands is also indicative of a broader reaction to neoliberalism and the need to focus on critical economic and environmental issues within which tourism is embedded. The authors conclude that Pacific Island states should embrace a revised metagovernance framework that places destination resilience and long-term sustainability within the context of the region's new policy trajectory.

In Chapter 4, Timothy deals with tourism mobility and transport issues in Pacific Island states, examining the vagaries of human mobility and travel. He distinguishes three main markets of global and intra-regional travel. The first of these is beach-based, sun-sea-sand tourists who often spend their holidays at seaside resorts. The second market is diasporic islanders who live abroad, either in other Pacific states or beyond, where travel to and within the region is big business and supports a travel infrastructure that might not otherwise be viable. The third market concerns arrivals from nations which are politically and socio-economically connected to Pacific Island states, such as the current free association status of several small states with former patron states within or outside the Pacific. This enquiry recognizes the number of mobility challenges that keep some countries from developing mainstream tourism sectors, including fragmented geography, physical isolation, small physical and demographic sizes, small economies of scale, political fragmentation, transportation irregularities and the effects of climate change. The discussion stresses that these factors significantly affect the development of tourism, the provision of transportation options and people's mobility within and between countries. The chapter acknowledges the enormous influence that the COVID-19 pandemic has had on transportation and access in the region.

In Chapter 5, Pratt and Kuilamu provide an overview of the economic contribution of tourism to Pacific Island economies and separately for Melanesia, Micronesia and Polynesia. The work tests the 'tourism-led growth hypothesis' (TLGH) for Pacific Island economies, stating that an expansion in international tourism paves the way for economic growth because international tourism increases foreign exchange earnings that contribute to capital goods, further stimulating economic growth and employment. Alternatively, economic growth can stimulate tourism income, the so-called 'growth-led tourism hypothesis' (GLTH). This can occur when a growing economy can create tourism-related business opportunities or when economies of scale, due to economic growth, decrease costs for the tourism sector, thus, making tourism more competitive. Using pre-pandemic data, the work tests whether the TLGH or the GLTH holds for different Pacific Island economies. The authors find that, for many of these countries, there are few direct corresponding relationships between tourism and economic growth; expectations include Palau, where tourism 'Granger causes' economic growth and economic growth, in turn, 'Granger causes' tourism. For Fiji, economic growth 'Granger causes' tourism, while for the Solomon Islands and Vanuatu, tourism 'Granger causes' economic growth. Pratt and Kuilamu emphasize that more restructuring should take place post-pandemic to enable tourism to be an engine for economic growth in the region.

In Chapter 6, Loh, Patovaki and Kumar evaluate crisis-related tourism recovery strategies in the Pacific Island countries, providing suggestions to sustain the tourism industry in the long-term, especially as tourism businesses and community livelihoods in the Pacific Island countries have been affected by the COVID-19 pandemic. The work deals with three case studies – Fiji, the Solomon Islands and Samoa – to develop a deeper understanding of the challenges faced by each country before and during the pandemic. The discussion asserts that, whilst the almost immediate response was to utilize foreign and government aid to keep economies afloat, a long-term sustainable approach to tourism development is needed, one which pursues responsible tourism consumption and production. The long-term remedies include participating in safe travel bubbles, augmenting the role of the private sector in the tourism economy, technologizing infrastructures and shifting toward regenerative tourism.

In Chapter 7, Harrison critically reviews the 13 self-governing Pacific Island states, examining the similarities and differences between the island states in terms of tourism visitation numbers, tourism intensity and tourism-related development prior to the outbreak of COVID-19 in early 2020. During the decade leading to the pandemic, the sector was growing quite significantly, though in some of these states more than others. This was noticeable in terms of some Pacific Island states displaying particular mass tourism traits. Focus is directed to those states which received the greatest number of tourists relative to their population (the Cook Islands, Palau and Niue) and to destinations with a similar number of tourists to residents (Fiji and Samoa), leading to an inspection of states which received far fewer tourists than their population (Vanuatu, Tonga and the FSM). The final sections of the chapter concern a critical reflection on mini-states with significantly fewer tourists (Nauru, Kiribati, Tuvalu and the Marshall Islands) and on the Solomon Islands, as being atypical to the other 12 Pacific Island states. Harrison asserts that, while fully fledged mass tourism did not exist in any of the countries surveyed, some characteristics were notably evident in a few states. The implication of this assessment is to recognize the tourism diversities and uniqueness of each Pacific Island country.

In Chapter 8, Chong and Stephenson decipher tourism's auspicious and inauspicious relationships with food and agricultural production in the Pacific Island states and especially in terms of the production of local food for tourism consumption. The work identifies the range of challenges that these countries face, such as climate change and unpredictable natural disasters, urbanization, westernization, rural migration and food security. The chapter draws reference to the case of Fiji, indicating sectoral linkages between tourism and particular local food and agricultural products but also acknowledges the prevalence of concerns related to product ownership, food choice and product quality. The importance of innovation, training and the acquisition of specific skills are seen as important ways forward. The second case illustration concerns the Solomon Islands, indicating the potential of agritourism and food tours in strengthening the sectoral linkages between tourism and the local food and agricultural sectors and identifying the need for capacity building in terms of developing professional expertise and advancing a strong food security framework. Chong and Stephenson indicate that as the relationship between tourism consumption and food/agricultural production in the Pacific Island region is generally inauspicious, the intersections between tourism, food and agriculture need to be rethought, particularly in light of lessons learnt as a consequence of the impacts of the COVID-19 pandemic.

Part II: tourism and island states in Melanesia

In Chapter 9, Halter examines Australian travel writing and notions of savagery in Melanesia. He notes that, although the types of tourists that have visited the Pacific Islands since the late nineteenth century have changed over time, these countries have been portrayed as static and unchanging and also imagined to be in a primordial and savage state. This chapter explores representations of savagery in the Melanesian region, with a particular focus on three Australian travellers: Beatrice Grimshaw, who visited Fiji in the 1900s; Jack McLaren, who visited the Solomon Islands in the 1920s; and Colin Simpson, who visited Vanuatu in the 1950s. The work investigates how representations of the Melanesian savage changed over time, from a figure employed in colonial narratives to encourage settlement and economic development to a marketing tool for the tourism and publishing industries. Travel accounts also show that tourists do not always reconcile generalized stereotypes with the reality that they encounter during their travels. Halter emphasizes that a historical approach to Pacific tourism is needed to understand the colonial roots of travel in the Pacific and the ways in which history continues to shape the contemporary tourism industry.

In Chapter 10, Mohamad Noor, Patovaki and Stephenson identify tourism development challenges in the Solomon Islands, especially insufficient infrastructure, ongoing political and ethnic conflicts, environmental threats, land disputes, economic challenges associated with child labour and the COVID-19 pandemic. The authors argue that there is potential to develop the tourism industry, particularly in terms of attracting lucrative markets from Australia and China, promoting a heritage tourism product that is associated with World War II nostalgia, hosting extensive dive tourism locations and developing sustainable forms of tourism in relation to cultural and ecological conservation. Attention is drawn to the United Nations Educational, Scientific and Cultural Organization (UNESCO) World Heritage Site (WHS) of East Rennell, which has been criticized for not fostering beneficial economic and cultural outcomes for the local community. Focus is also directed to Marovo Lagoon which has the potential to develop eco-friendly forms of tourism and be considered as a WHS, though there is a need for strong support from the central government and more cultural and political awareness from environmentalists and non-government organizations. The chapter concludes that, although there are substantial tourism development opportunities for the Solomon Islands, particularly in the post-COVID-19 context, any advancement must be done in an environmentally and culturally sustainable manner.

In Chapter 11, Movono and Stephenson produce a critique of tourism policy and related planning directives in Fiji, examining the challenges faced in terms of policy direction, implementation and enforcement. The chapter initially provides a critical assessment of the pro-tourism development policies that have been established in Fiji since the 1920s, though rapidly increasing since the 1960s. It is noted that significant tourism growth prior to the COVID-19 pandemic symbolizes Fiji's uncomfortable economic dependency on tourism, which was substantially brought to light in early 2020 because of international border closures to control the COVID-19 virus. Ultimately, Movono and Stephenson's enquiry argues that tourism policy frameworks must intersect at multi-national and regional levels to ensure that national tourism policies pursue the UN SDGs so that the sustainability of policy implementation is achieved in Fiji. The discussion asserts that, despite Fiji's government being open to the idea of a sustainable tourism agenda, there is still a real need to develop policies that are purposeful to the sustainable cause and not preoccupied with a pro-tourism development stance.

In Chapter 12, Orsua, Cheer and Blaer examine gender empowerment in tourism development within the context of studying female bungalow hosts in Vanuatu. The tourism sector's potential to contribute to the empowerment of women in the country is the central theme of this chapter, which is explored through an examination of the pivotal role that female empowerment plays in sustainable tourism development. The female bungalow entrepreneur in the province of Shefa is the focus of this exploratory study, which employs a qualitative approach to uncover the challenges that women face in participating in tourism micro-enterprise development as well as factors hindering women's ability to exercise agency over their lives. Micro-enterprise development in the form of bungalow accommodation holds enormous potential, though without cultural and institutional change, hurdles to female empowerment will prevail. The chapter uncovers the impacts of entrepreneurship on bungalow hosts' psyche and on social and familial relationships, financial stability and political standing within the village.

In Chapter 13, Trupp investigates the nexus between tourism, micro-entrepreneurship and handicrafts in the Solomon Islands. He emphasizes that, despite the development challenges faced by the country's tourism industry, handicraft producers and vendors can carve out their own niches by becoming entrepreneurial and commodifying some of their cultural features, such as the production of weavings, shell money and wood carvings. Trupp assesses the evolvement and status quo of Honiara's handicraft businesses, especially the economic benefits and socio-cultural characteristics. The author explains that the country has kept a relatively strong focus on locally crafted products and artworks, though the commodification process has meant that there have been attempts to adapt to tourism demand and to the broader processes of international exchange and globalization. From an economic perspective, interviewed micro-entrepreneurs have been relatively satisfied with their income and from selling crafts and arts prior to the pandemic. The final discussion, thus, contemplates the financial and industrial challenges facing handicraft micro-businesses during the pandemic.

In Chapter 14, Trau and Ballard focus on community management of cultural tourism in the context of the WHS of Chief Roi Mata's Domain (CRMD) in Vanuatu. The chapter addresses the intersecting processes of globalization and localization as they manifest in a cultural tourism business managed by Indigenous landowners. The Lelema community, distributed between the two settlements of Lelepa Island and Mangaliliu, on the mainland of Efate Island, owns and manages CRMD. This was Vanuatu's first WHS, which was inscribed in July 2008 as a continuing cultural landscape. Alongside multiple other forces, means and agents, and in the face of intensified global and local pressures, Lelema villagers seek to mobilize the World Heritage status of CRMD to improve both economic development opportunities (primarily through its associated local tourism enterprise, Roi Mata Cultural Tours) and local heritage conservation measures. The authors examine the Lelema experience of development and conservation, unpacking ambiguities between local customary norms, global development and conservation agendas. The work also highlights alternative local perspectives on the success and failure of community-led tourism and heritage initiatives in Vanuatu.

In Chapter 15, Sharma and Trupp assess the social media marketing performance of Fiji's hotel industry, which has been facing financial problems because of the global pandemic. The authors examine the hotel industry's 'social media marketing' (SSM) performance through the use of a 'digital marketing framework' (DMF) – before and during the COVID-19 pandemic. This study largely involves a netnography of social media content and engagement concerning a resort on the island of Denarau in Fiji. The study highlights the importance of eight DMF themes: 'attract', 'reach', 'volume', 'influence', 'engage', 'retain', 'learn', and 'relate back'. The work found that the resort had already invested extensively to attract and retain their users by providing ease of access and updating content and learning about user preferences and developing lasting relationships with their customers. During the pandemic, however, the resort's promotional efforts on social media platforms reflected its strategic shift toward the domestic market in which targeted social media strategies were implemented to retain and keep potential international guests engaged during the pandemic. Nonetheless, social media performance was inconsistent during this period, implicating the need to train hotel staff in SMM.

In Chapter 16, Mafi-Stephens, Kuilamu and Trupp assess the opportunities and challenges of ecotourism development in Fiji. The chapter's objective is to assesses the development and current situation of ecotourism initiatives in Fiji, examining public sector perspectives on the principles of ecotourism and how these principles have been implemented at the local level. Two case studies are introduced – an ecotourism resort and a National Marine Park – particularly to comprehend the implementation of ecotourism projects and challenges faced by such projects at local levels. In their implementation of eco-oriented initiatives, these two ventures have been quite successful in protecting the environment and encouraging local community participation. Nonetheless, the authors emphasize that there is a dilemma between Fiji's socio-cultural and environmental attributes as a sustainable island destination and the challenges in implementing ecotourism policies and practices.

Part III: tourism and island states in Micronesia

In Chapter 17, Tan and Arumugam assess tourism development challenges and opportunities in the Marshall Islands. Crucially, the chapter indicates how the islands have been influenced and determined by the burdensome legacy of nuclear colonization. Nonetheless, the chapter takes an optimistic approach despite past and ongoing challenges, including the ramifications of the COVID-19 pandemic to inbound tourism, through focusing on the importance of tourism development strategies. Attention is drawn to capacity building, developing tourism infrastructure, establishing public and private sector collaborations and creating productive marketing activities. The authors emphasize the advancement of 'high value and low impact' tourism, ecological-based tourism and the promotion of a positive and resilient destination image.

In Chapter 18, Vyas and Stephenson deal with the environmental challenges of tourism development in Palau. Until 2015, Palau was becoming a mass tourism destination, and there was a concern that negative impacts associated with overtourism could intensify, though this advancement ceased when tourist arrivals from China decreased rapidly from 2015 to 2019. Palau's geopolitical alignment with Taiwan seemingly had repercussions on the Chinese market. Nonetheless, the reduction in tourism numbers complemented the government's redirection toward the production of high-end tourism, characterized by low volume and high-quality experiences. However, as the chapter highlights, this form of tourism, especially luxury tourism, is also problematic to the environment. Consequently, Vyas and Stephenson emphasize the need to consider experiential and sustainable forms of tourism, those which are more nature-driven and conservation-based as opposed to those forms of tourism which are overly consumptive. The authors note that, despite the range of challenges and incongruities identified, the government of Palau has been pioneering in its approach to protect and conserve the environment, illustrated through such initiatives as the 'visitor pledge'. The chapter draws on the importance of advancing and promoting environmental citizenship behaviour among tourists but also among tourism industry employees and stakeholders.

In Chapter 19, Michalena, Hills and Samuwai focus on how the UN SDGs can be applied to tourism development in Kiribati, especially with regard to the advancement of 'thematic tourism' – notably energy tourism and marine-based tourism. The chapter infers that, although Kiribati has developed national development and tourism development plans, neither has had a significant effect on the development of its tourism sector. Through examining how the UN 2030 Agenda for Sustainable Development can provide an opportunity to reframe tourism development in Kiribati, the work implicates the broad and interlinked benefits across the social, economic and environment spheres. The authors discuss how thematic tourism in Kiribati can provide sustainable tourism development possibilities, especially when it is linked to its special energy status and its vast marine protected area.

In Chapter 20, Ali and Stephenson examine the challenges and potentialities of tourism development in Nauru but first draw attention to the concern of non-tourism. The authors assert that Australia's asylum and refugee policy strategically immobilizes such migrants to the status of non-tourists. The subsequent discussion explains how Nauru has the potential to transition from a non-tourist destination to a tourism destination in the future development of the 'visiting friends and relatives' (VFR) market. Nauru's history of phosphate mining also presents an opportunity to market Nauru as an industrial heritage tourism destination within the context of a unique physical landscape. The authors, too, recognize the potential to advance small-scale, high-quality fishing experiences in Nauru, as the region is renowned for game fish. As the work infers, pursuing appropriate forms of tourism development could counteract the common perception of Nauru as a non-tourism place, particularly through the projection of this Pacific Island state as a destination of tourism transformation and tourist activity.

In Chapter 21, Stephenson, Trupp and Shah deal with tourism development in the FSM with the intention of identifying a range of concerns: the complexities of land management and ownership, including difficulties in securing land for tourism development, and the problems of accessing the country, airline connectivity, internal mobility and poor transport infrastructure. The work identifies the problems of climate change on daily life and tourism development as well as challenges relating to the representation and interpretation of traditional heritage sites. Although the FSM is characterized by a rich cultural and diverse heritage, both in terms of tangible and intangible cultural resources, there are long-term concerns relating to the need to integrate cultural heritage tourism with heritage conservation and the natural environment. The chapter notes that prior to the pandemic, the FSM government acknowledged tourism as one of the main pillars of the country's economic development agenda, despite the fact that tourism growth was not significant. Although the discussion infers that this objective is purposeful, especially if the country is to lessen its economic dependence on the US, such forms of economic assistance enable greater freedom of mobility of FSM citizens. Nonetheless, as a consequence of the economic impacts associated with the COVID-19 pandemic, the authors stress the importance of developing new niche forms of sustainable tourism for the country.

Part IV: tourism and island states in Polynesia

In Chapter 22, Lovelock, Tuipoloa-Utuva and Carr focus on a critical assessment of the challenges of sustainable tourism planning in Samoa, including the impact of COVID-19, or KOVITI-19 as it is known in Samoa. Crucially, the authors include a discussion of the susceptibility of Samoa to external shocks that reflect the vulnerabilities of being a small island. However, the chapter initially provides an overview of the current state of the tourism sector in Samoa, especially transport networks, tourism attractions and accommodation, and the country's tourism planning framework and tourism vision. The chapter recognizes that, although tourism activities in Samoa are significantly operated by local people, pathways to sustainability are not always clear and communities' acceptance of (and capacities to manage) tourism are by no means universal. In order to illustrate some of the challenges and complexities around sustainable tourism development in Samoa, the work draws reference to the island of Apolima as an example of a peripheral setting in which potential tourism development must be aligned to social and infrastructural needs. The final section presents a case study concerning the importance of utilizing transportation networks to produce socially sustainable tourism experiences in which participation on island buses can encourage tourists to experience 'fa'a Samoa' ('Samoan way of life').

In Chapter 23, Cave and Fili Grover examine the main challenges to tourism development in Tonga and recommend solutions to the concerns identified. The enquiry scans the historical, cultural and environmental backstories that shape contemporary Tongan life and the development of its tourism economy. The discussion indicates that Tonga's main challenges concern its tourism markets and accessibility issues and tourism's impact on culture, tradition and the environment, including critical attention to the threat of climate change on Tongan livelihoods and biodiversity. The authors assert that international tourism development is at odds with cultural patterns of traditional behaviour embedded in Tongan society. The chapter produces a positive line of enquiry suggesting useful ways forward, including long-term programmes and methodologies that would be beneficial to understanding more of tourism's potential, especially in relation to market focused research, traditional practices research and ecological-based research.

In Chapter 24, Ali argues for the viability of Niue pursuing yoga (retreat) tourism as a complementary regenerative and degrowth form of tourism, especially in light of the mobility challenges associated with the global pandemic. The author conceptualizes yoga tourism as a type of niche sport tourism that promotes wellness experiences and a form of tourism that is appropriate to the size of the economy, limited land capacity and operational constraints of the destination. It is inferred that this form of tourism is aligned to the Niue National Strategic Plan (2016–2026), emphasizing the importance of preserving natural and cultural environments. As the discussion emphasizes, given Niue's invaluable coast as an environmental asset, sustainable sport tourism development in the form of yoga (retreat) tourism synchronizes with the island's natural therapeutic backdrop for those journeying in the pursuit to restore the 'mind, body and soul'. Protecting the country's natural assets by proposing yoga tourism as a special interest product and experience could be a resourceful way forward to delimit overtourism and stimulate regeneration, especially with sustainability at the helm of a post-COVID-19 recovery strategy. Importantly, the chapter evokes the limitations of Western-centric principles and practices when discussing sustainable development, regeneration and degrowth in non-Western settings.

In Chapter 25, Chaichi and Leong deconstruct the impacts of tourism development in the Cook Islands and identify key sustainable ways forward. The discussion observes that, prior to the COVID-19 pandemic, the country experienced significant tourism growth and was optimistic that this would continue, but the unforeseen phenomenon of 'undertourism' caused by the global pandemic lockdown has been challenging to the country's economy. The discussion acknowledges that, although the country needs to re-establish itself as a tourism destination, the prevailing question concerns the future direction that the Cook Islands should take. Whilst the government recognizes the importance of sustainability, with several important initiatives in place (e.g., a marine park, the Mana Tiaki Eco Certification Programme and the Sustainable Tourism Development Policy Framework), its pro-tourism stance would need to be re-addressed. By critically reflecting on the various socio-cultural impacts, human resource needs and environmental effects of tourism development, Chaichi and Leong stress that it is imperative that the country moves in a more sustainable direction.

In Chapter 26, Prideaux and Trupp examine Tuvalu's tourism industry, observing that the island has had little academic attention in tourism studies, which is probably because Tuvalu is one of the least visited nations in the world. The authors identify remoteness, inadequate infrastructure, lack of an appealing selling proposition, an image problem and the growing dangers posed by climate change as the country's key tourism development challenges. The chapter introduces the idea of undertourism, a problem that is the antithesis of overtourism problems faced by many well-known European tourism destinations. While acknowledging these problems, the work identifies opportunities for the country to participate in the global tourism industry, including a range of niche tourism opportunities that could be developed: 'off the beaten track' tourism, ecotourism, bucket list tourism, last chance tourism and climate change tourism. Evidence concerning the image problems faced by Tuvalu from climate change and isolation are drawn from a media scan of 23 stories that focus on aspects of travel to Tuvalu. The chapter demonstrates how application of tourism theory, in this case 'push-pull', 'comparative and competitive advantage' and 'periphery' can be used in identifying opportunities for tourism development in small island states.

In Chapter 27, the concluding chapter, Stephenson and Timothy expand on some of the research themes highlighted in the volume and also identify research trajectories which are thought to advance a critical conception of tourism within the wider study of the self-governing Pacific Island states. Consequently, this chapter assembles six thematic-based research trajectories: (1) tourism, sovereignty and geopolitics; (2) tourism, culture and heritage; (3) climate change, the environment and the 'blue economy'; (4) gender, frontline workers and social equality; (5) travel and tourism mobilities of Pacific Islanders; and (6) Indigenous methodologies and local diversities. The chapter also connects the work to issues raised in the previous contributing chapters, which helps to reflect on the volume and its interconnecting parts.

References

Akauola, L., L. Ilaiu and A. Samate (1980) 'The social and cultural impact of tourism in Tonga'. In: R. Crocombe and F. Rajotte (eds.) *Pacific Tourism as Islanders See it*, Suva, Fiji: Institute of Pacific Studies, University of the South Pacific, pp. 17–23.

Alexeyeff, K. (2016) 'Re-purposing paradise: Tourism, image and affect'. In: K. Alexeyeff and J. Taylor (eds.) *Touring Pacific Cultures*, Acton, ACT: ANU Press, pp. 403–421.

Alexeyeff, K. and J. Taylor (eds.) (2016) *Touring Pacific Cultures*, Acton, ACT: ANU Press.

Allen, M.G. (2018) *Resource Extraction and Contentious States: Mining and the Politics of Scale in the Pacific Islands*, Singapore: Palgrave Macmillan.

Ashman, M. (1975) 'Micronesia tastes tourism'. In: B.R. Finney and K.A. Watson (eds.) *A New Kind of Sugar: Tourism in the Pacific*, Honolulu, HI: East-West Technology and Development Institute, pp. 135–143.

Baldacchino, G. and A. Wivel (eds.) (2020) *Handbook on the Politics of Small States*, Cheltenham: Edward Elgar Publishing.

Berno, T. and N. Douglas (1998) 'Tourism in the South Pacific: A Polynesia/Melanesia discussion', *Asia Pacific Journal of Tourism Research*, 2(2): 65–73.

Bertram, G. (2018) 'Why does the Cook Island still need overseas aid?', *The Journal of Pacific History*, 53(1): 44–63.

Bertram, I.G. and R.F. Watters (1985) 'The MIRAB economy in South Pacific Microstates', *Pacific Viewpoint*, 26(3): 497–519.

Bolabola, C.A.B. (1980) 'The impact of tourism on Fijian woodcarving'. In: F. Rajotte and R. Crocombe (eds.) *Pacific Tourism: As Islanders See It*, Suva: Institute of Pacific Studies, University of the South Pacific, pp. 93–97.

Britton, S.G. (1980) 'The spatial organisation of tourism in a neo-colonial economy: A Fiji case study', *Pacific Viewpoint*, 21(2): 144–165.

Britton, S.G. (1982) 'The political economy of tourism in the third world', *Annals of Tourism Research*, 9(3): 331–358.

Britton, S.G. (1983) *Tourism and Underdevelopment in Fiji*, Canberra: Australian National University Press.

Buck, P. and C.M. Hall (1996) 'The Cook Islands'. In: C.M. Hall and S.J. Page (eds.) *Tourism in the Pacific: Issues and Cases*, London: International Thomson Business Press, pp. 219–234.

Burley, D.W. (2013) 'Fijian Polygenesis and the Melanesian/Polynesian divide', *Current Anthropology*, 54(4): 436–462.

Campbell, S. (2018) 'Tourism Pacific cultures', *Pacific Affairs*, 22 June, pp. 211–212.

Caneen, J.M. (2014) 'Tourism and cultural identity: The case of the Polynesian Cultural Center', *Athens Journal of Tourism*, 1(2): 101–120.

Cooper, C. and C.M. Hall (eds.) (2005) *Oceania: A Tourism Handbook*, Clevedon: Channel View Publisher.

Corbett, J., W. Veenendaal and L. Ugyel (2017) 'Why monarchy persists in small states: The cases of Tonga, Bhutan and Liechtenstein', *Democratization*, 24(4): 689–706.

Countryeconomy.com (2022) 'Countries data: Demographic and economy'. Available online at https://countryeconomy.com/countries (accessed 7 May 2022).

Craig-Smith, S.J. (2005) 'Global warming and tourism in Oceania'. In: C. Cooper and C. M. Hall (eds.) *Oceania: A Tourism Handbook*, Clevedon: Channel View Publications, pp. 353–361.

de Burlo, C. (1996) 'Vanuatu'. In: C.M. Hall and S.J. Page (eds.) *Tourism in the Pacific: Issues and Cases*, London: International Thomson Business Press, pp. 235–255.

d'Hauteserre, A.-M. (2005) 'Customary practices and tourism development in the French Pacific'. In: C. Cooper and C. M. Hall (eds.) *Oceania: A Tourism Handbook*, Clevedon: Channel View Publications, pp. 308–320.

de Kadt, E. (ed.) (1979) *Tourism, Passport to Development? Perspectives on the Social and Cultural Effects of Tourism in Developing Countries*, Oxford: Oxford University Press.

Douglas, B. (2010) 'Terra Australis to Oceania: Racial geography in the "fifth part of the world"', *The Journal of Pacific History*, 45(2): 179–210.

Douglas, N. (2004) 'Towards a history of tourism in Solomon Islands', *The Journal of Pacific Studies*, 26(1&2): 29–49.

Droessler, H. (2017) 'Colonialism by deferral: Samoa under the Tridominium, 1889–1899'. In: S. Rud and S. Ivarsson (eds.) *Rethinking the Colonial State (Political Power and Social Theory)*, Bingley: Emerald Publishing Limited, pp. 203–224.

Duncan, R. and H. Nakagawa (2006) *Obstacles to Economic Growth in Six Pacific Island Countries*, Washington, DC: World Bank.

ESCAP (United Nations Economic and Social Commission for Asia and the Pacific) (2022) *Ensuring Public Debt Sustainability in the Pacific Small Island Developing States*, March'. Available online at https://www.unescap.org/sites/default/d8files/event-documents/Issues%20paper%20-%20Ensuring%20Public%20Debt%20Sustainability_4Apr2022.pdf (accessed 12 June 2022).

Fagence, M. (1996) 'Planning issues in Pacific tourism'. In: C.M. Hall and S.J. Page (eds.) *Tourism in the Pacific: Issues and Cases*, London: International Thomson Business Press, pp. 91–108.

Farrell, B.H. (1979) 'Tourism's human conflicts: Cases from the Pacific', *Annals of Tourism Research*, 6(2): 122–136.

Finney, B. (1993) 'Rediscovering Polynesian navigation through experimental voyaging', *Journal of Navigation*, 46(3): 383–394.

Finney, B.R. and K.A. Watson (1975) *A New Kind of Sugar: Tourism in the Pacific*, Honolulu, HI: East-West Technology and Development Institute.

Firth, S. (2018) *Instability in the Pacific Islands: A Status Report*, Sydney: Lowy Institute.

Fong, P. (1980) 'Tourism and urbanization in Nausori'. In: F. Rajotte and R. Crocombe (eds.), *Pacific Tourism: As Islanders See It*, Suva: Institute of Pacific Studies, pp. 87–88.

Fontenoy, P.E. (1997) 'Ginseng, otter skins, and sandalwood: The conundrum of the China trade', *The Northern Mariner/Le Marin du Nord*, 7(1): 1–16.

Gibson, D. (2015) 'Community-based tourism in Fiji: A case study of Wayalailai Ecohaven Resort, Yasawa Island Group'. In: S. Pratt and D. Harrison (eds.), *Tourism in Pacific Islands*, Abingdon: Routleldge, pp. 118–133.

Greenwood, D.J. (1972) 'Tourism as an agent of change: A Spanish Basque Case', *Ethnology*, 11(1): 80–91.

Hall, C.M. and S.J. Page (eds.) (1996a) *Tourism in the Pacific: Issues and Cases*, London: International Thomson Business Press.

Hall, C.M. and S.J. Page (1996b) 'Australia's and New Zealand's role in Pacific tourism: Aid, trade and travel'. In: C.M. Hall and S.J. Page (eds.) *Tourism in the Pacific: Issues and Cases*, London: International Thomson Business Press, pp. 161–189.

Hall, C.M. and S.J. Page (1996c) 'Introduction: The context of tourism development in the South Pacific'. In: C.M. Hall and S.J. Page (eds.) *Tourism in the Pacific: Issues and Cases*, London: International Thomson Business Press, pp. 1–15.

Hanlon, D. (2009) 'The "sea of little lands": Examining Micronesia's place in "our sea of islands"', *The Contemporary Pacific*, 21(1): 91–100.

Harrison, D. (1994) 'Tourism, capitalism and development in less developed countries'. In: L. Sklair (ed.), *Capitalism and Development*, London: Routledge, pp. 232–257.

Harrison, D. (2003) *Pacific Island Tourism*, New York: Cognizant Communication Corporation.

Harrison, D. (2004a) 'Tourism in the Pacific islands', *Journal of Pacific Studies*, 26(1&2): 199–232.

Harrison, D. (2004b) 'Levuka, Fiji: Tourism, UNESCO and the world heritage list', *Journal of Pacific Studies*, 26(1–2): 199–232.

Harrison, D. and B. Prasad (2013) 'The contribution of tourism to the development of Fiji and other Pacific island countries'. In: C.A. Tisdell (ed.), *Handbook of Tourism Economics: Analysis, New Applications and Case Studies*, Singapore: World Scientific, pp. 741–761.

Hau'ofa, E. (1994) 'Our sea of islands', *Contemporary Pacific*, 6(1): 148–161.

Hess, J.S. (2016) 'A review of "Tourism in Pacific Islands: Current issues and future challenges", edited by Stephen Pratt and David Harrison', *Journal of Sustainable Tourism*, 24(7): 1059–1060.

Juswanto, W. and Z. Ali (2016) 'Economic growth and sustainable development in the Pacific island countries', No. 2016-6 (December), Tokyo: *Asian Development Bank Institute*. Available online from https://www.adb.org/sites/default/files/publication/219266/adbi-pb2016-6.pdf (accessed 13 May 2022).
Kabutaulaka, T.T. (2000) 'Rumble in the jungle: Land, culture and (un)sustainable logging in Solomon islands', *Culture and Sustainable Development in the Pacific*, 33: 88–97.
Kent, N. (1975) 'A new kind of sugar'. In: B.R. Finney and K.A. Watson (eds.) *A New Kind of Sugar: Tourism in the Pacific*, Honolulu, HI: East-West Technology and Development Institute, pp. 169–199.
King, B. (2005) 'Pacific Island tourism', *Tourism Management*, 26: 623–633.
Kiste, R.C., M. Kahn and R.C. Suggs (2022) 'Polynesian culture', *Encyclopedia Britannica*. Available online at https://www.britannica.com/place/Polynesia (accessed 5 June 2022).
Koth, B. (2015) 'Hosting bluewater sailors: A destination model for the Pacific Islands'. In: S. Pratt and D. Harrison (eds.) *Tourism in Pacific Islands*, Abingdon: Routledge, pp. 219–238.
Linnekin, J. and L. Poyer (eds.) (1990) *Cultural Identity and Ethnicity in the Pacific*, Honolulu, HI: University of Hawaii Press.
Lindstrom, L. (2016) 'Darkness and light in black and white: Travelling mission imagery from the new Hebrides'. In: K. Alexeyeff and J. Taylor (eds.) *Touring Pacific Cultures*, Acton ACT: ANU Press, pp. 33–57.
Lockhart, D.G., P.J. Schembri and D.W. Smith (eds.) (1993) *The Development Process in Small Island States*, London: Routledge.
Martins, K. (2020) 'Polynesian navigation and settlement of the Pacific', *World History Encyclopedia*. Available online at https://www.worldhistory.org/article/1586/polynesian-navigation--settlement-of-the-pacific/ (accessed 10 May 2022).
McGarry, D., S. Chanel and E. Samaglou (2021) 'Deserted islands: Pacific resorts struggle to survive a year without tourists', *The Guardian*, 2 April. Available online at https://www.theguardian.com/world/2021/apr/03/covid-coronavirus-deserted-islands-pacific-resorts-struggle-to-survive-a-year-without-tourists (accessed 20 November 2021).
McGavin, K. (2014) 'Being "nesian": Pacific islander identity in Australia', *The Contemporary Pacific*, 26(1): 126–154.
Mehmet, O. and M. Tahiroglu (2002) 'Growth and equity in microstates: Does size matter in development?' *International Journal of Social Economics*, 29(1/2): 152–162.
Meller, N. (1987) 'Pacific Island microstates', *Journal of International Affairs*, 41(1): 109–134.
Milne, S. (1992) 'Tourism and development in South Pacific microstates', *Annals of Tourism Research*, 19(2): 191–212.
Milne, S. (1996) 'Tourism marketing and computer reservation systems in the Pacific'. In: C.M. Hall and S.J. Page (eds.) *Tourism in the Pacific: Issues and Cases*, London: International Thomson Business Press, pp. 109–129.
Minerbi, L. (1996) 'Hawaii'. In: C.M. Hall and S.J. Page (eds.) *Tourism in the Pacific: Issues and Cases*, London: International Thomson Business Press, pp. 190–204.
Moulin, J.F. (2017) 'Touristic encounters: Imag(in)ing Tahiti and its performing arts'. In: K. Gillespie, S. Treloyn and D. Niles (eds.) *A Distinctive Voice in the Antipodes: Essays in Honour of Stephen A. Wild*, Canberra: Australian National University Press, pp. 267–306.
Nash, D. (1977) 'Tourism as a form of imperialism'. In: V. Smith (ed.), *Hosts and Guests: The Anthropology of Tourism*, Philadelphia, PA: University of Pennsylvania Press, pp. 33–38.
Nunn, P.D., L. Kumar, I. Eliot and R.F. McLean (2016) 'Classifying Pacific islands', *Geoscience Letters*, 3(1): 1–19.
Opeskin, O. and T. MacDermott (2009) 'Resources, population and migration in the Pacific connecting islands and rim'. *Asia Pacific Viewpoint*, 50(3): 353–373.
Pacha, A. (2022) 'Explained: What is the Pacific "ring of fire"?' *The Indian Express*, 19 January. Available online at https://indianexpress.com/article/explained/tonga-volcano-eruption-pacific-ring-of-fire-explained-7731351/ (accessed 17 May 2022).
Petersen, G. (2009) *Traditional Micronesian Societies: Adaptation, Integration, and Political Organization*, Honolulu, HI: University of Hawai'i Press.
Phipps, P. (2016) 'Performing indigenous sovereignties across the Pacific'. In: K. Alexeyeff and J. Taylor (eds.) *Touring Pacific Cultures*, Acton, ACT: ANU Press, pp. 245–265.
Plange, N. (1996) 'Fiji'. In: C.M. Hall and S.J. Page (eds.) *Tourism in the Pacific: Issues and Cases*, London: International Thomson Business Press, pp. 205–218.
Pratt, S. and D. Harrison (eds.) (2015) *Tourism and in Pacific Islands: Current Issues and Future Challenges*, Abingdon: Routledge.

Rajotte, F. and R. Crocombe (eds.) (1980) *Pacific Tourism: As Islanders See It*, Suva, Fiji: Institute of Pacific Studies, University of the South Pacific.

Ratter, B.M.W. (2018) *Geography of Small Islands: Outposts of Globalisation*, Cham: Springer Nature.

Remeikis, A. (2021) 'Pacific nations face "lost decade" due to economic cost of Covid', *The Guardian*, 29 September. Available online at https://www.theguardian.com/world/2021/sep/29/pacific-nations-face-lost-decade-due-to-economic-cost-of-covid (accessed 19 January 2022).

Ringer, G. (2004) 'Geographies of tourism and place in Micronesia: The "sleeping lady" awakes', *The Journal of Pacific Studies*, 26(1&2): 131–150.

Rudkin, B. and C.M. Hall (1996) 'Health and tourism in the Pacific'. In: C.M. Hall and S.J. Page (eds.) *Tourism in the Pacific: Issues and Cases*, London: International Thomson Business, pp. 130–145.

Russell, D. and J. Stabile (2003) 'Ecotourism in practice: Trekking the highlands of Makira Island, Solomon Islands'. In: D. Harrison (ed.), *Pacific Island Tourism*, New York: Cognizant Communication Corporation, pp. 38–57.

Samy, J. (1975) 'Crumbs from the table? The workers' share in tourism'. In: B.R. Finney and K.A. Watson (eds.) *A New Kind of Sugar: Tourism in the Pacific*, Honolulu, HI: East-West Technology and Development Institute, pp. 111–121.

Scheyvens, R. and E. Hughes (2015) 'Tourism and CSR in the Pacific'. In: S. Pratt and D. Harrison (eds.) *Tourism in Pacific Islands*, Abingdon: Routledge, pp. 134–147.

Shoji, S., R. Dahlgren and M. Nanzyo (1993) 'Terminology, concepts and geographic distribution of volcanic ash soils'. In: S. Shoji, M. Nanzyo and R. Dahlgren (eds.) *Volcanic Ash Soil – Genesis, Properties, and Utilization*, Developments in Soil Science 21, Amsterdam: Elsevier, pp. 1–5.

Sikivou, J. (1980) 'A conversation with two sword sellers'. In: F. Rajotte and R. Crocombe (eds.) *Pacific Tourism: As Islanders See It*, Suva, Fiji: Institute of Pacific Studies, University of the South Pacific, pp. 99–100.

Skinner, R.J. (1980) 'The impact of tourism on Niue'. In: F. Rajotte and R. Crocombe (eds.) *Pacific Tourism: As Islanders See It*, Suva, Fiji: Institute of Pacific Studies, The University of the South Pacific, pp. 60–64.

Smith, S.E. (2010) 'Uncharted waters: Has the Cook Islands become eligible for membership in the United Nations?', *New Zealand Journal of Public and International Law*, 8(2): 169–215.

Sönmez, S.F. (1998) 'Tourism, terrorism, and political instability', *Annals of Tourism Research*, 25(2): 416–456.

SPC (Pacific Community) (2022) 'Statistics by country', Statistics for Development Division, SPC. Available online at https://sdd.spc.int/all-countries (accessed 25 May 2022).

Stanton, M. (2003) 'Economics and tourism development on Easter Island'. In: D. Harrison (ed.) *Pacific Island Tourism*, New York: Cognizant Communication Corporation, pp. 110–124.

Steel, F. (2016) 'The cruise ship'. In: K. Alexeyeff and J. Taylor (eds.) *Touring Pacific Cultures*, Acton, ACT: ANU Press, pp. 61–71.

Stephenson, M.L. (2017) 'Deciphering international tourism development in the GCC region'. In: M.L. Stephenson and A. Al-Hamarneh (eds.), *International Tourism Development and the Gulf Cooperation Council States: Challenges and Opportunities*, London: Routledge, pp. 1–25.

Stephenson, M.L. and A. Al-Hamarneh (eds.) (2017) *International Tourism Development and the Gulf Cooperation Council States: Challenges and Opportunities*, London: Routledge.

Storlazzi, C., L. Erikson, S.B. Gingerich and C.I. Vloss (2021) 'The impact of sea-level rise and climate change on Pacific Ocean Atolls', Pacific Coastal and Marine Science Centre. Available online at https://www.usgs.gov/centers/pcmsc/science/impact-sea-level-rise-and-climate-change-pacific-ocean-atolls (accessed 2 June 2022).

Taylor, J. and K. Alexeyeff (2016) 'Departures and arrivals in touring Pacific cultures'. In: K. Alexeyeff and J. Taylor (eds.) *Touring Pacific Cultures*, Acton, ACT: ANU Press, pp. 1–28.

Teaiwa, T.L. (1996) 'Reviewed work: A new Oceania: Rediscovering our sea of islands by Eric Wadell, Vijay Naidu, Epei Hau'ofa', *The Contemporary Pacific*, 8(1): 214–217.

The Journal of Pacific Studies (2004) 'Edition of tourism in Pacific islands', Suva, Fiji: The University of the South Pacific, 26(1&2).

Thomas, N. (1989) 'The force of ethnology: Origins and significance of the Melanesia/Polynesia division', *Current Anthropology*, 30(1): 27–41.

Throsby, D. (2005) 'Pacific Island tourism: Reviews', *Pacific Economic Bulletin*, 20(1): 212–213.

Toatu, T. (2021) 'Unravelling the "Pacific paradox"-The case of Kiribati', *Pacific Economic Bulletin*, 16(1): 109–122.

Tong, P. and B. Tanentoa (1980) 'Urbanization, tourism and national environment (Kiribati)'. In: F. Rajotte and R. Crocombe (eds.), *Pacific Tourism: As Islanders See It*, Suva, Fiji: Institute of Pacific Studies, University of the South Pacific, pp. 127–132.

Townend, A. (2003) 'The strange death of the realm of New Zealand: The implications of a New Zealand Republic for the Cook Islands and Niue', *Victoria University of Wellington Law Review*, 34(3): 571–607.
Turner, L. and J. Ash (1975) *The Golden Hordes: International Tourism and the Pleasure Periphery*, London: Constable.
Twining-Ward, L. and T.S. Tuailemafua (2004) 'Small island tourism: Monitoring sustainable tourism development in Samoa', *The Journal of Pacific Studies*, 26(1&2): 77–103.
UNWTO (United Nations World Tourism Organization) (2021) *International Tourism Highlights*, 2020 edition, Madrid: UNWTO.
UNWTO (United Nations World Tourism Organization) (May 2022) *World Tourism Barometer*, 20(3), Madrid: UNWTO.
Urry, J. (2002) *The Tourist Gaze: Leisure and Travel in Contemporary Societies*, second edition, London: Sage.
West, F.J. and S. Foster (2020) 'Pacific islands', *Encyclopedia Britannica*, 17 November. Available online at https://www.britannica.com/place/Pacific-Islands (accessed 9 May 2022).
White, C.M. (2007) 'More authentic than thou: Authenticity and othering in Fiji tourism discourse', *Tourist Studies*, 7(1): 25–49.
Wilks, J. (2005) 'Destination risk management in Oceania'. In: C. Cooper and C.M. Hall (eds.) *Oceania: A Tourism Handbook*, Clevedon: Channel View Publications, pp. 335–352.
Wolf, F., W.L. Filho, P. Singh, N. Scherle, D. Reiser, J. Telesford, I.B. Miljković, P.H. Havea, C. Li, D. Surroop and M. Kovaleva (2021) 'Influences of climate change on tourism development in small Pacific island states', *Sustainability*, 13(8): 4223.
World Population Review (2022) *'Smallest countries 2022'*. Available online at https://worldpopulationreview.com/country-rankings/smallest-countries (accessed 13 May 2022).

2

TOURISM AND THE SUSTAINABLE DEVELOPMENT GOALS IN SMALL ISLAND STATES IN THE PACIFIC

Emma Hughes

Introduction

The United Nations (UN) Sustainable Development Goals (SDGs) are the central platform of the 2030 Agenda for Sustainable Development, a universal action plan for 'people, planet and prosperity' (United Nations 2015). Extensive stakeholder input into the development of the goals from civil society and Indigenous peoples, as well as governments and the private sector, extended the eight development priorities of the Millennium Development Goals to the final line up of 17 SDGs with 169 nested targets. Pacific Island nations took a leading role in negotiating a focus on the ocean as a stand-alone goal (SDG 14; PIFS 2015), and the goals have since been integrated within national policies and development strategies across the region. As a major economic sector in many Pacific Island states (SPTO 2019a; World Bank 2017), tourism has a key role to play in enabling the achievement of the goals in the region. Whilst research has emerged on the alignment of tourism with different goals and targets, greater discussion is needed on the localization of goals to ensure the implementation of the agenda responds to national priorities in practice. This chapter, thus, examines the relevance of the SDG goals and targets to tourism development priorities in the Pacific. In a post-COVID-19 context in which the ability to meet the SDGs is jeopardized (United Nations 2020), this will be all the more significant.

The following section introduces the place of tourism in Agenda 2030 and makes links between tourism and the SDGs, highlighting implications for the Pacific. The chapter then goes on to show how tourism planning in independent island states across the Pacific is responding to the challenge of addressing the SDGs through examining national development and tourism strategies and policies. It examines how specific strategies and tourism policies are linked directly or indirectly to the SDGs, where goals and targets are mapped or linked, and the challenges of implementation. The chapter goes on to identify to what extent local priorities and perspectives are reflected in the goals. The chapter concludes by considering how the SDGs can support tourism development in the Pacific and how local and regional approaches can inform sustainable tourism management.

 DOI: 10.4324/9780429019968-3

Tourism in 'Agenda 2030'

Businesses are positioned in a key role alongside governments and civil society in delivering on the SDG targets (Scheyvens et al. 2016), with tourism clearly specified as a tool for sustainable development within Agenda 2030. Three goals explicitly refer to engagement with the tourism sector: Target 8.9 seeks to 'promote sustainable tourism that creates jobs and promotes local culture and products'; Target 12b promotes the development and implementation of 'tools to monitor sustainable development impacts for sustainable tourism that creates jobs and promotes local culture and products'; and Target 14.7 aims to 'increase the economic benefits to small island development states and least developed countries from the sustainable use of marine resources, including through sustainable management of fisheries, aquaculture and tourism' (UNWTO 2015). The United Nations World Tourism Organization (UNWTO) identified the relevance and significance of tourism for all the goals, spanning renewable energy and sustainable infrastructure, preserving biodiversity and halting land degradation, enabling community development and empowering women, in addition to the potential for economic growth and development. The South Pacific Tourism Organisation (SPTO) notes issues of particular importance for the region, including employment and livelihood opportunities, climate change, water management and preserving marine resources and terrestrial ecosystems (SPTO 2017).

Tourism scholars have issued a call for critical analysis of how global tourism systems can achieve sustainable and inclusive development, taking into account power differentials, different worldviews and diverse methodologies and perspectives (Boluk et al. 2017). Research advocating for degrowth as an alternative paradigm for the industry (Higgins-Desbiolles et al. 2019) now holds particular resonance given the impact of COVID-19 on tourism. The next section provides an overview of the links between the SDGs and tourism and, in particular, the relevance to the Pacific context.

Tourism and the SDGs: critical issues for the Pacific

Tourism is the main driver of economic development in the Pacific, constituting a major source of income and employment for many Pacific Island countries (World Bank 2017). The sector contributed more than US$3.8 billion to the region in 2018 or 11.1% of gross domestic product (GDP) (SPTO 2019a). In 2019, prior to the outbreak of the COVID-19 pandemic, arrivals were forecasted to grow by an average of 3.3% reaching 2.7 million by 2024 (SPTO 2019b). The lead tourism destination in the region was Fiji, which accounted for 40% of the region's arrivals with all other destinations receiving a share of 10% or less. Other major destinations were the Cook Islands, Samoa, Palau and Vanuatu, with tourism being a significant employer for the Federated States of Micronesia, Tuvalu, Niue and Tonga. A focus on tourism as an income-generation mechanism, however, neglects to address the multi-dimensional nature of poverty, including wellbeing, human rights and empowerment as well as enabling secure livelihoods (Scheyvens and Hughes 2019). Indeed, development impacts resulting from tourism in the region have been 'muted' (Cheer et al. 2018). It is important, then, to examine how national tourism planning is responding to the challenge of addressing the SDGs and to what extent local and Indigenous perspectives are reflected in the goals. Key touch points between the SDGs and tourism in the Pacific are identified below, drawing on research from UNWTO (2015), SPTO (2017), Scheyvens (2018), and Cheer et al. (2018). These points are summarized in Table 2.1, identifying specific alignment with each of the SDGs and linked to the SDG themes of people, prosperity, planet, peace and partnership (UNDP 2015). Interconnections between the themes and goals are also assumed (Le Blanc et al. 2017; Pradhan et al. 2017).

Table 2.1 SDGs and tourism in the Pacific

SDG by theme	*Relevance to tourism and the Pacific*
People	
Goal 1. End poverty in all its forms everywhere.	• Income-generation supporting secure livelihoods and access to services and infrastructure. • Multiple vulnerabilities, e.g., seasonality of tourism, labour conditions, climate change and natural disasters, access to land.
Goal 2. End hunger, achieve food security and improved nutrition and promote sustainable agriculture.	• Local produce as a contributor to the securing of livelihoods, sustainable agriculture and fisheries, agritourism and good nutrition. • Heavy reliance on imported goods and impacted by climate change.
Goal 3. Ensure healthy lives and promote wellbeing for all at all ages.	• Improved health care services in tourist locations. • Impact of tourism on mental, spiritual and emotional wellbeing of residents and tourism workers.
Goal 4. Ensure inclusive and equitable quality education and promote lifelong learning opportunities for all.	• Accessible opportunities for vocational education. • Training for promotion and management positions.
Goal 5. Achieve gender equality and empower all women and girls.	• High proportion of women working in tourism, fisheries, food production and handicrafts. • Gender discrimination, low pay, insecure contracts, sex tourism and sexual harassment and lack of inclusion in policy and planning processes.
Goal 6. Ensure availability and sustainable management of water and sanitation for all.	• Access to water and sanitation. • Water usage and management, including wastewater and pollution control.
Prosperity	
Goal 7. Ensure access to affordable, reliable, sustainable and modern energy for all.	• Energy provision in challenging remote destinations. • Importance of renewable energy sources.
Goal 8. Promote sustained, inclusive and sustainable economic growth, full and productive employment and decent work for all.	• Decent work with fair pay, secure contracts, labour rights, career pathways and dignity. • Opportunities for small businesses and locally owned/managed operations. • Supporting and protecting local culture and heritage.
Goal 9. Build resilient infrastructure, promote inclusive and sustainable industrialization and foster innovation.	• Good infrastructure for tourism and local populations in challenging remote locations.
Goal 10. Reduce inequality within and among countries.	• Significant contributor to GDP. • Distribution of benefits between tourism and non-tourism regions. • Access to opportunities, finance and information for small businesses. • Economic leakages.
Goal 11. Make cities and human settlements inclusive, safe, resilient and sustainable.	• Urban infrastructure, transport, waste treatment and safety. • Protecting cultural and natural heritage.

(*Continued*)

Table 2.1 (Continued)

SDG by theme	*Relevance to tourism and the Pacific*
Planet	
Goal 12. Ensure sustainable consumption and production patterns.	• Pressures on the environment, energy and water equity. • Use of sustainable products; for example, sunscreens to protect reefs and use of recyclable materials. • Sourcing local products or using local materials.
Goal 13. Take urgent action to combat climate change and its impacts.	• Climate change impacts. • Severe weather events, erosion, sea level rise and pollution. • Importance of local and Indigenous knowledge.
Goal 14. Conserve and sustainably use the oceans, seas and marine resources for sustainable development.	• Importance of coastal and marine areas to tourism and community livelihoods. • Damage to coastal and marine environments from tourism development, cruise ships and waste disposal.
Goal 15. Protect, restore and promote sustainable use of terrestrial ecosystems, sustainably manage forests, combat desertification, halt and reverse land degradation and halt biodiversity loss.	• Threat to habitats, biodiversity and vulnerable species. • Conservation, protection and restoration of natural heritage areas, ecosystems and protected areas.
Peace and Partnership	
Goal 16. Promote peaceful and inclusive societies for sustainable development, provide access to justice for all and build effective, accountable and inclusive institutions at all levels.	• Importance of spiritual and cultural values, including their manifestation in tourism. • Protection of cultural heritage. • Opportunities for intercultural exchange.
Goal 17. Strengthen the means of implementation and revitalize the Global Partnership for Sustainable Development.	• Partnership between private sector, non-governmental organizations, government and civil society. • Policy coherence across sectors.

Source: Adapted from UNDP (2015: 132), UNWTO (2015), SPTO (2017), Scheyvens (2018), and Cheer et al. (2018).

Whilst each goal, therefore, has relevance for tourism in the Pacific, Pacific leaders continue to assert that climate change remains the 'single greatest threat to the welfare, security and livelihood of Pacific people' (PIFS 2018a). The fundamental challenge for tourism to meet the SDGs, then, becomes operating 'within the limits of the ecosystems of which we are a part' (Hall 2019: 1056). As has been pointed out, this may prove elusive for an industry focused on excessive consumption (Jones et al. 2017). Others suggest that what is required, instead, is a degrowth model which prioritizes the rights and interests of local communities (Higgins-Desbiolles et al. 2019: 16), coupled with a radical redistribution of power and wealth (Bianchi and de Man 2021: 365). Agenda 2030 can also be seen as misrepresenting realities in the Pacific, particularly regarding the importance of the connection between people, the land, sea and air (Cheer et al. 2018) and the sociocultural, spiritual and emotional value of land (Stumpf and Cheshire 2018). Stumpf and Cheshire (2018) show that the assumption that local land-holding systems hinder economic development fails to recognize the strength in informal institutions and the social, cultural and financial capital of land. They surmise that the realization of the SDGs must, thus, be based on an understanding of the interface between the SDGs and local realities. Examining priorities articulated in national development and tourism plans offers the opportunity to identify synergies and gaps between tourism planning in the Pacific and the UN agenda.

SDGs and Pacific Island tourism planning

This section analyses the national development plans and tourism strategies of the 13 Pacific Island nations reflected in this volume (Table 2.2). The documents span a date range from pre-2000 to looking forward to 2030 and range significantly in scope and length. The plans are in different stages of development and review and, therefore, also incorporate the SDGs to different degrees. The countries themselves include both established and emerging tourist destinations which necessarily determine the scale of tourism policies and strategies. Linkages with the SDGs are noted, and key priorities are highlighted. The analysis draws out similar themes across strategies and infers connections with the priorities and the intent of the SDGs. This also allows identification of national priorities less well-represented in the SDGs.

Tourism and the SDGs in 'national development strategies'

Most national plans target economic, social and environmental objectives and draw on cross-sectoral collaboration and partnerships with the private sector and civil society organizations, with the result that strategies all broadly align with the SDG goals and targets. Sustainability is expressed as central to every development agenda. Many recent plans link directly to specific SDGs and map national strategies to goals and targets, as identified in Table 2.2. Where existing strategies pre-date the SDGs, SDG localization efforts have focused on identifying linkages between the 2030 Global Agenda and national plans. Palau, for example, has developed national core indicators mapped against SDGs, mandated through Presidential Executive Order (No. 419), and Palau, Nauru and the Federated States of Micronesia have all completed Voluntary National Reviews.

Tourism is expected to contribute, in economic terms, to increases in GDP and employment and is also aligned to related infrastructure development, appearing almost exclusively within strategies for economic growth. High-level tourism strategies are, in this way, linked directly or by implication to the SDG 8 target of promoting sustainable tourism that creates jobs and the SDG 9 target of resilient infrastructure to support economic development. For example, plans for Fiji, the Solomon Islands and Kiribati specifically align with SDG 8.9, increasing tourism earnings as a percentage of GDP (Government of Fiji 2017; Government of Solomon Islands 2016; Government of Kiribati 2020). In emerging tourist destinations, the potential to develop niche tourism (Nauru Government n.d.), or ecotourism (Government of Tuvalu 2016) is identified as an economic strategy. Tourism is further referred to as a way to promote local culture and heritage and strengthen the agriculture sector (Government of Fiji 2017), as well as supporting the retention of cultural traditions and enabling sustainable environmental management (Government of Kiribati 2020). In the Cook Islands, tourism is acknowledged as a contributor to the decline in agriculture, negatively impacting food security (Government of the Cook Islands 2016: 39). Several plans refer to the importance of backward linkages. Vanuatu's plan refers to 'strengthening linkages between tourism, infrastructure, agriculture and industry' (Department of Strategic Policy, Planning and Aid Coordination 2016: 17), and the Federated States of Micronesia also sees the role of tourism as contributing to increasing production of traditional farming systems in its strategic plan (Government of the Federated States of Micronesia n.d.: 142). Individual tourism policies and strategies, which are analysed in the following section, focus on tourism linkages across all the SDGs.

Table 2.2 National and tourism strategies for 13 Pacific Island nations

Country	*National strategy*	*Alignment to SDGs*	*Tourism strategy*	*Alignment to SDGs*
Melanesia				
Fiji	5-Year and 20-Year National Development Plan. Transforming Fiji (Government of Fiji 2017).	National development targets and Key Performance Indicators for each sector are aligned to SDGs.	Fijian Tourism 2021 (Government of Fiji 2021).	28 strategies within nine thematic areas are linked to SDGs in an implementation framework covering all SDGs.
Solomon Islands	National Development Strategy 2016 to 2035 (Government of Solomon Islands 2016).	Aligned to all SDGs with specific targets mapped against 15 medium-term strategies.	National Tourism Development Strategy 2015–2019 (SPTO 2015).	Links sustainable tourism development to economic, environmental and social benefits.
Vanuatu	Vanuatu 2030 The People's Plan (Department of Strategic Policy, Planning and Aid Coordination 2016).	15 national SDGs organized into three pillars: society, environment and economy. The SDGs are embedded into the plan.	Vanuatu Sustainable Tourism Policy 2019–2030 (Government of Vanuatu 2019).	Developed own sustainable tourism policy goals. The policy has been developed to encompass economic, environmental, social and cultural objectives.
Micronesia				
Federated States of Micronesia	Strategic Development Plan 2004–2023 (Government of the Federated States of Micronesia n.d.).	N/A	National Tourism Policy Volume 1 Final Report (Government of the Federated States of Micronesia 2015).	Guided by principle of sustainable tourism to account for economic, social, cultural and environmental impacts.
Kiribati	Kiribati Development Plan 2020–23 (Government of Kiribati 2020).	All national indicators aligned with SDGs.	Ministry of Information, Communication, Transport and Tourism Development Strategic Plan 2020–2023 (MICTTD and SOE n.d.).	Four strategic objectives with a focus on strengthening sustainable tourism development.
Republic of the Marshall Islands	National Strategic Plan 2020–2030 (Government of the Marshall Islands 2020).	Links overall to SDGs through 10 major challenges.	Office of Commerce, Investment and Tourism Business Plan 2019–2021. (Office of Commerce, Investment and Tourism n.d.).	Work programmes linked to SDGs, specifically SDG 8.

(*Continued*)

Table 2.2 Continued

Country	*National strategy*	*Alignment to SDGs*	*Tourism strategy*	*Alignment to SDGs*
Nauru	Nauru Sustainable Development Strategy 2005–2025 (Nauru Government n.d.).	N/A	National Tourism Development Strategy 2015–2019.	*Not available*
Palau	Palau 2020 National Master Development Plan (Palau and Sagric International 1996).	N/A	Palau Responsible Tourism Framework 2017–2021 (Government of the Republic of Palau 2016).	Addresses economic, environmental and social objectives.
Polynesia				
Cook Islands	Te Kaveinga Nui National Sustainable Development Plan 2016–2020 (Government of the Cook Islands 2016).	16 sector-based development goals closely aligned to the Pacific Regional Framework and the SDGs.	Cook Islands Sustainable Tourism Development Policy Framework and Goals (Cook Islands Tourism Corporation 2020).	Seven national sustainable tourism goals supported by core 'kia orana values', focusing on culture, protection and sustainability.
Niue	Niue National Strategic Plan 2016–2026 (Government of Niue 2016).	Seven national development pillars focus on social and economic needs, culture and heritage values and protecting the environment.	Niue Responsible Tourism Policy (Government of Niue 2017).	Supports contribution to the SDGs with actions for environment, community and business/economic sustainability.
Samoa	Strategy for the Development of Samoa 2016/17–2019/20 (Government of Samoa 2016).	Aligned specifically with SDGs and SAMOA pathway.	Samoa Tourism Sector Plan 2014–2019 (Samoa Tourism Authority 2014).	Five programme areas underpinned by Sustainability Principles with three sustainable pillars, economic, cultural, and environmental.
Tonga	Tonga Strategic Development Framework 2015–2025 (Government of Tonga 2015).	Seven national outcomes and five pillars aligned to SDGs and SAMOA pathway.	Tonga Tourism Roadmap 2014–2018. Final Report (TRIP Consultants 2013).	Five focal areas with short and medium-term priorities to maximize economic, social and cultural wellbeing.
Tuvalu	Te Kakeega III National Strategy for Sustainable Development 2016–2020 (Government of Tuvalu 2016).	The development of the Strategy was framed by the SDGs and development measured against the SDGs as well as local factors.	National Tourism Development Strategy 2015–2019.	*Not available*

Source: Author's own compilation.

The SDGs and Pacific Island tourism policies

As laid out in national development strategies, tourism strategies and policies principally link tourism to economic development, aiming toward increased employment and the contribution of tourism to GDP. Nonetheless, tourism policies demonstrate a much greater emphasis on the imperative to achieve *sustainable* tourism development. Strategies emphasize the intersection between economic gain and social and environmental impacts, tempering the drive to grow tourism income with the urgency to protect the environment and cultural heritage. Whilst a number of policies employ strategies to attract high-value rather than high-volume tourists, Vanuatu's tourism policy specifies that high-value tourists are defined not only in terms of economic value but in terms of the social, cultural and environmental value they offer (Government of Vanuatu 2019). Policies also demonstrate the interlinkages between SDGs. For example, by linking tourism with sustainable agriculture, policies seek to address food security (SDG 2) as well as improve livelihoods (SDG 1) and sustainable production practices (SDG 12). Fiji identifies strengthening linkages between tourism sectors and the local economy as 'one of the best ways to enhance and retain the benefits of tourism' (Government of Fiji 2021: 47) (see Chapters 8 and 16), in which the resulting implementation framework aligns strategies to improve linkages between agriculture, aquaculture, and tourism to almost every SDG.

Approaches to ensure that tourism benefits the local population (SDG 1, 10) are repeated across a number of policies. The Solomon Islands specifies the responsibility to 'equitably address the needs of tourists, industry, the natural environment, but most importantly of the host communities' (SPTO 2015: 24), Vanuatu's policy seeks to enable 'equitable access to economic development' (Government of Vanuatu 2019: 13), and the Cook Islands identifies tourism's critical role in meeting goals of 'equity, income distribution and the alleviation of economic hardship' (Cook Islands Tourism Corporation 2020: 10). Palau links job creation, tourism development and industry to community benefits (Government of the Republic of Palau 2016: 6), whilst Samoa seeks to ensure that socioeconomic benefits are 'fairly distributed' (Samoa Tourism Authority 2014: 18). Both the Marshall Islands (Office of Commerce, Investment and Tourism n.d.), and the Federated States of Micronesia (Government of the Federated States of Micronesia 2015) use selection criteria to determine the community benefit of any new tourism projects, including assessing whether the initiative is socially inclusive, provides local employment, and shares economic benefits with host communities.

Alongside improved tourism income, social wellbeing (SDG 3) is a stated aim of the tourism policies of the Cook Islands, Fiji and Vanuatu. Consistent with references to gender in international reporting which focus on tourism as an 'income-generating opportunity' for women (UNWTO 2010 p.i.), a number of policies refer to economic opportunities for women in tourism. Several specifically focus on gender equity (SDG 5). Vanuatu's policy commits to gender equity in tourism businesses (Government of Vanuatu 2019) (see Chapter 12), and Kiribati's plan seeks gender equality in decision-making processes, whilst the Cook Islands also aspires to increase the percentage of tourism businesses owned or managed by women (Cook Islands Tourism Corporation 2020), perhaps signalling a shift toward 'gender responsible tourism' (Alarcón and Cole 2019: 904).

References to *decent* work (SDG 8) are limited. Discussions largely revolve around the challenges of seasonality's impact on employment stability and the limited management opportunities and career pathways available to local employees, with an attendant focus on marketing campaigns to increase tourism numbers in low seasons and improving access to training (SDG 4). Vanuatu's policy is notable in that it mandates 'fair employment opportunities and conditions

… particularly for all marginalised groups' (Government ofVanuatu 2019:19), along with equal employment training opportunities, occupational health and safety and fair wages. The Cook Islands' policy stresses the need for a 'strong and sustainable workforce' (Cook Islands Tourism Corporation 2020: 8), pointing out that additional data gathering is necessary to be able to understand and, therefore, improve worker experiences.

The significance of climate change (SDG 13) to tourism is at the forefront of recent tourism plans. Policies span mitigation and adaption including developing climate resilient or green infrastructure and renewable energy/energy efficiency plans (e.g., Fiji, Vanuatu and Samoa) and ensuring tourism activities are aligned to climate change adaptation strategies (e.g., Kiribati, Niue and Palau). Tonga (TRIP Consultants 2013) and Samoa (Samoa Tourism Authority 2014) outline risk and disaster management training for tourism providers. Several policies encompass broad mitigation strategies; for example, Fiji aims to develop a Sustainable Tourism Development Framework 'to mitigate and prevent the negative impact of tourism on the ecology, society and culture of Fiji' (Government of Fiji 2021: 52). In practice, this includes actions such as legislating environmental compliance and incentivizing sustainable practices through certification and accreditation schemes. SDG 14 is addressed via initiatives such as legislating extra protection for the marine environment, protecting vulnerable marine species and coastal resource management and monitoring, with Fiji, Niue and Samoa committing to establish and/or increase marine protected areas. Environmental protection is reflected across all goals and strategies. For example, Palau coordinates tourism development with the environment, labour, agriculture and fisheries sectors to ensure the protection of biodiversity and ecosystems (Government of the Republic of Palau 2016: 7).

Other strategies explicitly refer to the actions necessary to eliminate unsustainable practices, corresponding to SDG 12 and SDG 15. The Cook Islands includes targets and goals to reduce the waste generated by tourism activities (Cook Islands Tourism Corporation 2020), and Niue's Waste Action Plan promotes the use of low-waste, non-polluting products (Government of Niue 2017). Both Vanuatu and Palau encourage visitors to pledge responsibility to the islands during their visit and commit to respectful, appropriate behaviour (Government of Vanuatu 2019; Government of the Republic of Palau 2016) (see Chapter 18).

Implementation of SDGs

Governments play a key role in the implementation of goals, which, in line with SDG 17, is envisaged in partnership across the public, private and not-for-profit sectors. The reality for many independent island nations is that there are gaps between sectors and a lack of policy coherence (Hills et al. 2019). Further constraints for governments are imposed by the control vested in multi-national tourism operations, which are often resistant to regulation (Mowforth and Munt 2009). Although Agenda 2030 recognizes the role of governments in creating an enabling environment for business development, there is less emphasis on the need to regulate business to benefit communities and protect the environment (Scheyvens et al. 2016: 377). Further, business accountability is based on voluntary reporting, running the risk of 'SDG-washing' or SDG 'window dressing' (McCarton et al. 2021: 171). Indeed, we do not need to look far to see the gap between policy and practice; for instance, development on Fiji's Malolo Island halted in April 2019 after developers repeatedly flouted Environmental Impact Assessment requirements (Rovoi 2019). Nevertheless, examining the content of tourism policies is an important step in shifting our understanding of the nature of sustainable tourism and the implications for how tourism is developed and managed for island states in the Pacific. Pacific knowledge, based on local and Indigenous values, suggests that development initiatives have a greater chance of success when they are aligned to local priorities (Gegeo 1998) and take account of Indigenous development strategies (Hau'ofa 1994; Maiava and King 2007; Meo-Sewabu and Walsh-Tapiata 2012). Prior to the ratification of the SDGs, the SDG Pacific consultation group proposed that addressing the root causes of structural inequalities 'may require an alternative model of development other than a conventional "growth" model, to one that puts humans at the centre and is linked to justice' (PIFS 2015: 159). Accordingly, the following section reflects on the localization of the SDGs in the Pacific and what national approaches to tourism in the Pacific might add to this debate.

Localizing the SDGs in the Pacific

Strategies and policies reflect the need to localize goals and targets, foregrounding the place of locally developed goals founded on national values and principles. This acknowledges the need identified in Agenda 2030 to account for 'different national realities, capacities and levels of development and respecting national policies and priorities' (United Nations 2015, para. 55). For example, Palau's alignment with the SDGs is framed in terms of identifying relevance to national priorities and using local data as a starting point (Government of the Republic of Palau 2016), and Tuvalu's national strategy is based on 'Tuvalu's own development perceptions and needs' (Government of Tuvalu 2016: 3). The importance of localizing the SDGs supports findings from the Pacific consultation on the post-2015 development agenda, which highlight the need to account for national priorities to better reflect the Pacific context and ensure ownership of the SDGs (PIFS 2015: 110). A regional approach to addressing the SDGs builds on the involvement of Papua New Guinea, Palau and Nauru in the SDGs Open Working Group and continues to be supported through additional regional mechanisms, including the Framework for Pacific Regionalism (PIFS 2014), the Pacific Roadmap for Sustainable Development (PIFS 2018b) and, most recently, the Pacific Tourism Policy Framework (SPTO 2021). There are two tourism issues of particular significance which stand out for island states in the Pacific, namely climate change and the impact of tourism on the ocean and environment, and the protection of cultural heritage. This is encapsulated in the Framework for Pacific Regionalism, stating:

> Pacific peoples are the custodians of the world's largest, most peaceful and abundant ocean, its many islands and its rich diversity of cultures. We celebrate and draw strength from the culture and traditions, language, social values, and religious freedoms and beliefs that bind citizens and communities together, providing sustenance, social stability, and resilience. We acknowledge our shared responsibility for our significant terrestrial and oceanic resources, which provide livelihoods and opportunities for sustainable development.
>
> *(PIFS 2014: 1)*

Climate change is addressed in tourism policies via the priority accorded to environmental issues. In the Cook Islands, guardianship of the environment is written into core 'kia orana' values, underpinning the tourism policy (Cook Islands Tourism Corporation 2020), and the environment constitutes the most extensive section of responsible tourism actions in Niue's tourism policy (Government of Niue 2017). The localization of climate goals can be understood by reconceptualizing the *management* of marine and coastal ecosystems, fisheries, aquaculture and tourism as described in Agenda 2030 to *stewardship* of natural resources. Dame Meg Taylor, in her role as Ocean Commissioner, underscores the role of the Pacific Island countries as Big Ocean Stewardship States, quoting the Prime Minister of the Solomon Islands, Manasseh

Sogavare, who cautions that 'we cannot manage the oceans but instead need to manage the behaviour of people who use the ocean' (Taylor 2017). The concept of stewardship of resources, also expressed as guardianship or custodianship, is embedded within strategy and policy documents and reflects values and principles which understand conservation and the protection of natural resources as the foundation for the whole community and ecosystem, past, present and future. Kiribati's development plan is founded on a responsibility as 'custodians of our land, our oceans and our resources [to] ... harness these resources wisely' (Government of Kiribati 2020: 3). Tonga's development strategy is underpinned by the wise and prudent use of 'our inherited wealth' (Government of Tonga 2015: 17), and, in the Cook Islands, the value of Mana Tiaki refers to the responsibility for guardianship of the islands for future generations (Cook Islands Tourism Corporation 2020: 4). Recognizing how stewardship of resources is understood, rather than taking a resource management perspective, also leads to an expanded view of 'equity' in tourism. This is exemplified by Vanuatu's tourism policy which seeks 'generational equity' in tourism development by accounting for the impact on both present and future generations (Government of Vanuatu 2019: 18).

Culture is likewise differently represented in policy documents. In Agenda 2030, culture is embedded within other goals. With respect to tourism, SDG 8.9 advocates for sustainable tourism that promotes local culture and products, whilst SDG 4.7 calls for appreciating cultural diversity and the contribution of culture to sustainable development from an educational perspective. SDG 11.4 places emphasis on strengthening efforts to protect and safeguard the world's cultural and natural heritage in cities and human settlements. However, as Yap and Watene (2019) note, culture in the SDGs is seen as a means to achieve goals rather than valued in its own right, whilst, within Indigenous philosophies, culture is regarded as 'a dimension of sustainable development, indivisible from environmental, social and economic dimensions' (2019: 456). Within Pacific plans and policies, the contribution of culture to sustainable development is linked to all themes and goals with a focus on protecting and valuing tangible and intangible cultural assets. Paradoxically, culture constitutes a tourist attraction at the same time that it poses a threat to its integrity: opportunities to promote local culture create concerns around the pressure to 'monetize the performative aspects of islander culture' (Cheer et al. 2018: 2). However, policies show traditional knowledge is protected and simultaneously regarded as a means to address tourism development challenges. For example, strategies aim to protect unique cultural identity (Government of Fiji 2021: 54), promote local traditional knowledge and livelihood skills (Government of the Federated States of Micronesia 2015: 20), and safeguard traditional architecture and built heritage (Government of Vanuatu 2019: 9; Samoa Tourism Authority 2014: 18; TRIP Consultants 2013: 36). At the same time, tourism is seen as an opportunity to educate tourists and communities in environmental knowledge (Government of the Republic of Palau 2016: 7) and cultural heritage (Cook Islands Tourism Corporation 2020: 7). Visiting Friends and Family (VFR) tourism dominated arrivals in the pre-COVID era, namely in Kiribati, Samoa and Tonga, and this presents an opportunity to further embed such an educational approach across tourists and communities. Rather than focusing on tourism promoting local culture, therefore, greater strategic attention is needed to examine the ways in which culture can play a role in promoting sustainable tourism, informed by local and Indigenous approaches to tourism management.

Pacific Island tourism and the COVID-19 pandemic

Tourism destinations across the Pacific responded to increasing COVID-19 infections worldwide by closing their borders. With tourism numbers decimated, the economic contribution of tourism to island states in the Pacific all but halted, with economic contractions experienced across the region (Forward Insight and Strategy 2020). The UN Sustainable Development Group outlined the implications of the COVID-19 impact on tourism for progress on meeting SDGs, noting, in particular, the negative impact on addressing poverty and gender inequality. The impact on small island states and the cultural and social impact on communities are also noted as well as the opportunity for positive transformation of the impact of tourism on the environment, economy and society (United Nations 2020). Whilst change is likely to be incremental rather than transformational (Becken and Loehr 2022a), Pacific leaders have recognized the opportunity presented by COVID-19 to allow a rethinking of tourism development for the region. The then Secretary General to the Pacific Islands Forum, Dame Meg Taylor, highlighted the need to link COVID-19 recovery to the 2030 Agenda for Sustainable Development and to find new ways to address sustainable tourism in conjunction with climate change (Krishnamurthi 2020). The SPTO's regional sustainable tourism policy framework for the next ten years has since identified the COVID-19 pandemic as a catalyst toward a paradigm shift in sustainable tourism for the region, involving a model that is 'economically resilient, environmentally sustainable and socially inclusive', with the regional tourism sector seen as a champion and a steward for the environment in the Pacific (SPTO 2021: 5).

Research in progress is beginning to examine what transformation might mean for Pacific Islanders involved in tourism. Research examining the impact of COVID-19 on Indigenous people involved in tourism in Fiji the Cook Islands, Samoa, Vanuatu and the Solomon Islands showed that, although financial wellbeing significantly declined, embedded social systems fostered resilience and supported mental and emotional wellbeing (Scheyvens and Movono 2020). Evidence also demonstrates the significance of local and Indigenous knowledge in mitigating the social and economic impacts of COVID-19, for example, through customary cultivation and fishing practices (Movono et al. 2022) and extended bartering systems (Boccuzzi 2021:13). Reimagining tourism post-COVID-19 could include better terms and working conditions, more local ownership and greater diversification, pointing toward more sustainable and resilient forms of tourism for the future (Scheyvens and Movono 2020). Longer term, we can see some movement toward regenerative strategies (Becken and Loehr 2022b), with South Pacific governance through the SPTO indicative of a 'shift in underlying values that broadens pure commercial gains to consider wider aspects of wellbeing' (Becken and Loehr 2022a: 12).

Conclusion and research implications

This chapter has shown the relevance of tourism beyond the three tourism-focused SDG targets to each one of the goals and has identified the significance of goals for tourism in the Pacific. Particular challenges include creating sustainable livelihood opportunities supported by appropriate infrastructure and services whilst protecting cultural heritage and marine and terrestrial resources. Climate change impacts experienced by the islands underline the urgency of the pursuit of sustainable forms of tourism. Consistent with research advocating for a critical approach to viewing tourism as a panacea for sustainable development (Bricker 2018: 206) and calling for a more contextualized understanding of the goals (Stumpf and Cheshire 2018), this chapter suggests that more attention is needed to the localization of the goals in relation to the tourism sector. Further, the need is highlighted for Indigenous values and culture to be incorporated in our understanding sustainable development (Yap and Watene 2019: 456).

Looking at how goals are contextualized within national development plans and tourism policies has shown that localizing the goals can be effective in accounting for local priorities and values. Pacific policies focus on balancing economic, social, environmental and cultural gains for which enhancing the wellbeing of the population is intimately connected with protecting the environment and cultural heritage. Examining strategies and actions in closer detail, thus, demonstrates where Pacific Island priorities are not as well-reflected in the SDGs, namely the role of cultural values and traditional knowledge in promoting sustainable tourism alongside the reconceptualization of environmental management as environmental stewardship. Cultural perspectives inform strategies and policies formulated from a position as stewards with a responsibility to protect cultural and environmental heritage for future generations. Any future focus on tourism planning in island states of the Pacific will need to be revisited in the context of the consequences of the pandemic, but it is clear that the use of local, traditional and Indigenous knowledge can constitute key solutions to the challenges of tourism development. Initial lessons learned suggest that future policies will only see strengthened links between climate, tourism and health. Greater attention to national contexts, consequently, offers huge potential to inform the global sustainability agenda.

Regional perspectives provide further insight into the importance of localizing the SDGs. The Pacific Roadmap articulates the SDGs as an opportunity to demonstrate leadership, advocate regional priorities and tell the 'region's sustainable development story' (PIFS 2018b: 6–7). Regional leadership is provided through the Pacific Island Forum Secretariat (PIFS) in addition to other bodies including the Pacific SDGs Taskforce, the Council of Regional Organisations of the Pacific and the SPTO. The SPTO plays a key role at a regional level in brokering 'international enablers and regional/national values' (Becken and Loehr 2022a: 16). This is clearly seen in Pacific 2030, which, in conjunction with the Pacific Tourism Statistics Strategy, foregrounds the SDGs but stresses the importance of local control over tourism development and monitoring progress using regionally developed measures (SPTO 2021: 15).

These findings present the opportunity for future research to explore the mechanisms for transferability of tourism-focused learnings both across the Pacific and at a global level. Further research is needed on the ways in which islands can (and do) learn from one another in establishing good practices, connecting across governments and ministries and advocating for sustainable tourism practices. Paying attention to Pacific perspectives of the goals, and their translation into locally relevant targets, will allow us to extend our understanding of the opportunities for Agenda 2030 to balance people, prosperity and planet in practice.

References

Alarcón, D. M. and S. Cole (2019) 'No sustainability for tourism without gender equality', *Journal of Sustainable Tourism*, 27(7): 903–919.

Becken, S. and J. Loehr (2022a) 'Tourism governance and enabling drivers for intensifying climate action', *Journal of Sustainable Tourism*, Online First. https://doi.org/10.1080/09669582.2022.2032099.

Becken, S. and J. Loehr (2022b) 'Asia–Pacific tourism futures emerging from COVID-19 recovery responses and implications for sustainability', *Journal of Tourism Futures*, Online First. https://doi.org/10.1108/JTF-05-2021-0131.

Bianchi, R.V. and F. de Man (2021) 'Tourism, inclusive growth and decent work: A political economy critique', *Journal of Sustainable Tourism*, 29(2–3): 353–371.

Boccuzzi, E. (2021) *The Future of Work for Women in the Pacific islands*, The Asia Foundation. Available online at https://asiafoundation.org/wp-content/uploads/2021/02/The-Future-of-Work-for-Women-in-the-Pacific-Islands.updateMarch1.pdf (accessed 18 January 2022).

Boluk, K., C. T. Cavaliere and F. Higgins-Desbiolles (2017) 'Critical thinking to realize sustainability in tourism systems: Reflecting on the 2030 sustainable development goals', *Journal of Sustainable Tourism*, 25(9): 1201–1204.

Bricker, K. (2018) 'Positioning sustainable tourism: Humble placement of a complex enterprise', *Journal of Park and Recreation Administration*, 36(1): 205–211.

Cheer, J. M., S. Pratt, D. Tolkach, A. Bailey, S. Taumoepeau and A. Movono (2018) 'Tourism in Pacific island countries: A status quo round-up', *Asia and the Pacific Policy Studies*, 5(3): 442–461.

Cook Islands Tourism Corporation (2020) *Protecting Our Future: Cook Islands Sustainable Tourism Development Policy Framework and Goals*, Government of the Cook Islands. Available online at https://policycook-islands.files.wordpress.com/2017/05/2016_ci-sustainable-tourism-development-policy-framework-goals.pdf (accessed 23 January 2022).

Department of Strategic Policy, Planning and Aid Coordination (2016) *Vanuatu, 2030: The People's Plan: National Sustainable Development Plan 2016 to 2030*, November, Port Vila, Vanuatu: Department of Strategic Policy, Planning and Aid Coordination. Available online at https://www.gov.vu/images/publications/Vanuatu2030-EN-FINAL-sf.pdf (accessed 23 January 2022).

Forward Insight and Strategy (2020) *Pacific Tourism: Covid-19 Impact and Recovery*, Sector Status Report: Phase 1B, Ministry of Foreign Affairs and Trade with the South Pacific Tourism Organisation.

Gegeo, D. (1998) 'Indigenous knowledge and empowerment: Rural development examined from within', *The Contemporary Pacific*, *10*(2): 289–315.

Government of Fiji (2017) *5-year 20-year National Development Plan: Transforming Fiji*, November, Republic of Fiji: Ministry of Economy. Available online at https://www.fiji.gov.fj/getattachment/15b0ba03-825e-47f7-bf69-094ad33004dd/5-Year---20-Year-NATIONAL-DEVELOPMENT-PLAN.aspx (accessed 23 January 2022)

Government of Fiji (2021) *Fijian Tourism 2021*. Republic of Fiji: Government of Fiji. Available online at http://mitt.gov.fj/wp-content/uploads/2019/04/FT2021.pdf (accessed 23 January 2022).

Government of Kiribati (2020) *Kiribati Development Plan 2020–2023*. Available online at https://policy.asiapacificenergy.org/sites/default/files/Kiribati%20Development%20Plan%202020-2023.pdf (accessed 27 February 2022).

Government of Niue (2016) *Niue National Strategic Plan 2016–2026*, Prepared for the Office of the Premier by P. Talagi, D. Siataga, P. Kapaga, K. Vaha, F. Sioneholo, M. Marsh and P. Salatielu. Available online at https://www.theprif.org/sites/default/files/2020-08/Niue%20National%20Strategic%20Plan%202016-2026_0.pdf (accessed 23 January 2022).

Government of Niue (2017) *Niue Responsible Tourism Policy*. Alofi: Government of Niue.

Government of Samoa (2016) *Samoa: Strategy for the development of Samoa 2016/17-2019/20: 'Accelerating Sustainable Development and Broadening Opportunities for All'*, Ministry of Finance. Available online at https://policy.asiapacificenergy.org/node/2834 (accessed 23 January 2022).

Government of Solomon Islands (2016) *National Development Strategy 2016–2035: Improving the Social Economic Livelihoods of all Solomon Islanders*, Honiara, Solomon Islands: Ministry of Development Planning and Aid Coordination. Available online at https://solomonislands-data.sprep.org/system/files/National%20Development %20Strategy.pdf (accessed 23 January 2022).

Government of Tonga (2015) *Tonga Strategic Development Framework 2015–2025. A More Progressive Tonga: Enhancing Our inheritance*, May, Ministry of Finance and National Planning. Available online at http://extwprlegs1.fao.org/docs/pdf/ton168846.pdf (accessed 23 January 2022)

Government of Tuvalu (2016) *National Strategy for Sustainable Development 2016 to 2020*, Suva, Fiji: Pasifika Communications Ltd. Available online at https://www.adb.org/sites/default/files/linked-documents/cobp-tuv-2017-2019-ld-02.pdf (accessed 23 January 2022).

Government of Vanuatu (2019) *Vanuatu Sustainable Tourism Policy 2019–2030*. Available online at https://tourism.gov.vu/images/DoT-Documents/Policies/SUSTAINABLE_TOURISM_POLICY_2019-2030_New.pdf (accessed 27 February 2022).

Government of the Cook Islands (2016) *Te Kaveinga Nui – National Sustainable Development Plan 2016–2020*, January, Central Policy and Planning Office, Office of the Prime Minister. Available online at https://cookislands-data.sprep.org/system/files/cobp-coo-2017-2019-ld-01.pdf (accessed 23 January 2021).

Government of the Federated States of Micronesia (2015) *National Tourism Policy, Volume 1*, Final Report, June. Federated States of Micronesia. Available online at https://www.fsmstatistics.fm/wp-content/uploads/2019/10/4-FSM-National-Tourism-Policy-Vol-1a_W-Foreword_8July.pdf (accessed 21 January 2022).

Government of the Federated States of Micronesia (n.d.) *Federated States of Micronesia's Strategic Development Plan (2004–2023) – The Next 20 Years: Achieving Economic Growth and Self-Reliance (Volume 1: Policies and Strategies for Development)*, Federated States of Micronesia. Available online at https://fsm-data.sprep.org/system/files/ccbp-fsm-2015-2017-sd-02.pdf (accessed 23 January 2022).

Government of the Marshall Islands (2020) *National Strategic Plan 2020–2030*, Majuro, Republic of the Marshall Islands: Economic Policy, Planning and Statistics Office. Available online at https://www.rmieppso.org/eppso_files/nsp/NSP_2020_2030.pdf (accessed 23 January 2022).

Government of the Republic of Palau (2016) *Palau Responsible Tourism Policy Framework: Ensuring a Pristine Paradise, Palau. Palau for Everyone 2017–2021*, December, Republic of Palau: Ministry of National Resources, Environment and Tourism. Available online at https://www.palaugov.pw/wp-content/uploads/2017/04/Final_Palau-Responsible-Tourism-Framework1.pdf (accessed 21 January 2022).

Hall, C. M. (2019) 'Constructing sustainable tourism development: The 2030 agenda and the managerial ecology of sustainable tourism', *Journal of Sustainable Tourism*, 27(7): 1044–1060.

Hau'ofa, E. (1994) 'Our sea of islands', *The Contemporary Pacific*, 6(1): 148–161.

Higgins-Desbiolles, F., S. Carnicelli, C. Krolikowski, G. Wijesinghe and K. Boluk (2019) 'Degrowing tourism: Rethinking tourism', *Journal of Sustainable Tourism*, 27(12): 1926–1944.

Hills, J., S. Bala, A. Solofa, P. Dunstan, M. Fischer and D. Hayes (2019) 'The disjuncture between regional ocean priorities and development assistance in the South Pacific', *Marine Policy, 107*. Online First. https://doi.org/10.1016/j.marpol.2019.01.009.

Jones, P., D. Hillier and D. Comfort (2017) 'The sustainable development goals and the tourism and hospitality industry', *Athens Journal of Tourism*, 4(1): 7–18.

Krishnamurthi, S. (2020) 'Eco-tourism is the way of the post-covid future for the "blue" Pacific', Pacific Media Centre. Available online at https://www.scoop.co.nz/stories /HL2007/S00153/eco-tourism-is-the-way-of-the-post-covid-future-for-the-blue-pacific.htm (accessed 21 January 2021).

Le Blanc, D., C. Freire and M. Vierros (2017) *Mapping the Linkages between Oceans and Other Sustainable Development Goals. A Preliminary Exploration*, New York: UN. Available online at https://www-un-ilibrary-org.ezproxy.massey.ac.nz/economic-and-social-development/mapping-the-linkages-between-oceans-and-other-sustainable-development-goals_3adc8369-en (accessed 19 October 2019).

Maiava, S. and T. King (2007) 'Pacific Indigenous development and post-intentional realities'. In: A. Ziai (ed.), *Exploring Post-Development: Theory and Practice, Problems and Perspectives*, London: Routledge, pp. 83–98.

McCarton, L., S. O'Hogain and A. Reid (2021) 'Sustainable development goals (SDGs)'. In: L. McCarton, S. O'Hogain and A. Reid (eds.), *The Worth of Water: Designing Climate Resilient Rainwater Harvesting Systems*, Cham: Springer, pp. 159–172.

Meo-Sewabu, L. and W. Walsh-Tapiata (2012) 'Global declaration and village discourses: Social policy and Indigenous wellbeing', *AlterNative: An International Journal of Indigenous Peoples*, 8(3): 305–317.

MICTTD and SOE (n.d.) *Ministry Strategic Plan 2020–2023*, Betio, Tarawa, Republic of Kiribati: Ministry of Information, Communication, Transport and Tourism Development (MICTTD) and State-Owned Enterprises. Available online at https://www.micttd.gov.ki/sites/default/files/FINAL%20MICTTD%20MSP%202020-2023%20PRINT2.pdf (accessed 23 January 2022).

Movono, A., R. Scheyvens and S. Auckram (2022) 'Silver linings around dark clouds: Tourism, Covid-19 and a return to traditional values, villages and the vanua', *Asia Pacific Viewpoint*, Online First. https://doi.org/10.1111/apv.12340.

Mowforth, M. and I. Munt (2009) *Tourism and Sustainability: Development, Globalisation and New Tourism in the Third World* (3rd ed.), New York: Taylor and Francis.

Nauru Government (n.d.) *Nauru: National Sustainable Development Strategy 2005–2025 (Partnerships for Quality of Life)*, Nauru: Development Planning and Policy Division, Ministry of Finance and Economic Planning, Nauru Government. Available online at https://nauru-data.sprep.org/system/files/nauru_development_strategy_2025_en_ 2005.pdf (accessed 23 January 2022).

Office of Commerce, Investment and Tourism (n.d.) *Business Plan 2019–2021*, Majuro, Republic of the Marshall Islands. Available online at https://www.ioes.ucla.edu/wp-content/uploads/OCIT-Business-Plan-2019-2021.pdf (accessed 23 January 2022).

Palau and SAGRIC International (1996) *Palau National Master Development Plan Issues, Options and Strategies for Palau's Development*. Final Report, Adelaide: SAGRIC.

PIFS (Pacific Islands Forum Secretariat) (2014) *Framework for Pacific Regionalism*, Suva, Fiji: PIFS.

PIFS (Pacific Islands Forum Secretariat) (2015) *Pacific Regional MDGs Tracking Report*, Suva, Fiji: PIFS.

PIFS (Pacific Islands Forum Secretariat) (2018a) *The BOE Declaration on Regional Security*, Suva, Fiji: PIFS. Available online at https://www.forumsec.org/2018/09/05/boe-declaration-on-regional-security/ (accessed 7 May 2022).

PIFS (Pacific Islands Forum Secretariat) (2018b) *The Pacific Roadmap for Sustainable Development, Prepared by Pacific Sustainable Development Goals Taskforce*, Suva, Fiji: PIFS.

Pradhan, P., L. Costa, D. Rybski, W. Lucht and J. P. Kropp (2017) 'A systematic study of sustainable development goal (SDG) interactions', *Earth's Future*, 5(11): 1169–1179.

Rovoi, C. (2019) 'Fiji Landowners hail rejection of Malolo project', *Radio New Zealand*. Available online at https://www.radionz.co.nz/international/pacific-news/386685/fiji-landowners-hail-rejection-of-malolo-project (accessed 12 April 2019).

Samoa Tourism Authority (2014) *Samoa Tourism Sector Plan 2014–2019*, Government of Samoa. Available online at https://pafpnet.spc.int/attachments/article/684/Samoa-Tourism-Sector-Plan-2014-2019.pdf (accessed 23 January 2022).

Scheyvens, R. (2018) 'Linking tourism to the sustainable development goals: A geographical perspective', *Tourism Geographies*, 20(2): 341–342.

Scheyvens, R., G. Banks and E. Hughes (2016) 'The private sector and the SDGs: The need to move beyond "business as usual"', *Sustainable Development*, 24(6): 371–382.

Scheyvens, R. and E. Hughes (2019) 'Can tourism help to "end poverty in all its forms everywhere"? The challenge of tourism addressing SDG1', *Journal of Sustainable Tourism*, 27(7): 1061–1079.

Scheyvens, R. and A. Movono (2020) *Development in a World of Disorder: Tourism, COVID-19 and the Adaptivity of South Pacific People*, Palmerston North, NZ: Massey University.

SPTO (South Pacific Tourism Organisation) (2015) *The Solomon Islands National Tourism Development Strategy 2015–2019: A Strategy for Growth (Final Report)*, February, TRIP consultants within the framework of the Pacific Regional Tourism Capacity Building Programme, which is implemented by the South Pacific Tourism Organisation, Suva, Fiji. Available online at http://macbio-pacific.info/wp-content/uploads/2017/08/National-Tourism-Strategy-2015.pdf (accessed 23 January 2022).

SPTO (South Pacific Tourism Organisation) (2017) *Tourism and the Sustainable Development Goals*, Suva, Fiji. Available online at https://consumer.southpacific islands. travel/tourism-and-the-sustainable-development-goals/ (accessed 19 October 2019).

SPTO (South Pacific Tourism Organisation) (2019a) *2018 Annual Visitor Arrivals Report*, Suva, Fiji. Available online at https://pic.or.jp/ja/wp-content/uploads/2019/07/2018-Annual-Visitor-Arrivals-ReportF.pdf (accessed 19 October 2019).

SPTO (South Pacific Tourism Organisation) (2019b) '*SPTO Releases 2019–2024 Pacific Tourism Forecast*', Suva, Fiji. Available online at https://corporate.south pacificislands.Travel/spto-releases-2019-2024-pacific-tourism-forecast/ (accessed 29 January 2021).

SPTO (Pacific Tourism Organisation) (2021) *Pacific 2030: Sustainable Tourism Policy Framework*, Suva, Fiji. Available online at https://southpacificislands.travel/wp-content/uploads/2021/07/Pacific-Sustainable-Tourism-Policy-Framework.pdf (accessed 27 February 2022).

Stumpf, T. and C. Cheshire (2018) 'The land has voice: Understanding the land tenure–sustainable tourism development nexus in Micronesia', *Journal of Sustainable Tourism*, 27(7): 957–973.

Taylor, M. (2017) 'A sea of islands: How a regional group of Pacific states is working to achieve SDG 14', *United Nations Chronicle*, 54(2): 19–22.

TRIP Consultants (2013) *Tonga Tourism Sector Roadmap 2014–2018 Final Report*, August. Available online at http://macbio-pacific.info/wp-content/uploads/2017/08/Tonga-Tourism-Roadmap-2014-2018.pdf (accessed 23 January 2021).

UNDP (United Nations Development Program) (2015) *Human Development Report 2015: Work for Human Development*, New York: UNDP.

United Nations (2015) *Transforming our World: The 2030 Agenda for Sustainable Development*. Available online at https://sustainabledevelopment.un.org/post2015 /transformingourworld (accessed 13 April 2019).

United Nations (2020) *Policy Brief: COVID-19 and Transforming Tourism*. Available online at https://unsdg.un.org/resources/policy-brief-covid-19-and-transforming-tourism (accessed 29 January 2021).

UNWTO (United Nations World Tourism Organization) (2010) *Global Report on Women in Tourism. Preliminary findings*, Madrid: UNWTO.

UNWTO (United Nations World Tourism Organization) (2015) *Tourism and the Sustainable Development Goals*, Madrid: UNWTO.

World Bank (2017) *Pacific Possible: Long-term Economic Opportunities and Challenges for Pacific Island Countries*, Washington, DC: World Bank Group. Available online at http://documents.worldbank.org/curated/en/168951503668157320/Pacific-Possible-long-term-economic-opportunities-and-challenges-for-Pacific-Island-Countries (accessed 19 October 2019).

Yap, M. L. M. and K. Watene (2019) 'The sustainable development goals (SDGs) and indigenous peoples: Another missed opportunity?' *Journal of Human Development and Capabilities*, 20(4): 451–467.

3

TOURISM AND THE NEW REGIONAL GOVERNANCE OF 'PACIFIC LARGE OCEAN STATES'

From governance and development to metagovernance and resilience

C. Michael Hall and Alberto Amore

Introduction

The imagery of Pacific Islands as 'romantic idylls' (Hall 2009a; Harrison 2004) is a consolidated feature of the global tourism industry. This imagery was generated not by the island communities that the representation was meant to describe but by the mercantile and colonial interests that were seeking to open up the Pacific for god, empire and economics. The integration of the Pacific into global capitalism and, hence, governance structures is, therefore, very much a part of colonial and post-colonial networks and structures that linger to the present day and continue to shape economic, social and political realities for the Pacific Islands (Cheer et al. 2018). Arguably, tourism is only one of numerous waves of economic development that have served to tie the interests of Pacific Island communities to those of international capital and global politics (Yang 2011).

International tourism represents a key economic driver for the development of Pacific Island states, as it contributes to infrastructural development, employment, foreign currency exchange and trade balance (Gössling 2003). Like many island nations, the countries of the Pacific are generally highly dependent on tourism for their economic well-being and employment generation (Hall 2015). According to the estimates from the World Travel Tourism Council (WTTC), tourism directly and indirectly contributed to 34.7% of Vanuatu's gross domestic product (GDP) in 2019, with international visitor revenues accounting for 67.3% of the exports (WTTC 2020a). Similarly, tourism accounted to 34% of Fiji's GDP and nearly 46.9% of the exports in 2019 (WTTC 2020b), while tourism expenditure in Hawaii, in the same year, totalled US$16.8 billion and was anticipated to reach US$20.2 billion by 2022 (Department of Business, Economic Development and Tourism 2019). Nevertheless, 'tourism has long been recognized as having both positive and negative externalities in island developments' (Hall 2010: 246).

Overreliance on international tourism markets places the economies of Pacific Island countries and territories under substantial pressure from the state of the global economic and political environment. In the case of the Cook Islands, for example, nearly 83% of international tour-

DOI: 10.4324/9780429019968-4

ists in 2018 came from Australia and New Zealand (MFEM 2019). This, in turn, exacerbates destination vulnerability to economic and social triggers in generating tourism markets upon which Pacific Island states have little to no control. Moreover, the relative geographical remoteness of these destinations, the relatively elastic nature of market demand and the substantial overseas ownership in the hospitality sector exacerbate the imbalance between developed countries and Pacific Island states (Tourism Concern 2009). The case of Vanuatu is emblematic, with Carnival Cruises having been offered discounted port access fees (ADB 2018), while Pacific states have long struggled to maintain aviation linkages, often only doing so with substantial financial difficulty (Guthrie 2011, 2013). Finally, Pacific Island states currently face increasing environmental vulnerability as result of climate change and sea-level rise (IPCC 2014). The latter is a major problem for the islands of Kiribati, with increased flooding and sea incursions polluting water wells and devastating the already vulnerable agriculture of the 33 atolls (Bowers 2017; Hall 2015).

Pacific Island states face major challenges and opportunities in framing policy making to tackle economic and environmental vulnerabilities in increasingly uncertain times. As Hall (2015: 55) states, island states 'have also become both actual and symbolic representations of the central challenge of sustainable development to reconcile human demands with the limits of natural resources'. This chapter addresses both global and inherent characteristics of Pacific Island states to highlight the governance shortcomings in the current policy environment. It does so by looking at tourism-relevant perspectives such as global trade, hyper-neoliberalism, climate change, biodiversity loss and the increased frequency and magnitude of natural disasters.

This chapter delves into the current political dimension of tourism in the Pacific Island states. It does so by illustrating the current landscape and regime with respect to key economic, political and ecological dimensions (Amore et al. 2018). The inclusion of the ecological dimension is necessary, as it reflects the vulnerability and resilience of Pacific Island states and 'their intrinsic abilities to maintain, resume, or adaptively change in face of external disturbances' (Amore et al. 2018: 237). The chapter first discusses the phenomenon of tourism and globalization in Pacific Island states by addressing current environmental, economic and political issues in the region. The chapter then raises concerns on the sustainable tourism development rhetoric in international and national policy making and their implications for Pacific Island states. This contribution provides a revised appraisal of Pacific large ocean states from a metagovernance and resilience perspective. The chapter concludes with a discussion and conclusion on the future of tourism in Pacific Island states and the need to rethink approaches for economically sustainable development in a carbon-constrained world.

Globalization, tourism and the new regionalism in Pacific Island states

Contemporary tourism is an expression of voluntary human mobility in time and space (Hall 2005). In his view, the contemporary shifts in globalization and mobility 'not only have implications for tourism but also for a wide range of human activities, as well as ideas of accessibility, extensibility, distance and proximity' (Hall 2005: 130). Tourism is both a cause and a consequence of globalization and continually impacts the economic and political spheres at local and national levels (Mowforth and Munt 1998). Focusing on Pacific Island states, accounts in critical geography conceive global tourism as an exploitative form of neocolonialism (Britton 1982; Scheyvens and Momsen 2008) that further perpetuates socio-economic and spatial development inequalities. Or, as Milne (1997) described it, a 'vicious circle' of dependency that reinforces the centre-periphery relationships that surround island development. As a result, Pacific Island states are subject to the so-called 'Greek Squeeze' (Souty 2002), with large tour operators often putting pressure on local businesses and suppliers to increase profit or otherwise excluding them by developing enclave-type resorts. However, it is also important to note that the development of tourism and related infrastructure is also often tied into political pressures from states, such as Australia, China and New Zealand, that seek to both extend and/or maintain their sphere of influence (Connolly 2016; Dearden 2008; Wallis 2017). For example, Beijing's Belt and Road Initiative (BRI) exacerbated a longstanding concern about the power shift from the West to China among Western states but has also served to incorporate a number of Pacific Island states into the debate (Pan et al. 2018).

This situation, therefore, means that the governance of tourism relationships and connectivity cannot readily be seen as being isolated from broader issues of regional institutional architecture (Tarte 2014). The Pacific tourism and aviation bodies, for example, are part of the Council of Regional Organisations in the Pacific that also includes the Pacific Islands Forum Secretariat (PIFS), the Secretariat of the Pacific Community (SPC), the University of the South Pacific (USP), the Secretariat of the Pacific Regional Environment Program (SPREP) and the Forum Fisheries Agency (FFA). Significantly, countries such as Fiji have been proposing a new 'island-centred' regionalism (Maclellan 2015). Assertions of a new Pacific regionalism can be seen in the strengthening of the Pacific Small Island Developing States (PSIDS) caucus at the United Nations (UN) and its formal integration into the Asia group within the UN system, which has created a significant platform for Pacific states to promote their interests internationally. This is matched by the promotion of a new Framework on Pacific Regionalism by the PIFS (2014). Significantly, Pacific Island countries are placing the oceans at the centre of regional frameworks and, increasingly, identifying themselves as 'large ocean countries' (Maclellan 2015) rather than small island states. For example, the Pacific Oceans Alliance was launched by the Forum Secretariat and SPREP in 2014 to address concerns over ocean conservation, climate change and sustainable development. Interestingly, Maclellan (2015) also notes that there is increased criticism from within the Pacific of neoliberal policies for economic growth.

Since the early 1990s, Australia, New Zealand and the United States have strongly advocated a neoliberal ideology of trade liberalization to promote economic growth in the islands of which tourism is a part (Maclellan 2015; Williams and Blaiklock 2018). Advocacy for the 'Washington Consensus' has supported the corporatization and privatization of state-owned enterprises, cuts in public sector employment, the introduction of value-added taxes, the commoditization of culture and nature and policies to promote greater foreign investment in key industries, including tourism (Overton and Murray 2018; Murray et al. 2018; Schilcher 2007; Tabani 2017). This has been reinforced by transnational organizations such as the International Monetary Fund (IMF) and the World Bank, in addition to the multilateral and bilateral aid and policy donors, that have dominated policy advice on development in the Pacific with a heavy emphasis on neoliberal discourses on 'growth', 'efficiency', 'reform', and 'governance' (Anderson and Lee 2010). Indeed, Addinsall et al. (2015) even observe that culture and tradition have often been cited as impediments to growth, as they do not fit the dominant neoliberal approaches to economic growth in the region. Nevertheless, the emergence of the new regionalism in the Pacific means that there is a 'growing interest in new paradigms suitable for small island developing states such as "green growth", the blue (ocean) economy, and the traditional (village-centred) economy' (Maclellan 2015: 5, emphasis in the original).

The rise of new regionalism in the Pacific in recent years is the result of 'a new regional political architecture' (Leslie and Wild 2018: 20) that culminated with the establishment of the Framework for Pacific Regionalism in 2013. Within this enhanced framework, Pacific states seek to collectively coalesce in adhering to the Sustainable Development Goals (SDGs) and cooperate in mitigating the impacts of climate change in the region. The increasing exposure to natural hazards puts Pacific Island state governments in a position in which the instances of Indigenous people and the civic society (Leslie and Wild 2018) around ecological resilience prevail over the hegemonic discourse of economic growth. In the words of the then President of Kiribati, Anote Tong, Pacific Island states 'see the value of replicating communal systems at the regional and international levels as crucial to our advancement as a region and as a people' (Tong 2015: n.p.).

The almost absolute dominance of hyper-neoliberal discourses that are hegemonic, market driven and market led (Bosman and Dredge 2011) has become a key element in the narrative and operationalization of tourism governance in the Pacific since the 1990s. Hyper-neoliberalism exists in situations in which neoliberalism is taken to extremes, especially with respect to austerity reforms in the public administration and implementation of market-obeying policies (Amore and Hall 2017). According to Dredge and Jamal (2013: 560), hyper-neoliberalism occurs when 'profound structural changes are taking place in the way governments govern and public-private partnerships become the preferred mechanism to deal with a variety of public policy issues'. Indeed, in the Pacific Island context, Overton and Murray (2018) discuss the situation of 'inverse sovereignty', whereby island states and their small and stretched bureaucracies have to deal with complex and often burdensome donor reporting requirements, management systems, consultative meetings and strategic priorities. Nevertheless, the emergence of the new Pacific regionalism indicates an emerging counter-narrative to the previously dominant discourse, particularly with respect to climate change, fisheries policy and assertions of political sovereignty. For example, Prime Minister Bainimarama (2015) of Fiji has highlighted climate change as being at the heart of tensions between Australia and the Pacific Islands:

> As we see it, Australia and New Zealand have been put to the test on climate change and been found wanting. It should be no surprise that we have formed the view that at the very least, their position as full members of our island nation Forum needs to be questioned, reexamined and redefined. They simply do not represent our interests as we face this critical matter of survival.

Similarly, at the 2014 UN Climate Summit, the Foreign Minister of the Marshall Islands, Tony de Brum, stated:

> Probably one of the most frustrating events of the past year for Pacific islanders is Australia's strange behavior when it comes to climate change … Australia is a member of the Pacific Islands Forum and Australia is a Pacific island, a big island, but a Pacific island. It must recognize that it has a responsibility.
>
> *(quoted in O'Malley 2014)*

The focus on crises, such as that presented by climate change, is often integral to the paradigm of hyper-neoliberalism within the frame of crisis-prone capitalism (Klein 2015). This approach conceives public policy, and tourism policy making in particular, as manifestations of the contemporary depoliticized environment in which policy interventions are limited by the 'rationality' of a narrow range of economic thinking (Amore and Hall 2017). Hyper-neoliberalism can, therefore, potentially dominate the institutional and socio-technical system and frame the policy agenda that directly and indirectly affects tourism destinations and tourism agencies and sets destinations on particular development trajectories (Hall 2016). Amore and Hall (2017) suggest that hyper-neoliberalism is often reinforced by, if not a direct consequence of, crisis-prone capitalism, and they argue that because socio-technical systems and institutions are already set on certain economic and policy trajectories, 'the hyperneoliberal doctrine will only further exacerbate the propensity to further crises and failures' (Amore and Hall 2017: 10). For example, the case of Nauru is emblematic of decades of economic crises and market-driven policies that ultimately undermined the fragile ecosystem of the island in total disregard of its finite resources (Klein 2015) (see Chapter 20). Similarly, in 2019, in the Cook Islands, the manager of the world's largest marine park lost her job for backing a sea mining moratorium in opposition to the Cook Islands' government's desire to proceed with mining exploration, saying it wants to be at 'the frontier of the new gold rush' and could be ready to start seabed mining within five years (Doherty 2019). It is, therefore, highly relevant for the governance and metagovernance of tourism in the Pacific Islands that the new regionalism focuses on developing new responses to such crises. Partly due to the failure of Organisation for Economic Co-operation and Development (OECD) nations to recognize the 'special and differential' status of vulnerable island states in regional trade policy, Pacific Island governments are increasingly looking to new Asian markets and partners and have been advancing a range of sub-regional initiatives on trade and labour mobility (Maclellan 2015). Nevertheless, as the following discussion indicates, climate change and other regional challenges provide a significant focus for tourism-related governance in the Pacific in light of the new regionalism paradigm (Leslie and Wild 2018; Vince et al. 2017).

Climate change and biodiversity loss

Climate change poses a serious threat to the future of Pacific Island states (Gössling et al. 2009). Many low-lying small island countries are extremely vulnerable to sea-level rise as a result of coral bleaching events and ocean acidification (IPCC 2014). Damage to the coral reefs and mangroves that serve to reduce wave impact and tidal surges will make islands more vulnerable to coastal erosion, flooding and damage to their fresh water supplies (Hall 2015). Many island population centres are, therefore, subject to increased risk levels from high magnitude weather events. According to the Alliance of Small Island States' (AOSIS) Declaration on Climate Change, which is supported by nearly all Pacific Island states and territories, 'climate change poses the most serious threat to our survival and viability, and … undermines our efforts to achieve sustainable development goals and threatens our very existence' (AOSIS 2009: 1).

There is no doubt that international tourism, which is so important to the regional economy, contributes to climate change (Hall 2010; Hall et al. 2015; Scott et al. 2010). The first estimate based on figures from 2005 found that travel and tourism accounted for around 4.9% of global carbon dioxide (CO^2) emissions, with transport alone accounting for 1.77 million metric tons of CO^2 equivalent ($MtCO^2$) (UNWTO, WMO and UNEP 2008). However, this figure did not include any estimate of the effects of radiative forcing (Hall et al. 2013). More recently, and based on a lifecycle assessment, Lenzen et al. (2018) estimate that the global carbon footprint of travel and tourism was around 8% of all emissions, a situation that reinforces the potential vulnerability of businesses and destinations that are dependent on air travel for their efforts to reduce emissions (Scott et al. 2016a, 2016b).

Some accounts argue that poverty reduction through tourism somehow compensates environmental externalities (e.g., UNTWO 2007a). However, the total economic contribution of tourism in the least developed countries in 2016 is insignificant when compared to the estimated economic losses as a result of climate change (UNFCCC 2019; UNWTO 2017). The trade-off between environmental externalities and tourism development underpinned the pro-growth rhetoric of the past 30 years by overlooking the very finite resource stock of Pacific Island states. Evidence from Kiribati shows how sea-level rise represents a serious threat to the 33 atolls (Bowers 2017), yet the Kiribati government continued to pursue a tourism strategy that reiterated the paradigm of economic development as a flywheel to environmental and social sustainability (Government of Kiribati 2016). In the Cook Islands, the hospitality industry contributed much to the depletion of the already limited water resources and vulnerable ecosystems over the years, with the government now coordinating with the private sector to establish practices of social and environmental sustainability in tourism (Evans 2019) (see Chapter 25). Like many low-lying states, Tuvalu also faces increased vulnerability from climate change in the forms of sea-level rise, ocean acidification and water shortages (Gössling et al. 2009) (see Chapter 26). The media exposure on the impacts of climate change on small island destinations has grown exponentially, with public awareness seeking to understand social and environmental vulnerabilities behind the mass tourism idylls of islands in the Pacific (Hall 2010).

Resource use, population growth, development patterns and urbanization in island states in the Pacific are beyond sustainable levels, where increased tourism pressure hinders the carrying capacity of these islands with potentially irreparable consequences to the natural environment and biodiversity (Hall 2010). Subsequently, balancing tourism and the effects of climate change presents a significant policy challenge. Presently, this appears to be managed under the notion that, while tourism is a major source of emissions because of the aviation component, the overall low emissions of the islands by international measures provide justification for further expansion. However, the tensions created by the relationship between tourism and climate change will clearly need to be managed in the long term. Maclellan (2015: 7) notes that as Pacific Island 'countries move toward negotiation of a global climate treaty and forge new Sustainable Development Goals, regional institutions will find it increasingly difficult to paper over contested visions for the future'.

Hazards and disasters

Due to their small size and disproportionate share of the population living in hazard-prone coastal areas, the Pacific Islands are particularly vulnerable to natural disasters (Campbell 2014). Moreover, the El Niño–Southern Oscillation severely affects the agriculture and fisheries of the Pacific Island states, thereby affecting food security. This weather-related phenomenon also affects the appeal of Pacific Island states as tourist destinations, as increased precipitation and chances of tropical storms can affect perceptions of destinations in the region (Scott et al. 2012). Nevertheless, it could be argued that there is no such thing as a natural disaster (Squires and Hartman 2006), as human activity contributes much to the vulnerability of destinations against natural hazards. Samoa, Saint Lucia, Grenada, Vanuatu, Tonga and Maldives are among the top 12 countries with the highest economic losses on capital stock, in relative terms, due to natural disasters from 1970 to 2006. Low-lying islands, such as those of Kiribati and Tuvalu, which are extremely vulnerable to storm surge and sea-level rise, have chronically limited freshwater resources, low annual rainfall and shallow water tables, which have necessitated water shipments at times of drought as well as the purchase of desalinization plants (Hall 2015).

Governance failures and destination vulnerability

There have been multi-scaled governance failures with respect to the ecological future. This can be seen, for example, at international and regional levels on issues such as tourism, climate change, education and alleviation of poverty. The UN World Tourism Organization (UNWTO 2007a: n.p.) is committed 'to seek balanced and equitable policies to encourage both responsible energy-related consumption as well as anti-poverty operational patterns', which translates into meaning that carbon emissions from long-haul trips should not inhibit travel and tourism to economically and environmentally fragile island destinations (UNWTO 2007b). Nevertheless, as noted above, there is increasing concern in the Pacific as to how development and emissions and climate change trajectories can be reconciled. In the case of Kiribati, for instance, Mallin (2018: 214) notes that the promotion of a 'sinking nation paradigm' by successive governments has constrained the political vision of the country in the long term. Instead, 'adaptation' has permeated the economic development agenda in which budgetary requirements are put ahead of the needs and desires of island communities. This policy paradigm, in turn, has facilitated the emergence of highly environmentally problematic deep-sea mineral initiatives.

A good extent of governance frictions over ecological futures is the response to the hegemonic rhetoric around sustainability, growth and the alleged role of tourism as a catalyst for development in the region (d'Hauteserre 2010). One statement about tourism that is often taken for granted is that it fosters economic development and poverty alleviation (Hall 2007). However, the tendency among tourism stakeholders is to overlook aspects such as greenhouse gas emissions, carbon offsetting and mitigation costs (Gössling et al. 2009). Instead, most policy makers and communities appear to opt for hard, short-term adaptation measures instead of long-term solutions to adapt communities to the inevitable vulnerabilities of climate change (Hall 2015; IPCC 2014).

A second governance issue concerns the myth of the so-called green economy (Hall and Amore 2016). Moseley (2003: 21) argues that 'green' practices, such as ecotourism, are often questionable because it relies 'on tourists making extravagant demands on fossil fuels by travelling thousands of miles for the experience'. Moreover, there is rising criticism as to whether the green economy in tourism actually leads to efficient resource use. In particular, the nexus between local policies and global processes has proved to be incoherent in the context of geographically remote islands, as policies commonly consider carbon offsetting and environmental protection at the level of the destination rather than the wider context of the trip (Hall et al. 2013). To date, there is limited research on the complex policy conundrum of tourism and the environment in island destinations. Hall (2015) is among the few providing an overview of the socio-ecological and socio-political landscape of small island destinations and emphasizes that, while most of Pacific Island states 'are low contributors to greenhouse emissions on per capita basis, [they] bear the brunt of many of the effects of climate change' (Hall 2015: 62). Accordingly, Dornan et al. (2018: 421) express: 'Climate-related challenges and responses will therefore likely shape development policymaking in the coming decades'. Importantly, in the case of the Pacific, the policy response also needs to consider the significance of the large ocean context and the coastal interface, i.e., the blue economy, and the role that this plays not only with tourism but also with sustainable development in the region overall (Keen et al. 2018).

Nevertheless, despite criticisms of the green growth approach, it is significant to note that elements of the green growth discourse have become very significant in the Pacific, especially for countries such as Fiji and Vanuatu. Dornan et al. (2018), for example, highlight that green growth discourse in the Pacific Islands has been shaped by regional and national debates about appropriate development policy and the 'failure' of neoliberal policies. In fact:

> Amidst concern that orthodox prescriptions for development focused too narrowly on free markets and economic growth, green growth was proposed as an 'alternative' model; one that was more in tune with a 'Pacific' balancing of the economy with social and ecological stewardship.
>
> *(2018: 419)*

Sound sustainable tourism policies and practices will define current and future destination trajectories for Pacific Island states. The globalizing reach of contemporary tourism has further manifested inner vulnerabilities as a result of the COVID-19 pandemic. Transnational hospitality corporations and the cruise industry do not represent a valid option for remote island destinations. On the contrary, it should be the prerogative of Pacific Island states to enable the conditions for environmentally and socially responsible forms of local entrepreneurship. The shift toward destination metagovernance should challenge the prevailing rhetoric of market-driven tourism recovery and policy response to COVID-19. Pacific Island states should restructure tourism development to align to the principles of steady state and sufficiency and degrowth (Hall 2009b). The current pandemic, thus, encourages a progressive rethinking of tourism for Pacific Island states, calling for a bold and much-needed focus on 'the environmental, social and economic limits of the ecosystems' (Hall and Seyfi 2021: 233).

Conclusions and research implications

Over the decades, neoliberal policies have framed the development of tourism in the Pacific Island states while also enhancing the economic, environmental and social vulnerabilities of these destinations. Yet, despite clear signs of policy failures, the current hyper-neoliberal shift in tourism policy and adaptation to crises calls for a timely rethinking of tourism in the Pacific Island states. This chapter has sought to place some of the issues of tourism governance in the Pacific Islands within a wider metagovernance perspective (Amore and Hall 2016). Such an approach encourages looking beyond the narrow confines of an explicit 'tourism policy' to understand the broader context within which tourism-related governance is made. In the Pacific, this is especially important because of the emergence of new narratives of islanders and sustainability that serve to challenge hyper-neoliberal processes of governance that currently dominate tourism policy thinking. There is a clear need for sustainable forms of development that are both resilient to the changing climate and which do not exacerbate climate change (AOSIS 2014). The issue as to whether tourism in the Pacific, in its present form, can ever accomplish genuine sustainable development and resilience is open to question. Nevertheless, in the wider metagovernance context, it is clear that the adoption of large ocean state terminology and advocacy for green growth represent a change in orientation with respect to development strategies. Some Pacific actors are using the term 'green growth' to advocate for an 'alternative development model' that supports the 'traditional economy'. Accordingly, Dornan et al. (2018: 421) state: 'Pacific Island actors interpret or vernacularise international messages concerning green growth in ways that are consistent with their worldviews and agendas'.

A significant dimension in sustainable development thinking is the 'islands of sustainability approach' (IOS) (Deschenes and Chertow 2004; Hall 2015; Wallner et al. 1996). This conceptualization of sustainability addresses not only the interactions between the economic system and the ecosphere but also the structural aspects of the anthropogenic system, notably economic diversity and economic connectedness (Wallner et al. 1996). In the case of the Pacific, it is notable that the frame of thinking with respect to the 'boundary' of island states has come to include their oceanic territories. Governments and regional agencies of the Pacific Islands are strengthening their commitment to sustainable oceans management, including tourism dimensions, through increasingly proactive policies and programmes, although the capacity to implement them is, at times, limited because of a lack of resources and the market power of transnational operations. Nevertheless, the conceptual framework of the blue economy has been found to be useful for structuring evaluations of practice and helping to reveal missing ingredients necessary for the sustainable development of oceans (Bennett 2018; Keen et al. 2018). However, marine policy makers in large ocean/small island developing states (SIDS) need to be more vigilant to wider political economic agendas and narratives when considering options for the ocean and coastal governance (Mallin 2018).

According to Hall et al. (2018), a destination is resilient when stakeholders:

- are aware of the vulnerabilities and of the likely impact of potential hazards;
- embark on redevelopment paths that benefit the local community at large;
- engage in networked and collaborative forms of destination planning;
- reframe the metagovernance of destinations;
- operate predominantly at the regional and local level; and
- reflexively and willingly learn from previous crises and reduce destination vulnerability for the future.

Such a goal would appear to mesh well with the reorientation of the Pacific Island governance agenda toward new regionalism and a growing commitment to climate change mitigation and adaptation, moving beyond Western green growth and neoliberal 'solutions' to a more nuanced localized perspective. Nevertheless, manufactured islands of vulnerability and resilience can reinforce unwanted stereotypes that can constrain the political visions of Pacific Island states into hegemonic – and uneven – relationships between centre and periphery. The diversity of island geographies illustrates that vulnerability and resilience are neither opposites nor independent or objective variables, and it is best supported by sustainable island development endeavours when accepted as being subjective, contextualized and nuanced (Kelman, 2018). The shift toward resilience and metagovernance, therefore, potentially frames a new trajectory for tourism and development studies (and research) in the Pacific. Arguably, the present experience with respect to the importance of metagovernance of tourism and the reframing of Pacific SIDS as large ocean states reinforces Wallner et al.'s (1996) observation that islands of sustainability can be regarded as 'troublemakers', infiltrating the whole unsustainable system, acting as new cells of development and enabling a rethinking of governance in a changing world.

References

ADB (Asian Development Bank) (2018) *Tourism as a Driver of Growth in the Pacific: A Pathway to Growth and Prosperity*, Manila: Asian Development Bank.

Addinsall, C., K. Glencross, P. Scherrer, B. Weiler and D. Nichols (2015) 'Agroecology and sustainable rural livelihoods: A conceptual framework to guide development projects in the Pacific Islands', *Agroecology and Sustainable Food Systems*, 39(6): 691–723.

Amore, A. and C.M. Hall (2016) 'From governance to meta-governance in tourism? Re-incorporating politics, interests and values in the analysis of tourism governance', *Tourism Recreation Research*, 41(2): 109–122.

Amore, A. and C.M. Hall (2017) 'National and urban public policy in tourism: Towards the emergence of a hyperneoliberal script?', *International Journal of Tourism Policy*, 7(1): 4–22.

Amore, A., G. Prayag and C.M. Hall (2018) 'Conceptualizing destination resilience from a multilevel perspective', *Tourism Review International*, 22(3–4): 235–250.

Anderson, T. and G. Lee (2010) 'Introduction: Understanding Melanesian customary land'. In: T. Anderson and G. Lee (eds.) *Defence of Melanesian Customary Land*, Sydney, NSW: Aid/Watch, pp. 2–4.

AOSIS (Alliance of Small Island States) (2009) *Declaration on Climate Change 2009*, New York: AOSIS High-level Summit on Climate Change, American Museum of American History. Available online at https://sustainabledevelopment.un.org/content/documents/1566AOSISSummitDeclarationSept21FINAL.pdf (accessed 8 March 2019).

AOSIS (Alliance of Small Island States) (2014) 'Intervention on climate change', 11th Session of Open Working Group on Sustainable Development, 8 May. Available online at https://sustainabledevelopment.un.org/content/documents/9358aosis.pdf (accessed 8 March 2019).

Bainimarama, V. (2015) 'Building an inclusive and independent institution for Pacific islanders, speech at meeting on agreement to institutionalize PIDF', Suva: Government of Fiji, 6 May.

Bennett, N.J. (2018) 'Navigating a just and inclusive path towards sustainable oceans', *Marine Policy, 97*: 139–146.

Bosman, C. and D. Dredge (2011) 'Histories of placemaking in the gold coast city: The neoliberal norm, the state story and the community narrative', *Urban Research Program, Research Paper 33*, April, Brisbane, QLD: Griffith University.

Bowers, M. (2017) 'Waiting for the tide to turn: Kiribati's fight for survival', *The Guardian*, 23 October. Available online at https://www.theguardian.com/world/2017/oct/23/waiting-for-the-tide-to-turn-kiribatis-fight-for-survival (accessed 9 March 2019).

Britton, S.G. (1982) 'The political economy of tourism in the third world', *Annals of Tourism Research*, 9(3): 331–358.

Campbell, J.R. (2014) 'Climate-change migration in the Pacific', *The Contemporary Pacific*, 26(1): 1–28.

Cheer, J.M., S. Pratt, D. Tolkach, A. Bailey, S. Taumoepeau and A. Movono (2018) 'Tourism in Pacific island countries: A status quo round-up', *Asia and the Pacific Policy Studies*, 5(3): 442–461.

Connolly, P.J. (2016) 'Engaging China's new foreign policy in the South Pacific', *Australian Journal of International Affairs*, 70(5): 484–505.

d'Hauteserre, A.M. (2010) 'Government policies and indigenous tourism in New Caledonia', *Asia Pacific Journal of Tourism Research*, 15(3): 285–303.

Dearden, S.J. (2008) 'EU aid policy towards the Pacific ACPs', *Journal of International Development*, 20(2): 205–217.

Department of Business, Economic Development and Tourism (2019) *Outlook for the Economy: 1st Quarter 2019 Report*. Available online at http://dbedt.hawaii.gov/economic/qser/outlook-economy/ (accessed 9 March 2019).

Deschenes, P.J. and M. Chertow (2004) 'An island approach to industrial ecology: Towards sustainability in the island context', *Journal of Environmental Planning and Management*, 47(2): 201–217.

Doherty, B. (2019) 'Cook Islands: Manager of world's biggest marine park says she lost job for backing sea mining moratorium', *The Guardian*, 19 October. Available online at https://www.theguardian.com/world/2019/oct/20/cook-islands-manager-of-worlds-biggest-marine-park-says-she-lost-job-for-backing-sea-mining-moratorium (accessed 9 November 2019).

Dornan, M., W. Morgan, T. Newton, T. Cain and S. Tarte (2018) 'What's in a term? "green growth" and the "blue-green economy" in the Pacific islands', *Asia and the Pacific Policy Studies*, 5(3): 408–425.

Dredge, D. and T. Jamal (2013) 'Mobilities on the Gold Coast, Australia: Implications for destination governance and sustainable tourism', *Journal of Sustainable Tourism*, 21(4): 557–579.

Evans, M. (2019) 'Paradise, polluted: Cook Islands tries to clean up its tourism sector', *Mongabay*, 23 September. Available online at https://news.mongabay.com/2019/09/paradise-polluted-cook-islands-tries-to-clean-up-its-tourism-sector/ (accessed 25 October 2019).

Gössling, S. (2003) *Tourism and Development in Tropical Islands: Political Ecology Perspectives*, Aldershot: Edward Elgar.

Gössling, S., C.M. Hall and D. Scott (2009) 'The challenges of tourism as a development strategy in an era of global climate change'. In: E. Palosou (ed.), *Rethinking Development in a Carbon-Constrained World: Development Cooperation and Climate Change*, Helsinki: Ministry of Foreign Affairs, pp. 100–119.

Government of Kiribati (2016) *Kiribati 20 Year Vision 2016–2036*, Tarawa: Government of Kiribati.

Guthrie, K. (2011) 'The role of civil aviation safety and security in the economic development of Pacific island countries', *The Journal of Social, Political, and Economic Studies*, 36(2) 218–248.

Guthrie, K. (2013) 'Aviation regionalism in the Pacific: A history', *The Journal of Pacific History*, 48(3): 294–308.

Hall, C.M. (2005) *Tourism: Rethinking the Social Science of Mobility*, Harlow: Prentice Hall.

Hall, C.M. (2007) 'Pro-poor tourism: Do tourism exchanges benefit primarily the countries of the South?', *Current Issues in Tourism*, 10(2–3): 111–118.

Hall, C.M. (2009a) 'Heritage tourism in the Pacific: Modernity, myth, and identity'. In: D.J. Timothy and G. Nyaupane (eds.) *Cultural Heritage and Tourism in the Developing World*, London: Routledge, pp. 108–125.

Hall, C.M. (2009b) 'Degrowing tourism: Décroissance, sustainable consumption and steady-state tourism', *Anatolia*, 20(1): 46–61.

Hall, C.M. (2010) 'Island destinations: A natural laboratory for tourism: Introduction', *Asia Pacific Journal of Tourism Research*, 15(3): 245–249.

Hall, C.M. (2015) 'Global change, islands and sustainable development: Islands of sustainability or analogues of the challenge of sustainable development?' In: M. Redclift and D. Springett (eds.) *The Routledge International Handbook of Sustainable Development*, Abingdon: Routledge, pp. 55–73.

Hall, C.M. (2016) 'Intervening in academic interventions: Framing social marketing's potential for successful sustainable tourism behavioural change', *Journal of Sustainable Tourism*, 24(3): 52–72.

Hall, C.M., B. Amelung, S. Cohen, E. Eijgelaar, S. Gössling, J. Higham, R. Leemans, P. Peeters, Y. Ram and D. Scott (2015) 'On climate change skepticism and denial in tourism', *Journal of Sustainable Tourism*, *23*(1): 4–25.

Hall, C.M. and A. Amore (2016) 'Turismo, sostenibilità e crescita verde: Green economy o una semplice pennellata di verde?' In: A. Pecoraro Scanio (ed.) *Turismo Sostenibile*, Rome: Aracne Editrice, pp. 145–188.

Hall, C.M., G. Prayag and A. Amore (2018) *Tourism and Resilience: Individual, Organisational and Destination Perspectives*, Clevedon: Channel View Publications.

Hall, C.M., D. Scott and S. Gössling (2013) 'The primacy of climate change for sustainable international tourism', *Sustainable Development*, 21(2): 112–121.

Hall, C.M. and S. Seyfi (2021) 'COVID-19 pandemic, tourism and degrowth'. In: C.M. Hall, L. Lundmark and J. Zhang (eds.) *Degrowth and Tourism: New Perspectives on Tourism Entrepreneurship, Destinations and Policy*, Abingdon: Routledge, pp. 220–238.

Harrison, D. (2004) 'Editor's introduction: Tourism in Pacific islands', *Journal of Pacific Studies*, *26*(1–2): 1–28.

IPCC (International Panel on Climate Change) (2014) 'Small islands'. In: T.F. Stocker, D. Qin, G.K. Plattner, M. Tignor, S.K. Allen, J. Boschung, A. Nauels, Y. Xia, V. Bex and P. Midgley (eds.) *Climate Change 2014: Impacts, Adaptation and Vulnerability, Part B: Regional Aspects*. Contribution of Working Group II to the Fifth Assessment Report of the Intergovernmental Panel on Climate Change, Cambridge: Cambridge University Press, pp. 1613–1655.

Keen, M.R., A.M. Schwarz and L. Wini-Simeon (2018) 'Towards defining the blue economy: Practical lessons from Pacific Ocean governance', *Marine Policy*, 88: 333–341.

Kelman, I. (2018) 'Islands of vulnerability and resilience: Manufactured stereotypes?' *Area*, 52(1): 6–13.

Klein, N. (2015) *This Changes Everything: Capitalism vs. The Climate*, London: Penguin.

Lenzen, M., S. Ya-Yen, F. Futu, T. Yuan-Peng, G. Arne and M. Arunima (2018) 'The carbon footprint of global tourism', *Nature Climate Change*, 8(6): 522–528.

Leslie, H. and K. Wild (2018) 'Post-hegemonic regionalism in Oceania: Examining the development potential of the new framework for Pacific regionalism', *The Pacific Review*, 31(1): 20–37.

Maclellan, N. (2015) 'Transforming the regional architecture: New players and challenges for the Pacific islands', *Asia-Pacific Issues*, No. 118, August, Honolulu, HI: East-West Centre.

Mallin, M.A.F. (2018) 'From sea-level rise to seabed grabbing: The political economy of climate change in Kiribati', *Marine Policy*, 97: 244–252.

Milne, S. (1997) 'Tourism, dependency and South Pacific microstates: Beyond the vicious cycle?' In: D.G. Lockhart and D. Drakakis-Smith (eds.) *Island Tourism: Trends and Prospects*, London: Pinter, pp. 281–301.

MFEM (Ministry of Finance and Economic Management) (2019) *Cook Islands Statistical Bulletin: Migration Statistics 2019*, Avarua: Cook Islands Ministry of Finance and Economic Management.

Moseley, M.J. (2003) *Rural Development: Principles and Practice*, London: Sage.

Mowforth, M. and I. Munt (1998) *Tourism and Sustainability: Tourism in the New World*, London: Routledge.

Murray, W.E., J. Overton, G. Prinsen, T.A.J. Ulu and N. Wrighton (2018) *Aid, Ownership and Development: The Inverse Sovereignty Effect in the Pacific Islands*, Abingdon: Routledge.

O'Malley, N. (2014) 'Australia is a Pacific island - It has a responsibility', *Sydney Morning Herald*, 21 September 2014. Available online at https://www.smh.com.au/world/australia-is-a-pacific-island--it-has-a-responsibility-20140921-10jwdw.html (accessed 24 March 2022).

Overton, J. and W.E. Murray (2018) 'Migration, education and marginality: Networks and strategies in the Pacific islands'. In: S. Pelc and M. Koderman (eds.) *Nature, Tourism and Ethnicity as Drivers of (De) Marginalization: Insights to Marginality From Perspective of Sustainability and Development*, Cham: Springer, pp. 215–233.

PIFS (Pacific Island Forum Secretariat) (2014) *The Framework for Pacific Regionalism*, Suva: PIFS.

Pan, C., M. Clarke and S. Loy-Wilson (2018) 'Local agency and complex power shifts in the era of belt and road: Perceptions of Chinese aid in the South Pacific', *Journal of Contemporary China*, 28(117): 385–399.

Scheyvens, R. and J.H. Momsen (2008) 'Tourism and poverty reduction: Issues for small island states', *Tourism Geographies*, 10(1): 22–41.

Schilcher, D. (2007) 'Growth versus equity: The continuum of pro-poor tourism and neoliberal governance', *Current Issues in Tourism*, 10(2–3): 166–193.

Scott, D., C.M. Hall and S. Gössling (2012) *Tourism and Climate Change*, Abingdon: Routledge.
Scott, D., C.M. Hall and S. Gössling (2016a) 'A review of the IPCC 5th assessment and implications for tourism sector climate resilience and decarbonization', *Journal of Sustainable Tourism*, 24(1): 8–30.
Scott, D., S. Gössling, C.M. Hall and P. Peeters (2016b) 'Can tourism be part of the decarbonized global economy?: The costs and risks of carbon reduction pathways', *Journal of Sustainable Tourism*, 24(1): 52–72.
Scott, D., P. Peeters and S. Gössling (2010) 'Can tourism deliver its 'aspirational emission reduction targets?', *Journal of Sustainable Tourism*, 18(3): 393–408.
Souty, F. (2002) *Passport to Progress: Competition Challenges for World Tourism and Global Anticompetitive Practices in the Tourism Industry*, Madrid: UNWTO.
Squires, G. and C.W. Hartman (2006) *There Is No Such Thing as a Natural Disaster: Race, Class, and Hurricane Katrina*, London: Routledge.
Tabani, M. (2017) 'Development, tourism and commodification of cultures in Vanuatu'. In: E. Gnecchi-Ruscone and A. Pain (eds.) *Tides of Innovation in Oceania: Value, Materiality and Place*, Canberra, ACT: Australian National University Press, pp. 225–260.
Tarte, S. (2014) 'Regionalism and changing regional order in the Pacific islands', *Asia and the Pacific Policy Studies*, 1(2): 312–324.
Tong, A. (2015) 'Charting its own course: A paradigm shift in Pacific diplomacy'. In: G. Fry and S. Tarte (eds.) *The New Pacific Diplomacy*, Acton, ACT: ANU Press, pp. 21–26.
Tourism Concern (2009) *Putting Tourism to Rights: A Challenge to Human Rights Abuses in the Tourism Industry*, London: Tourism Concern.
UNFCCC (United Nations Framework of Convention on Climate Change) (2019) *UN climate Change Annual Report 2018*, Bonn: UNFCCC.
UNWTO (United Nations World Tourism Organization) (2007a) 'Tourism will Contribute to Solutions for Global Climate Change and Poverty Challenges' (press release), Madrid: UNWTO.
UNWTO (United Nations World Tourism Organization) (2007b) *From Davos to Bali – A Tourism Contribution to the Challenge of Climate Change*, Madrid: UNWTO.
UNWTO (United Nations World Tourism Organization) (2017) *Tourism for Sustainable Development in Least Developed Countries: Leveraging Resources for Sustainable Tourism with the Enhanced Integrated Framework*, Madrid: UNWTO.
UNWTO (United Nations World Tourism Organisation), WMO (World Meteorological Organization) and UNEP (United Nations Environment Programme) (2008) *Climate Change and Tourism: Responding to Global Challenges*, Madrid: UNWTO.
Vince, J., E. Brierley, S. Stevenson and P. Dunstan (2017) 'Ocean governance in the South Pacific region: Progress and plans for action', *Marine Policy*, 79: 40–45.
Wallis, J. (2017) *Pacific Power?: Australia's Strategy in the Pacific Islands*, Clayton, VIC: Melbourne University Publishing.
Wallner, H.P., M. Narodoslawsky and F. Moser (1996) 'Islands of sustainability: A bottom-up approach towards sustainable development', *Environment and Planning A: Economy and Space*, 28(10): 1763–1778.
Williams, C. and A. Blaiklock (2018) 'Human rights informed the sustainable development goals, but are they lost in New Zealand's neoliberal aid program?' In: G. MacNaughton and D.F. Frey (eds.) *Economic and Social Rights in a Neoliberal World*, Cambridge: Cambridge University Press, pp. 236–260.
WTTC (World Travel and Tourism Council) (2020a) *Travel and Tourism Economic Impact 2018: Vanuatu*, London: WTTC.
WTTC (World Travel and Tourism Council) (2020b) *Travel and Tourism Economic Impact 2018: Fiji*, London: WTTC.
Yang, J. (2011) *The Pacific Islands in China's Grand Strategy: Small States, Big Games*, New York (NY): Palgrave Macmillan.

4

TOURISM MOBILITY AND TRANSPORT

Issues and developments in Pacific Island states

Dallen J. Timothy

Introduction

The Pacific Islanders have long been a mobile people. Their historical seafaring prowess facilitated much inter-island travel, trade and fishing. However, during the past two centuries of modernization, several geographical and economic conditions have challenged their regional mobility and the ability of tourism to flourish in some countries. The Pacific Islands exude a touristic appeal like no other region on Earth. Several countries are widely regarded as tropical paradises, and many have become thriving tourism destinations, largely in the leisure-oriented sun, sea and sand marketplace. Since World War II, tourism has become one of the most pervasive industries in the Pacific region, and many countries rely on its dividends for their economic well-being. However, given the region's unique geographical, demographic and economic characteristics, the success of tourism within the Pacific Islands depends on efficient transportation and adequate accessibility, perhaps more than in any other insular region of the world. The manifold issues associated with accessibility and its challenges are the focus of this chapter, as it examines the mobility patterns, transportation dilemmas and opportunities in the sovereign and self-governing states of the Pacific Islands. Its large size, heavily dispersed geography, small populations and perceptively small economies of scale both limit tourism's possibilities and promote its uniqueness as a magnet for tourism.

The chapter first examines transportation mobility in the Pacific region from a general perspective. This section describes mobility and transportation patterns in the Pacific Islands at large and then at national levels throughout the region. The chapter then elucidates many of the challenges and opportunities for mobility in the region pertaining to characteristics such as highly fragmented geographies, isolation, physical size and topography, small populations, unbalanced economies of scale, political fragmentation and irregular and unstable transportation networks. While these erect significant development barriers in many cases, some of them simultaneously provide a stimulus for tourism demand. Finally, the chapter considers other mobility issues, such as freedom of travel and visa regulations, emerging air transportation networks and the effects of the COVID-19 pandemic on regional travel.

DOI: 10.4324/9780429019968-5

Transportation mobility in the Pacific Island states

The mobility regime of the Pacific Islands is synonymous with long-haul transportation, even within most individual states. While maritime transportation has played an important role in the region for centuries, today, air transportation dominates human mobility for Pacific Islanders and global tourists. Contemporary maritime mobility in the region has been geared toward the cruise sector, as relatively few water-based options for long-distance travel currently exist, and even within states, maritime services remain inadequate. The next two sections discuss general mobility patterns from global/regional and subnational perspectives.

General patterns of global and regional mobility

As other chapters in this volume have highlighted, the Pacific Islands are appealing global destinations for several reasons, but this section describes two primary markets. The first market is sun, sea and sand (SSS) seekers. Many of the island states in the region are home to beautiful beaches, a pleasant climate, coral reefs, fascinating cultures and/or water sports. This has long attracted a leisure vacation-oriented market, and many island resorts have developed in response to this demand (Duval 2016; Rao 2002). From a mobilities perspective, the second global market comes from the islands' diasporic populations. There are hundreds of thousands of Pacific Islanders scattered throughout the world. For example, there are more than 57,000 native Tongans or people of Tongan descent in the United States, more than 36,000 in Australia, and more than 80,000 in New Zealand (Australian Bureau of Statistics 2015; Taumoefolau 2015). According to the 2018 New Zealand census, besides the Indigenous Māori, Samoans (182,721) were the largest group of Pacific Islanders living in New Zealand. These were followed by Tongans (82,389), Cook Islander Māori (80,532) and Niueans (30,867), with other large populations of Fijians and Tuvaluans also living in the country (Stats NZ 2019). Some of these high numbers, such as Samoa, Niue and the Cook Islands, reflect those countries' former colonial and current free association status with New Zealand. These sizable numbers have significant implications for global and regional travel. While Pacific Islanders are scattered throughout the world, these specific New Zealand figures illustrate the diversity of Pacific populations overseas. Like many other diasporic populations, Pacific peoples are avid return travellers who desire to visit their home countries for holiday, familial, religious and cultural reasons (Coles and Timothy 2004; Scheyvens 2007). Accordingly, there is high demand for travel between clusters of diasporic populations abroad and their homelands. While much of this demand is regional (i.e., from within the Pacific area), much of it derives from further afield.

In addition to climatic, beachfront and diasporic stimuli, there is a third reason for the travel connectedness between several island states and other countries: political ties. For instance, the Cook Islands are well linked to New Zealand because the two countries are politically connected through 'free association'. The same relationship exists between Niue and New Zealand, and New Zealand is the only international origin and destination of flights to and from Niue. The Marshall Islands, Palau and all four states in the Federated States of Micronesia (FSM), have international transport connections to the United States at Guam and Honolulu, not only because of their geographic proximity but also because of their former protectorate status and current free association with the United States.

The main mode of transportation throughout the Pacific and from other regions is, predictably, air travel (Duval 2005, 2007; Taumoepeau 2015, 2016a). Each sovereign state in the region has at least one international airport, although not all of these have flight connections beyond the Pacific (Table 4.1). The states with the best interconnections to localities outside the Pacific

Table 4.1 International airports in the Pacific Island states, 2020

Country	*International Airport(s)*	*Main International Connections*	*Max. Runway Length (metres)*
Cook Islands	Rarotonga International Airport	Australia, New Zealand, Tahiti, US	2,328
Federated States of Micronesia	Chuuk International Airport	Japan, Kiribati, Marshall Islands, Nauru, Papua New Guinea (PNG), US	1,831
	Kosrae International Airport	Marshall Islands, Nauru, US	1,753
	Pohnpei International Airport	Kiribati, Marshall Islands, Nauru, PNG, US	1,829
	Yap International Airport	Guam (US)	1,829
Fiji	Nadi International Airport	Australia, China, Hong Kong, Japan, New Zealand, most Pacific states/islands, PNG, Singapore, Taiwan, US	3,273
	Nausori International Airport	Australia, New Zealand, Samoa, Tonga, Tuvalu, Vanuatu	1,868
Kiribati	Bonriki International Airport	Australia, Fiji, FSM, Marshall Islands, Nauru, Solomon Islands, Tuvalu	2,011
	Cassidy International Airport	Fiji, US	2,103
Marshall Islands	Amata Kabua International Airport	Kiribati, Nauru, FSM, US	2,407
Nauru	Nauru International Airport	Australia, Fiji, FSM, Kiribati, Marshall Islands and Solomon Islands	2,150
Niue	Niue International Airport	New Zealand	2,335
Palau	Roman Tmetuchl International Airport	FSM, Japan, Macau, Philippines, South Korea, Taiwan, Guam (US)	2,195
Samoa	Faleolo International Airport	Australia, Fiji, New Zealand, Tonga, American Samoa (US)	3,000
Solomon Islands	Honiara International Airport	Australia, Fiji, Kiribati, Nauru, PNG, Vanuatu	2,200
Tonga	Fua'amotu International Airport	Australia, Fiji, New Zealand, Samoa	2,681
	Vava'u International Airport	Fiji	1,705
Tuvalu	Funafuti International Airport	Fiji, Kiribati	1,524
Vanuatu	Bauerfield International Airport	Australia, Fiji, New Caledonia, New Zealand, PNG, Solomon Islands, Vanuatu	2,600
	Santo-Pekoa International Airport	Australia, New Caledonia	1,988

Source: Author's own compilation

Islands are Fiji and Palau. The best connected countries within the insular region are the FSM, Fiji, Kiribati and Vanuatu. The least connected countries are the Marshall Islands, Niue and Tuvalu. All the island states, however, are either directly or indirectly linked to larger transportation centres where passengers can transfer to other destinations.

During the past half century, air transportation in the Pacific has been rather volatile, with many unsuccessful efforts to restructure airlines and develop new routes (as discussed later). Likewise, there has been relatively little competition between air carriers, owing largely to small demand, long distances from major markets, high operation costs and strict regulatory policies, all of which dissuade airlines from competing on a large scale in a large region (ADB 2007). In spite of these challenges, several airports in the region have undergone expansions and improvements in recent years to accommodate expected eventual tourism growth by accommodating larger aircrafts and more frequent landings. The Marshall Islands' Amata Kabua International Airport was improved between 2007 and 2009 with resurfaced runways and expanded aprons to enable larger aircraft to land. This came about, in part, through pressure by Japan Airlines (JAL), which had planned to start charter flights to Majuro in 2006 (ADB 2007). Despite the expansion, which was funded by the United States government, the Japanese company only flew nine charter flights in 2007 and 2008 but ceased operating owing to a lack of demand and a change in JAL leadership that had different destination priorities (Johnson 2009).

In an effort to grow Fiji's tourism industry, upgrades were made in 2018 to lengthen and widen the runway at Nausori International Airport near Suva to accommodate larger airplanes from more international locations and to improve domestic services. The work was completed in November 2021 (Chute 2022) and, thus, Nausori is planned to be another major Pacific gateway airport. In 2016, Air New Zealand and Qantas ceased flying to Vanuatu's Bauerfield International Airport because of safety concerns over its unmaintained runway. This brought about an emergency project funded by the World Bank to rebuild the runway for the sake of safety and to facilitate non-stop flights from Asia and North America (RNZ 2019a). This Port Vila project helped initiate efforts to upgrade a handful of smaller domestic airports throughout Vanuatu as well. Several other Pacific airports have recently undergone runway expansions or other improvements, and new terminals are currently being built or are planned at several airports to accommodate expected increases in international travel, although this will undoubtedly depend on the extent to which tourism markets will revive once international mobility resumes following COVID-19 lockdowns.

Besides air travel, water-based transportation is the second obvious means of human mobility. While there are few global-scale ship transport services, beyond merchant vessels, to carry passengers to the Pacific Islands from areas such as Europe, Africa or North America, the cruise sector had become increasingly popular in the Pacific Islands (see Dowling and Weeden 2017). Following World War II, as trans-oceanic air travel began to develop rapidly, ship-based long-distance travel declined. This decline stimulated the re-invention or re-branding of the transport ship industry, giving birth to the modern cruise sector. Nonetheless, there are several physical characteristics that prevent cruise tourism from developing on some islands (these will be discussed later in the chapter), but the sector had been thriving in many areas until the COVID-19 pandemic, which was responsible for the collapse of the cruise sector in March and April 2020. As Table 4.2 illustrates, during normal times, Fiji and Vanuatu are significant cruise destinations (Treloar and Hall 2005), while others, such as the Marshall Islands, Nauru and Tuvalu, traditionally have received few if any cruise arrivals.

The popularity of some island states over others as cruise destinations is often a result of physical geography and location. Some islands have less suitable deep-water ports than others. Coral atolls frequently require expansive dredging works to enable ships to access port moorings. This is costly and environmentally unsound. In most cases, ships are required to drop anchor offshore, with guests being tendered into port, which is not uncommon in other parts of the world as well. This means that ship size and island size can affect a destination's popularity as a cruise port of call. Likewise, in this region, locations that were already popular tourist

Table 4.2 Major cruise ports in the countries of the Pacific, schedule for 2019

Country	*Cruise Ports*	*Number of Scheduled Cruise Arrivals for 2019*	*Other Comments*
Cook Islands	Avarua (Rarotonga)	8	Cruise ships occasionally call at Aitutaki
	Aitutaki	n/a	
Federated States of Micronesia	Chuuk	n/a	Cruise ships occasionally call at Chuuk, Yap and Pohnpei
	Pohnpei	n/a	
	Yap	n/a	
Fiji	Dravuni	16	
	Lautoka	16	
	Port Denarau	19	
	Savusavu	4	
	Suva	42	
Kiribati	Fanning Island	n/a	Cruise ships occasionally call at Fanning Island, Christmas Island and Tarawa
	Kiritimati (Christmas Island)	n/a	
	Tarawa	n/a	
Marshall Islands	Majuro	n/a	
Nauru	No cruise port	0	
Niue	Alofi	3	
Palau	Koror	n/a	Cruise ships occasionally call at Koror
Samoa	Apia	4	
Solomon Islands	Gizo	n/a	Cruise ships occasionally call at Gizo
	Guadalcanal	15	
Tonga	Nukualofa	8	
	Vava'u	10	
Tuvalu	Funafuti	n/a	
Vanuatu	Champagne Bay	10	
	Luganville	5	
	Mystery Island	62	
	Pentecost Island	1	
	Port Vato	0	
	Port Vila	68	

Source: Adapted from data from Crew Center (2019); WhatsinPort (2019); Worldwide Cruise Centres (2019).

destinations have also tended to be popular cruise destinations. Vanuatu and Fiji, in Table 4.2, are illustrative of this. Islands that offer unique cultural or natural experiences typically receive more cruises than those which might only have beaches. Finally, proximity to other islands appears to be an important variable that makes islands popular destinations. A 'critical mass' of islands, or a relatively compact archipelago, can determine cruise popularity. Although some cruises call at isolated localities, they more commonly stop at ports that are reasonably proximal to other ports, given time and cost constraints. The examples of Vanuatu and Fiji in Table 4.2 illustrates this point well.

The worldwide pandemic hit the cruise industry severely, including outbreaks of the Coronavirus, causing many companies to cancel voyages and ships to return to their ports of origin as the virus spread among passengers and crew (Parker 2020). Although the Pacific Islands were among the most favoured cruise destinations before the pandemic, the sector suffered tremendously from the effects of COVID-19. Many governments disallowed cruise ship dockings, and consumers regarded the ships as super-spreaders of the disease (Ng 2020).

General patterns of national mobility

With the exceptions of Nauru and Niue, which are single-island states, the countries in the region comprise national archipelagos, or groups of islands. Inter-island mobility in each country has long been of major significance for government functions, medical care, religious adherence, fishing, trade and commerce, attending school, business travel and return visits (VFR) by residents who have emigrated. While the majority of regional airlines serve a limited number of intra-regional destinations, their primary focus is providing domestic services (Table 4.3). Many countries in the region have their own domestic airlines, some of which also fly to limited intra-regional destinations. Domestic airlines usually function as a government-funded public service to support domestic routes, even among some of the far-flung and least populated islands (Taumoepeau 2015). Some national island groups are relatively compact (e.g., Samoa), whereas others (e.g., Kiribati) are extremely dispersed. Kiribati's geography stretches nearly

Table 4.3 Domestic and international airlines based in the Pacific Island states, 2019

Country	*National Airlines*	*Domestic*	*International*	*International Destinations*
Cook Islands	Air Rarotonga	Yes	Yes (charters)	Tahiti, Niue, Tonga, Samoa
FSM	Caroline Islands Air	Yes	No	–
Fiji	Northern Air	Yes	No	–
	Pacific Island Air	Yes	No	–
	Fiji Airlines/Fiji Link	Yes	Yes	Samoa, Tonga, Tuvalu, Vanuatu
	Fiji Airways	No	Yes	Services 13 countries in the Pacific, plus Hong Kong, Singapore, US
Kiribati	Air Kiribati	Yes	Yes	Tuvalu
	Coral Sun Airways	Yes	No	–
Marshall Islands	Air Marshall Islands	Yes	No	–
Nauru	Nauru Airlines	No	Yes	Australia, Fiji, FSM, Kiribati, Marshall Islands, Solomon Islands
Niue	–	–	–	–
Palau	Belau Air	Yes	No	–
Samoa	Samoa Airways	Yes	Yes	Australia, New Zealand, American Samoa (US)
	Talofa Airways	No	Yes	Tonga and American Samoa (US)
Solomon Islands	Solomon Airlines	Yes	Yes	Australia, Fiji, Kiribati, PNG, Vanuatu
Tonga	Real Tonga	Yes	Yes	Samoa
Tuvalu	–	–	–	–
Vanuatu	Air Vanuatu	Yes	Yes	Australia, Fiji, New Caledonia, New Zealand, PNG, Solomon Islands

Source: Author's own compilation from individual airlines.

3,000 kilometres between the country's western Gilbert Islands and its easternmost Line Islands, creating significant challenges for that country's transportation network. In fact, even Kiribati's national airline, Air Kiribati, does not fly between the country's western islands, including the capital Tarawa, to its easternmost archipelago. Some domestic connections, such as flights between Tarawa and Christmas Island must take place in an international transit space via Fiji. Likewise, although Fiji is one of the main gateways and hubs in the region, travelling between Fiji and the Cook Islands, for example, currently requires transiting through New Zealand, and in doing so, passengers lose another vacation day because of the International Date Line. The challenges to air-based mobility are discussed in more depth later in the chapter.

Most countries in the region have also developed at least a few boat services and water taxis between islands as well as land transportation with public buses and taxi services, although this also depends on the size of the islands and their populations (Treloar and Hall 2005). In a few countries, there are no inter-island sea transportation options. Several countries have no publicly operated land transportation either, but local entrepreneurs have set up businesses to utilize mini-buses or cars for local transportation. Some of these services are formalized with regularly scheduled departures and arrivals, while others are more ad hoc as transport needs dictate. Domestically, Tuvalu is the least accessible country in the region. There are no domestic inter-island flights, and only one cargo vessel delivers people to other islands periodically. Even land transport is limited, with only a handful of taxis and private mini-buses providing services around Funafuti. Most of the other countries have private mini-bus services and taxis along with car rentals. While the local concept of 'ride sharing' has long existed with local buses and vans, the Western notion of ride sharing (i.e., Uber, Lyft) has not caught on in the majority of the countries, and it will likely be some time before it becomes legal or popular.

Pacific mobility challenges and opportunities

The islands' location, proximity to markets, physical size, colonial legacies and other variables have determined accessibility and created a wide range of challenges to transport mobility (Kissling 1984).

Fragmented geography

The region's most obvious geographic characteristic is its insularity and fragmented geography, as previously noted. Comprised of thousands of islands, islets and atolls, the Pacific Islands are scattered over hundreds of millions of square kilometres, with vast distances separating them from their neighbours. Even within some individual countries (e.g., Kiribati and the FSM), distances stretch into the thousands of kilometres between island groups. This scattering of landmass and large distances necessitate the use of aircraft as the primary mode of inter-island travel, at least within the most sprawling countries (Collison and Spears 2010; Duval 2005). Many of the populated islands in the most fragmented states are served by air carriers. For example, the Solomon Islands, which is home to six major islands and nearly 1,000 minor islands, has 36 airports. Vanuatu has 31 airports to serve half of its inhabited islands. The Cook Islands is home to 11 airports, Fiji has 28, and the Marshall Islands has 15 official airports (Central Intelligence Agency 2019). While relatively few of these airports have paved runways, they are a crucial part of each country's infrastructure, as they serve extremely dispersed populations, and airlines are typically government owned and heavily subsidized at a financial loss to the state (Taumoepeau 2015). Some airports and airport improvements have been funded by international aid agencies or foreign business interests, and several regional airlines have wealthy state shareholders from Asia and Europe.

The great distances between islands within a country affect marine transportation considerably. Most notably is that boat-based travel cannot easily be developed when domestic distances are measured in the hundreds or thousands of kilometres, which means that air transport is the most practical and pervasive mode. The lack of inter-island sea transportation has hindered the sustainable development of many island nations in the region, especially for those that lack larger air links with other states in the region (Bola 2017). For development purposes, water transportation could be an alternative to air access. However, water-based transportation in the most isolated and dispersed states tends to be dangerous and prone to disasters, as the January 2018 sinking of an over-loaded ferryboat in Kiribati demonstrates. Nearly all the 102 passengers on board perished (Reuters 2019). In July 2009, a similar incident occurred in Kiribati, killing the majority of the passengers and crew members on board. Likewise, in August 2009, Tonga's new ferryboat Princess Ashika sank in a storm, claiming many lives while the ferry was providing transport services between the capital and the Ha'apai island group (BBC News 2011).

In the cruise sector, the Pacific geography results in four significant differences from other island cruise destinations. First, owing to vast distances, Pacific cruises tend to be longer in duration compared to those in the Caribbean or Mediterranean. The shortest South Pacific cruises typically last 8–10 days, while the shortest Caribbean cruises are 2–3 days. Many Pacific cruises last more than a month, with some lasting as long as 50–55 days. Second, unlike other popular island destination regions, where cruises frequently visit multiple countries, the maritime spread between islands in the Pacific usually limits the number of countries, islands or ports where an individual cruise can call. Even the 40–50-day cruises focus on only two or three countries. Third, Pacific cruises generally spend more days on the open sea than cruises do in other popular island regions. Finally, Pacific cruises are generally more expensive than those in many other regions owing to their longer sailings and the vast distances travelled

It is not yet certain how the Pacific cruise ship industry will fully emerge from the COVID-19 crisis, what Pacific routes will be resurrected and how the structure and nature of cruise ship itineraries will evolve. It is anticipated that controlled shore excursions and reduced passenger numbers will be projected as important components of the cruise experience, especially as a consequence of the need to strengthen the health and safety directives of the cruise ship (Holland et al. 2021). The market could change, initially, to accommodate more 'small elite cruises with well-off passengers' (Connell and Taulealo 2021)

Isolation

Related to fragmentation, many of the Pacific nations are among the most isolated on the planet, which makes developing transportation linkages difficult or inefficient at best. This affects socio-economic development in many ways (Bola 2017), not just in terms of transporting people but also in conveying adequate supplies of fresh food (Berno 2011). Several countries have only one or two international air connections per week, and these are often only to one other destination, which makes visiting these places cumbersome and expensive. This is largely due to small populations and, therefore, small demand for air travel. Niue's only air connection is to New Zealand, with only two flights per week. Tuvalu is connected to Kiribati by one flight per week and Fiji three times per week, making Niue and Tuvalu among the world's most isolated countries. In addition to having limited tourism assets and resources, several countries' isolation has limited their ability to develop a thriving tourism industry. Tuvalu, for example, is the least visited country on the earth, having received only 2,000 tourists in 2016, many of whom came to see the effects of climate change from the rising oceans (Lemelin and Whipp 2019; RNZ 2017) (see Chapter 26). Isolation also has the effect of causing tourists to have to remain longer in the destination than they might prefer. This is particularly the case in the smallest states, which often receive the least frequent flights and have the fewest tourism assets and most limited tourism infrastructure (McElroy 2003). In the past, the once weekly service in and out of Nauru meant that visitors to that country were required to remain there an entire week despite a lack of activities and attractions to keep them occupied (Fagence 1997). This led to a great deal of dissatisfaction among Nauru's small tourism market and prevented many from visiting at all.

Related to isolation and fragmentation is the notion that some countries in the Pacific have been over-dependent on a single tourist market or on a small handful of markets. Prior to the immobilization of tourism and travel in March 2020 because of the COVID-19 pandemic, the Cook Islands and Niue were largely visited by tourists from New Zealand, which reflects their unique political relationship and transportation access. These two countries have been relatively successful in building and maintaining this relationship with New Zealand. However, not all dependency relationships with single tourism markets have been so successful. A recent example of this is Palau's over-dependence on the Chinese tourist market (see Chapter 18). In 2017, approximately half of all of Palau's tourists were from mainland China. However, because of Palau's growing diplomatic relationship with Taiwan, China severed tourism ties with Palau that year by declaring the country an 'illegal destination', prohibiting package tourists from visiting, and cancelling flights between China and Palau (RNZ 2018a, 2018b). At the time of writing, these flights had not yet resumed, and relations between the two countries remained strained.

Physical size, topography and small population

In common with many other microstates, the topography and small size of many Pacific Islands precludes them from accommodating large-scale aircraft from distant locations (Timothy 2001a, 2001b, 2021). Small and unstable landmasses frequently necessitate dredging and massive land reclamation projects to accommodate runways and other transportation infrastructure. Likewise, some islands lack deep-water ports or are even unable to maintain small ports, which obviously limits the sizes of passenger watercraft and cruise ships that can be tendered.

Microstates are defined not only by their small geographic sizes but also by their small populations. Many Pacific Island states have disproportionately small populations, which has implications not only for public services, social welfare and education but also for tourism and transportation. The smallest population is that of Niue with an estimated 1,645 inhabitants (Worldometer 2022a), while the largest is Fiji with approximately 907,829 inhabitants (Wordometer 2022b). The other demographically smallest states include the Cook Islands, the FSM, Kiribati, the Marshall Islands, Nauru, Palau, Tonga and Tuvalu. The populations of Fiji, Samoa, the Solomon Islands, and Vanuatu are relatively larger by island standards and are able to maintain more public services and accessibility. Except for Nauru and Niue, which are single-island states, these countries' populations are scattered throughout hundreds of islands, and yet many of the least populated islands receive air service from their national airlines. For example, the island of Futuna, Vanuatu, is serviced by Air Vanuatu twice a week from nearby airports, despite the island's small population of 613 people in five villages. Providing this level of service for even the most sparsely populated islands is common throughout the largest archipelagos and is often more a matter of national pride than it is economic pragmatism (Taumoepeau 2015).

Economies of scale

All the geographic and demographic challenges outlined so far significantly determine the tourism-based 'economies of scale' (EoS) in the Pacific region. EoS suggests that increased production will lower the price of output and result in cost savings for producers, suppliers and consumers. For example, producing thousands of pieces of a specific consumer product (e.g., books or items of clothing) will typically be more cost-effective than producing only a few hundred of the same item. Thus, the cost per item is reduced, resulting in considerable savings for consumers. In addition to cost alone, larger-scale production and consumption also mean a more varied selection in style and quality. In the context of tourism, EoS means that there must be sufficient demand to warrant large-scale production, such as public investments in infrastructure and transportation development (Abulibdeh 2019). Small demand (e.g., small travelling populations) and few competitive suppliers (e.g., airlines and cruise companies) keep prices high and diversity low. Small EoS contribute to the general lack of accessibility and make travel costs prohibitively expensive in many cases (Connell 2006; Taumoepeau 2016b; Vanderpool-Wallace 2018).

In a study of Australians in Vanuatu, Cassidy and Brown (2010) found that the main reason for not visiting an outer island was the high cost of travelling there. From the perspective of access and mobility, in the Pacific, insufficient EoS typically means a lack of adequate transportation connections to other islands and major source markets. It also translates into an under-developed infrastructure, inadequate facilities, and irregular services. One case in point is the dominance of unpaved landing strips, rather than airports and paved runways, on the vast majority of islands. It also translates into infrequent flights between certain origins and destinations, smaller aircraft, and limited-use landing facilities (Taumoepeau 2015). For example, some airports have night-lights and intricate operating systems to guide incoming flights. For many Pacific Island countries, however, this is not possible, owing to EoS; there simply is insufficient demand to afford and maintain such luxuries. As Kissling (2016: 41) notes, 'Given the high cost of such systems, only airports with high traffic volumes and major airlines moving thousands of passengers on numerous flights could justify acquiring and using this equipment on a regular basis'. EoS affects even local transportation. Between islands with small populations, there may be no regular ferryboat services, and on the least populated islands, regardless of their physical size, there is commonly a lack of public transportation, including buses and taxis. In some cases, this also reflects the islands' small physical size, which is easy to navigate on foot or bicycle.

Perhaps the most observable travel pattern in the entire region that derives from the EoS, scattered geography, isolation, and small physical size has been the prominent role of Fiji as a regional air transport hub. This was particularly the case until the outbreak of the COVID-19 pandemic, where Australia and New Zealand functioned as major gateways to the Pacific Islands on a global scale as well as Fiji with Fiji Airways through Nadi International Airport, servicing Kiribati, Nauru, Samoa, Tonga, the Solomon Islands, Tuvalu, and Vanuatu from North America, Asia, Europe, Australia and New Zealand. Fiji has also been the main hub for intra-Pacific air travel for all but three countries of Micronesia (the FSM, Palau and the Marshall Islands) and New Zealand's free-associated states (Niue and the Cook Islands).

Political fragmentation and connectedness

In addition to being physically fragmented, the Pacific region is politically fragmented, with many islands or archipelagos forming independent states, while others are dependents of larger developed states. Colonial legacies and current colonial control have created conditions of dependency, even for the area's independent states, and have a bearing on the sorts of collaboration that can take place to develop mobility regimes and transportation networks. EoS come into play in this regard. For instance, for many years, there were no direct flights between the Marshall Islands and neighbouring Kiribati; passengers had to travel via Fiji. Air Kiribati, thus, planned to start a new route between Kiribati and the Marshall Islands by the end of 2018. Unfortunately, at the time of writing, despite continued negotiations, this new connection had not yet begun. This is, in part, a result of the colonial history of the region, working together with small EoS. In this case, the collaboration between Kiribati and the Marshall Islands was hindered logistically because Kiribati uses New Zealand's aviation regulations, while the Marshall Islands uses the United States regulations, which are inharmonious and challenging (RNZ 2018c).

Given the relatively recent independence of Palau, the Marshall Islands and the FSM from US trusteeship and the influence of EoS and colonial dependency, these countries have yet to establish their own aviation laws and policies. Instead, they are connected to US flight regulations through the Federal Aviation Administration (FAA), and their Air Route Traffic Control Centers (ARTCCs) are located in Honolulu, Hawaii, or Oakland, California, despite their considerable distance from Hawaii and the continental US (AirNav.com 2019).

Another political perspective with recent currency is the vying for influence in the Pacific between China and Taiwan. For many years, Taiwan has wielded considerable influence and power in the Pacific region, particularly with international aid to countries such as Kiribati, Tuvalu, the Marshall Islands, Nauru, Palau, the Solomon Islands and Nauru. Some of these countries have also been among the world's few to recognize the sovereignty of Taiwan. In a major blow to Taiwan, in September 2019, two of its biggest supporters, the Solomon Islands and Kiribati, discontinued their recognition of Taiwan and transferred their diplomatic relations and trade support to China. In response, the Marshall Islands, Nauru, Palau and Tuvalu reaffirmed their commitment to Taiwan. Negotiations have also been held between China and a handful of states (e.g., Vanuatu) to develop military relationships that could enable China to establish military bases in the region, much to the chagrin of Australia, New Zealand and the US (Hollingsworth 2019). Thus, the US is keen to encourage the Pacific states to continue supporting the interests of Taiwan and Australia over those of China. This diplomatic row in the region is largely connected to Chinese and Taiwanese promises of aid and investments in the island states, much of which includes transportation and tourism infrastructure, such as new roads, port facilities and airport expansions. Recent negotiations have taken place between China and the Solomon Islands for the Chinese to build a new 'tourism hub' on the island of Guadalcanal, including a new airport for easier access. In August 2019, similar negotiations took place between China and Samoa to build a new port in that country, which has the potential to become a Chinese military installation (Barrett 2019). China's incursion into the region has stimulated considerable interest in Australia, Japan, New Zealand and the US to expand their own interests and diplomatic efforts in the region.

Transport irregularity, inconsistency and instability

One of the most glaring challenges facing many of the islands, from a global perspective at least, is inconsistent and unreliable marine and air services. Schedules frequently change, challenging the growth of international tourism (Taumoepeau 2015); travellers may be stuck on an outer island without a way to return to the main island to connect with an international flight. In Kiribati, for example, Air Kiribati frequently cancels flights to and from the country's outer islands or over-sells flights without providing alternative transport options. In addition to possessing a well-established hospitality sector with adequate services and experience options, successful tourism requires regularly scheduled and reliable transportation to meet tourists' goals and connect with flights to other destinations. For some of the least developed countries, this is a salient problem that affects international demand for their tourism products.

Another challenge concerns frequent bankruptcies, airline turnovers and corporate restructurings. Something that nearly all domestic airlines in the region have in common is financial insolvency (Taumoepeau 2015). There is a long list of regional airlines that have shuttered their operations in recent years. Air Fiji ended operations in 2009, owing to financial mismanagement. Royal Tonga Airlines ceased operations in 2004, owing to funding challenges, and Vanuatu's Vanair declared bankruptcy in 2005 because of mismanagement. Palau Pacific Airways ceased operating in 2018 when China (Palau's largest foreign market) declared Palau an 'illegal destination', owing to its close relations with Taiwan, as noted earlier (RNZ 2018a). One of the most often cited examples of this airline insolvency in the Pacific is the airline(s) of Nauru (Connell 2006). Air Nauru began operating in 1970 to link the microstate with the outside world. Owing to mismanagement, conflict within the company, high operation costs, insolvency and a lack of consistent scheduling, the airline's only airplane was seized in Australia by creditors in 2005 while parked at Melbourne Airport. This left Nauru and Kiribati without international air service for almost two years. However, Nauru was later able to acquire another aircraft, and the country re-established the airline under the name of Our Airline in October 2006 (Taumeopeau 2015). In 2014, the airline was once again re-branded as Nauru Airlines, which is entirely state owned, and operates several 373-300s with plans to expand its services throughout the Pacific (Blue Swan 2017). However, during the height of the border closures during the COVID-19 pandemic, Nauru Airlines, like many other airlines in the region, was reduced to providing essential services relating to charter trips for repatriation of citizens and medical treatment as well as freight services (Nauru Bulletin 2021).

Political and economic instability has sometimes severed access by eliminating air services, especially in the case of political unrest, airline disputes and bankruptcies and economic upheavals faced by island states. For example, a pilot strike at Air Nauru in 1988 effectively ended the only airline service between New Zealand and Niue for some time (Milne and Nowosielski 1997). Similarly, in 2014, air service between Fiji and the Solomon Islands was severed for a time because of a dispute between Fiji Airways and Solomon Airlines in which the Solomon Islands and Solomon Airlines were concerned about Fiji Airways trying to usurp one of the weekly services between the two countries (Economist 2014). In the Solomon Islands in November 2021, political unrest in the capital city of Honiara disrupted repatriation flights which were scheduled to return nationals and students from overseas (Eddie 2021).

Climate change

One of the most salient challenges in the Pacific is global warming. Several island states are beginning to suffer the consequences of rising ocean levels, most notably Kiribati, Tuvalu and the Marshall Islands. In Kiribati, several villages are already under water on a regular basis, and seawalls no longer keep the rising waters out. This is a significant concern in itself for population stability and food security, for example, but it raises specific challenges for human mobility and transportation. Bonriki International Airport (Tarawa, Kiribati) lies only 3 metres above sea level. Cassidy International Airport (Christmas Island, Kiribati) and Amata Kabua International Airport (Marshall Islands) are only 2 metres above sea level. Rising oceans have clear implications for basic transportation and long-distance mobility in several island countries.

Climate change is responsible for increasing numbers and intensity of weather events and coastal erosion. On January 4, 2019, a severe tropical storm dropped 367 millimetres (mm) of rain in 72 hours with large storm surges (2–3 metres) that affected air and sea transport in Kiribati and the Marshall Islands. Rising ocean levels and coastal erosion are other concerns that will continue to affect transportation, mobility, shipping and tourism development in many Pacific Island states

(Hall 2017) (see Chapters 2, 3, 19, 21, 22, 26 and 27). Coastal volatility makes port development and maintenance increasingly difficult, which will directly affect maritime travel and the cruise sector. Coastal erosion endangers the very resources upon which many island states depend for their tourism livelihoods, including countries such as the Cook Islands, Fiji and Vanuatu.

The benefits of these challenges

Many of these same challenges underlay some of the charm and allure of the Pacific nations. For instance, first, while small size profoundly affects transportation EoS, it may simultaneously create a fascination factor that appeals to many (Timothy 2001a, 2021). Four of the ten smallest countries in the world in area and population are in the Pacific region (the Marshall Islands, Nauru, Niue and Tuvalu). According to Baum (1997), visiting small islands is a significant novelty for people from large countries. For them, it breaks up the monotony of metropolitan living. Smallness may also equal quaintness, which adds a unique personality that may be lacking in large metropolitan countries, something Scheyvens and Momsen (2008) acknowledge in suggesting that miniature size can create its own niche markets. Secondly, the remoteness of some island states renders them exotic and romantic in a 'Robinson Crusoesque' sort of way (Scheyvens and Momsen 2008). Their isolation may also render them culturally and ecologically unique because they have developed in culturally distinct ways different from other island states, and their natural environments have remained largely intact (Baum 1997; Butler 1993; Sufrauj 2011).

Political separateness was pointed out by Baum (1997) and Timothy (2001a) as a potential benefit for tourism in island microstates. Through their exercise of sovereignty, they have developed their own personalities, culturally and politically (Butler 1993). Political individuality and remoteness also work together to create a strong sense of nationalism and solidarity among the citizenry, which can increase their political strength in developing tourism or rejecting it (Scheyvens and Momsen 2008). This is a key characteristic of many Pacific Island states, which have had to adapt to external forces that are far beyond their control, such as global economic downturns, political turmoil, climate change and an evolving marketplace. Being at the mercy of globalization forces and other countries' decisions, whether bad or good, most Pacific states have demonstrated remarkable resilience in many respects regarding mobility and transportation. The failure of some state airlines might reflect the weaknesses of small markets and small-scale sovereignty. Nevertheless, the Pacific states' determination to resurrect bankrupt carriers, continue developing global marketing plans and collaborating with foreign powers, such as China, Taiwan, Japan, Australia, the US and New Zealand, to expand accessibility for themselves and for tourists demonstrates a level of resilience incommensurate with their small demographic and geographic sizes.

Border-crossing regulations and visa requirements

Despite their critical role in the travel experience and in facilitating transnational mobility, one element of travel mobility that is often overlooked by tourism scholars is border-crossing regulations and visa requirements (Timothy 2001c, 2017; Whyte 2008). Visa requirements, especially reciprocal ones, are often reflective of bilateral relationships between states. Poor socio-political and economic relations between countries often translate into stricter visa requirements between discordant state parties, while visa requirements are frequently waived or are minimal between states that share benevolent relationships. Likewise, Whyte's (2008) study suggests that people's ability to travel to other countries is often predicated upon the level of affluence and openness of their own home country. This means that the majority of the world's population has relatively few visa-free travel privileges.

The Pacific states diverge considerably in their visa requirements for foreign nationals. Visitors, regardless of nationality or citizenship, are welcome to visit the Cook Islands, the FSM, Samoa, and Niue without a visa. Tuvalu requires visas of all nationalities outside the European Union, but these are provided on arrival at Funafuti. Some countries in the region provide visas on arrival for the majority of nationalities. Fears of ethnic 'others' is less of a concern for some of the geographically isolated Pacific states, as illustrated in Samoa's case, where Iranians are granted a three-month visa-free entry on arrival. Most countries have a mixed visa policy, with many nationalities not needing a visa, visas being issued on arrival or requiring visas ahead of time. Nauru has a strict arrival policy that requires almost all nationalities to acquire a visa in advance. This is, in part, a means of skirting negative media scrutiny regarding the country's holding of asylum seekers. Journalists and other media representatives must seek special permission, and their actions are carefully monitored (Greenfield and Westbrook 2019) (see Chapter 20). Throughout the Pacific, the need to acquire a tourist visa is usually based upon length of stay, purpose of visit and the nature of bilateral relations.

From an intra-regional perspective, most Pacific Islanders are relatively free to travel throughout the region (Table 4.4). Although several countries require visas from other Pacific citizens, these endorsements are almost always issued on arrival. Nauru is the only country to require a visa ahead of time for its neighbours from Kiribati. These visa-free or visa-on-arrival policies

Table 4.4 Intra-regional visa policies for Pacific Island states, 2020

Destination country	*No visa required for citizens of:*	*Visa on arrival for citizens of:*	*Visa required ahead of time for citizens of:*
Cook Islands*	All Pacific Island countries	–	–
Federated States of Micronesia	All Pacific Island countries	–	–
Fiji	All Pacific Island countries	–	–
Kiribati	All Pacific Island countries	–	–
Marshall Islands	FSM and Palau	Cook Islands, Fiji, Kiribati, Nauru, Niue, Samoa, Solomon Islands, Tonga, Tuvalu and Vanuatu	–
Nauru	–	Cook Islands, Fiji, Marshall Islands, FSM, Niue, Palau, Samoa, Solomon Islands, Tonga, Tuvalu and Vanuatu	Kiribati
Niue*	All Pacific Island countries	–	–
Palau	FSM and Marshall Islands	Cook Islands, Fiji, Kiribati, Nauru, Niue, Samoa, Solomon Islands, Tonga, Tuvalu and Vanuatu	–
Samoa	All Pacific Island countries	–	–
Solomon Islands	–	All Pacific Island countries	–
Tonga	–	All Pacific Island countries	–
Tuvalu	–	All Pacific Island countries	–
Vanuatu	–	All Pacific Island countries	–

Source: Author's own compilation

Note: * Cook Islanders and Niueans are citizens of New Zealand; visa policies for New Zealanders also apply.

have the potential to facilitate increased inter-island mobility for leisure, business and commerce and VFR purposes. In light of this potential, in the years leading to the outbreak of the COVID-19 pandemic, there has been increased interest among national airlines to improve connectivity in the region (see RNZ 2018d; 2018e). Real Tonga Airlines started direct flights between Tonga and Samoa early in 2018 (RNZ 2018e). At the beginning of 2020, Air Vanuatu had four aircraft for its international and domestic routes. The airline's plan to acquire 13 new aircraft over the next decade (including four in 2020) was indicative of Vanuatu's interest in becoming increasingly connected to other countries of the region and to facilitate more frequent intra-regional travel (RNZ 2019b). Joint venture negotiations between Real Tonga Airlines and Nauru Airlines focused on increasing direct routing between Tonga and other countries of the region, especially Australia (RNZ 2018d).

COVID-19, travel immobility and closed borders

Nothing has affected travel mobility in the Pacific since World War II more than the 2020–2022 COVID-19 pandemic. The crisis has had devastating effects on global tourism everywhere, yet the poorest and smallest countries appear to have been especially hard hit, particularly those who depend on tourism for a large portion of their economic well-being. In the Pacific region, the geographical characteristics that simultaneously create the islands' appeal and restrict their mobility (e.g., isolation, fragmentation, inaccessibility and small size) appear to have benefitted some of them in the pandemic. According to the Center for Strategic and International Studies (Shen 2020), the isolation of the Pacific Islands, together with their fast responses and mobility restrictions (closed borders) at the outset of the pandemic, enabled them to minimize infections and transmissions. However, by the end of January 2022, the number of cases concerning the transmission of the COVID-19 virus was significant in Fiji (62,855 total cases), Kiribati (460), Palau (1,633) and the Solomon Islands (1,183), especially as there were attempts by particular government authorities to ease borders to allow for more tourism and mobility, as was the case with Fiji and Kiribati (Gunia 2022).

Although travel restrictions and border controls helped to minimize infections, they also caused an economic crisis, sending tourist arrivals plummeting (Shen 2020). Almost all sea vessels ceased operations, contributing to increased commodity prices and maritime inaccessibility between islands in the region. Most international flights were cancelled to and within the region. They only began operating to the region since 2021 and in an inconsistent manner, particularly in the form of travel bubbles from New Zealand and Australia. In late January 2021, a travel bubble was also established between New Zealand and the Cook Islands, which reduced the Cook Islands' isolation factor considerably (Curran 2021). Some countries attempted to open up for tourism in mid-2020 but re-closed their borders owing to increased infection rates. Many of the planned airport expansions described earlier in this chapter and other efforts to grow tourism were necessarily halted, but there is a common sense that these efforts will resume once infection rates have declined significantly and as countries begin to re-open for international arrivals and departures.

Conclusion and research implications

The same geography that inspires people to visit the Pacific Islands also raises some of the region's most formidable challenges to tourism development and human mobility. The widespread dispersion of islands and their isolation means that air services have now become the most crucial means of inter-island and intra-regional travel, leaving maritime transportation far behind in its wake. The success of tourism in the Pacific is entirely dependent on the aviation industry and air access (Boneberg 2017; Duval 2005; Taumoepeau 2015), with the least reachable countries also being among the least visited in the world.

Future research needs to consider core-periphery development options that address incongruous regional transportation policies, uneven development and the physical and geopolitical characteristics of the region. Prospective research ought to consider alternative tourism and transportation technologies that are not only ecologically friendlier but also provide physical access in ways that have heretofore not been considered feasible in the Pacific region. Research also needs to consider the limitations of implementing innovations that might bridge the gaps in the Pacific Islands' mobility and resolve the effects of the COVID-19 pandemic on mobilities and transportation, especially addressing how these gaps and effects might be mitigated better in the future.

The chapter has indicated how this region is simultaneously plagued with small EoS, translating into a high level of airline bankruptcies, restructurings, and inter-government conflicts. The region has witnessed the emergence of numerous national air carriers that continue to provide inefficient services as a public utility but also represent a significant fiscal loss to national budgets. Nevertheless, there continues to be a pattern of under-supply and over-demand for inter-island transportation, which means that increasing numbers of airlines are appearing in the marketplace, they are finally beginning to collaborate and they are flying to an increasingly diverse range of destinations. Low-cost carriers have developed to satisfy some of this growing demand, to increase the mobility of less-affluent regional travellers and to cater to more budget-conscious holiday-makers (Taumoepeau 2016b). However, their reach has largely been to and from the destinations that are already popular holiday hotspots, such as the Cook Islands, Vanuatu and Samoa, from the larger market source countries that already dominate the region: Australia and New Zealand.

References

Abulibdeh, A. (2019) 'Urban transportation and tourism in the MENA region'. In: D.J. Timothy (ed.), *Routledge Handbook on tourism in the Middle East and North Africa*, London: Routledge, pp. 272–289.

ADB (Asian Development Bank) (2007) *Ocean Voyages: Aviation in the Pacific*. Manila: Asian Development Bank.

AirNav.com (2019) 'FAA information effective 05 December 2019'. Available online at https://www.airnav.com/airport/PTRO (accessed 7 December 2019).

Australian Bureau of Statistics (2015) 'Migration, Australia, 2011–12 and 2012–13'. Available online at http://www.abs.gov.au/AUSSTATS/abs@.nsf/DetailsPage/3412.02011-12%20and% 202012-13?OpenDocument (accessed 20 March 2019).

Barrett, J. (2019) 'Sink or swim: Chinese port plans put Pacific back in play', *Reuters World News*, 6 August. Available online at https://www.reuters.com/article/us-pacific-samoa-china-insight/sink-or-swim-chinese-port-plans-put-pacific-back-in-play-idUSKCN1UX01I (accessed 2 November 2019).

Baum, T. (1997) 'The fascination of islands: A tourist perspective'. In: D.G. Lockhart and D. Drakakis-Smith (eds.), *Island Tourism: Trends and Prospects*, London: Pinter, pp. 21–35.

BBC News (2011) 'Four found guilty at Tonga ferry disaster trial', 1 April. Available online at https://www.bbc.com/news/world-asia-pacific-12930463 (accessed 1 February 2021).

Berno, T. (2011) 'Sustainability on a plate: Linking agriculture and food in the Fiji Islands tourism industry'. In: R.M. Torres and J.H. Momsen (eds.), *Tourism and Agriculture: New Geographies of Consumption, Production and Rural Restructuring*. London: Routledge, pp. 87–103.

Blue Swan (2017) 'Air Nauru is a small fish in a very big pond', *The Blue Swan Daily*, 15 March. Available online at https://blueswandaily.com/air-nauru-is-a-small-fish-in-a-very-big-pond/ (accessed 22 March 2019).

Bola, A. (2017) 'Potential for sustainable sea transport: A case study of the Southern Lomaiviti, Fiji Islands', *Marine Policy*, 75: 260–270.

Boneberg, T. (2017) *Air Connectivity on Samoa: Chances and Risks*, Munich: GRIN.

Butler, R.W. (1993) 'Tourism development in small islands: Past influences and future directions'. In: D.G. Lockhart, D. Drakakis-Smith and J. Schembri (eds.) *The Development Process in Small Island States*, London: Routledge, pp. 71–91.

Cassidy, F. and L. Brown (2010) 'Determinants of small Pacific island tourism: A Vanuatu study', *Asia Pacific Journal of Tourism Research*, 15(2): 143–153.

Central Intelligence Agency (2019) *World Factbook*, Washington, DC: Central Intelligence Agency. Available online at https://www.cia.gov/library/publications/the-world-factbook/ (accessed 30 January 2019).

Chute, I. (2022) 'Runaway upgrade nears completion', *Fiji Times*, 9 January. Available online from https://www.fijitimes.com/runaway-upgrade-nears-completion/ (accessed 24 March 2022).

Coles, T. and D.J. Timothy (eds.) (2004) *Tourism, Diasporas and Space*, London: Routledge.

Collison, F.M. and D.L. Spears (2010) 'Marketing cultural and heritage tourism: The Marshall Islands', *International Journal of Culture, Tourism and Hospitality Research*, 4(2): 130–142.

Connell, J. (2006) 'Nauru: The first failed Pacific state?' *The Round Table*, 95(383): 47–63.

Connell, J. and T. Taulealo (2021) 'Island tourism and COVID-19 in Vanuatu and Samoa: An unfolding crisis', *Small States and Territories*, 4(1): 105–124.

Crew Center (2019) 'Port schedules 2019'. Available online at www.crew-center.com (accessed 13 February 2019).

Curran, A. (2021) 'Air New Zealand operates first travel bubble flight from the Cook Islands'. Available online at https://simpleflying.com/air-new-zealand-cook-islands/ (accessed 5 April 2021).

Dowling, R.K. and C. Weeden (eds.) (2017) *Cruise Ship Tourism* (2nd ed.). Wallingford: CABI.

Duval, D.T. (2005) 'Tourism and air transport in Oceania'. In: C. Cooper and C.M. Hall (eds.) *Oceania: A Tourism Handbook*. Clevedon: Channel View Publications, pp. 321–334.

Duval, D.T. (2007) *Tourism and Transport: Modes, Networks and Flows*. Clevedon: Channel View Publications.

Duval, D.T. (2016) 'Scale and scope of commercial air transport in the Asia Pacific'. In: D.T. Duval (ed.), *Air Transport in the Asia Pacific*. London: Routledge, pp. 1–10.

Economist (2014) 'Talks on Fiji-Solomon Islands air dispute', *The Economist*, 29 October. Available online at https://country.eiu.com/article.aspx?articleid=1172443101&Country=Fiji&topic=Politics&subtopic=Fo_3 (accessed 22 November 2019).

Eddie, L. (2021) 'Flights disrupted by unrest', *Solomon Star*, 8 December. Available online at https://www.solomonstarnews.com/flights-disrupted-by-unrest/ (accessed 3 April 2022).

Fagence, M. (1997) 'An uncertain future for tourism in microstates: The case of Nauru', *Tourism Management*, 18(6): 385–392.

Greenfield, C. and T. Westbrook (2019) 'New Zealand TV journalist held by Nauru policy', *Euronews*, 9 December. Available online at https://www.euronews.com/2018/09/04/new-zealand-tv-journalist-held-by-nauru-police (accessed 26 March 2022).

Gunia, A. (2022) 'A COVID-free Pacific nation opened its border a crack: The virus came rushing in', *Time*, 31 January. Available online at https://time.com/6143 260/covid-19-pacific-islands-kiribati/ (accessed 26 March 2022).

Hall, C.M. (2017) 'Climate change and its impacts on coastal tourism: Regional assessments, gaps and issues'. In: A. Jones and M. Phillips (eds.) *Global Climate Change and Coastal Tourism: Recognizing problems, Managing Solutions and Future Expectations*, Wallingford: CABI, pp. 48–61.

Holland, J., T. Mazzarol, G.N. Soutar, S. Tapsall and W.A. Elliott (2021) 'Cruising through a pandemic: The impact of COVID-19 on intentions to cruise', *Transportation Research Interdisciplinary Perspectives*, 9: 1–15.

Hollingsworth, J. (2019) 'Why China is challenging Australia for influence over the Pacific Islands', *CNN*, 22 July. Available online at https://www.cnn.com/2019/07/22/asia/china-australia-pacific-investment-intl-hnk/index.html (accessed 3 November 2019).

Johnson, G. (2009) 'Japan Airlines flights to Marshalls slow', *Marianas Variety*, 9 March. Available online at http://www.mvariety.com/community-bulletin-sp-595/15230-japan-airlines-flights-to-marshalls-slow (accessed 3 December 2019).

Kissling, C. (1984) *Transport and Communication for Pacific Microstates: Issues in Organisation and Management*, Suva: University of the South Pacific.

Kissling, C. (2016) 'Networks enabling air transport services in the South Pacific: 40 years of change'. In: D.T. Duval (ed.), *Air Transport in the Asia Pacific*, London: Routledge, pp. 33–51.

Lemelin, H. and P. Whipp (2019) 'Last chance tourism: A decade in review'. In: D.J. Timothy (ed.), *Handbook of Globalisation and Tourism*, Cheltenham: Edward Elgar, pp. 316–322.

McElroy, J.L. (2003) 'Tourism development in small islands across the world', *Geografiska Annaler, Series B: Human Geography*, 85(4): 231–242.

Milne, S. and L. Nowosielski (1997) 'Travel distribution technologies and sustainable tourism development: The case of South Pacific microstates', *Journal of Sustainable Tourism*, 5(2): 131–150.

Nauru Bulletin (2021) 'Leaders urge all to get COVID vaccine', 25 May, Republic of Nauru. Available online at http://naurugov.nr/media/145713/nauru_bulletin__02_ 25may2021__227_.pdf (accessed 26 March 2022).

Ng, R. (2020) 'Why cruise ships are setting sail again as COVID-19 rages', *National Geographic*, 10 December. Available online at https://www.nationalgeographic.com/travel/2020/12/how-cruise-ships-are-sailing-again-during-coronavirus/ (accessed 1 February 2021).

Parker, B. (2020) 'Fresh blow for cruise industry as passenger on ship in the South Pacific tests positive', *The Telegraph*, 3 August. Available online at https://www.telegraph.co.uk/travel/ cruises/news/coronavirus-covid-19-passengers-on-cruise-ship-sailing-south-pacific/ (accessed 1 February 2021).

Rao, M. (2002) 'Challenges and issues for tourism in the South Pacific island states: The case of the Fiji Islands', *Tourism Economics*, 8(4): 401–429.

Reuters (2019) 'Kiribati ferry that sank, killing 95, not licensed to take passengers, inquiry finds', *The Guardian*, 8 October. Available online at https://www.theguardian.com/world/2019/oct/08/kiribati-ferry-that-sank-killing-95-not-licensed-to-take-passengers-inquiry-finds (accessed 1 February 2021).

RNZ (Radio New Zealand) (2017) 'Tourists could help Tuvalu', *RadioNZ*, 20 September. Available online at https://www.radionz.co.nz/international/programmes/datelinepacific/audio/201859221/tourists-could-help-tuvalu (accessed 16 February 2019).

RNZ (Radio New Zealand) (2018a) 'Palau tourism industry "suffering greatly" from China ban', *RadioNZ*, 3 September. Available online at https://www.radionz.co.nz/international/pacific-news/365556/palau-tourism-industry-suffering-greatly-from-china-ban (accessed 16 February 2019).

RNZ (Radio New Zealand) (2018b) 'Palau's tourism revenue grows despite China's ban', *RadioNZ*, 1 November. Available online at https://www.radionz.co.nz/international/pacific-news/369937/palau-s-tourism-revenue-grows-despite-china-s-ban (accessed 16 February 2019).

RNZ (Radio New Zealand) (2018c) 'Air Kiribati plans Majuro Flights', *RadioNZ*, 27 August. Available online at https://www.radionz.co.nz/international/pacific-news/364992/air-kiribati-plans-majuro-flights (accessed 20 February 2019).

RNZ (Radio New Zealand) (2018d) 'Nauru and Tonga airlines discuss joint venture', *RadioNZ*, 11 December. Available online at https://www.radionz.co.nz/international/pacific-news/377991/nauru-and-tonga-airlines-discuss-joint-venture (accessed 10 April 2019).

RNZ (Radio New Zealand) (2018e) 'Tonga prepares for direct flights to Samoa', *RadioNZ*, 5 February. Available online at https://www.radionz.co.nz/international/pacific-news/349710/tonga-prepares-for-direct-flights-to-samoa (accessed 16 February 2019).

RNZ (Radio New Zealand) (2019a) 'Upgrade of troubled Vanuatu airport nears completion', *RadioNZ*, 25 February. Available online at https://www.rnz.co.nz/international/pacific-news/383313/upgrade-of-troubled-vanuatu-airport-nears-completion (accessed 3 December 2019).

RNZ (Radio New Zealand) (2019b) 'Air Vanuatu to acquire 13 new aircraft by 2030', *RadioNZ*, 30 January. Available online at https://www.radionz.co.nz/international/pacific-news/381234/air-vanuatu-to-acquire-13-new-aircraft-by-2030 (accessed 5 February 2019).

Scheyvens, R. (2007) 'Poor cousins no more: Valuing the development potential of domestic and diaspora tourism', *Progress in Development Studies*, 7(4): 307–325.

Scheyvens, R. and J. Momsen (2008) 'Tourism in small island states: From vulnerability to strengths', *Journal of Sustainable Tourism*, 16(5): 491–510.

Shen, K. (2020) 'The economic costs of the pandemic for the Pacific islands', 9 September 2020. Available online at https://www.csis.org/blogs/new-perspectives-asia/economic-costs-pandemic-pacific-islands (accessed 5 April 2021).

Stats NZ. (2019) '2018 census population stats', New Zealand Government. Available online at https://www.stats.govt.nz/topics/population (accessed 16 October 2019).

Sufrauj, S. (2011) 'Islandness and remoteness as resources: Evidence from the tourism performance of small remote island economies (SRIE)', *European Journal of Tourism, Hospitality and Recreation*, 2(1): 19–41.

Taumoefolau, M. (2015) 'Tongans—Facts and figures', *Te Ara - the Encyclopedia of New Zealand*. Available online at https://teara.govt.nz/en/tongans/page-5 (accessed 22 March 2019).

Taumoepeau, S. (2015) 'Air transportation and tourism linkages in the South Pacific islands'. In: S. Pratt and D. Harrison (eds.) *Tourism in Pacific Islands: Current Issues and Future Challenges*, Abingdon: Routledge, pp. 183–195.

Taumoepeau, S. (2016a) 'Low cost carriers in Asia and the Pacific'. In: S. Gross and M. Lück (eds.) *The Low Cost Carrier Worldwide*, London: Routledge, pp. 113–138.

Taumoepeau, S. (2016b) 'Suitability of the low-cost airline model in the South Pacific region'. In: D.T. Duval (ed.), *Air Transport in the Asia Pacific*, London: Routledge, pp. 93–112.

Timothy, D.J. (2001a) 'Benefits and costs of smallness and peripheral location in tourism: Saint Pierre et Miquelon (France)', *Tourism Recreation Research*, 26(3): 61–70.

Timothy, D.J. (2001b) 'Postage stamps, microstates and tourism', *Tourism Recreation Research*, 26(3): 85–88.

Timothy, D.J. (2001c) *Tourism and Political Boundaries*, London: Routledge.

Timothy, D.J. (2017) 'Passport'. In: L.L. Lowry (ed.), *The SAGE International Encyclopedia of Travel and Tourism*, Thousand Oaks, CA: Sage, pp. 939–940.

Timothy, D.J. (2021) *Tourism in European Microstates and Dependencies: Geopolitics, Scale and Resource Limitations*, Wallingford: CABI.

Treloar, P. and C.M. Hall (2005) 'Tourism in the Pacific islands'. In: C. Cooper and C.M. Hall (eds.) *Oceania: A Tourism Handbook*, Clevedon: Channel View Publications, pp. 173–294.

Vanderpool-Wallace, V. (2018) 'Island travel transportation'. In: M. McLeod and R. Croes (eds.) *Tourism Management in Warm-Water Island Destinations*, Wallingford: CABI, pp. 11–26.

WhatsinPort (2019) 'Cruise guide to 1200 ports of call'. Available online at www.whatsinport.com (accessed 13 February 2019).

Whyte, B. (2008) 'Visa-free travel privileges: An exploratory geographical analysis', *Tourism Geographies*, 10(2): 127–149.

Worldometer (2022a) 'Niue population'. Available online at https://www. worldometers.info/world-population/niue-population/ (accessed 26 March 2022).

Worldometer (2022b) 'Fiji population'. Available online at https://www.worldometers.info/world-population/fiji-population/ (accessed 26 March 2022).

Worldwide Cruise Centres (2019) 'Ports'. Available online at https://www.worldwidecruisecentres.com.au/ports (accessed 15 October 2019).

5

TOURISM AND THE PACIFIC ISLAND ECONOMIES

Outcomes and implications

Stephen Pratt and Marika Kuilamu

Introduction

Tourism is a promoter of economic growth (Carmignani and Moyle 2018). Tourism's potential to increase economic activity, create and sustain jobs, attract investment, contribute to an economy's balance of payments, and reduce poverty is highlighted in almost every academic paper on tourism economics (Lin et al. 2019) and introductory textbook on tourism (Cooper *et al.* 2008). For Pacific Island states, which have a comparative advantage in 'good weather', 'beautiful scenery', 'distinctive cultures', and 'friendly locals', tourism provides a comparative advantage that has grown into a leading economic sector. Pratt (2013) notes that from colonial times, the South Pacific has a ubiquitous image of a tropical paradise with untouched scenery, swaying palm trees, warm crystal-clear waters, and golden sandy beaches, coupled with happy and welcoming islanders. This tourism-induced image has been reinforced through modern destination marketing organization campaigns; however, the image is changing given the international media attention to climate change in the region (Conrich and Smith 2018).

Tourism is a key sector in the Pacific Island states because there are few other economic alternatives (Pratt 2015a). This chapter provides an overview of the economic contribution of tourism to Pacific Island economies. Many scholars divide Pacific peoples into Polynesia, Melanesia, and Micronesia, although this taxonomy is not straightforward and is seen as contentious by some (see Chapter 1). For this chapter, these three regions will be reviewed. Micronesia will cover the Federated States of Micronesia (FSM), Kiribati, the Marshall Islands, Nauru, and Palau. Melanesia is represented by Fiji, the Solomon Islands, and Vanuatu, while the Polynesian countries and territories under investigation include the Cook Islands, Niue, Samoa, Tonga, and Tuvalu. These 13 Pacific Island economies are 'small island developing states' (SIDS) as classified by the United Nations. As noted by other scholars (Briguglio 1995; Hampton and Christensen 2007), the characteristics of SIDS form obstacles to economic growth. These obstacles include small size, remoteness, and environmental vulnerability and have 'migration, remittances, aid, and bureaucracy' (MIRAB) characteristics (Bertram and Watters 1985). These barriers to growth affect not only economic growth in general but also the pattern and magnitude of tourism development. Despite these commonalities, there is also a significant degree of heterogeneity among SIDS (McElroy 2003, 2006). Pacific Island states have little other choices than to

DOI: 10.4324/9780429019968-6

seek further tourism development (Cheer et al. 2018). Yet, the economic benefits from tourism outlined above are not necessarily guaranteed. Both internal and external factors will affect the degree to which tourism benefits remain in the economy.

This chapter initially provides an overview of the economic contribution of tourism to Pacific Island economies, especially in the context the pre-COVID-19 period. Through a review of recent secondary data outlining the types of tourism found in the Pacific and tourism's contribution to 'gross domestic product' (GDP), and other economic indicators of tourism to the Pacific overall, as well as separately for Micronesia, Melanesia, and Polynesia, an understanding of the economic contribution of tourism to Pacific Island economies will be provided. Second, the chapter will test the 'tourism-led growth hypothesis' (TLGH) for Pacific Island economies in Micronesia, Melanesia, and Polynesia. Through this analysis, we can determine the extent to which tourism drives the economies of these 'Pacific Island countries' (PICs) and territories. Lastly, we provide recommendations on how economic benefits provided by tourism can be strengthened in the Pacific. This will be useful to policy-makers and practitioners to ensure tourism benefits are disbursed widely and remain in the Pacific.

The economic contribution of tourism to Pacific Island economies

The economic contribution of tourism to Pacific Island economies depends on both external and internal factors affecting both tourism and the wider economies. Several external factors that have impacted tourism in Pacific Island states include geopolitical drivers, the emergence of China, natural disasters, and political instability. Cheer et al. (2018) highlight the implications of geopolitical drivers for the Pacific Islands for tourism. The People's Republic of China has been seen to exert its influence in the Pacific. Along with increased investments in infrastructure, both tourism and non-tourism, and other development projects, such as the development of public infrastructure and utilities, industry export development, and interventions in health and education, the growing importance of China as a source market for international tourism arrivals can influence the economic contribution of tourism in the Pacific. For example, in 2017, China banned state-run package tours from visiting Palau, resulting in Chinese visitor arrivals dropping to 58,000 in 2017 from a high of 87,000 in 2015 (ABC News 2018). This move was perceived as retribution from Palau's diplomatic ties with the Republic of China (Taiwan), resulting in empty hotel rooms and cancelled flights (see Chapter 18).

More generally, the increasing disposable income of Chinese citizens has created a growing middle class that is travelling internationally. Outbound tourism growth in recent decades is unprecedented. Chinese outbound tourists increased from 3.74 million in 1993 to 10.47 million in 2000 (Kim et al. 2005), increasing further to 98.19 million in 2013 (Fu 2014). Pacific Island economies hope to benefit from this increased travel. In 2016, Chinese tourists to the Pacific comprised 0.14% of total Chinese outbound tourism (approximately 140,000) (Cheer et al. 2018). Put another way, the Chinese market comprised, on average, 7.3% of all international tourists to the Pacific region. However, there are large differences by country. In 2016, the Chinese made up 47% of visitors to Palau, but, as noted above, this market is under threat. The burning question will be whether Pacific Island states can take advantage of this potential Chinese market, both in terms of being able to meet their needs (Vada-Pareti 2015) and ensuring that the types of tourism offered to the Chinese market are managed sustainably. In 2019, the environmental damage created by a Chinese-backed resort developer on Malolo, Fiji caught the attention of New Zealand media. On investigation, the construction of the 350-*bure* resort started without the required permits, destroying the local fishing grounds and mangroves, which will cost millions of dollars to restore (Reid and Jennings 2019).

Other external factors that significantly affect the economic contribution of tourism to the Pacific Islands are significant weather events. These include tsunamis and cyclones. SIDS are particularly vulnerable to environmental disasters (Pratt 2015a) Recent severe weather events in the Pacific that have negatively affected tourism include the Samoa earthquake and tsunami (September 2009); Severe Tropical Cyclone Pat in Aitutaki, Cook Islands (February 2010); Cyclone Pam in Vanuatu (March 2015); Cyclone Winston in Fiji (February 2016); Severe Tropical Cyclone Harold in the Solomon Islands, Vanuatu, Fiji, and Tonga (April 2020); Severe Tropical Cyclone Yasa in Fiji (December 2020); Severe Tropical Cyclone Ana in Fiji (January 2021); Severe Tropical Cyclone Cody in Fiji (January 2022); and Severe Tropical Cyclone Dovi in Vanuatu and New Caledonia (February 2022). These severe weather events not only deter tourists from visiting these Pacific Islands but also damage hotel inventory and transportation infrastructure that need repairing or rebuilding. For example, the government of Vanuatu estimated that the total economic damages caused by Cyclone Pam was approximately 64% of GDP (VUV48.6 billion/US$416.45 million) with the tourism sector being the second most damaged, after housing (20% of total damage costs) (ILO 2015). Tropical Cyclone Winston in Fiji caused FJ$1.29 billion (US$0.6 billion) in damages (i.e., destroyed physical assets) with damages to the tourism and hospitality sector totalling FJ$76.1 million (World Trade Organization 2019). In the north of Viti Levu, the Rakiraki Tanoa Hotel was in the direct path of Tropical Cyclone Winston. Three-quarters of the property were damaged. The hotel was closed for three months. The hotel operated at 20% capacity for the following three months (ITC 2018).

Political and social tensions have also negatively affected tourism in several Pacific Island states. Harrison and Pratt (2010) outline the effects of the political coups in Fiji where international tourist arrivals decreased 26% and international tourism receipts dropped 21% after the 1987 coup. There was a similar pattern after the 2000 coup when international tourist arrivals and international tourism receipts decreased by 28% and 29%, respectively. In contrast, the 2006 coup resulted in only a 1% decrease in arrivals and a 5% decrease in receipts, as Fiji tourism stakeholders became better at managing their response to political crises with improved marketing. Tourists to Fiji, as a whole, also realized they were not a target of the coups, which took place in the capital, Suva, on the other side of Viti Levu, away from the tourist hubs of the Coral Coast, Nadi, and the island groups of Mamanucas and the Yasawas. The Solomon Islands was also the site of ethnic violence between 1998 and 2003, which saw fighting between militants from Guadalcanal Island and the neighbouring Province of Malaita. About 200 people were killed in the violence with many more displaced. In 1997, the Solomon Islands welcomed almost 16,000 international tourists. This dropped to 2,400 in 1999 (ABC News 2017). The recent rioting against the government in Chinatown in Honiara, Solomon Islands in November 2021 was also a costly affair (see Chapter 10), adding to the country's economic troubles associated with impacts of the COVID-19 pandemic on the country's tourism industry.

While these external factors influence the economic contribution of tourism to the Pacific Islands, the volume and composition of tourism also influence the economic contribution of tourism. Source markets for Pacific states are driven partly by relative distance (McKercher and Lew 2003) and partly by historical and colonial ties (Harrison and Pratt 2013). As such, Australia is the largest source market for Melanesian countries, while New Zealand is important for the Polynesian tourist destinations, and Asia provides the most international arrivals for Micronesian tourist destinations. There are exceptions such as the FSM and the Marshall Islands where the United States (US) market is dominant because of the compact agreement between the US and those Pacific States.

Total direct international tourist receipts (expenditure) are a function of three variables: the total number of international tourists, the average tourist expenditure per day, and the length of stay. These direct expenditures are tourists' direct expenditures on tourism-oriented products such as accommodation, transportation, and food and beverages (Khoshkhoo et al. 2017). The direct expenditures lead to a series of successive or indirect economic impacts through the supply chain. The secondary sectors feature firms that provide goods and services to the tourism-oriented sectors, such as agriculture and manufactured goods, that are supplied to hotels and restaurants. The strength of these linkages between the tourism sectors and their supply chain determines the size of the multiplier effect (Pratt 2015b). Another factor that determines the size of the multipliers includes the local share of goods and services purchased by tourists, vis-a-vis the import component. The degree to which capital and labour are provided by the host economy will also contribute to the size of the multipliers. The majority of labour tends to be provided by residents, although not significantly at the senior management level. Nonetheless, major capital investments are often foreign owned, though not in all instances. Fiji's National Providence Fund owns Fiji Marriott Momi Bay Resort, Grand Pacific Hotel, Holiday Inn Suva Hotel, and Intercontinental Fiji Golf Resort & Spa. These hotels are operated under management contracts. Hence, the capital/labour ratio of these sectors and their ownership is important as is the overall ownership of the tourism operations. Foreign ownership of tourism products may result in profit being repatriated to the overseas headquarters rather than reinvested in the host economy (Pratt 2015a).

In Wanhill's (1994) review of tourism multiplier, he notes that the tourism income multiplier for island economies ranges between 0.39 and 1.59 with an average of 0.85. Although somewhat dated, Milne (1992) estimates the tourism income multiplier for five Pacific Island economies: Niue (0.35), Kiribati (0.37), Tonga (0.42), the Cook Islands (0.43) and Vanuatu (0.56). The tourism income multipliers show the economic contribution in the form of local income based on $1 spent in the economy. This means that Niue retains just 35 cents for every $1 a tourist spends on the island.

Just prior to the COVID-19 pandemic, the South Pacific Tourism Organisation (2019) reported several indices to measure the economic contribution of tourism in the 13 Pacific Island economies under investigation. Table 5.1 shows international tourism receipts, receipts per capita, receipts per international tourist arrivals, and several employment measures. These 13 Pacific Island economies accrued a total of $US2.16 billion in international tourism receipts in 2018. Fiji's share is by far the largest at 43.1% while Vanuatu and the Cook Islands' share is 13.0% and 11.7%, respectively. At the other end of the spectrum, Nauru gained US$8.30 million, and Tuvalu earned US$8.21 million in international tourism receipts in 2018 (SPTO 2019). Therefore, for such small island states, it is comprehensible that the economic repercussions of subsequent border closures from March 2020 have been a serious concern for island governments, especially as these states often lack opportunities to economically diversify. Therefore, reliance on a dominant industry such as tourism can be a significant economic risk.

The average tourism receipts per capita in 2018 across the 13 Pacific Island economies is US$2,710. There are large differences across economies. Palau's tourism receipts per capita are US$9,496 while Kiribati's is only US$119 per capita. International tourism receipts or expenditure per tourist averages US$1,878 across the 13 island states. Interestingly there is less variability among Pacific Island economies for this measure. The highest receipts per arrival are Tuvalu at US$3,008 while the lowest is Niue at US$837. Tourism earnings as a percentage of GDP is a broad measure which shows the size of the tourism sectors to the overall economy. The Cook Islands has the largest tourism sector compared to the remainder of the economy as tourism earnings are almost 87% of GDP. Two other Polynesian countries, Niue and Samoa, have significant tourism sectors as measured by tourism earnings as a percentage of GDP at 41.0% and 30.4%, respectively (SPTO 2019).

Table 5.1 Tourism receipts and employment

	2018 Receipts (US$ Million)	*2018 Receipts per Capita(US$)*	*2018 Receipts per Arrival (US$)*	*2018 Tourism Earnings as % of GDP*	*No. Tourism Employees*	*Tourism Employees as % of Total*
Micronesia						
FSM	$52.26	$464	$2,721	15.80	1,802	5.67
Kiribati	$13.84	$119	$2,028	7.86	2,100	15.50
Marshall Islands	$13.20	$226	$1,952	6.76	605	5.50
Nauru	$3.30	$309	$1,099	2.8	77	2.53
Palau	$170.04	$9,496	$1,600	21.00	2,690	44.83
Melanesia						
Fiji	$931.98	$1,055	$1,071	18.41	41 338	35.50
Solomon Islands	$81.51	$125	$2,925	7.59	6,400	11.00
Vanuatu	$281.51	$962	$2,434	22.1	15 000	34.51
Polynesia						
Cook Islands	$253.14	$14,450	$1,500	86.99	2336	34.39
Niue	$8.3	$5,123	$837	41.00	226	32.29
Samoa	$248.19	$1,265	$1,480	30.42	5,158	12.54
Tonga	$94.95	$920	$1,757	25.35	7,100	21.20
Tuvalu	$8.21	$713	$3,008	19.55	87	2.25

Source: Adapted from SPTO (2019 Table R14 and R15).

A total of 84,969 persons were employed in the tourism sectors in the 13 PICs. An average of 6,536 persons were employed in the tourism sectors. On an absolute level, Fiji tourism sectors employed more than 41,000 people, while an estimated 15,000 were employed in Vanuatu's tourism sectors. As a percentage of total employment, tourism employees comprised, on average, about 20% of total employees in these 13 economies. Palau, Fiji, and the Cook Islands had the highest proportion of tourism employees in their countries: Palau – 44.8%, Fiji – 35.5%, and the Cook Islands – 34.4%. Both Vanuatu and Niue have a relatively high proportion of tourism employees (SPTO 2019).

The analysis above shows the heterogeneity of tourism in the Pacific Island economies. This has implications for the different stakeholders: communities, governments, and the tourism industry. The more tourism intensive the economy, the larger the economic impact but also the larger the socio-cultural and environmental impacts. Those economies with larger tourism sectors can generate potentially more benefits, including taxes, which benefit the government, but those same governments will have a greater responsibility to manage and contain the negative impacts of tourism through taxation or legislation. This will also require resources to monitor, evaluate, and enforce sustainable tourism policies. A larger tourism sector means more employment opportunities for the local communities, but working in the tourism sector often involves long, tiring work. This can impact the social relations with the family and village. The next section examines the tourism-led growth hypothesis in the context of these 13 Pacific Island economies.

Tourism-led growth hypothesis (TLGH)

One commonly used technique to test the relationship between economic growth and tourism growth is Granger's approach to causality (Granger 1969), which assesses the significance of the lagged values of the independent variable (tourism in the case of the TLGH) on an autoregressive model of a dependent variable (income/GDP in the case of the TLGH). If the lagged tourism variables are significant, tourism is said to 'Granger cause' economic growth. In the growth-led tourism hypothesis (GLTH), Granger causality runs the other way, from economic growth to tourism. The work of Balaguer and Cantavella-Jorda (2002) is often cited as the first study to investigate the TLGH, which adapted the theory of export-led growth hypothesis, which posits that expanding exports can contribute to economic growth. In the national accounting framework, tourism receipts are categorized as exports of services, hence, the extension of investigating export-led growth hypothesis to the tourism-led growth hypothesis. Because that work was published around 20 years ago, there has been an explosion of research looking at whether the TLGH holds in different contexts and over different time periods. While, as a general rule, the tourism-led growth hypothesis holds, Brida et al. (2016) note that results are not uniform. There are contradictory results at times. The authors highlight cases in which, in the short run, there is a unidirectional Granger causality running from tourism growth to economic growth (South Africa: Akinboade and Braimoh 2010; Taiwan: Chen and Chiou-Wei 2009), cases in which there is bidirectional Granger causality (Greece: Dritsakis 2004b; Lebanon: Tang and Abosedra 2014) and cases in which there is unidirectional Granger causality running from economic growth to tourism growth (African countries: Lee and Chang 2008). In those 95 reviewed studies (Brida et al. 2016), a long-run co-integrating relationship is found in all the studies. The TLGH is validated in many studies (Antigua and Bermuda: Schubert et al. 2011; Spain: Balaguer and Cantavella-Jorda 2002). The GLTH has been validated in other studies (for example, North Cyprus: Katircioğlu 2010), while a bidirectional long-run relationship has been found in other economies (for example, Jamaica: Amaghionyeodiwe 2012; Aruba: Ridderstaat et al. 2014).

Narayan et al. (2010) tested the TLGH for four PICs: Fiji, the Solomon Islands, Tonga, and Papua New Guinea (PNG), using annual data from 1988 to 2004. They find that, in the short run, 'real GDP Granger causes tourism exports and in the long run, tourism exports through the lagged error correction term Granger cause real GDP' (2010: 180). On average, across the four PICs, a 1% increase in tourism exports increases GDP by 0.24% in the short run and 0.72% in the long run. This chapter differs from Narayan et al. (2010) in several ways. First, as their data set ranged from 1988 to 2004, we extend the period of data from 1980 to 2017 (depending on the availability of data) and, thus, increase the number of annual observations from 16 to 38 (in some cases). Second, we extend the coverage of Pacific Island states. Narayan et al., covered Fiji, PNG, the Solomon Islands, and Tonga. We extend the analysis by including other Pacific States. As well as the aforementioned, we cover Melanesia (Fiji, the Solomon Islands, and Vanuatu), Micronesia (the FSM, Kiribati, the Marshall Islands, Nauru, and Palau), and Polynesia (the Cook Islands, Niue, Samoa, Tonga, and Tuvalu). Third, we use international tourist arrivals instead of tourism receipts. International tourist arrivals are available for a wider range of PICs because of data scarcity of tourism receipts in several PICs. Tourism arrivals have been used as a measure for tourism in TLGH analyses in a variety of studies (Brida and Giuliani 2013; Jayathilake 2013; Katircioğlu 2011). Further, because tourism receipt data are published as part of the balance of payments accounts, these data are less accurate than arrival counts (Sinclair 1998). Tourism receipts and tourism arrivals are highly correlated (Neumayer 2004). However, Castro-Nuño et al. (2013) note in their meta-analysis that the extent to which tourism contributes to economic growth is higher when tourism expenditures are used as a measurement for tourism as opposed to tourist arrivals.

Methodological approach

To test the tourism-led growth hypothesis and the growth-led tourism hypothesis, we undertake the standard econometric procedures by first testing for the stationarity of the tourism and income variable. We then test for the optimal lag structure and perform the Granger causality tests (Granger 1969). Tourism is said to 'Granger cause' economic growth if the past values of tourism help predict the future values of economic growth. Economic growth is said to 'Granger cause' tourism if the past values of economic growth predict the future values of tourism. It should be noted that Granger causality is not true causality but a type of statistical test. Further, if tourism affects economic growth through a third variable, then this will not be detected through Granger causality. The TLGH takes the general form:

$$Y_t = \gamma_{0j} + \sum_{i=1}^{p} \delta_j Y_{t-i} + \sum_{i=0}^{q} \beta_j' TOUR_{t-i} + \varepsilon_{jt} \tag{1}$$

GLTH takes the form:

$$TOUR_t = \gamma_{0j} + \sum_{i=1}^{p} \delta_j TOUR_{t-i} + \sum_{i=0}^{q} \beta_j' Y_{t-i} + \varepsilon_{jt} \tag{2}$$

where Y_t is income and $TOUR_t$ is tourism; β and δ are coefficients, γ is a constant, $j = 1$ to k, ε_{it} is a vector of the error terms which exhibit white noise characteristics, and p and q are optimal lag orders, where p are lags for the dependent variable and q are lags for the independent variables.

Data

The variable to represent tourism is international tourist arrivals. Tourism data is generally sparse in the Pacific (Harrison and Pratt 2015), so using international tourism receipts is often not feasible. The data for international tourism arrivals is sourced predominantly from the South Pacific Tourism Organisation (SPTO) (SPTO 2017). These data were supplemented with data from the individual countries' statistical offices. The income variable is measured by the country's real GDP index, where 2010 = 100. Because we need to assess the relationship between income and tourism, the data set is limited by the availability of data for both the tourism and income variables. In other words, one series might be longer than another, but we can only use the maximum of the shorter data series. As with Carmignani and Moyle (2019), and given large cross-country differences in arrivals and GDP, it is customary to take the natural logarithms of the variables for a smoother distribution. Table 5.2 shows the length of the data series for each PIC under analysis.

Results

Unit roots test

If variables in these types of regression models are non-stationary, the regression analysis can result in artificially high goodness-of-fit measures and result in spurious relationships (Dritsakis 2004a). To overcome this problem, we test for stationarity and transform the variables by differencing, if necessary, until stationarity is achieved. We implement the Augmented Dickey-Fuller (ADF) test to test for unit roots (Harris 1992). The unit root tests were conducted to examine the integration orders (unit roots) of both the tourism and income variables. Table 5.3 shows that, for international tourism arrivals, the variables become stationary when first differenced for all countries except Kiribati, which is stationary at I(0). While for the GDP of most countries

Table 5.2 Data series for Pacific Island countries

Country	*Start Year*	*End Year*	*Annual Observations*
Micronesia			
FSM	1996	2017	22
Kiribati	1980	2017	38
Marshall Islands	1998	2017	20
Nauru	2007	2014	8
Palau	2000	2017	18
Melanesia			
Fiji	1980	2017	38
Solomon Islands	1985	2017	33
Vanuatu	1985	2017	33
Polynesia			
Cook Islands	2001	2017	17
Niue	1997	2011	15
Samoa	1985	2017	33
Tonga	1985	2017	33
Tuvalu	1990	2017	28

Source: Authors' own compilation.

Table 5.3 Unit roots tests for tourism and income

	Tourism (International Tourism Arrivals)			*Income (Real GDP)*		
ADF	*Level*	*First Difference*	*Stationarity*	*Level*	*First Difference*	*Stationarity*
Micronesia						
FSM	0.690	0.0007***	I(1)	0.0155**	0.0117**	I(0)
Kiribati	0.0082***	0.0027***	I(0)	0.9995	0.0000***	I(1)
Marshall Islands	0.268	0.0010***	I(1)	0.8303	0.0062***	I(1)
Nauru	0.160	0.0007***	I(1)	0.2620	0.1925	I(2)
Palau	0.759	0.0236**	I(1)	0.2814	0.0213**	I(1)
Melanesia						
Fiji	0.930	0.0000***	I(1)	0.9687	0.0000***	I(1)
Solomon Islands	0.732	0.0000***	I(1)	0.7374	0.0159**	I(1)
Vanuatu	0.798	0.0022***	I(1)	0.9680	0.0503*	I(1)
Polynesia						
Cook Islands	0.504	0.0006***	I(1)	0.9827	0.5022	I(2)
Niue	0.877	0.0018***	I(1)	0.2009	0.0639*	I(1)
Samoa	0.906	0.0094***	I(1)	0.9620	0.0106**	I(1)
Tonga	0.477	0.0000***	I(1)	0.9207	0.0005***	I(1)
Tuvalu	0.771	0.0003***	I(1)	0.7451	0.0018***	I(1)

Notes. (1) ***, **, and * denote rejection of the null hypothesis based on the critical values from at the 1%, 5%, and 10% significance level, respectively. The null hypotheses of the ADF test are the series contains unit roots, whereas the null hypothesis for the KPSS test is series is without unit roots. ADF = Augmented Dickey-Fuller.
Source: Authors' own compilation.

is stationary when first differenced, Nauru and the Cook Islands are stationary when second differenced, and the FSM is stationary at levels.

Next, the most appropriate lag length for the model is determined. As Goh and Wong (2014) note, short-lag lengths may lead to incorrect specification, but longer lag lengths will decrease the degrees of freedom which will be problematic in a study with a relatively small sample size for some of these PICs. Based on several indices, such as the Akaike information criterion and Schwarz information criterion (Cavanaugh and Neath 2019), the VAR lag order selection criteria suggests a lag of one as the optimal lag length for all the countries except for the FSM where the criteria suggest an optimal lag of length two. With only eight observations for Nauru, this country is dropped from further analysis.

Table 5.4 shows the pairwise Granger causality tests by geographic region. The first thing that can be noted from Table 5.4 is that, for many PICs, there are no co-integrating relationships between tourism and economic growth. The growth trajectory of tourism and the growth

Table 5.4 Pairwise Granger causality tests

Economy	*Null Hypothesis*	*Optimal Lag Length^*	*Obs.*	*F-Statistic*	*Prob.*
Micronesia					
FSM	Tourism does not Granger cause economic growth	2	19	1.476	0.262
	Economic growth does not Granger cause tourism			0.251	0.782
Kiribati	Tourism does not Granger cause economic growth	1	36	0.478	0.494
	Economic growth does not Granger cause tourism			0.218	0.644
Marshall Islands	Tourism does not Granger cause economic growth	1	18	0.028	0.869
	Economic Growth does not Granger cause tourism			0.003	0.958
Nauru	Tourism does not Granger cause economic growth				
Palau	Economic growth does Granger cause tourism	1	16	3.787	0.074*
	Tourism does Granger cause economic growth			3.383	0.089*
Melanesia					
Fiji	Tourism does not Granger cause economic growth	1	36	0.064	0.802
	Economic growth does Granger cause tourism			3.403	0.074*
Solomon Islands	Tourism does Granger cause economic growth	1	31	3.697	0.065*
	Tourism does not Granger cause economic growth			1.299	0.264
Vanuatu	Economic growth does Granger cause tourism	1	31	5.676	0.024**
	Tourism does not Granger cause economic growth			0.794	0.381
Polynesia					
Cook Islands	Tourism does not Granger cause economic growth	1	14	0.853	0.376
	Economic growth does not Granger cause tourism			0.753	0.404
Niue	Tourism does not Granger cause economic growth	1	9	1.529	0.263
	Tourism does not Granger cause economic growth			0.094	0.769
Samoa	Economic growth does not Granger cause tourism	1	31	0.765	0.389
	Tourism does not Granger cause economic growth			0.262	0.613
Tonga	Tourism does not Granger cause economic growth	1	31	0.246	0.624
	Economic growth does not Granger cause tourism			0.353	0.557
Tuvalu	Tourism does not Granger cause economic growth	1	26	0.105	0.749
	Tourism does not Granger cause economic growth			0.776	0.388

^ Based on Schwarz information criterion (Koehler and Murphree 1988).
Obs. = Observations
***, ** and * denote rejection of the null hypothesis based on the critical values from at the 1%, 5%, and 10% significance level, respectively.
Source: Authors' own compilation.

trajectory of the national income do not correspond. The exception in Micronesia is Palau, where it is found that tourism 'Granger causes' economic growth and economic growth in turn 'Granger causes' tourism. For the Melanesia countries, there are unidirectional relationships between tourism and economic growth. These unidirectional relationships occur in opposition directions, depending on the country. For Fiji, economic growth 'Granger causes' tourism, while tourism 'Granger causes' economic growth for the Solomon Islands and Vanuatu. Surprisingly for Polynesia, as with most of Micronesia, tourism and economic growth move independently.

The impact of COVID-19 on tourism in Pacific Island states

The analysis above summarizes the situation in the pre-COVID-19 era. From March 2020, with the outbreak of the COVID-19 pandemic and its fast transmission, governments of the Pacific Island states quickly moved to close international borders to isolate their vulnerable populations. PIC governments realized that, with the communal lifestyle of Pacific Islanders with their extended families and the limited resources of their health systems, an outbreak of the deadly virus would have significant fatal ramifications among Pacific Island residents. As a result, several Pacific Island states remained some of the only few countries to report no cases. Table 5.5 shows that Samoa, the Cook Islands, the FSM, Kiribati, Nauru, Niue, Tokelau, Tonga, and Tuvalu did not register any COVID-19 cases. Until February 2021, all of the other PICs were COVID-19-contained, meaning any cases being brought into the country were identified in the quarantine period.

Pacific Island states did avoid the health impacts of the pandemic until later in 2021. By March 2022, however, all the 13 independent states did have COVID-19 cases, though in the case of the Marshall Islands there were only four transmissions and no deaths to-date, and in Niue, there were just five transmissions and no deaths. This contrast sharply to six states that have been significantly impacted, most notably Fiji, with 64,394 total cases and 834 deaths as of 30 March 2022, the Solomon Islands with 11,174 cases and 133 deaths, Tonga with 6,423 cases and 6 deaths, Palau with 4,024 cases and 6 deaths, Vanuatu with 3,386 cases and 2 deaths,

Table 5.5 COVID-19 cases and deaths (as of 17 February 2021)

Pacific Country / Territory	*Past 14 Days*	*Past 28 Days*	*Past 42 Days*	*Total Cases*	*Total Deaths*
Cook Islands				0	0
FSM				0	0
Fiji	0	1	3	56	2
Kiribati				0	0
Marshall Islands	0	0	0	4	0
Nauru				0	0
Niue				0	0
Palau				0	0
Samoa	1	2	2	4	0
Solomon Islands	1	1	1	18	0
Tonga				0	0
Tuvalu				0	0
Vanuatu	0	0	0	1	0
Total	**2**	**4**	**6**	**83**	**2**

Source: Adapted from SPC (2021).

and Kiribati with 3,065 cases and 13 deaths (SPC 2022). Nonetheless, even by early 2021, the collapse of international tourism was still devastating. The World Bank (2021) forecasts that the cumulative output loss over 2020–22 is estimated to be around 10% of its 2019 level. Given the reliance of many Pacific Island states on tourism, many livelihoods throughout the region have been severely affected by COVID-19. National airlines grounded their fleets with the exception of repatriation and cargo flights. Hotels and resorts either closed altogether or have stayed open in the hope of breaking even by attracting the domestic market.

The drastic drop in international tourism arrivals in 2020 illustrates the extent of tourism's impact. Although an incomplete data set (SPTO official, pers. comm. 22 February 2021), Table 5.6 shows the extent of the cessation of tourism. A number of Pacific Island states have not allowed any more tourists in the two quarters following the travel embargoes in March 2020. Kiribati, Palau, Samoa, the Solomon Islands, Tonga, and Vanuatu are among this group. Fiji has also started to accept some international tourists, although these tourists are subject to certain requirements. One scheme is Fiji's Blue Lane initiative, whereby yachts and pleasure craft can dock at Port Denarau in Nadi (Fiji) (Figure 5.1) once they can certify that they have met certain criteria, including a 14-day quarantine at sea (RNZ 2020). Fiji's Minister for Tourism claimed the first yacht to enter Fiji on this scheme spent nearly $US20,000 in the local economy. Another scheme that is operating in Fiji, despite the ban on conventional international tourism is the Luxury Vacation in Paradise (VIP) initiative that aims to attract high net worth individuals. Under this scheme, travellers can fly to Fiji and quarantine on an approved private island resort. The VIP Lane refers to a 'bubble', whereby there is a safe travel pathway from the point of origin, through the mode of transportation to the port of arrival and transfers to a designated private resort. Movement is contained with the bubble, and there is minimal contact with the local Fijian population to minimize any contagion. There are clear guidelines for quarantine and testing for travellers and those who come in contact with them.

Despite initiatives to restart tourism, the reality on the ground has been devastating. The International Finance Corporation (IFC) conducted a survey of 3,596 tourism and non-tourism

Table 5.6 International tourism arrivals in 2020

Pacific Country	*Quarter 1*	*Quarter 2*	*Quarter 3*	*Total*
Cook Islands	24,728	98	221	25,047
FSM	0	0	0	0
Fiji	139,701	62	333	140,096
Kiribati	1,409	0	0	1,409
Marshall Islands	0	0	0	0
Nauru	0	0	0	0
Niue	0	0	0	0
Palau	18,174	0	0	18,174
Samoa	20,485	0	0	20,485
Solomon Islands	4,080	0	0	4,080
Tonga	8,932	0	0	8,932
Tuvalu	0	0	0	0
Vanuatu	21,965	0	0	21,965
Total	**239,474**	**160**	**554**	**240,188**

Note: Table based on data received from the countries. For the COVID-19 impact period, some have recorded zero tourists, while others have not submitted data.
Source: Adapted from SPTO (2021).

Figure 5.1 Port Denarau, Fiji. Source: Kosita Butratana, October 2019.

Fijian businesses from 28 April to 15 May 2020 and found that half of all tourism businesses were hibernating or fully closed at the time of the survey. This compared to only one in five (19%) of non-tourism businesses who were hibernating or closed. More than 500 businesses, including 29% of tourism businesses and 11% of non-tourism businesses, predicted they would be bankrupt by November 2020 if the situation did not change (which, in hindsight, it has not). It is the 'micro-, small-, and medium-size enterprises' (MSMEs) that have suffered disproportionately. Tourism MSMEs have lost seven times more income than non-tourism MSMEs. This equates to a median loss of FJ$21,000 (currently US$9,464) for each tourism MSME compared to FJ$3,000 (US$1,352) loss for each non-tourism MSME. Large tourism businesses have lost a median of FJ$500,000 (US$225,339), twice as much as large businesses in the non-tourism sectors ($FJ250,000/US$112, 669) (IFC 2020).

However, Scheyvens and Movono (2020) paint a more optimistic view of COVID-19 in the Pacific in their research investigating the impact of the subsequent economic slowdown on the wellbeing of Pacific Island communities. Despite almost 90% of respondents reporting living in households that have faced significant reductions in income and almost 85% of tourism business owners reporting a major decline in earnings, losing three-quarters or more of their usual income, the authors report that some residents have called the impacts of COVID-19 a 'blessing', 'silver lining', and a 'wakeup call'. To offset the economic hardships of formal employment income loss, residents report an increase in subsistence farming and fishing, trading or bartering goods, and selling other goods and services. Many commentators have seen this health disaster as a chance to reset tourism economies to be more equitable and just, spreading the economic benefits to a wider group of stakeholders while minimizing the negative economic impacts of tourism.

Discussion

Since colonial days to the present, the Pacific has been aggregated into a uniform picture of a tropical paradise. This image enticed tourists to visit providing an economic boost to these island economies. However, as can be seen from a review of the economic contribution of tourism to 13 Pacific Island states, and affirmed in the TLGH analysis, the economic impact of tourism on these SIDS varies widely. The economic contribution of tourism to different island nations depends on both internal and external factors. In general, tourism has had more importance in the Melanesian countries. The proximity to the main markets of Australia and New Zealand and being relatively larger in population and natural resources has meant that tourism represents a large part of the Fiji and Vanuatu economies. The Solomon Islands has the potential to grow its tourism but has been plagued by poor governance, corruption, and a reputation for being unsafe. In Micronesia, the FSM and Palau could be considered to have 'tourism' economies. Much of the contribution of tourism in this sub-region is a result of US political ties, but Palau's tourism sector could be jeopardized by political tension with China. Polynesia is also dependent on tourism, particularly the Cook Islands and Niue. Being located further into the Pacific Ocean, accessibility will always be an issue.

Results from the analysis show that the TLGH does not hold in many PICs. This means that past values of tourism growth cannot explain current changes in economic growth, the exception being Palau. In other words, changes in tourism in the previous year do not directly correspond with changes in economic growth in the current year. Depending on the Pacific Island economy, there are several reasons for this. Several characteristics of these SIDS may mean that tourism growth and economic growth may not correspond. Bertram and Watters (1985) note several of these characteristics and termed these island nations MIRAB states. This implies that these SIDS rely heavily on remittances, foreign aid, and large public sectors. As such, these other economic stimuli may conflate the relationship between tourism and economic growth. Other characteristics of SIDS pose obstacles to economic growth. One of these is their small size, both in terms of populations and sometimes geographic areas. Thus, there is a limited supply of the factors of production. Economies have a high level of subsistence households based around the farming of small family-size plots of land as well as fishing. There is little manufacturing or value-added, meaning that a significant amount of goods needs to be imported, particularly capital goods. There are limited possibilities for import substitution. As a result of these characteristics, a lot of what is consumed by tourists cannot be produced locally in either sufficient quality or quantity (Sharpley and Ussi 2014). A small population results in small domestic markets. As such, it is difficult for domestic industries to leverage economies of scale to take advantage of lower average costs. These are reasons for the weak linkages and low multipliers that exist between the tourism sector and other sectors in the economy.

Regardless of the number of tourist arrivals, the key to economic prosperity from tourism is to retain tourism expenditures through maximizing the linkages and minimizing the leakages. This is easier said than done, though the following section recommends ways in which this can happen.

Recommendations

There are several opportunities and tourism typologies that Pacific Island economies could develop to further benefit economically, especially to reencourage and redevelop tourism following a long period of undertourism. For Pacific Island economies to achieve sustainable tourism growth, appropriate investment and political commitment will need to be devoted to tourism product development as well as cost-effective tourism promotion and marketing.

Marine-based resources are a dominant feature of all Pacific Islands. Commoditizing this resource through tourism needs to be done in a way that improves the local communities' livelihoods. Pacific Island economies need to make use of the unique marine resources they have. Tonga and Niue are endowed with humpback whales visiting every year. Rather than gaining more tourists and putting pressure on resources, prices could be increased to gain more revenues from these activities. Other Pacific Island economies have exceptional scuba diving and snorkelling sites that include World War II wrecks, for instance in Vanuatu and Fiji. Unspoilt fishing opportunities can be found in the Cook Islands, Kiribati, and Tuvalu. While some of these types of tourism occur presently, more could be made of them.

Melanesian countries and territories that are relatively close to the major markets of Australia and New Zealand could benefit from rising cruise ship tourism if they can provide sufficient infrastructure and minimize leakages. While this is a viable option, cruise tourism in the Pacific needs close monitoring and significant investment in transportation infrastructure to minimize the negative impacts on the local communities (Cheer 2016). Cruise ship tourism will also need to ensure that passengers are confident that health and safety measures are strictly in place, especially given the problems encountered by cruise ship companies in dealing with COVID-19 transmissions as well as to ensure that host communities also feel safe in terms of receiving virus-free tourists. Accordingly, restoring confidence in the cruise ship market is an important economic priority for Pacific Island states (see Chapter 4). In terms of encouraging and developing other markets in the future, several states are well-known for their sporting heritage, particularly rugby, and this could be enhanced to a greater degree. The Pacific Games are held every four years, attracting spectators as well as participants from around the Pacific. Sporting events and festivals could be held more regularly so that these events could become regulars on the international touring calendar. This could attract not only international tourists in the form of spectators but also competing athletes and friends and relatives.

For the geographically larger Pacific Island economies, developing agritourism would be a good strategy to pursue. Such economies where this would be applicable are Fiji, Samoa, Tonga, and Vanuatu. There are several forms of agritourism. Adapting working farms to become tourist attractions that offer farm stays, interactive cooking demonstrations, and immersive fruit and vegetable harvesting activities for tourists are several options. Farm-to-table initiatives have the benefits of providing direct links between the agricultural sector and the tourism industry, hence, strengthening the supply chain and helping to minimize leakages. Strengthening links between the tourism and the agriculture sector is not an option but a requirement for Pacific Island states if they are to realize fully the benefits that can be accrued from tourism. This will require reviewing policies that empower and support local farmers, providing them with the right technology, skills, and business knowledge to be able to supply quality goods consistently. Governments must provide incentives to encourage businesses to buy local products; for example, buy Fiji Made initiatives. Import substitution will retain tourist dollars and reduce leakages.

Strengthening the linkages between tourism and the education sector is another way to ensure the tourist dollar remains in PICs. Hotels might hire foreign labourers at the construction stage because their wages are lower. They then hire managers and directors from overseas because Pacific residents may not have the qualifications, skills, and experience to manage international chain businesses. More can be achieved if the government, training institutions, and the tourism sector meet more often and have proper policies in place to ensure people are trained according to the needs of the industry. Strengthening linkages between tourism and local communities is also vital. Through community-based tourism initiatives, locals can invest their resources (land) and start small tourism enterprises. These businesses can employ fellow residents and provide alternative livelihoods for others. Different groups in the village (men, women, youth) can contribute or provide services or goods and benefit from the business. This can be developed by utilizing tourism as a tool to improve a community's business goal. It provides income to members of the community, and it helps in other aspects of their daily existence.

Determining the type of tourism development Pacific Island states want will also have an impact on strengthening the links between these key sectors. Embracing conventional or mass tourism will mean accommodating multinational or international branded hotels, which leads to high importation to cater to the demands of these tourists. This will exacerbate socio-economic concerns, too. Opting for niche markets such as ecotourism, backpacker, dive, and adventure tourism will mean fewer guests, more interaction with the local people, and higher retention of the tourist dollar.

Conclusions and research implications

The chapter provides an overview of the economic contribution of tourism to those Pacific Island economies which represent independent states. This was achieved by studying secondary data on tourism's contribution to GDP and several other economic indicators with a special focus on Micronesia, Melanesia, and Polynesia. Using secondary data, the TLGH for economies in Micronesia, Melanesia, and Polynesia was tested. The results show a bidirectional relationship between economic growth and tourism growth for Palau, that economic growth 'Granger causes' tourism growth for Fiji and Vanuatu, and that tourism growth 'Granger causes' economic growth for the Solomon Islands. For the other Pacific Island economies, there is no direct predictive relationship between economic growth and tourism growth. Despite tourism being an important economic pillar for most of these Pacific Island economies, the growth trajectories of these indicators differ.

Tourism is still a good strategy when promoting economic growth for PICs. Despite the drawbacks resulting from weak linkages and low multipliers, PICs can still benefit economically from tourism by strengthening links between the tourism sector and the agricultural and education sectors as well as with the local communities. With the right support from government and tourism stakeholders, tourism can bring about much needed economic development for PICs, particularly as the COVID-19 pandemic has brought about economic uncertainty.

In terms of research implications, more resources are needed to collect and disseminate economic tourism statistics. Some progress has been made with the New Zealand Government's Ministry of Foreign Affairs and Trade and International Finance Corporation funding the New Zealand Tourism Research Institute (https://www.nztri.org.nz/projects-pacific-tourism-data-initiative) to undertake regular international visitor surveys throughout Pacific Island economies to capture tourist behaviour and attitudes of international visitors to the region. While this is a start, it will provide only the direct economic contribution of tourism in the various island economies. To estimate the direct, indirect, and induced economic impacts of tourism, countries need an input–output table coupled with a tourism satellite account. At present, only Fiji has an input–output table that would enable analysts to estimate the economy-wide impact of tourism in Fiji (Oum and Singh 2019). Other research that would be useful to policy-makers in the Pacific would include an assessment of the types of tourism that could contribute to sustainable livelihoods and the quality of life of Pacific peoples, especially within the context of an economic need to not only kick-start tourism due to significant tourism inactivity since March 2020 but also to ensure that tourism revenues productively benefit local communities and economies.

References

ABC News (2017) 'Solomon Islands at a crossroads as Australian-led assistance mission bids farewell'. Available online at https://www.abc.net.au/news/2017-06-29/solomon-islands-at-a-crossroads-as-australian-led-mission-ends/8661532 (accessed 8 May 2019).

ABC News (2018) 'China's "tourist ban" leaves Palau struggling to fill hotels and an airline in limbo'. Available online at https://www.abc.net.au/news/2018-08-26/china-tourist-ban-leaves-palau-tourism-in-peril/10160020 (accessed 8 May 2019).

Akinboade, O. A., and L. A. Braimoh (2010) 'International tourism and economic development in South Africa: A Granger causality test', *International Journal of Tourism Research*, 12(2): 149–163.

Amaghionyeodiwe, L. A. (2012) 'Research note: A causality analysis of tourism as a long-run economic growth factor in Jamaica', *Tourism Economics*, 18(5): 1125–1133.

Balaguer, J., and M. Cantavella-Jorda (2002) 'Tourism as a long-run economic growth factor: The Spanish case', *Applied Economics*, 34(7): 877–884.

Bertram, I. G., and R. F. Watters (1985) 'The MIRAB economy in South Pacific Microstates', *Pacific Viewpoint*, 26(3): 497–519.

Brida, J. G., I. Cortes-Jimenez, and M. Pulina (2016) 'Has the tourism-led growth hypothesis been validated? A literature review', *Current Issues in Tourism*, 19(5): 394–430.

Brida, J. G., and D. Giuliani (2013) 'Empirical assessment of the tourism-led growth hypothesis: The case of the Tirol - Südtirol - Trentino Europaregion', *Tourism Economics*, 19(4): 745–760.

Briguglio, L. (1995) 'Small island developing states and their economic vulnerabilities', *World Development*, 23(9): 1615–1632.

Carmignani, F., and C.-L. Moyle (2019) 'Tourism and the output gap', *Journal of Travel Research*, 58(4): 608–621.

Castro-Nuño, M., J. A. Molina-Toucedo, and M. P. Pablo-Romero (2013) 'Tourism and GDP: A meta-analysis of panel data studies', *Journal of Travel Research*, 52(6): 745–758.

Cavanaugh, J. E., and A. A. Neath (2019) 'The Akaike information criterion: Background, derivation, oroperties, application, interpretation, and refinements', *WIREs Computational Statistics*, 11(3): e1460. https://doi.org/10.1002/wics.1460.

Cheer, J. (2016) 'Cruise tourism in a remote small island–high yield and low impact?' In: R. Dowling, and C. Weeden (eds.) *Handbook of Cruise Ship Tourism*, Oxfordshire: CABI, pp. 408–423.

Cheer, J. M., S. Pratt, D. Tolkach, A. Bailey, S. Taumoepeau, and A. Movono (2018) 'Tourism in Pacific island countries: A status quo round-up', *Asia and the Pacific Policy Studies*, 5(3): 442–461.

Chen, C.-F., and S. Z. Chiou-Wei (2009) 'Tourism expansion, tourism uncertainty and economic growth: New evidence from Taiwan and Korea', *Tourism Management*, 30(6): 812–818.

Conrich, I., and R. Smith (2018) 'Between reality and myth: Small islands changed and re-imagined in an age of global warming', *Post Script, suppl. Special Issue: Islands and Film*, 37(2/3): 40–52.

Cooper, C., J. Fletcher, A. Fyall, D. Gilbert, and S. Wanhill (2008) *Tourism: Principles and Practice* (4th edition), Upper Saddle River, NJ: Financial Times Prentice Hall.

Dritsakis, N. (2004a) 'Tourism as a long-run economic growth factor: An empirical investigation for Greece using causality analysis', *Tourism Economics*, 10(3): 305–316.

Dritsakis, N. (2004b) 'Cointegration analysis of German and British tourism demand for Greece', *Tourism Management*, 25(1): 111–119.

Fu, R. (2014) 'Rise of the China outbound tourism', *China Internet Watch*, 12 September. Available online at https://www.chinainternetwatch.com/8832/outbound-travelers/ (accessed 26 September 2017).

Goh, S. K., and K. N. Wong (2014) 'Could inward FDI offset the substitution effect of outward FDI on domestic investment? Evidence from Malaysia', *Prague Economic Papers*, 23(4): 413–425.

Granger, C. W. J. (1969) 'Investigating causal relations by econometric models and cross-spectral methods', *Econometrica*, 37(3): 424–438.

Hampton, M. P., and J. Christensen (2007) 'Competing industries in islands: A new tourism approach', *Annals of Tourism Research*, 34(4): 998–1020.

Harris, R. (1992) 'Testing for unit roots using the Augmented Dickey-Fuller test: Some issues relating to the size, power and the lag structure of the test', *Economics Letters*, 38(4): 381–386.

Harrison, D., and S. Pratt (2010) 'Political change and tourism: Coups in Fiji'. In: R. Butler and W. Suntikul (eds.) *Tourism and Political Change*, Oxford: Goodfellow Publishers, pp. 160–174.

Harrison, D., and S. Pratt (2013) 'Tourism in Pacific island countries'. In: C. Cooper (ed.), *Contemporary Tourism Reviews*, Oxford: Goodfellows Publishers, pp. 1–24.

Harrison, D., and S. Pratt (2015) 'Tourism in Pacific island countries: Current issues and future challenges'. In: S. Pratt and D. Harrison (eds.) *Tourism in Pacific Islands: Current Issues and Future Challenges*, Abingdon: Routledge, pp. 3–21.

IFC (International Finance Corporation) (2020) *Fiji COVID-19 Business Survey: Tourism Focus*, July, World Bank Group. Available online at https://www.ifc.org/wps/wcm/connect/ 4fc358f9-5b07-4580-a28c-8d24bfaf9c63/Fiji+COVID-19+Business+Survey+Results+-+Tourism+Focus+Final. pdf?MOD= AJPERES&CVID=ndnpJrE (accessed 23 February 2021).

ILO (International Labour Organisation) (2015) 'Cyclone PAM causes devastating impact on employment and livelihoods', 3 April. Available online at https://www.ilo.org/suva/public-information/WCMS_368560/lang--en/index.htm (accessed 30 January 2020).

ITC (International Training Centre) (2018) 'Tropical Cyclone Winston: Impact on Fiji's labour market and enterprises: Promotion of youth employment in fragile settings', Turin, Italy. Available online at https://fragilestates.itcilo.org/2018/04/09/tropical-cyclone-winston-impact-fijis-labour-market-enterprises/ (accessed 30 January 2020).

Jayathilake, P. B. (2013) 'Tourism and economic growth in Sri Lanka: Evidence from cointegration and causality analysis', *International Journal of Business, Economics and Law*, 2(2): 22–27.

Katircioğlu, S. T. (2011) 'Tourism and growth in Singapore: New extension from bounds test to level relationships and conditional Granger causality tests', *The Singapore Economic Review*, 56(3): 441–453.

Katircioğlu, S.T. (2010) 'International tourism, higher education and economic growth: The case of North Cyprus', *The World Economy*, 33(12): 1955–1972.

Khoshkhoo, M. H. I., V. Alizadeh, and S. Pratt (2017) 'The economic contribution of tourism in Iran: An input-output approach', *Tourism Analysis*, 22(3): 435–441.

Kim, S. S., Y. Guo, and J. Agrusa (2005) 'Preference and positioning analyses of overseas destinations by mainland Chinese outbound pleasure tourists', *Journal of Travel Research*, 44(2): 212–220.

Koehler, A. B., and E. S. Murphree (1988) 'A comparison of the Akaike and Schwarz criteria for selecting model order', *Journal of the Royal Statistical Society: Series C (Applied Statistics)*, 37(2): 187–195.

Lee, C.-C., and C.-P. Chang (2008) 'Tourism development and economic growth: A closer look at panels', *Tourism Management*, 29(1): 180–192.

Lin, V. S., Y. Yang, and G. Li (2019) 'Where can tourism-led growth and economy-driven tourism growth occur?', *Journal of Travel Research*, 58(5): 760–773.

McElroy, J. L. (2003) 'Tourism development in small islands across the world', *Geografiska Annaler*, 85(B): 231–242.

McElroy, J. L. (2006) 'Small island tourist economies across the life cycle', *Asia Pacific Viewpoint*, 47(1): 61–77.

McKercher, B., and A. A. Lew (2003) 'Distance decay and the impact of effective tourism exclusion zones on international travel flows', *Journal of Travel Research*, 42(2): 159–165.

Milne, S. (1992) 'Tourism and development in South Pacific microstates', *Annals of Tourism Research*, 19(2): 191–212.

Narayan, P. K., S. Narayan, A. Prasad, and B. C. Prasad (2010) 'Tourism and economic growth: A panel data analysis for Pacific island countries', *Tourism Economics*, 16(1): 169–183.

Neumayer, E. (2004) 'The impact of political violence on tourism: Dynamic cross-national estimation', *Journal of Conflict Resolution*, 48(2): 259–281.

Oum, S., and R. Singh (2019) 'A new computable general equilibrium model for the Fiji Economy', *School of Economics Working Paper Series*, Suva, Fiji: University of the South Pacific. Available online at https://www.usp.ac.fj/fileadmin/files/schools/ssed/economics/working_papers/2019/Wroking_Paper-A_New_Computable_General_Equilibrium_Model_for_the_Fiji_ Economy.pdf. (accessed 22 May 2019).

Pratt, S. (2013) 'Same, same but different: Perceptions of South Pacific destinations among Australian travellers', *Journal of Travel and Tourism Marketing*, 30(6): 595–609.

Pratt, S. (2015a) 'The economic impact of tourism in SIDS', *Annals of Tourism Research*, 52: 148–160.

Pratt, S. (2015b) 'Potential economic contribution of regional tourism development in China: A comparative analysis', *International Journal of Tourism Research*, 17(3): 303–312.

Reid, M., and M. Jennings (2019) 'Fiji revokes big Chinese resort's rights', *Newsroom*, 9 April. Available online at https://www.newsroom.co.nz/fiji-revokes-chinese-resorts-rights (accessed 8 May 2019).

Ridderstaat, J., R. Croes, and P. Nijkamp (2014) 'Tourism and long-run economic growth in Aruba', *International Journal of Tourism Research*, 16(5): 472–487.

RNZ (Radio New Zealand) (2020) *'Fiji tourism minister praises start of "blue lane" initiative'*. Available online at https://www.rnz.co.nz/international/pacific-news/422235/fiji-tourism-minister-praises-start-of-blue-lane-initiative (accessed 23 February 2021).

Scheyvens, R., and A. Movono (2020) 'Development in a world of disorder: Tourism, COVID-19 and the adaptivity of South Pacific people', *Institute of Development Studies Working Paper Series*, October, Palmerston North, New Zealand, Massey Univeristy: Institute of Development Studies. Available online at https://mro.massey.ac.nz/bitstream/handle/10179/15742/IDS%20Working%20Paper%20-%20Tourism%20Covid%20Pacific%20-%20final.pdf?sequence=2&isAllowed=y (accessed 23 February 2020).

Schubert, S. F., J. G. Brida, and W. A. Risso (2011) 'The impacts of international tourism demand on economic growth of small economies dependent on tourism', *Tourism Management*, 32(2): 377–385.

Sharpley, R., and M. Ussi (2014) 'Tourism and governance in small island developing states (SIDS): The case of Zanzibar', *International Journal of Tourism Research*, 16(1): 87–96.

Sinclair, M. T. (1998) 'Tourism and economic development: A survey', *Journal of Development Studies*, 34(5): 1–51.

SPC (Pacific Community) (2021) 'COVID-19: Pacific community updates', Noumea, New Caledonia. Available online at https://www.spc.int/updates/blog/2021/02/covid-19-pacific-community-updates (accessed 23 February 2021).

SPC (Pacific Community) (2022) 'COVID-19: Pacific community updates', Noumea, New Caledonia. Available online at https://www.spc.int/updates/blog/2022/03/covid-19-pacific-community-updates (accessed 2 April 2022).

SPTO (South Pacific Tourism Orgaisation) (2017) *Annual Review of Tourist Arrivals in Pacific Island Countries 2016*, Suva, Fiji: South Pacific Orgainsation. Available online at https://corporate.southpacificislands.travel/wp-content/uploads/2017/02/2016-Annual-Visitor-Arrivals-ReviewF.pdf (accessed 26 September 2017).

SPTO (South Pacific Tourism Organisation) (2019) *Annual Review of Tourist Arrivals in Pacific Island Countries 2018*, Suva, Fiji: South Pacific Orgainsation. Available online at https://pic.or.jp/ja/wp-content/uploads/2019/07/2018-Annual-Visitor-Arrivals-berReportF.pdf. (accessed 5 January 2020).

Tang, C. F., and S. Abosedra (2014) 'Small sample evidence on the tourism-led growth hypothesis in Lebanon', *Current Issues in Tourism*, 17(3): 234–246.

Vada-Pareti, S. (2015) 'The Chinese are coming – Is Fiji ready? A study of Chinese tourists to Fiji', *The Journal of Pacific Studies*, 35(3): 145–167.

Wanhill, S. (1994) 'The measurement of tourist income multipliers', *Tourism Management*, 15(4): 281–283.

World Bank (2021) *Global Economic Prospects*, World Bank Group. Available online at https://www.worldbank.org/En/Publication/Global-Economic-Prospects (accessed 23 February 2021).

World Trade Organization (2019) 'Natural disasters and trade symposium'. Available online at https://www.wto.org/english/tratop_e/devel_e/study_1_pacific_country_annex_18_april_draft_final.pdf (accessed 30 January 2020).

6

THE COVID-19 PANDEMIC AND THE SOUTH PACIFIC

Evaluating crisis-related tourism recovery strategies

Evelyn G.Y. Loh, Jefferson Patovaki, and Navneet Nimesh Kumar

Introduction

The livelihoods of individuals across Pacific Island states or countries (PICs) have been socially and economically affected by the COVID-19 pandemic since March 2020. Consequently, governments of PICs have been seeking ways to curb the spread of the virus but should arguably be putting long-term recovery strategies into action, as it is believed that the pandemic will continue to spread over the next decade or indefinitely until mass populations have been *effectively* inoculated (Bliss et al. 2021). The chapter draws critical attention to discussions which comprehend the readiness of PICs in becoming self-sustaining nations and being able to effectively advance their approach toward tourism development in the post-COVID-19 context. PICs have significantly focused on immediate concerns and capital injection approaches into the hardest hit areas. However, more attention is arguably required in academic inquiries concerning the importance and utilization of crisis management and long-term recovery strategies in tourism development, particularly within the context of PICs. Therefore, this chapter has three main objectives: (1) to provide an overview of the tourism industry background; (2) to outline and discuss the short-to-immediate term COVID-19 economic remedies by governmental institutions and partner countries/agencies; and (3) to propose ways forward in creating new foundations in the tourism industry that could serve the long-term interest of local governments, communities, tourists, and tourism operators.

Given these objectives in place, three island states were selected based on the MIRAB-SITE model, with Fiji selected as part of the 'small island tourism economies' (SITE) category and the Solomon Islands selected to represent the category of 'migration, remittances, aid, and bureaucracy' (MIRAB). Finally, Samoa is selected as a moderate representation, as it does not represent a SITE nor a MIRAB state. These cases develop insights from lessons learnt in each PIC before and during the pandemic, thereby deliberating long-term tourism development strategies for a post-COVID-19 recovery. These discussions lead to two resounding themes: (1) the need for tourism crisis management plans to be effectively initiated and formally established; and (2) the need to advocate for responsible and sustainable forms of tourism consumption.

DOI: 10.4324/9780429019968-7

Tourism economy classifications and the role of foreign aid in the PICs

The tourism economy in PICs can be categorized into SITE and MIRAB states (Movono et al. 2015). These authors further note that the former comprises countries with tourism as a major contributor to national income for which the contribution of travel and tourism to the country's gross domestic product (GDP) is more than 30%, whilst the latter concerns states which experience lower tourism income with a GDP contribution lower than 10%. Accordingly, in the pre-pandemic year of 2019, the Cook Islands recorded the highest tourism revenue, with at least a 60% contribution to its GDP in 2019, followed by Vanuatu (35.8%), Fiji (32%), Samoa (24%), Tonga (18.5%), and the Solomon Islands (9.3%) (Global Economy, n.d.; WTTC 2021). These GDP figures implicate the Cook Islands, Fiji, and Vanuatu as SITEs and the Solomon Islands as a MIRAB nation. Samoa and Tonga would neither be classified as a SITE nor a MIRAB nation as their travel and tourism GDP contributions are between 10% to 30%.

This spectrum of tourism's contribution to GDP offers insights on how these nations have fared in their respective tourism development agendas. Tourism is identified as a key area for economic growth in the PICs, with each country possessing unique tourism branding characteristics. For instance, Fiji is known as a luxury holiday destination targeted toward couples and families (Tourism Fiji 2013). Small private tourism operators in the Solomon Islands found that heritage tourism derived from historical World War II sites and relics represents a competitive advantage component (Kitching 2014), and thus is a key identifying characteristic. On the other hand, Kiribati attracts angling tourists to Kiritimati (*Christmas*) Island – categorized as a form of sports tourism (FAO 2010). Destination branding is an imperative strategic marketing technique which drives tourists to visit destinations and enables effective destination positioning supported by efficient resource allocation (Piggot et al. 2004). However, destination branding would not be effective without appropriate infrastructure development, particularly in geographical areas where tourists often visit. Given that island states often struggle to fund these developments, despite their political independence, it is common for these nations to become economically dependent on foreign aid to support such developments that can be then utilized to drive tourism growth (Alisjahbana et al. 2020).

The provision of foreign aid sources is linked to different purposes in each PIC. For instance, the Cook Islands receives grants from the New Zealand government for infrastructural development such as irrigation systems, road networks, and telecommunications aimed to supplement tourism growth (Bertram 2018). Foreign aid in the South Pacific is also directed toward sustainability in terms of environmental health, social welfare, and economic independence. For example, the Australian Government Department of Foreign Affairs and Trade (DFAT n.d.) facilitate programmes toward strengthening the governance of climate change in the Republic of the Marshall Islands. In the case of Niue, the provision of foreign aid by New Zealand focuses on developing niche and luxury tourism in a sustainable manner in order to avoid mass tourism (Watson 2019). On the other hand, foreign aid from China is arguably focused on the extension of political ideology for which development funds are largely channelled to the PICs which support the 'One China' policy (Dornan and Pryke 2017). For instance, China assisted Vanuatu in a range of infrastructural projects involving the construction of Santo Island wharf, a major airport runway extension, convention centre, stadium, and major roads (Gorman 2019).

Historically, this region received tremendous financial aid in the past decade. Australia is often hailed for being the most significant donor in the region, but there are also other donors, including China, Japan, New Zealand, and the United States (US) (Dziedzic 2018). From 2000 to 2012, the World Bank (n.d.-a) recorded an increase of 180%, from US$387 million to US$1.1 billion, of foreign aid received by the PICs. According to Dziedzic (2018), in 2016, foreign aid for the PICs dipped to its lowest amount for the first time in six years. However, foreign aid channelled into the PICs increased again from 2016 to 2018 by 35% (World Bank, n.d.-a). The economic challenges caused by the COVID-19 pandemic have driven PICs to rely on the support of partner countries, as will be indicated in the cases of Fiji, Samoa, and the Solomon Islands.

Undoubtedly, foreign aid can drive regional competitiveness for the tourism industry and provide entrepreneurial opportunities to local communities. Despite significant foreign aid received, the PICs are also some of the poorest nations in the world, with about 25% of the population living below the poverty line, based on pre-pandemic data (United Nations 2019). However, governmental institutions in the PICs have been criticized for being inefficient in terms of financial resource management (Hassal 2018). Although some PICs have been able to effectively expand their tourism development agenda in the past few decades, together with the support of foreign aid, there should also be serious considerations to improve economic independence through long-term strategic measures to sustain the tourism economies of PICs and the livelihoods of the local community and to preserve the natural environment.

These reflections lead to the presentation of three Pacific Island cases based on the MIRAB-SITE model criteria. Fiji, Samoa, and the Solomon Islands are presented to illustrate their respective tourism agendas, prevailing social and political issues, and economic sustenance during the COVID-19 pandemic. These cases provide insights into varying tourism development approaches, which have been significantly challenged by the pandemic. The lessons learnt from these case scenarios could help to comprehend ways in which sustainable tourism could be advanced in PICs in the post-COVID-19 context.

Fiji

Tourism development in the pre-pandemic era

Fiji is categorized as a SITE, as the nation's travel and tourism industry contributed to 32% of the total GDP in 2019 (WTTC 2021). In 2017 and 2018, the tourism industry was the third highest waged employer in Fiji, preceded by retail and wholesale, and manufacturing (United Nations Pacific 2020). Although tourism in Fiji displayed an upward trend in tourist arrivals, the industry was affected by past political unrests and complex land ownership structures and leases (Fletcher and Morakabati 2008). On average, prior to the pandemic 80% of Fijian households had at least one family member participating in the tourism economy (Scheyvens and Hughes 2019). However, Fiji faced significant socio-economic issues pertaining to tourism employment in which women dominated lower-level jobs (Naidu et al. 2013) and worked significantly in the informal sector (Scheyvens and Hughes 2019).

Tourism growth in Fiji advanced the accommodation sector in terms of multinational ownership, dominated by such corporations as Accor, Hilton, Intercontinental, Radisson, and Starwood-Sheraton (Klauss 2011). In recent years leading to the pandemic, Fiji witnessed significant increases in tourism receipts from Australia, New Zealand the US, and India, as the Fijian government strengthened its marketing initiatives, public policy, and infrastructure (Koya 2018). From 2013 to 2019, there was a growth rate of 26% for arrivals from China, indicating significant potential to target this market in the post-pandemic era (World Bank n.d.-b). Consequently, there has been a notable reliance on mass tourism and foreign investment to build and sustain the tourism economy. Given the nation's long-term vision aligned to the 2017–2036 National Development Plan, tourism development in Fiji began to shift its commitment toward environmental sustainability (ADB 2019). This agenda is driven by the needs of local communities and the increasing preferences of tourists for sustainable experiences (Tyllianakis et al. 2019). Environmental 'preservation' can take place along with the strengthening of an ecotourism agenda in Fiji (Kundra et al. 2021). Ecotourism could be an alternative development option to mass tourism, as it advocates for environmental conservation and reduces environmental damage (Western 1993), and there have been significant discussions on ecotourism development in Fiji (Kundra *et al.* 2021; Pratt 2013) (see Chapter 16). However, making ecotourism a mainstream type of tourism can be challenging when Fiji has historically targeted and planned for resort-based tourism. Inevitably, the pandemic unfolded a tourism crisis which, arguably, led to the realization that resort-type tourism is not a sustainable way forward, especially as people had become over-dependent on this industry.

A tourism crisis response

Similar to other tourism economies, Fiji's tourism industry was significantly affected since the start of the pandemic (see Chapters 11 and 15). In March 2020, the Fijian government announced the closure if its national borders, meaning international tourists could no longer enter the country (Dean 2020). Public protection came at a cost when the tourism-reliant state witnessed a decrease of 83.6% in visitor arrivals for 2020 compared to 2019, and tourism 'micro, small, and medium enterprises' (MSMEs) were the most vulnerable, as they lost seven times more revenue compared to businesses with no tourism connections (IFC 2020).

Witnessing the impact of the COVID-19 virus on the economy, the government started to work closely with national and international agencies to introduce cash-based aid to the public. A stimulus package of FJ$1 billion (US$467 million) was provided for the country to overcome (or confront) the health crisis (Xinhua 2020a). International partner agencies, namely the Asian Development Bank (ADB) and World Bank, donated financial support of FJ$100 million (US$46 million) and FJ$5.5 million (US$2.6 million), respectively (United Nations Pacific 2020). Australia and New Zealand also provided direct financial support of FJ$152.5 million (US$71.2 million) and NZ$40 million (US$27 million) respectively (DFAT 2021; New Zealand Foreign Affairs and Trade 2021). Subsequently, the pandemic inevitably increased the dependence of PICs on partner countries, and Fiji became particularly dependent on foreign cash-based aid.

Tax measures were also implemented to enable economic relief to tourism businesses for the short term. These included the abolishment of the Service Turnover Tax (STT) of 6%, reduction of the Environment and Climate Adaptation Levy (ECAL) from 10% to 5%, reduction of duty levied on alcohol by 50% for tourism businesses, and reduction of airport departure tax (KPMG 2020). The STT and ECAL are taxes imposed on tourism-related businesses, including accommodation establishments, transportation, and tour and recreational services (PwC 2022). The government also provided COVID-19 flexible concessional loans to MSMEs which can be directly injected into business operations or used to fund capital (Government of the Republic of Fiji 2020).

The tourism-reliant economy also launched the 12-month 'Love Our Locals' (LOL) campaign to stimulate spending from domestic tourism, which began in June 2020 and became a success when increased occupancy rates and resumption of work for tourism operators amid the pandemic were observed (RNZ 2020a; Tuimaisala 2020). Fiji also began to realign their tourism strategies and set forth the following plans: (1) establishing a travel bubble with its main tourist groups, Australians and New Zealanders; (2) relaunching its national airline and ocean passage for yachts; and (3) developing an online channel – 'PayNowStayLater' to encourage potential tourists to pay first and allow them to enjoy their stays when international borders reopen (Xinhua 2020b). The Fijian Prime Minister also noted that the tourism industry will require significant consumption to drive the economy, pointing to a regional travel bubble with Australia and New Zealand as an answer to the tourism crisis faced (Office of the Prime Minister 2021). In December 2021, Fiji eagerly reopened its international borders to resuscitate tourism as a main contributor to its GDP (CNN Travel 2021). Tourism Fiji launched a celebrity-endorsement campaign known as 'open for happiness', targeted toward tourist markets from Australia, New Zealand, the United Kingdom (UK), and the US (Branding in Asia 2021). The COVID-19 financial policies and tourism campaigns can be initially perceived as helpful solutions to remedy the tourism crisis in the short run; however, there should arguably be an emphasis to adapt to the changing landscape of the tourism business environment instead of reliance on cash-based aid and the legacy of mass forms of tourism. While short-term measures could keep the economy afloat, Fiji is losing sight of the bigger picture, which is to build more resilient yet sustainable tourism that could buffer against a tourism crisis.

Samoa

Tourism prior to the pandemic

Historically, Samoa has received a significant number of tourists who are emigrants returning to their home country with the purpose of visiting friends and relatives (VFRs) (Scheyvens 2007). Based on a four-year average from 2014 to 2018, statistics published by the Samoan Tourism Authority (STA) indicate that around 35% of tourist arrivals are VFR tourists (STA 2019), mainly from New Zealand and Australia. Leisure tourists still top the list, comprising an average of 38% of tourist arrivals within the same period (STA 2019). Samoa's tourism industry represented a moderate proportion of 24% of the country's GDP in 2019 (Global Economy, n.d.), indicating that Samoa is neither categorized as a SITE nor a MIRAB state. In the same year, its tourism employment represented approximately 15% of the total employment (Parliament of Samoa 2019), whilst employment in the agricultural sector represents approximately five times more than tourism industry employment (Government of Samoa 2018). Although tourism has been a key contributor to national income, there are limited employment opportunities compared to the agricultural sector.

Based on 2018 statistics, approximately 18.8% of Samoa's population lives below the 'national poverty line' (ADB 2020a). This is despite attempts by the Samoan government to reduce poverty through advancing an economic framework to strengthen structural and economic linkages between tourism and the agricultural sectors. Prior to the pandemic, more than two-thirds of accommodation establishments in Samoa utilized local food and beverage producers (STA 2014). These linkages created employment opportunities and empowered local communities to generate a self-sustaining income. This self-sustaining economic model for tourism development is arguably crucial as the country is challenged by its limited natural and human resources, inability to achieve economies of scale, and lack of international exposure because it is geographically isolated (Cheer et al. 2018).

Samoa has been cautious not to over-rely on tourism income but, rather, focus on economic diversification (Guardian 2021). To support this strategy, Samoa pursued a 'low volume, high yield' tourism policy with an emphasis on nature and cultural preservation integral to the tourism experience (Scheyvens 2005; Twining-Ward and Butler 2002). Landowners can practice the guiding principles of sustainability by ensuring that land is developed and maintained for tourism activities, for instance, through beach *fale* accommodation (Scheyvens 2005). This form of lodging prioritized guests' experiences of local cultural elements such as food, Sunday observances, and community sociability (Scheyvens 2006), thereby indicating the close affinity that Samoans have with their culture and the pride they possess in sharing their traditional hospitality with foreign tourists (Scheyvens 2005). There have been a range of activities available whereby cultural brokers, such as crafts persons, can connect with tourists to share their knowledge concerning the production of *siapo*, a traditional barkcloth (Ford et al. 2019).

The Samoan community is also known for its collectivist values, which is a key component in its social structure and 'way of life' (Government of Samoa 2020). Together with its staunch cultural practices, Samoa poses a location that could potentially advance its cultural and natural heritage tourism products (see Chapter 22). Nonetheless, this type of community and cultural-based strategy will require significant intervention from tourism sector authorities to educate local Samoans in creating authentic tourism experiences and increasing Samoa's visibility as a cultural heritage tourism destination. While this strategy can be a sustainable way forward in the post-COVID-19 context, its success will depend on the government's willingness to invest in such community-centric developments and ability to garner the public's support to participate as well.

COVID-19 and the tourism economy: ramifications and responses

Since the start of the pandemic, the Samoan government has taken stringent lockdown measures as a response to the 'still lingering' effects of the 2019 measles outbreak and the historical experience of the 1918 influenza pandemic, which were detrimental to the Samoan population (Tahana 2020). Due to the COVID-19 pandemic, 68% of Samoans reported loss of income due to layoffs, business closures, reduced working hours, and changes to the terms of employment (United Nations in Cook Islands, Niue, Samoa, and Tokelau 2020). This economic impact is considered substantial even though tourism was not the country's main economic source. Samoa's tourism industry significantly relies on the informal sector, comprising food hawkers, micro-scale sellers, and equipment rentals, which have become the most vulnerable type of employment when tourism activities ceased (Connell 2020). Although informal employment in tourism is often a means to ascertain additional income, informal workers are often at risk due to poorly regulated remuneration systems, inconducive work environments, and inadequate skills development (Uguz and Ismet 2016). Moreover, women micro-entrepreneurs have experienced financial losses and thus resorted to agricultural activities for the purpose of small-scale selling and consumption (United Nations in Samoa 2020). As cash was difficult to attain, some of the communities in Samoa have turned to the barter system, leveraging on social media's reach. For instance, the 'Le Barter' Facebook page was established to allow the exchange of goods and services amongst its members (Siutaia 2020). This once again reflects the collectivist values upheld by the Samoan community in which individuals and families can depend on their respective communities for shared resources. In other cases, communities have also relied on subsistence agriculture and fishery for survival, which, however, led to an over-supply of local produce, driving the prices down (Connell and Taulealo 2021).

To buffer against the economic downturn, the Samoan government introduced financial aid to the public, including partial withdrawals from pension funds, transfers to those who were financially vulnerable or have experienced job losses, and three-month loan repayment subsidies for those applied under commercial banks and Government Guarantee Schemes (ADB 2020b). Apart from relocating hospitality staff to other industries that remain in operation during the pandemic, the Samoan government also received cash-based aid from the ADB amounting to US$23.6 million to compensate retrenched hospitality staff and US$29.88 million for assisting those who are financially affected by the pandemic (United Nations Samoa in 2020). On a longer-term basis, the Samoan government recognized the need to upskill and reskill the hospitality workforce and improve digital literacy to empower women entrepreneurs (United Nations in Samoa 2020). While micro-entrepreneurs are encouraged to increase their digital presence to cater to a wider local and tourist market, it is also important for MSME-governing bodies to coach the usage of technology and provide the necessary equipment support to existing and potential micro-business operators.

The case of Samoa presents a rather unique scenario in comparison to Fiji. As noted in the earlier section of this case, Samoa has been treading carefully with its tourism development by focusing on cultural heritage and avoiding mass tourism (see Chapter 22). Although cash-based aid seems inevitable in this tourism crisis, Samoa has exhibited the importance of community closeness for survival. However, there also needs to be significant human resource development, particularly in paving the way for the empowerment of women and improved digital literacy in the tourism industry.

Solomon Islands

Pre-COVID-19: tourism development deterrents and growth measures

The Solomon Islands make an attractive 'sun, sea, and sand' destination, but the tourism industry has encountered challenges since the 1990s, notably natural disasters and political instability (Fugui and Wate 1994). Ethnic tensions between the Malaita and Guadalcanal groups in 1998 resulted in decades of social unrest and significantly marred community relationships, affecting growth of the tourism industry (Allen 2012). In November 2021, social unrest reappeared in the form of rioting in Honiara's Chinatown by islanders who opposed the government change in diplomatic allegiances from Taiwan to China (Donald 2022). Being categorized as a MIRAB state, travel and tourism in the Solomon Islands represented a GDP of 9.3% in 2019 (WTTC 2021). The nation's economy is largely reliant on harnessing natural resources such as logging, mining, fishing, and agriculture, implying that the utilization of land and other natural resources are an important economic avenue (Monson 2017).

The tourism industry in the Solomon Islands witnessed a lack of local community participation. For instance, 90% of household livelihoods in villages nearby tourism locations are sustained by subsistence farming and fisheries, generating cash through small-scale enterprises (Diedrich and Aswani 2016). This suggests that exposure of local communities to a tourism economy is limited, questioning the extent to which tourism revenue is distributed throughout local communities. The Solomon Islands government recognizes the challenges of tourism development concerning the absence of effective policies and procedures to aid the industry's growth, focusing on the need to promote stakeholder partnerships and create financial systems that encourage local participation (Nolan 2020). The creation of clear intersectoral linkages between tourism and the agriculture could encourage more local community representation in the tourism sector (see Chapter 8), as was indicated earlier in the case of Samoa.

Prior to the pandemic, the Solomon Islands' cruise tourism witnessed a boom, encouraging tourists to visit post-war sites and relics (Gibson et al. 2021). The direct economic impact from cruise tourism was AU$0.6 million (US$0.43 million), of which 65% was diverted to the government as income generated from port fees (IFC 2016), indicating the government's large degree of control over cruise tourism revenue. However, to encourage higher participation of local tourism operators, the Solomon Islands government also needs to consider the division of revenue that is economically attractive for local operators. This calls for the need to revisit the local benefits of cruise ship tourism and ways in which revenue can trickle down to local communities and businesses.

Education investment has been identified as a key area for human resource development in the Solomon Islands, which could strengthen the formal economy such as tourism (DFAT 2014). Given the patriarchal nature of the Solomon Islands society, gender disparity is a significant concern, as women are less likely than men to receive a full, formal education (United Nations Women, n.d.). This leads to other prevailing issues such as poor economic inclusion, lack of job opportunities, and financial insecurity (DFAT 2014). Despite these challenges that ensued prior to the outbreak of the COVID-19 virus, there was an emerging realization by partner countries and agencies that community-driven tourism can take place by fostering gender equity and economic independence. In 2018, DFAT provided funding of AU$3 million (US$2.2 million) for an economic empowerment project to redevelop several market districts in the Solomon Islands, including markets in Auki (Malaita Province), Gizo (Western Province), and Honiara (Guadalcanal) (Fleming and Tabualevu 2018). These authors also highlighted that this project, known as the Market4Change (M4C), was established in collaboration with the organization 'United Nations (UN) Women' to enable female empowerment through economic inclusion and financial independence in businesses.

It is evident that the Solomon Islands requires significant collective effort between such key stakeholders as tourism operators, local communities, and government bodies to progressively advance a clear and strategic direction for tourism development. The tourism development agenda of the Solomon Islands represents a contrast to the path taken by Fiji and Samoa, which possess a clearer vision with regard to placing emphasis on destination identity and tourism branding, and have even begun identifying ways to advance their respective strategic directions toward the tourism industry. Undoubtedly, the Solomon Islands possesses vast natural and heritage assets to advance a sustainable tourism agenda and, hence, need to begin developing the necessary action plans (see Chapter 10).

Impact of COVID-19 and coping mechanisms

The tourism industry in the Solomon Islands faced tremendous pressure since the start of the pandemic. Businesses of major hotels reduced to a 10% occupancy rate and others have closed (Leni 2020). The country's marketing arm, Tourism Solomons, reported a loss of about SI$350 million (US$43.5 million) in tourism revenue in 2020 (Mamu 2021). Employees affected by layoffs were allowed a one-time withdrawal of SI$5,000 (US$620) from the Solomon Islands National Provident Fund (SINPF) (Nanau and Labu-Nanau 2021). Additionally, each member of Parliament would receive SI$250,000 (US$31,000) to assist repatriating constituents to return home (Kusapa 2020). The prime minister also announced a government 'economic stimulus package' (ESP) of US$37.5 million, drawing from financial sources such as bonds, loans, and foreign donors (RNZ 2020b). The purpose of the ESP is to stimulate the rural and national economy and introduce measures such as tax relief, interest payment relief, and electricity tariff subsidies (ADB 2020b). However, the challenge of the government is to reach out to scattered populations, particularly those in remote locations. Other strategies utilized to soften the economic impact concerned the government's move to pursue a domestic travel campaign initiated in 2021 (Fly Solomons 2021). Also, the government and the International Finance Corporation (IFC) launched an investor guide for tourism investments in land projects (Solomon Islands Government 2021). The implication is to stimulate the tourism and hospitality industry after a long period of inactivity. However, opening investable land for tourism development is both a political and social challenge given the complexities in land ownership in the Solomon Islands (McDougall 2005).

Similar to the cases of Fiji and Samoa, the government of the Solomon Islands also saw capital injection as a solution to sustain the economy temporarily. However, given the larger degree of governmental control over financial resources (as indicated above), there is a need to question the efficiency and effectiveness of the distribution of these resources to tourism businesses, particularly the MSMEs and communities that have been significantly affected by the pandemic. For the tourism economy to develop, there should be serious considerations to empower the private sector to enable innovation and entrepreneurial opportunities, increasing employment opportunities overall. However, international tourism mobility would need to become more fluid, and international tourism demand would need to fully recover.

Contemplating long-term remedies for sustainable tourism

Recent scholarly discussions on post-pandemic tourism recovery strategies have debated issues surrounding remedial actions to 'restart' tourism and whether to reopen nations for international tourism or to abet domestic tourism (Benjamin et al. 2020; Lee 2021). Plans for bilateral quarantine-free or eased-quarantine travel corridors have been initiated. In late 2021, the Cook Islands created a travel bubble with New Zealand, though it was not continuous due to virus outbreaks (Downes 2021). Tourism Solomons and Solomon Airlines were also keen on pursuing an intra-regional travel bubble within the Pacific Islands, which did not materialize when the Solomon Islands was no longer COVID-free (Lehmann n.d.; Solomon Times 2020). Travel bubbles can be a feasible solution to keeping the tourism economy afloat in the short run and attract VFR tourists or seasonal workers seeking to return home (Connell and Taulealo 2021). This strategy can sustain economic needs and give confidence to individuals to travel internationally. However, increased COVID-19 cases can be challenging to PICs which have strained and vulnerable health and medical services.

PICs should look inward for solutions to recover from this crisis. Samoan government's domestic tourism campaign in late 2020 was not successful because of limited coordination amongst tourism operators, implementation of unfriendly travel policies, poor promotions strategy, and over-estimation of customer demands (Samoa Observer 2020). Accordingly, there is a need to conduct detailed market research and adequate forecasting. After all, the Pacific Islands have been positioned to attract affluent international visitors (Pearlman 2021). Understanding domestic tourist behaviour will need to be prioritized, focusing on national demographics and differentiated market segments pertaining to disposable income for domestic tourism consumption, preferred locations and activities, and choice of accommodation facilities. Tourism authorities also play a leading role in bringing together public and private tourism operators to implement impactful marketing strategies and communicate with policymakers, enabling successful tourism campaigns. With governmental support, the private sector can be empowered to drive sustainable tourism industry competitiveness and, at the same time, generate profits that encourage local and international investments (Everett et al. 2018).

There should also be considerations by governmental agencies to develop a tourism 'crisis management plan' (CMP), involving a set of policies and procedures to reduce negative financial impacts on the industry during a crisis, especially one with long-term implications. Drawing inspirations from Blake and Sinclair (2003) on the recovery response from the September 11 incident by the Travel Industry Recovery Coalition, a tourism CMP should aim to include tax deductibles for consumers and tourism employees, loan schemes, and infrastructural upgrades as well as upskilling and reskilling workforce training programmes to prepare the tourism workforce for a post-crisis business environment. This means that a CMP requires detailed planning amongst tourism stakeholders, including tourism governing bodies, financial institutions, tourism operators and, most importantly, community representatives. However, these strategies require implementation at different phases, given the varying levels of infrastructural and human resource development in each PIC.

During the pandemic, technology became an essential tool for convenient and 'touchless' spending and engaging public safety through contact tracing. Furthermore, the post-pandemic consumer is likely to be far more familiar with digital applications and driven to use digital platforms because of greater convenience and accessibility to tourism products and services (Hajro et al. 2021). Technology presents tremendous opportunities in allowing eased travel processes for international tourists. For instance, the Cook Islands launched the *CookSafe+* application, a compatible contact tracing system with New Zealand's COVID tracer application, which enables eased and safer travel processes between the two countries (RNZ 2021). While challenges to connectivity in remote areas of the PICs are still prevalent, there are opportunities for the PICs to further develop technological infrastructure to gain confidence of the post-lockdown tourist.

With drastic changes to the tourism landscape, Connell and Taulealo (2021) argue that the emergence of 'new normal' tourism would likely take place when the pandemic ends. This begs the crucial question: 'Is the way forward actually "the way back"?' Prior to the pandemic, tourism presented a range of disadvantages, notably pollution, social disparities, and unfavourable working conditions caused by capitalistic forms of mass tourism and consumption (Cave and Dredge 2020). These authors added that the pandemic sheds light on the shortcomings of mass tourism, suggesting that regenerative forms of tourism which are community-inclusive, authentic, and responsible should become more prevalent. The need for community-inclusivity also resonates from pre-pandemic times, as was seen in the cases of Samoa and Solomon Islands, for which there needs to be higher inclusion and equality of women in the tourism economy.

Regenerative tourism could enable sustainable developments – environmentally, socially, and economically – shifting away from a focus on monetary gains. It concentrates on the notion of the scarcity of resources and the need to sustain the natural and social environments through transformative measures (McEnhill et al. 2020) (see Chapter 24). Hence, there is a need to focus on the protection of natural and heritage sites and, at the same time, enable the economic participation of local communities as well as advocating for fair wages regardless of social backgrounds (Christ 2017). In this sense, 'transformation' can be seen to embrace an emancipatory vision for the host and destination community. This strategy can be a sensible way forward, particularly for Fiji, where mass tourism for far too long has been the nation's economic stronghold. By avoiding a fixation on profit-maximation, regenerative tourism implicates that a PIC will need to take more ownership of financial resource management to drive economic growth in a way that is holistic and socially inclusive for host communities (Duxbury et al. 2021). The lack of ownership (political or otherwise) of financial resource management is evident given the over-reliance on financial aid received for economic development and the dependence of tourism revenue from specified countries. Hence, there is a need for effective leadership and governance to improve social protection and financial assistance, particularly for the most vulnerable communities.

With a regenerative tourism agenda, PICs will need to reset their marketing strategies. This structural change can begin by considering social, green, and ethical marketing. These strategies possess different approaches but with similar underpinnings, involving social responsibility and social awareness. Taking the first strategy, social marketing aims to change the behaviour of target groups by promoting social responsibility or awareness (Kotler and Keller 2016). This can be an effective strategy for Fiji, where tourism has been more resort-based and distanced from the local communities. Hence, social marketing can be used to target both locals and tourists to 're-brand' into regenerative tourism by communicating the importance of *responsible consump-*

tion, whereas green marketing is an approach in which products and services are modified to reduce the effects of consumption on the environment (Polonsky 1994). This could represent an effective strategy for Samoa, where advocacy for small-scale tourism and consumption could preserve its natural and cultural heritage alongside its community-based tourism identity. This strategy will require effective leadership in green tourism management and eventually allow Samoa to achieve its strategic objective of becoming a sustainable tourism destination. Whereas ethical marketing, a strategy whereby social responsibility is a fundamental element in addressing social issues and consumer protection (Ferrell et al. 2013), can be used to expand the social marketing approach for the Solomon Islands. Community-led tourism development will be able to advance this strategy by providing better opportunities to local marginalized communities to participate in the tourism economy. This will, in turn, lead to enhanced economic inclusion, which is particularly important for the betterment of the Solomon Islands' marginalized communities.

Conclusion and research implications

As nations begin to recover from the perils of the pandemic, it would be interesting to witness how the tourism industry in the region will evolve. Will there be serious considerations to shift into a new agenda that focuses on regenerative tourism and seek a diversified approach? Or will PICs return to their comfort zones and continue with familiar tourism practices that prevailed in the pre-COVID-19 era? Responding to crises can be a challenge, as some crises are short-lived, whereas others, such as the COVID-19 pandemic, exist over a longer period than most expected. Proactive crisis management strategies will require high levels of coordination amongst governments, operators, and local communities of host nations. This poses a challenge, especially when there are many poorly socially networked tourism operators (Basili et al. 2014). There needs to be unity in terms of planning and decision-making, as well as ensuring that tourism recovery strategies and innovative solutions are feasible for all stakeholders.

These discussions on the current tourism crisis faced by PICs lead back to the question of whether the way forward for post-pandemic tourism is essentially 'the way back'. When Fiji launched the 'open for happiness' tourism campaign in December 2021, branding the country as a romantic destination, its celebrity-branded advertisement brought about criticisms that the campaign was centred around 'white saviourism' and an unequal socio-economic status between the tourist and the host (Girma 2021), especially when it implicated that affluent Caucasian people could revive the nation's tourism industry. Concerningly, Girma (2021) points out that tourism branding makes a significant impact in communication to a global audience, citing the need to attract tourists that are mindful with their consumption and respectful of local cultures. This example serves as a reminder that the shift toward regenerative tourism and community-driven tourism starts with communicating the right message to locals and tourists. Hence, tourism marketers in PICs play a vital role in understanding these tourism concepts to be able to restart tourism on the right track.

The lessons drawn from this chapter can also function as fundamentals for future research and development, particularly for small and marginalized tourism-dependent communities. As indicated, the pandemic significantly changed the tourism landscape, and this calls for the need to embrace new forms of tourism that are mindful and responsible. Nevertheless, there should be a thorough analysis of the different perspectives of tourism stakeholders including PIC governments, foreign aid donors, tourism authorities, local communities, hoteliers, tourism operators, small and micro-entrepreneurs, and tourists. The suggestions put forth in this chapter are based on academic inquiries, which can be further strengthened through an action-based research approach which engages public opinion. This type of research could be conducted to revamp existing frameworks and construct new ways of thinking and, indeed, new knowledge (Small 1995), which can be a powerful method to collaboratively develop tourism and work closely with Pacific Island communities on tourism transformation processes. Sustainable tourism is not a new concept. Although there have been efforts in PICs to protect and conserve the natural environment and empower marginalized communities and women in the workforce, this chapter posits that there is room for improvement, especially to achieve the long-term sustainability of the tourism industry.

References

Allen, M. G. (2012) 'Land, identity and conflict on Guadalcanal, Solomon Islands', *Australian Geographer*, 43(2): 163–180.

ADB (Asian Development Bank) (2019) *Fiji, 2019–2023 —Achieving Sustained, Inclusive, Private Sector-led Growth*. Available online at https://www.adb.org/sites/default/files/institutional-document/495256/cps-fij-2019-2023.pdf (accessed 20 August 2021).

ADB (Asian Development Bank) (2020a) *Basic Statistics, Asia and the Pacific*. Available online at https://data.adb.org/dataset/basic-statistics-asia-and-pacific (accessed 16 April 2021).

ADB (Asian Development Bank) (2020b) *Pacific Economic Monitor December 2020*. Available online at https://www.adb.org/publications/pacific-economic-monitor-december-2020 (accessed 7 May 2022).

Alisjahbana, A. S., H. Hahm and H. A. Malik (2020) *Leveraging Ocean Resources for Sustainable Development of Small Islands Developing States: Asia-Pacific Countries with Special Needs Development Report 2020*, Bangkok, Thailand: United Nations Publications, pp. 35–66.

Basili, A., W. Liguori and F. Palumbo (2014) 'NFC smart tourist card: Combining mobile and contactless technologies towards a smart tourist experience'. In: *23rd International WETICE Conference, IEEE*, pp. 249–254.

Benjamin, S., A. Dillette and D. H. Alderman (2020) '"We can't return to normal": Committing to tourism equity in the post-pandemic age', *Tourism Geographies*, 22(3): 476–483.

Bertram, G. (2018) 'Why does the Cook Islands still need overseas aid?', *The Journal of Pacific History*, 53(1): 44–63.

Blake, A. and M. T. Sinclair (2003) 'Tourism crisis management: US response to September 11', *Annals of Tourism Research*, 30(4): 813–832.

Bliss, C., L. Musikanski, R. Phillips and L. Davidson (2021) 'When will the pandemic end? Suggestions for US communities to manage well-being in the face of COVID-19', *International Journal of Community Well-Being*, 4(3): 299–313.

Branding in Asia (2021) 'Rebel Wilson's boat lands on the shores of Fiji in "open for happiness" tourism film', 3 December. Available online at https://www.brandinginasia.com/rebel-wilsons-boat-lands-on-the-shores-of-fiji-in-open-for-happiness-tourism-film/ (accessed 17 January 2021).

Cave, J. and D. Dredge (2020) 'Regenerative tourism needs diverse economic practices', *Tourism Geographies*, 22(3): 503–513.

Cheer, J. M., S. Pratt, D. Tolkach, A. Bailey, S. Taumoepeau and A. Movono (2018) 'Tourism in Pacific island countries: A status quo round-up', *Asia and the Pacific Policy Studies*, 5(3): 442–461.

Christ, C. (2017) '6 Ways to be a more sustainable traveller', *National Geographic*, 12 July. Available online at https://www.nationalgeographic.com/travel/article/sustainable-travel-tips#:~:text=The%20three%20pillars%20of%20sustainable,communities%20(ranging%20from%20upholding%20the (accessed 23 December 2021).

CNN Travel (2021) 'Fiji reopens to tourism', 2 December. Available online at https://edition.cnn.com/travel/article/fiji-reopening-tourism-intl-hnk/index.html (accessed 23 August 2021).

Connell, J. (2020) 'Blue ocean tourism in Asia and the Pacific: Trends and directions before the coronavirus crisis', *ADBI Working Paper Series*, Tokyo, Japan: Asian Development Bank Institute. Available online at https://www.adb.org/publications/blue-ocean-tourism-asia-pacific-trends-directions-before-coronavirus-crisis (accessed 15 August).

Connell, J. and T. Taulealo (2021) 'Island tourism and COVID-19 in Vanuatu and Samoa: An unfolding crisis', *Small States and Territories*, 4(1): 105–124.

Dean, M. R. U. (2020) 'COVID-19 and Fiji: A case study', *Oceania*, 90(S1): 96–106.

DFAT (Department of Foreign Affairs and Trade) (n.d.) 'Development assistance in the Republic of Marshall Islands', Australian Government. Available online at https://www.dfat.gov.au/geo/republic-of-marshall-islands/development-assistance/development-assistance-in-the-republic-of-the-marshall-islands (accessed 1 July 2021).

DFAT (Department of Foreign Affairs and Trade) (2014) *Investment Design: Solomon Islands Education Sector Program 2*, Australian Government. Available online at https://www.dfat.gov.au/sites/default/files/solomon-islands-education-sector-program-2-investment-design.pdf (accessed 8 August 2021).

DFAT (Department of Foreign Affairs and Trade) (2021) 'Australia's development partnership with Fiji', Australian Government. Available online at https://www.dfat.gov.au/geo/fiji/development-assistance/development-assistance-in-fiji (accessed 7 July 2021).

Diedrich, A. and S. Aswani (2016) 'Exploring the potential impacts of tourism development on social and ecological change in the Solomon Islands', *Ambio*, 45(7): 808–818.

Donald, R. (2022) 'Analysts point to logging and mining to explain Solomon Islands unrest', *Mongabay News*, 13 January. Available online at https://news.mongabay.com/2022/01/analysts-point-to-logging-and-mining-to-explain-solomon-islands-unrest/ (accessed 19 January 2022).

Dornan, M. and J. Pryke (2017) 'Foreign aid to the Pacific: Trends and developments in the twenty-first century', *Asia and the Pacific Policy Studies*, 4(3): 386–404.

Downes, S. (2021) 'The overseas destinations we'll have the best chances of visiting in 2022', *Stuff*, 30 December. Available online at https://www.stuff.co.nz/travel/kiwi-traveller/300487822/the-overseas-destinations-well-have-the-best-chances-of-visiting-in-2022 (accessed 1 January 2022).

Duxbury, N., F. E. Bakas, T. Vinagre de Castro and S. Silva (2021) 'Creative tourism development models towards sustainable and regenerative tourism', *Sustainability*, 13(1): 2. Online first DOI: 10.3390/su13010002.

Dziedzic, S. (2018) 'Which country gives the most aid to Pacific island nations? The answer might surprise you', *ABC News*, 9 August. Available online at https://www.abc.net.au/news/2018-08-09/aid-to-pacific-island-nations/10082702?nw=0 (accessed 5 May 2021).

Everett, H., D. Simpson and S. Wayne (2018) 'Tourism as a driver of growth in the Pacific: A pathway to growth and prosperity for Pacific Island Countries', *The Pacific Private Sector Development Initiative*, No. 2, June, Manila, Philippines: Asian Development Bank.

Ferrell, O. C., V. L. Crittenden, L. Ferrell and W. F. Crittenden (2013) 'Theoretical development in ethical marketing decision making', *AMS Review*, 3(2): 51–60.

Fleming, F. and M. Tabualevu (2018) 'UN women markets for change midterm review report'. Available online at https://www.dfat.gov.au/sites/default/files/markets-for-change-independent-mid-term-review-2018.pdf (accessed 15 May 2021).

Fletcher, J. and Y. Morakabati (2008) 'Tourism activity, terrorism and political instability within the Commonwealth: The cases of Fiji and Kenya', *International Journal of Tourism Research*, 10(6): 537–556.

Fly Solomons (2021) '"Let's keep our country moving" as national tourism effort Iumi Tugeda Holidays expands', 6 July. Available online at https://www.flysolomons.com/about-us/news/general/lets-keep-our-country-moving-iumi-tugeda-holidays-expansion-july-2021 (accessed 15 August 2021).

FAO (Food and Agriculture Organization of the United Nations) (2010) *National Fishery Sector Overview, Kiribati*. Available online at https://www.fao.org/fishery/docs/DOCUMENT/fcp/en/FI_CP_KI.pdf (accessed 21 May 2021).

Ford, A., A. Carr, N. Mildwaters, D. Fonotti and G. Jackmond (2019) 'Promoting cultural heritage for sustainable tourism development: Samoa', *Analysis and Policy Observatory*, New Zealand Institute for Pacific Research. Available online at https://apo.org.au/sites/default/files/resource-files/2019-06/apo-nid242511.pdf (accessed 20 April 2020).

Fugui, J. M. and M. Wate (1994) 'Solomon Islands in review: Issues and events', *The Contemporary Pacific*, 6(2): 457–463.

Gibson, D., E. Yai and S. Pratt (2021) 'Journeying into the past to discover the potential for WWII dark tourism in the Solomon Islands', *Current Issues in Tourism*, 25(14): 2285–2302.

Girma, L. L. (2021) 'Tourism Fiji's shameless new campaign is a reminder for marketers everywhere', *Skift*, 7 December. Available online at https://skift.com/2021/12/07/tourism-fijis-shameless-new-campaign-is-a-reminder-for-marketers-everywhere/ (accessed 2 January 2022).

Global Economy (n.d.) 'Samoa: International tourism revenue, percent of GDP'. Available online at https://www.theglobaleconomy.com/Samoa/international_tourism_revenue_to_GDP/ (accessed 16 April 2021).

Gorman, C. (2019) 'Rumor has it, China in Vanuatu', *The University of Texas at Austin,* October 31. Available online at http://sites.utexas.edu/climatesecurity/2019/10/31/rumor-has-it-china-in-vanuatu/ (accessed 9 April 2022).

Government of Samoa (2018) *Ministry of Commerce and Industry Labour: Sector Profiles.* Available online at https://www.mcil.gov.ws/storage/2018/07/Sector-Profiles-bookletweb.pdf (accessed 16 April 2021).

Government of Samoa (2020) 'Second voluntary national review on the implementation of sustainable development goals to ensure improved quality of life for all'. Available online at: https://spccfpstore1.blob.core.windows.net/digitallibrary-docs/files/88/88697faa2e4c8ef0f27f023532e4f25c.pdf?sv=2015-12-11&sr=b&sig=5KP6g5hcJclDG7ERd%2BNL%2FX229vTIVS8QZoCycVyp2EI%3D&se=2023-02-14T06%3A00%3A20Z&sp=r&rscc=public%2C%20max-age%3D864000%2C%20max-stale%3D86400&rsct=application%2Fpdf&rscd=inline%3B%20filename%3D%22Samoa_VNR_2020.pdf%22 (accessed 15 January 2022).

Government of the Republic of Fiji (2020) 'Fijian government unveils COVID-19 concessional loan packages for Fijian micro, small and medium enterprises'. Available online at https://www.fiji.gov.fj/getattachment/bba34afb-b4f6-4e6b-90f6-b0392ae4a18f/CONCESSIONAL-LOAN-PACKAGES-FOR-MSMEs.aspx (accessed 10 July 2021).

Guardian (2021) '"Job-killer of the century": Economies of Pacific islands face collapse over COVID-19', *The Guardian,* 4 April. Available online at https://www.theguardian.com/world/2021/apr/03/covid-coronavirus-deserted-islands-pacific-resorts-struggle-to-survive-a-year-without-tourists (accessed 16 April 2021).

Hajro, N., K. Hjartar, P. Jenkins and B. Vieira (2021) 'What's next for digital consumers', *McKinsey.* Available online at https://www.mckinsey.com/business-functions/mckinsey-digital/our-insights/whats-next-for-digital-consumers (accessed 23 December 2021).

Hassal, G. (2018) 'Special Issue on public sector enhancement in Pacific Island states', *Asia Pacific Journal of Public Administration,* 40(4): 207–211.

IFC (International Finance Corporation) (2016) 'Assessment of the economic impact of cruise tourism in Papua New Guinea and Solomon Islands', *World Bank Group.* Available online at https://www.ifc.org/wps/wcm/connect/region__ext_content/ifc_external_corporate_site/east+asia+and+the+pacific/resources/assessment+of+the+economic+impact+of+cruise+tourism+in+papua+new+guinea+and+solomon+islands (accessed: 11 July 2021).

IFC (International Finance Corporation) (2020) 'Fiji COVID-19 business survey: Tourism focus–Ministry of Commerce trade tourism and transport', *World Bank Group.* Available online at https://www.ifc.org/wps/wcm/connect/region__ext_content/ifc_external_corporate_site/east+asia+and+the+pacific/resources/fiji+covid-19+business+survey+-+tourism+focus (accessed 12 July 2021).

Kitching, C. (2014) 'From idyllic beaches to tanks and dog tags: How a local man turned WW2 battlefield into open-air museum in the Solomon Islands', *Daily Mail,* 20 October. Available online at https://www.dailymail.co.uk/travel/travel_news/article-2800059/solomon-islands-world-war-two-relics-draw-tourists.html (accessed 21 May 2021).

Klauss, L. (2011) 'Sustainable tourism and climate change in the Pacific Island region', GTZ Sector Project, *Secretariat of the Pacific Regional Environment Programme.* Available online at https://www.sprep.org/att/irc/ecopies/pacific_region/673.pdf (accessed 17 December 2021).

Koya, F. S. (2018) 'Positive development in Fiji's tourism industry', *The Sun,* 13 March. Available online at https://fijisun.com.fj/2018/03/13/positive-development-in-fijis-tourism-industry/ (accessed 19 December 2021).

Kotler, P. and K. L. Keller (2016) *Marketing Management,* Harlow, England: Pearson Education Limited.

KPMG (2020) *Fiji Budget 2020/2021,* KPMG International Limited, 17 July. Available online at https://assets.kpmg/content/dam/kpmg/us/pdf/2021/03/tnf-fiji-mar18-2021.pdf (accessed 14 January 2022).

Kundra, S., M. Alam and M. A. Alam (2021) 'How do political coups disrupt Fiji's tourism? Impact assessment on ecotourism at Koroyanitu National Heritage Park (KNHP), Abaca', *Heliyon,* 7(5): 20–25.

Kusapa, J. (2020) 'MPS silent over $250K repatriation fund', *Islands Sun,* 14 April. Available online at https://theislandsun.com.sb/mps-silent-over-250k-repatriation-fund/ (accessed 15 June 2021).

Lee, C. G. (2021) 'Tourism-led growth hypothesis: International tourism versus domestic tourism-evidence from China', *International Journal of Tourism Research,* 23(5): 881–890.

Lehmann, F. (n.d.) *Supporting Sustainable Tourism Development in Least Developed Countries amid the COVID-19 Recovery,* Geneva, Switzerland: Executive Secretariat for the Enhanced Integrated Framework (EIF) at the WTO (World Trade Organization). Available online at https://enhancedif.org/en/system/files/uploads/eif-policybrief-tourism_final_screen.pdf (accessed 23 April 2022).

Leni, Z. (2020) 'Tourism business feels the brunt of COVID-19', *Solomon Islands Chamber of Commerce and Industry*, 7 April. Available online at https://www.solomonchamber.com.sb/news-reports/posts/2020/april/global-pandemic-casts-shadow-over-local-tourism-industry/ (accessed 15 May 2021).

Mamu, M. (2021) 'Big loss', *Solomon Star*, 8 March. Available online at: https://www.solomonstarnews.com/big-loss/ (accessed 16 May 2021).

McDougall, D. (2005) 'The unintended consequences of clarification: Development, disputing, and the dynamics of community in Ranongga, Solomon Islands', *Ethnohistory*, 52(1): 81–109.

McEnhill, L., E. S. Jorgensen and S. Urlich (2020) 'Paying it forward and back: Regenerative tourism as part of place', Centre of Excellence for Sustainable Tourism Internal Report 2020/101, Lincoln University.

Monson, R. (2017) 'The politics of property: Gender, land and political authority in Solomon Islands'. In: S. McDonnell, G. A. Matthew and F. Colin (eds.) *Kastom, Property and Ideology: Lands Transformations in Melanesia*, Canberra, Australia: Australia National University Press, pp. 383–403.

Movono, A., S. Pratt and D. Harrison (2015) 'Adapting and reacting to tourism development: A tale of two villages on Fiji's Coral Coast'. In: D. Harrison and S. Pratt (eds.) *Tourism in Pacific Island Countries: Current Issues and Future Challenges*, Abingdon: Routledge, pp. 125–141.

Naidu, V., A. Matadradra and M. Sahib and J. Osborne (2013) *Fiji: The Challenges and Opportunities of Diversity*, London, UK: Minority Rights Group International.

Nanau, G. and M. Labu-Nanau (2021) *The Solomon Islands' Social Policy Response to Covid-19: Between Wantok and Economic Stimulus Package*, Bremen, Germany: University of Bremen.

New Zealand Foreign Affairs and Trade (2021) 'Support for Fiji's 2021 COVID-19 outbreak', 21 June. Available online at https://www.mfat.govt.nz/en/aid-and-development/humanitarian-action/support-for-fijis-2021-covid-19-outbreak/ (accessed 8 July 2021).

Nolan, J. (2020) 'Governor General highlights challenges faced by tourism sector', *Solomon Times*, 17 March. Available online at https://www.solomontimes.com/news/governor-general-highlights-challenges-faced-by-tourism-sector/9622 (accessed 2 June 2021).

Office of the Prime Minister (2021) 'Price Minister Josaia Voreqe Bainimarama's address at the 23rd Fiji Excellence in Tourism Awards', 19 March. Available online at http://www.pmoffice.gov.fj/prime-minister-josaia-voreqe-bainimaramas-address-at-the-23rd-fiji-excellence-in-tourism-awards-19032021/ (accessed 25 October 2021).

Parliament of Samoa (2019) *Samoa Tourism Authority 2018–2019 Annual Report*, Samoa Tourism Authority. Available online at https://www.palemene.ws/wp-content/uploads/Samoa-Tourism-09.07.2020-Final-English-AR-2018-2019-Vaine.pdf (accessed 20 April 2021).

Pearlman, J. (2021) 'Billionaires vacation in South Pacific islands as region struggles over pandemic curbs', *The Straits Times*, 25 January. Available online at https://www.straitstimes.com/asia/billionaires-vacation-in-south-pacific-islands-as-region-struggles-over-pandemic-curbs (accessed 9 September 2021).

Piggot, R., N. Morgan and A. Pritchard (2004) 'New Zealand and the Lord of the Rings: Leveraging public and media relations'. In: N. Morgan, A. Pritchard and R. Pride (eds.) *Destination Branding: Creating the Unique Destination Proposition*, Massachusetts, US: Elsevier Butterworth-Heineman, pp. 207–225.

Pratt, S. (2013) 'Minimising food miles: Issues and outcomes in an ecotourism venture in Fiji', *Journal of Sustainable Tourism*, 21(8): 1148–1165.

Polonsky, M. J. (1994) 'An introduction to green marketing', *Electronic Green Journal*, 1(2).

PwC (PricewaterhouseCoopers) (2022) 'Fiji: Corporate – Other taxes'. Available online at https://taxsummaries.pwc.com/fiji/corporate/other-taxes (accessed 7 August 2021).

RNZ (Radio New Zealand) (2020a) 'Fijians urged to support local tourism', 9 July. Available online at https://www.rnz.co.nz/international/pacific-news/420821/fijians-urged-to-support-local-tourism (accessed 11 July 2021).

RNZ (Radio New Zealand) (2020b) 'Details of Solomon Islands economic package revealed by PM', 5 May. Available online at https://www.rnz.co.nz/international/pacific-news/415823/details-of-solomon-islands-economic-package-revealed-by-pm (accessed 12 September 2021).

RNZ (Radio New Zealand) (2021), 'Cook Islands contact tracing app to help travel with NZ', 19 April. Available online at https://www.rnz.co.nz/international/pacific-news/439133/cook-islands-contact-tracing-app-to-help-travel-with-nz (accessed 23 November 2021).

Samoa Observer (2020) 'Domestic tourism a fine idea; but it must be realised', 20 September. Available online at https://www.samoaobserver.ws/category/editorial/71129 (accessed 14 May 2021).

Scheyvens, R. (2005) 'The growth of beach fale accommodation in Samoa: Doing tourism the Samoan way', *Centre for Indigenous Governance and Development Working Paper Series, 3*, New Zealand: Massey University.

Scheyvens, R. (2006) 'Sun, sand, and beach fale: Benefiting from backpackers – The Samoan way', *Tourism Recreation Research*, 31(3): 75–86.

Scheyvens, R. (2007) 'Poor cousins no more: Valuing the development potential of domestic and diaspora tourism', *Progress in Development Studies*, 7(4): 307–325.

Scheyvens, R. and E. Hughes (2019) 'Can tourism help to "end poverty in all its forms everywhere"? The challenge of tourism addressing SDG1', *Journal of Sustainable Tourism*, 27(7): 1061–1079.

Siutaia, H. (2020) 'Le Barter trading platform glimpse of the past', *Samoan Observer*, 1 May. Available online at https://www.samoaobserver.ws/category/samoa/62224?utm_content=buffer9c695&utm_medium=social&utm_source=facebook.com&utm_campaign=buffer&fbclid=IwAR2eJgwoyuD_cgkOqJoNwl9H2SyMguZi8qZfp0AyNJWxf50tYcq6EmUnNbg (accessed 13 May 2021).

Small, S.A. (1995) 'Action-oriented research: Models and methods', *Journal of Marriage and the Family*, 57(4): 941–955.

Solomon Islands Government (2021) 'Solomon Islands launches tourism guides for investors and governments', 8 November. Available online at https://solomons.gov.sb/solomon-islands-lunches-tourism-guides-for-investors-and-governments/ (accessed 3 December 2021).

Solomon Times (2020) 'Travel Solomons and Solomons Air push for intra-regional travel bubble', 9 July. Available online at https://www.solomontimes.com/news/travel-solomons-and-solomon-airlines-push-for-intraregional-travel-bubble/10006 (accessed 7 June 2021).

STA (Samoa Tourism Authority) (2014) *Samoa Tourism Sector Plan*. Available online at: https://pafpnet.spc.int/attachments/article/684/Samoa-Tourism-Sector-Plan-2014-2019.pdf (accessed 18 April 2021).

STA (Samoa Tourism Authority) (2019) 'Total visitor arrivals by purpose/year'. Available online at https://www.samoatourism.org/articles/254/total-visitor-arrivals-by-countryyear (accessed 20 April 2021).

Tahana, J. (2020) 'A surge in poverty as tourism jobs in Samoa disappear', *Radio New Zealand*, 7 May. Available online at https://www.rnz.co.nz/international/pacific-news/416094/a-surge-in-poverty-as-tourism-jobs-in-samoa-disappear (accessed 18 April 2021).

Tourism Fiji (2013) 'Tourism Fiji unveils new brand around the world', *Cision PR Newswire*, 10 October. Available online at https://www.prnewswire.com/news-releases/tourism-fiji-unveils-new-brand-around-the-world-227209801.html (accessed 21 May 2021).

Tuimaisala, L. (2020) '"Love Our Locals" campaign successful', *The Fiji Sun*, 7 September. Available online at https://fijisun.com.fj/2020/09/07/love-our-locals-campaign-successful/ (accessed 27 May 2021).

Twining-Ward, L. and R. Butler (2002) 'Implementing STD on a small island: Development and use of sustainable tourism development indicators in Samoa', *Journal of Sustainable Tourism*, 10(5): 363–387.

Tyllianakis, E., G. Grilli, D. Gibson, S. Ferrini, H. Conejo-Watt and T. Luisetti (2019) 'Policy options to achieve culturally-aware and environmentally-sustainable tourism in Fiji', *Marine Pollution Bulletin*, 148: 107–115.

Uguz, S. C. and K. Ismet (2016) 'Informal employment in tourism'. In: C. Avcikurt, M. Dinu, N. Hacioglu, R. Efe, A. Soykan and N. Tetik (eds.) *Global Issues and Trends in Tourism*, Sofia, Bulgaria: St Kliment Ohridski University Press, pp. 518–526.

United Nations (2019) 'A quarter of Pacific islanders live below "Basic Needs Poverty Lines", top UN development forum hears', 10 July. Available online at https://www.un.org/development/desa/en/news/sustainable/hlpf-2019-pacific-islands-forum.html (accessed 3 July 2021).

United Nations in Samoa (2020) 'COVID-19: Socio-Economic Response Plan', *United Nations Sustainable Development Group*. Available online at https://unsdg.un.org/resources/covid-19-socio-economic-response-plan-samoa (accessed 15 August 2021).

United Nations Pacific (2020) *Socio-Economic Impact Assessment of COVID-19 in Fiji*, Suva, Fiji: United Nations Pacific.

United Nations in Cook Islands, Niue, Samoa, and Tokelau (2020) 'COVID-19 survey: Majority of Samoans lost income, struggle with debt', 13 August. Available online at https://samoa.un.org/en/87292-covid-19-survey-majority-samoans-lost-income-struggle-debt (accessed 14 August 2021).

United Nations Women (n.d.) *Asia and Solomon Islands: The Pacific*. Available online at https://asiapacific.unwomen.org/en/countries/fiji/co/solomon-islands (accessed 8 August 2021).

Watson, R. (2019) *Tourism development in Niue and the impact of New Zealand's aid*, Master of Arts Thesis, Dunedin, New Zealand: University of Otago.

Western, D. (1993) 'Defining ecotourism'. In: K. Lindberg and D. E. Hawkins (eds.) *Ecotourism: A Guide for Planners and Managers*, 7–11, Vermont, USA: Ecotourism Society, pp. 7–11.

World Bank (n.d.a) 'Net official development assistance received (current US$) – Pacific island small states'. Available online at https://data.worldbank.org/indicator/DT.ODA.ODAT.CD?locations=S2 (accessed 25 March 2022).

World Bank (n.d.b) 'International tourism, number of arrivals – Fiji'. Available online at https://data.worldbank.org/indicator/ST.INT.ARVL?locations=FJ (accessed 25 October 2021).

WTTC (World Travel and Tourism Council) (2021) *Annual Research: Key Highlights*. Available online at https://wttc.org/Research/Economic-Impact (accessed 13 July 2021).

Xinhua (2020a) 'Fiji announces stimulus package to help fight COVID-19', 17 July. Available online at http://www.xinhuanet.com/english/2020-07/17/c_139220750.htm (accessed 29 July 2021).

Xinhua (2020b) 'Fiji strives to restart shrinking tourism amid COVID-19 pandemic', 5 July. Available online at http://www.xinhuanet.com/english/2020-07/05/c_139189524.htm (accessed 11 July 2021).

7

MASS TOURISM IN SMALL PACIFIC ISLAND STATES

A Critical Review of Tourism Development in the Pre-COVID era

David Harrison

Introduction

Mass tourism undoubtedly poses challenges to some destinations, and sometimes its negative impacts have been considerable, as highlighted, for example, by local protests in Venice and Barcelona (Sharpley and Harrison 2018: 232–233). However, it has also brought widespread economic and other benefits. Whilst critics may legitimately argue for controls and restrictions, few advocate a return to a non-tourist past, especially in light of the socio-economic challenges that the tourism industry and destination host communities have been facing due to international border closures and the global pandemic lockdown. If anything, the apprehension of tourism's impact in the developing world is even more immediate – whether too much impact or too little. From the author's analysis of the available data (UNWTO: 2019a), of the 30 states where the contribution of tourism to gross domestic product (GDP) is greatest, no less than 28 are islands. This chapter focuses on tourism in 13 small Pacific Island states that are considered to be self-governing, analyzing their similarities and differences. In the context of global tourism, which registered more than 1,460 million tourists in 2019 (UNWTO 2021:2), the arrivals recorded for the Pacific Island states may well be insignificant. Nevertheless, even a few tourists may have a considerable impact, and, in this, as in other matters, Pacific Island states can be a barometer for the rest of the world, demonstrating miniature problems and solutions relating to global tourism. In the pages that follow, attention will be paid to those island countries which received the greatest number of tourists relative to their population (the Cook Islands, Palau and Niue) within the context of the pre-COVID-19 era, but other continua here and elsewhere relating to mass tourism will also be considered. This segment of the discussion will be followed by a focus on those destinations with a similar number of tourists to residents (Fiji and Samoa), leading to an inspection of island states which received far fewer tourists than their population (Vanuatu, Tonga and the Federated States of Micronesia (FSM)). The work then reflects on mini-states with significantly fewer tourists (Nauru, Kiribati, Tuvalu and the Marshall Islands), with final attention to the Solomon Islands, which is considered to be atypical to the rest of the islands.

 DOI: 10.4324/9780429019968-8

Tourism development in small Pacific Island states

Prior to the outbreak of the COVID-19 pandemic in early 2020, international tourism growth was fairly robust. The 13 island states attracted more than 1.5 million arrivals in 2018 (Table 7.1), modestly increasing to a peak of more than 1.6 million in 2019 (SPTO 2021). When tourists arrive in the Pacific Island states, they tend to stay in accommodation that, by and large, is small in scale and locally owned (Table 7.2), though foreign ownership is greater in such countries as Fiji, Palau and Vanuatu.

Destinations with historically many more tourists than residents: The Cook Islands, Niue and Palau

Prior to the pandemic the Cook Islands, Niue and Palau received considerably more tourists than their populations. The Cook Islands had, by far, the greatest ratio, with a tourism intensity rate of 961.7 (Table 7.3), and tourism contributed nearly 70% to the state's GDP (Syme-Buchanan 2019). Tourism facilities are mostly around the coastline of Rarotonga, the main island and, to a lesser extent, on the island of Aitutaki. At first sight, the Cook Islands, which is in free association with New Zealand, has been a stereotypical mass tourist economy, but, in five key respects, it has not conformed to this type of classification. First, tourism facilities are overwhelmingly owned by Cook Islanders (including some residents in New Zealand), and second, the units are generally small in scale, with an average of fewer than five rooms (Table 7.2), ranging from luxury properties owned or franchised to the (Cook Islands) Pacific Resort Hotel Group, to individual properties for rent to returning Cook Islanders. Third, arrivals have been relatively constant throughout the year (SPTO 2019), and fourth, tourist facilities are physically apart from the Cook Island centres of population. Consequently, the immediate impact of tourism on residents has been limited. Finally, there has been little social distance between tourists (which include Cook Islanders visiting friends and relatives) and local residents (many of whom visit New Zealand on a regular basis) (Berno 2003).

Table 7.1 Tourist arrivals for 2018 and characteristics of Pacific Island states

Country	*A* *2018 annual total arrivals*	*B* *Population*	*C* *Area (sq. miles)*	*D* *Tourist intensity rate = A/B x 100*	*Cruise ship visitors*
Cook Islands	168,760	17,548	91	961.7	–
FSM	19,207	105,544	270	18.2	–
Fiji	870,309	889,953	7,095	97.8	187,890
Kiribati	6,824	117,606	313	5.8	232
Marshall Islands	6,761	58,791	70	11.5	1,908
Nauru	3,000 (est.)	10,756	8	27.9	–
Niue	9,500 (est.)	1,615	104	588.2	2,331
Palau	106,273	18,008	196	590.1	788
Samoa	167,651	197,097	1,093	85.1	4,350
Solomon Islands	27,866	669,823	10,639	4.2	4,984
Tonga	54,046	104,494	289	51.7	23,260
Tuvalu	2,729	11,646	10	23.4	–
Vanuatu	115,634	299,882	4,707	38.6	234,567

Source: Adapted from SPTO (2019: appendices 1 and 4), UNWTO (2019b) and Worldometers (2019).

Table 7.2 Accommodation units, number of rooms and average number of rooms per accommodation unit for the latest years for Pacific Island states

Country	*Year – latest update*	*Number of accommodation units*	*Number of rooms*	*Average number of rooms per accommodation unit*
Cook Islands	2019	805	3,300	4.10
FSM	2015	29	–	–
Fiji	2019	423	12,888	30.47
Kiribati	2021*	37	351	9.49
Marshall Islands	2016	12	281	23.42
Nauru	2015	8	–	–
Niue	2019	39	197	5.05
Palau	2018	118	2,409	20.42
Samoa	2021*	119	2,268	19.06
Solomon Islands	2019	181	1,991	11.00
Tonga	2021*	356	1,300	3.65
Tuvalu	2016	10	–	–
Vanuatu	2018	867	1,722	1.99
Total Number		3004	26,707	8.89

Source: Adapted from SPTO (2021: 14).
Notes: *provisional data

Table 7.3 The Cook Islands, Palau and Niue: tourist numbers, intensity rates and select indicators for 2018

Country	*Tourists*	*Tourist intensity rate (2018)*	*Tourism employment as % of all employed (2018)*	*Tourism receipts as % of GDP (2018)*
Cook Islands	168, 760	961.7	34.4	86.99
Palau	106, 273	590.1	21.0	21.0
Niue	9,500 (est)	588.2	32.3	–

Source: Adapted from SPTO (2019: 13–14).

Tourism and remittances from islanders living in New Zealand (numbering more than 60,000) have brought considerable prosperity to the Cook Islands, which could be classified a developed economy – a prospect raising mixed emotions (Roy 2017). So 'developed' are they that hoteliers have had to obtain labour from elsewhere, notably Fiji and the Philippines. Tourism might be blamed for the loss of a slower way of life or for sexualizing 'traditional' songs and dances when performed for tourists, but any perceived subservience (which is rare) is countered by Cook Island humour, where the butt of the joke is frequently the white female tourist (Alexeyeff 2008). The rapid growth of tourism in the Cook Islands has had a negative effect on water supply and sewage (Burns and Cleverdon 1995; Hajkowicz and Okotai 2005: 14; Taylor 2001: 80). Aid from New Zealand and China largely remedied the situation (Zhang 2015), but there were concerns that further increases to tourism in Rarotonga could become unsustainable (Government of the Cook Islands 2018: 25–26; Syme-Buchanan 2017) (see Chapter 25). Therefore, from an environmental point of view, the worldwide pandemic provided a useful respite to the environmental pressures of tourism, consumption and the airline industry (Harvey 2020).

Tourism in Palau differed noticeably from that of the Cook Islands. At 590.1, its tourist intensity rate was lower, as it covers a wider area, even though tourists were more than six times the population (Table 7.3). The land mass of Palau is more than twice the size of the Cook Islands, with a population numbering around 18,000 (Table 7.1), and is largely concentrated on eight islands. In addition, accommodation units average more than 20 rooms (Table 7.2) and are more likely to be owned by overseas investors, especially from Japan. However, because of the need to change planes to reach Palau, the cost of holidays (and the character of Palau's tourists) has been different from neighbouring Guam. Therefore, in Palau tourists are traditionally wealthier with more free time, while tourism to Guam has tended to involve lower costs (Carlile 2000: 429). At various times, Palau has been a popular diving destination for tourists from Japan, Taiwan, South Korea, the United States (US), and Europe. In 2009, as a consequence of concern over tourism's pressure on the environment, the government announced the creation of the world's first shark sanctuary and took the lead in the regional initiative on climate change (the Micronesia Challenge) (United Nations 2019).

The emergence of Chinese tourism gave such efforts a new impetus. In 2013, arrivals from China were fewer than a thousand, but in 2015, with Chinese government support (Cheng 2016; Chung 2018), they rose to more than 91,000 (Chung 2018). It was a mixed blessing, as Chinese tourists were accused by locals of being loud, creating litter, damaging coral, and generally showing disrespect to the environment (HKEJ 2015) – so much so that the government reduced flights to the island by half. It also moved to minimize tourism's negative impacts by introducing the Palau pledge to protect the environment, to be signed on entry by tourists to the country (see Chapter 18), later to be followed by an accreditation programme for tourism businesses (Hamdi 2019). The influence of the pledge is debatable because, in pressurizing Palau to switch its allegiance from Taiwan, the Chinese government withdrew its support for tourism. Consequently, by the end of 2017, Chinese visitors declined by almost 60,000 tourists (Beldi 2018), leading to empty hotels, the abandonment of many projects and a crisis in the economy. An influx of *mass* tourists was averted, but the need for tourism continued, preferably such 'niche' tourists as divers, 'soft' ecotourists and others interested, for example, in archaeological attractions (Fitzpatrick and Kanai 2001: 42). Nonetheless, prior to the global pandemic, there were elements of mass tourism found in Palau: high level of foreign ownership and control, high ratio of tourists to the local population and the tourist sector serviced largely by immigrants, especially Filipinos; though their impact may be reduced by the Palauan tendency to absorb newcomers into the traditional family structure (Nero et al. 2000: 34). In 2017, the industry contributed more than 80% of the country's exports (World Bank 2019) and provided nearly 20% of employment in 2018 (SPTO 2019). However, low-cost mass tourism led to the overconsumption of reef fish and damage to reefs (Morrison 2012: 235; Wabnitz et al. 2018), and social pressures arose from the (alleged) behaviour of Chinese tourists.

Niue is one of the smallest territories in the South Pacific, with few natural resources and a population of only 1,600, most of whom practise subsistence agriculture. However, Niue is also heavily reliant on employment in the government sector and depends on overseas aid from New Zealand, with which Niue enjoys free association (and of which all Niueans have citizenship). Tourism has been linked to future prosperity. In 2018, Niue received just under 10,000 visitors, representing six times its population (Table 7.1). These tourists were primarily divers exploring the numerous inlets and caves, tourists aiming to watch dolphins and whales and some of the 20,000 islanders who are resident in New Zealand visiting friends and relatives. Up to 40% of all arrivals could be categorized as repeat visits (Watson 2018:107). Accommodation on the island reflects differences among tourists: the 55-room Scenic Matavai Resort, built with money from the New Zealand Aid Programme, caters to the wealthy, whereas most of the remaining accommodation units are budget-priced guest houses owned and/or operated by Niueans, sometimes with their New Zealand partners. Despite the small numbers, tourism has been important to Niue, contributing to 32% of employment in 2018 (SPTO 2019: 14). Arguably, tourism will not make the island self-sufficient, though the economic impact of the pandemic would be frustrating some Niueans who have been affected by the rapid decline in numbers, irrespective of the resilience of people from the Pacific Islands to be able adapt to many crisis scenarios (see Khor et al. 2016). Yet sustainable and regenerative tourism options may have potential (see Chapter 24).

Despite the number of tourists relative to the population, mass tourism did not fully reach Niue. The Scenic Mativai resort is the only international chain, thus, locals largely exercise control over the industry, and many tourists are in the 'visiting friends and relatives' (VFR) category. Anecdotal evidence suggests that most tourists are seen as friends, at least potentially, and are often involved in island life (Watson 2018: 74–79). Furthermore, because most visitors have come from New Zealand, which many Niueans visited regularly, there has been little social distance between them. Realistically, islanders recognize that without tourism the island has no future. Nonetheless, in the Cook Islands and Niue there have been elements of mass tourism. Tourist numbers relative to the local population were undoubtedly high, but tourism's impacts were reduced by a high percentage of VFR tourists, relatively little social distance between tourist and host, and a high level of local ownership of tourist facilities. In Palau, by contrast, there has been more overseas investment and concern over lower class tourists who damage the environment, though the *decline* in Chinese visitors and the loss of tourists during the pandemic lockdown have made this problem less potent, at least for the present.

Destinations with much the same numbers of tourists as residents: Fiji and Samoa

Fiji and Samoa received approximately the same numbers of tourists as their respective populations, where, in 2018, tourists contributed 49% and 60%, respectively, of all exports (World Bank 2019) but differed in marked respects from each other. With its strong economy and population of almost 900,000 people, Fiji is the gateway to the South Pacific. Except for brief periods of domestic unrest, the trend over the last two decades has been upward. In 2018, there were 870,309 tourists (SPTO 2019) rising to 968,926 in 2019 (excluding cruise ship tourists) (SPTO 2021: 9) in contrast to 389,000 in 1998 (WTO 2001). Tourists are predominantly families coming to Fiji from Australia or New Zealand for leisure purposes (SPTO 2019). It was estimated that Fiji received around US$963 million in international tourism receipts in 2019, which had increased significantly from 2010, which stood at around US$634 million (UNWTO 2021:19).

Because of Fiji's relatively large land area, the tourist intensity rate in 2018 was 97.8 (Table 7.4). However, most tourism has been centred on the west (Coral) coast of Viti Levu and the small

Table 7.4 Fiji and Samoa: tourist numbers, intensity rates and select indicators

Country	*Tourists (2018)*	*Tourist intensity rate*	*Tourism employment as a % of all employed*	*Tourism receipts as % of GDP (2018 estimates)*
Fiji	870,309	97.8	35.5 (2018)	18.41
Samoa	167,651	85.1	12.5 (2015)	30.42

Source: Adapted from SPTO (2019: 13–14).

islands of the Mamanucas and Yasawas to the northwest and is normally constant throughout much of the year. Transnational companies (TNCs) have operated more than half the rooms in the premium and high categories (Harrison and Pratt 2015: 10). In *some* respect, Fiji was a reasonable example of mass tourism. It has had a distinctly expatriate orientation and has been highly capitalistic, and (for the region) received many tourists, mostly visiting for leisure purposes (Harrison and Sharpley 2018). The outer islands feature a few boutique resorts and TNCs with expatriate senior management dominating most of the middle and high-range sector. By contrast, lower-range hotels, catering to budget international or local tourists, are likely to be owned and operated by Fijians of Indian descent. The exception is small-scale, indigenous, outer island budget 'ecotourist' enterprises owned and operated by ethnic Fijians, whose primary motivation is to meet such requirements as school expenses and church obligations (Gibson 2012).

It is worth reiterating that in the pre-pandemic period, the ratio of tourists to the total population was low in Fiji. However, Pratt et al. (2016) warn that tourism can disrupt social patterns, especially as some traditional aspects of Fijian culture have reportedly declined, for example, yam cultivation, and that more (and less healthy) processed food is purchased, increasing the risk of diabetes (see Chapter 8). Some villagers, especially elders, also claim that women and young men working in hotels defer less to (male) traditional authority. Less controversially, there is concern that young villagers leave school early to take up unskilled jobs in tourism rather than continuing their studies (Movono et al. 2015: 107–112).

In some respects, Samoa is similar to Niue but considerably larger and more populated, with almost 200,000 people, receiving almost 167,651 tourists in 2018 (SPTO 2019: appendix 1; see Table 7.1) with a modest increase to 173,920 in 2019 (SPTO 2021). Estimated data suggests that international tourist receipts reached US$199 million in 2019 compared to US$132 million in 2010 (UNWTO 2021). Like Fiji, arrivals have been predominantly from New Zealand and Australia. As with Niue, considerably more Samoans live in New Zealand than Samoa, and just less than half of all tourists were returning Samoans – i.e., the VFR market (SPTO 2019: appendix 3). As they tend to visit throughout the year, seasonality is reduced, while social distance is minimized and more local food is consumed. However, returning Samoans can also be problematic as their relatives in Samoa often incurred extra expenditure, and may be weary of hearing how good life in New Zealand can be. Indeed, to avoid family tensions many VFR tourists opt for budget accommodation (Taufatofua and Craig-Smith 2010: 95–96). By contrast, more affluent overseas visitors have selected segregated beachfront accommodation (Samoa Tourism Authority 2014: 11), and the more adventurous have patronized locally run beach *fales* (Scheyvens 2006). Unlike Fiji, Samoa's tourism is largely locally owned and under local control, and more than three-quarters of accommodation falls under the 'standard' or 'budget' category. Two notable foreign investors are Fiji-owned Tanoa Tusitala (Raicola 2009) and the Taumeasina Island Resort, owned by a partnership based in Papua New Guinea (Harrison and Prasad 2013: 749; Loop Pacific 2018), though the government would like to see more, especially in the 'deluxe' sector.

Destinations with far less tourists than residents: Vanuatu, Tonga and the Federated States of Micronesia

Vanuatu, Tonga and the FSM all received far fewer tourists than their populations (Table 7.5). However, pre-pandemic tourism contributed considerably to their exports, 71% for Vanuatu, 24% for the FSM and 48% for Tonga (World Bank 2019). While some 20%–25% of the population of Vanuatu has been formally engaged with the cash economy (Trau 2012: 154), the remainder being involved in agriculture, which has met the needs of most of the population

Table 7.5 Vanuatu, Tonga and the FSM: tourist numbers, intensity rates and select indicators

Country	*Tourists (2018)*	*Tourist intensity rate*	*Employment in tourism as % of all employed*	*Tourism receipts as a % of GDP (2018 estimates)*
Vanuatu	115,634	38.6	34.5 (2018)	22.1
Tonga	54,046	51.7	21.2 (2018)	25.4
FSM	19,207	18.2	5.7 (2014)	15.8

Source: Adapted from SPTO (2019: 13–14).

(Scheyvens and Russell 2013: 6). Tourism, though, is important, and in 2018 there were 115,634 visitors predominantly from Australia, New Zealand and other Pacific Islands (SPTO 2019: appendix 3). There were also more than 230,000 cruise ship arrivals in 2018 (SPTO 2019: appendix 4), whose economic contribution to Pacific Island countries has long been problematic (Scheyvens and Russell 2013: 35–43). Travel and tourism contributed an estimated 22.1% to GDP and 34.5% of all employment in 2018 (Table 7.5). With aid from the World Bank and New Zealand, the country had hoped to quadruple arrivals over the next decade (RNZ 2019). Up to 2019, Pacific Island states were largely optimistic and somewhat ambitious to extend tourism growth into the 2020s and beyond.

Tourism in Vanuatu has been heavily concentrated in Efate, especially in and around the capital city of Port Vila, an area that hosts 80% of tourists who arrive by air. Travel by air to other islands is expensive (Cassidy and Brown 2010); however, the islands of Tanna and Santo have attracted tourists for some years (Scheyvens and Russell 2013:27). As in Fiji and Palau, overseas-owned hotels and resorts have dominated the top and middle of the market, while ni-Vanuatu have little representation in the tourism sector, tending to occupy only the lowest positions in hospitality. They have suffered much from the lack of clarity over custom land ownership (Haccius 2011), and land leases in Efate, in particular, have been inadequately regulated (Stefanova 2008: 2). However, they may run guest houses that cater to local and some regional visitors (Scheyvens and Russell 2013: 30) or rent out bungalows (see Chapter 12). Finally, there is some evidence that when ni-Vanuatu *are* entrepreneurs, their aim may be to enhance their status within the community rather than to make a profit (de Burlo 2003: 74–76). Consequently, there has been a clear gulf between ni-Vanuatu and expatriate owners, mirrored in the social and cultural divide between tourists and local residents who tended to interact only in the marketplace. The socio-cultural impacts of tourism in Vanuatu have been debated at length, and questions have emerged: have such traditions as land diving in Pentecost been commercialized by tourism (Cheer et al. 2013; de Burlo 1996)? Have *kastom* villages on Tanna remained 'authentic' (Robinson and Connell 2008)? The issue of 'authenticity' has also been relevant in other Pacific Islands, but of more importance is the need for ni-Vanuatu to be incorporated into hotel management and the tourism sector generally. This should arguably be a post-pandemic objective. However, the gulf between tourist and host has been unacceptably wide. Resort-based 'mass' tourism is likely to 'remain fundamental to the industry's survival' (World Bank 2008 quoted in Scheyvens and Russell 2013: 35), but local involvement must be part of the solution. Ni-Vanuatu must see tourism as theirs rather than an expatriate incursion.

The Kingdom of Tonga has a population of only 104,494 (Table 7.1), and the economy is based on agriculture, fishing, and remittances from Tongans resident overseas, with generous contributions of overseas aid, especially from New Zealand. Its tourism industry, though small even by Pacific Island standards, is, nevertheless, an important source of foreign currency and local employment. In 2018, it was estimated that Tonga's tourism receipts contributed 25.4%

of GDP and tourism employment represented 21.2% of total employment, with tourist arrivals standing at 54,046 and a tourist intensity rate of 51.7 (Table 7.5). Arrivals from New Zealand were the greatest at 45%, followed by Australia (21%) and the US (14%) (SPTO 2018: 37–38). Furthermore, from earlier data available, about half of all visitors represent the VFR category, and only around 37% of tourists were non-Tongan holidaymakers (Ministry of Tourism 2009). According to the International Finance Corporation, Tonga has suffered from several key constraints: a failure by government to integrate policies, limited institutional capacity, regulatory barriers to investment, inadequate planning and promotion, poor management of the visitor experience, insufficient human resource capacity and poor destination marketing (Bartlett 2010). Others apprehensively note the increasing involvement of China in Tonga's economy (Perry 2019). The only international chain in Tonga is the Tanoa International Dateline Hotel, part of Fiji's Reddy group, which caters largely to business tourists. Most other hotels, significantly located on Nuku'alofa, where the capital city of Tongatapu is located, and are in the middle or lower rung of accommodation units, while small boutique resorts, often owned and run by expatriates, can be found on the outer islands. For holidaymakers, the major attraction has undoubtedly been whale watching (Orams 2013) (see Chapter 23). Such visitors are far from the typical mass tourist but, nevertheless, their expenditure has to be balanced against the need to import much of their daily requirements.

Composed of Chuuk, Kosrae, Pohnpei and Yap, with a capital in Pohnpei, the FSM is undeveloped, with a per capita income of around US$3,586 in 2019, with a slight decrease to $3,565 in 2020 (Countryeconomy.com 2020). The FSM has a population of just over 100,00 and a land mass of only 270 square miles (Table 7.1). The FSM is in a Compact of Free Association with the US, signed in 1982, and has relied heavily of subsidies from the US the 1980s (see Friberg et al. 2006). Over the past four decades, tourism has remained remarkably static, probably the result of distance from the main Asian markets and the high air fares (Hezel 2017: 117). In 1978, the number of visitors was 23,111 (Hezel 2017: 12) and by 2018 the number was only 19,207 (SPTO 2019: appendix 1). Most were from the US or Japan, usually staying in small, locally owned hotels, most of which are in Pohnpei, followed by Chuuk, Yap and Kosrae. Yap Pacific Dive Resort, owned by a Chinese company, is one of the few foreign-owned hotels in the region. Despite this lack of growth, in 2018, tourism contributed 24% to total exports with tourism employment (SPTO 2019; World Bank 2019). The future is unlikely to see much change, though if the US were to decide to reduce its commitments, China is waiting to extend its influence in the region.

Mini-states with few tourists: Nauru, Kiribati, Tuvalu and the Marshall Islands

The four island states in this category have received few tourists and pre-pandemic tourism has been relatively insignificant in their economies (Table 7.6). Nauru has a population of a little under 11,000 (Table 7.1). From the late 1970s through to the 1980s, Nauru became extremely wealthy through phosphate mining until phosphate became depleted. Gross over-spending and financial incompetence frittered away the fortunes obtained through mining, and Nauru turned to money laundering. It also became a tax haven, but nothing was able to prevent bankruptcy. In effect, Nauru became a failed state (Connell 2006). During the 2000s, Nauru profited by playing China and Taiwan against each other, and it also received financial assistance from Russia and Australia. Notoriously, in the early 2000s and again since 2012, in return for aid, it agreed to be an offshore immigration detention centre for Australia. There is a clear irony here: a tropical island in the Pacific acting as a prison for would-be immigrants unwelcome in Australia

Table 7.6 Nauru, Kiribati, Tuvalu and the Marshall Islands: tourist numbers, intensity rates and select indicators

Country	*Tourists (2018)*	*Tourist intensity rate*	*Tourism employment as % of all employed*	*Tourism receipts as % of GDP (2018 estimates)*
Nauru	3,000	23.4	2.5 (2014)	–
Kiribati	6,824	5.8	15.5 (2017)	7.9
Tuvalu	2,729	23.4	2.3 (2016)	19.6
Marshall Islands	6,761	11.5	5.5 (2015)	6.8

Source: Adapted from SPTO (2019: 13–14).

(Sanggaran and Zion 2016). Tourists and asylum-seekers, hotels and detention centres, hoteliers and wardens: the parallels are both uncomfortable and striking. Unsurprisingly, tourism is insignificant in Nauru (see Chapter 20), a situation likely to continue.

Independent since 1979, Kiribati is composed of 33 islands across 5 million square kilometres and in three groups: the Gilbert group (the most populous), the Phoenix group and the Line Islands. It produces mainly copra and fish and, as a stereotypical MIRAB economy, relies on migration, overseas remittances (including from seamen working on international ships), aid and bureaucracy (Borovnik 2006; Tisdell 2014). However, perhaps the island group's main (albeit unwelcome) claim to fame (along with Tuvalu) is its vulnerability to climate change. Already parts of Tarawa have succumbed to saltwater invasion of wells. As a consequence, relocation of the population is a necessity rather than an option (Bedford and Bedford 2010), and the island's president has already purchased land in Fiji for future resettlement (Hermann and Kempf 2017). Arrivals have vacillated considerably, but in 2018 the country received 6,824 overnight visitors (SPTO 2019: appendix 1), of whom less than half (2,044) visited for leisure purposes. A further 2,191 were business visitors and 152 were classified as visiting friends and relatives. In addition, 232 cruise ship visitors went to the island in 2018, a decline from 4,478 in 2014 (SPTO 2019: appendix 3 and 4). Accommodation in Kiribati is basic and, unsurprisingly, none of the major hotel chains are found in the state. Arrivals have been consistent throughout the year, and past statistics indicate that leisure visitors went primarily on fishing packages to Kiritimati atoll in the northern Line Islands, whereas business visitors went to Tarawa, the capital, in the Gilbert group (Government of Kiribati 2013), a division likely to continue today. Other specialist interests include battlefield tours of Tarawa (scene of one of the bloodiest battles of the Second World War), bird watching and surfing around Kiritimati atoll. Compared to Nauru, Kiribati has more to offer tourists, but tourism development is (again) constrained by distance to and within the island state, poor infrastructure, limited accommodation and poor communications (see Chapter 19). Over the long term, successful tourism in Kiribati is unlikely, though imminent disaster has highlighted 'a last chance to see' approach and the attraction of numerous scientists to the area.

Tuvalu has a population of a little more than 11,000 and a land area of 10 square miles (Table 7.1). Like Kiribati, it is threatened by climate change. Agriculture and fishing are important to the economy, and most families are involved in subsistence agriculture. As with Kiribati, other important sources of income are remittances from sailors or Tuvaluans living overseas, but overseas aid, especially from New Zealand, is crucial and contributes nearly 90% to the economy (Wallace 2018). In 2018, there were 2,729 visitors, primarily from Australia, New Zealand, Japan and Fiji, but only 740 were recorded as visiting for leisure purposes, mostly staying on the island area of Funafuti. Half were on business, reportedly to discuss climate change and conservation and attend conferences. Despite contributing around 20% of the country's GDP (SPTO 2019: 13–14) and with a rather insignificant level of tourism employment as a percentage the employed workforce (Table 7.6), tourism in Tuvalu has been relatively undeveloped (see Chapter 26). There is only one hotel, the government owned Vaiaku Lagi, and several locally owned guest houses and motels (Commonwealth Network 2019). Like other island states, there have been problems of accessibility, a lack of infrastructure and inadequacy of human resources. Furthermore, constraints of distance and a lack of an internal market have mitigated against reduced air fares. Without many more tourists, overseas investment is highly unlikely.

The Marshall Islands was taken over by the US in 1944 with a Compact of Free Association signed in 1986 (and amended in 2004), which gave the US sole responsibility for defence and islanders the right to emigrate and work in the US. It also involved hundreds of millions of dollars as recompense for the nuclear tests (U.S. Department of State 2019). Agriculture and fishing are now the main products of the state, but income from the US is crucial to the economy. Tourism in the Marshall Islands has remained fairly static in the decade preceding the pandemic, with arrivals in 2010 standing somewhere around 5,000 (UNWTO 2021: 19), increasing in 2018 to a mere 6,761, with only a minority of holidaymakers (mainly from the US and other Pacific Islands). Tourism, then, accounted for 5.5% of employment and 6.8% of GDP (SPTO 2019) (Table 7.6). Despite this poor record, the Marshall Islands has considerable potential: the country is stable, the weather is good, the beaches are excellent and the marine environment is outstanding, with numerous dramatic wrecks from World War II attracting divers (Pemberton and Kapak 2003: 38). There is also a potential for cultural and heritage tourism (Collison and Spears 2010: 136–137) (see Chapter 17). However, as with other islands, geographical remoteness, limited air access, untrained human resources, and high travel costs prevent further development (Collinson and Spears 2010: 138). Until (or if?) air access becomes cheaper, progress is unlikely to occur.

Island state with few tourists and notable challenges: Solomon Islands

The Solomon Islands is atypical of other islands discussed in this chapter, covering an area of 10.6 thousand square miles (Table 7.1). During World War II it was the scene of the battle of Guadalcanal, a major confrontation between Japanese and allied forces (August 1942–January 1943). Self-government was obtained in 1976 and independence in 1978, but considerable ethnic diversity led to much unrest, and from 1998 until 2003 the fledgling state was virtually ungovernable. In July 2003, responding to a government request for intervention, and nominally under the Pacific Forum, the first police and troops of the Regional Assistance Mission to Solomon Islands (RAMSI) arrived, and the multinational (but largely Australian) military presence was to remain in the Solomon Islands until 2017 (though civilians in RAMSI continued in the country) (Fraenkel 2017). Unsurprisingly, few tourists went to the Solomon Islands in periods of unrest.

However, arrivals increased from 11,482 in 2006 to 27,865 in 2018, and tourism contributed some 10% to the economy. Of the visitors, 40% were Australian and 20% other Pacific Islanders, followed by New Zealand, the US and China. Importantly, 25% (7,093) were on business and those visiting friends and relatives amounted to a further 4,163 (15%). Only 8,955 (32%) were leisure visitors and another 7,655 (27%) fell into the 'other' category (SPTO 2019: appendix 3). Overall, the country had a very low tourism intensity rate (Table 7.7), and tourism impacts were more likely to occur in those few places where tourists gather, such as in the capital city of Honiara. The Solomon Islands clearly has 'attractions'. There are excellent opportunities for diving, especially World War II sites, and many distinctive cultures can be experienced. In addi-

Table 7.7 The Solomon Islands: tourist numbers, intensity rate and select indicators

Tourists 2018	*Tourist intensity rate*	*Employment in tourism as % of total employed (2017)*	*Tourism receipts as % of GDP (2017)*
27,866	4.2	11.0	7.59

Source: Adapted from SPTO (2019: 13–14).

tion, East Rennell, the southernmost island in the state and the largest uplifted coral atoll in the world, became a World Heritage site in 1998 (see Chapter 10). However, outside the main urban centres, there is no regular electricity, roads are poor, water is unsafe, and public transport is inadequate. Poverty, low levels of education standards amongst the young population and crime in urban areas all prevail in the Solomon Islands (Evans 2017). Until these problems are addressed, it is unlikely that Solomons will quickly grow after the pandemic has lifted, though its tourism potential is significant given its many attractions.

Conclusion and research implications

This chapter has focused on 13 independent island states of the Pacific and has attempted to analyze their similarities and differences within the context of the pre-pandemic era. The emphasis, in particular, was on the volume of international tourists relative to the size of the population, aiming to assess how differently island states exhibited characteristics of mass tourism. Detailed discussions of tourism's impact in Pacific Island countries can be found elsewhere (Harrison and Pratt 2015; Everett et al. 2018), but it can be concluded from this survey that, while fully fledged mass tourism did not exist in any of the island states surveyed, *some* characteristics were notably evident in a few states.

The Cook Islands, Niue and Palau, in particular, all received high numbers of tourists relative to their population, and tourism was the main contributor to their GDP. However, in the Cook Islands and Niue, tourism's impacts were lessened by extensive local ownership of accommodation, its small scale, and the lack of social distance between 'host' and 'guest' who have long been exposed to each other's cultures, much of which they have in common. In Palau, by contrast, there has been more foreign ownership of hotels and a greater reliance on immigrant labour, as well as a recent history of reliance on (and awareness of) mass tourism – although the government has become more sensitive to the potential of 'high quality and low quantity' tourism market (see Chapter 18). Samoa and Fiji received approximately the same numbers of tourists as their population, but, whereas transnational hotels and non-indigenous owners have been major players in Fiji tourism, they are almost absent from Samoa, where hotels are largely locally owned. In both societies, too, the physical separation of hotels from villages has served to mitigate tourism's impacts, and Samoan tourism has been characterized by a strong VFR content and the presence of locally owned beach *fales* for more adventurous tourists.

Although Vanuatu, Tonga and the FSM all received few tourists relative to their population size, tourism has been economically important, especially in Vanuatu, because of limited employment alternatives. Tourism there is traditionally dominated by expatriates and heavily concentrated around Port Vila, and ni-Vanuatu have been largely excluded from participation except as unskilled service workers. By contrast, tourism in Tonga has been largely locally owned and run (except for boutique resorts) but remains undeveloped: tourists normally visit primarily for whale watching, and a high percentage are Tongans resident overseas visit friends and relatives. Tourism in the FSM has similarly been undeveloped. In Nauru, Kiribati, Tuvalu and the Marshall Islands, tourism has also been of minor significance. Most visitors to the 'doomed island' of Nauru (Kendall 2009) were on business, though some engaged in game fishing, scuba diving, or visiting a World War II site. More literally, Kiribati is also doomed, and its geographical isolation, poor communications, inadequate infrastructure and accommodation and lack of human resources make its future bleak indeed. Tuvalu has the same problems: equally threatened by climate change and visited primarily by business visitors, it has attractive topographical features but (like several other small island states surveyed) will continue to rely on aid from New Zealand. Similarly, despite tourist attractions in the Marshall Islands, tourism there, too, stagnated. Geographical isolation, limited air access and high travel costs prevent further development and ensure its reliance on continued financial support from the US. Finally, the Solomon Islands is a special case. It has a varied topography, distinctive cultures, and offers considerable opportunities for diving, including sunken ships from World War II. However, a decade of civil unrest from the early 1990s, poor infrastructure and roads outside the urban centres, unsafe water and inadequate health services and crimes against the person in urban areas have prevented successful tourism development.

Clearly, for many Pacific Island communities, tourism can be a path to prosperity, but its benefits are often unequally distributed. Future policies should, first, aim to reduce leakages from these island economies and develop closer links with the agricultural sector (see Chapter 8). This suggestion is not new, and efforts are frequently made to engage local farmers. However, piecemeal publicity is no substitute for thorough-going agricultural reform. Second, most societies reviewed have traditions of land tenure that differ from Western systems, and tourism activities have been affected (and investments curtailed), for example, in the Solomon Islands (Sofield 2003), Vanuatu (Stefanova: 2008) and Fiji (Harrison 1997: 173–176). In Fiji, government policies seem to have resolved the situation, but potential conflict continues to exist, both there and, especially, in the Solomon Islands and Vanuatu. Such land issues can be resolved, but only if Indigenous people believe tourism development is good for them as well as for expatriate investors. More generally, whether in states where tourism is little developed, as in the FSM or the Solomon Islands, or where it is already extensive, as in the Cook Islands, Palau or Samoa, tourism must be seen to benefit the many rather than the few. Third, as the review shows, aid is crucial to the survival of most states surveyed. Without support from New Zealand, in particular, most island states discussed would struggle to exist. New Zealand and the US (along with other aid agencies) may consider their assistance as (at least partial) recompense for past wrongs. However, Western governments are also aware that the expansion of Chinese tourism reflects growing Chinese interest in the region and that these small states have a vote at the United Nations. Finally, although climate change has been little discussed in this review, its effects on small Pacific Islands are considerable (see Chapters 2, 3, 11, 17, 19, 21, 22, 23, 26 and 27). However, its causes are not universally accepted (Fair 2018) and, even if these countries succeed in reducing their own carbon footprint, such efforts are globally insignificant.

More importantly, all Pacific Island countries actively seek more tourists, who inevitably travel by air. Perhaps such island nations as Kiribati and Tuvalu are best seen as canaries in the global mine, warning us of future disaster? If we in the West heed the warning, what would that mean for tourism to these paradise islands? Consequently, this conclusion has raised concerns and questions that need to be systematically researched. The COVID-19 pandemic, however, offers an unexpected opportunity to critically review pre-pandemic tourism practices and activities, helping to place into context those concerns and challenges which have affected individual Pacific Island states. Whether any actions are taken by governments, organizations and communities to address some of the common problems and issues that have been noted in this chapter, one fact remains almost certain: tourism will resume, and the wheels of commerce and capitalism will start to roll.

Acknowledgements

With David Harrison's endorsement in January 2021, the Handbook's editor, Marcus Stephenson, has made some minor updates to this chapter, especially for any reference published in 2021.

References

Alexeyeff, K. (2008) 'Are you being served? Sex, humour and globalisation in the Cook Islands', *Anthropological Forum*, 18(3): 287–293.

Bartlett, J. (2010) *Tonga Tourism Diagnostic - Presentation of Key Findings for Stakeholder Verification*, 11–12 June, Nuku'alofa: Tonga National Tourism Forum.

Bedford, R. and C. Bedford (2010) 'International migration and climate change: A post-Copenhagen perspective on options for Kiribati and Tuvalu'. In: B. Burson (ed.) *Climate Change and Migration: South Pacific Perspectives*, Wellington, New Zealand: Institute of Policy Studies, pp. 5–134.

Beldi, L. (2018) 'China's "tourist ban" leaves Palau struggling to fill hotels and an airline in limbo', *ABC News*, 28 August. Available online at https://www.abc.net.au/news/2018-08-26/china-tourist-ban-leaves-palau-tourism-in-peril/10160020 (29 January 2021).

Berno, T. (2003) 'Local control and the sustainability of tourism in the Cook Islands'. In: D. Harrison (ed.) *Pacific Island Tourism*, New York: Cognizant Communication Corporation, pp. 94–109.

Borovnik, M. (2006) 'Working overseas: Seafarers' remittances and their distribution in Kiribati', *Asia Pacific Viewpoint*, 47(1): 151–161.

Burns, P. and R. Cleverdon (1995) 'Destination on the edge? The case of the Cook Islands'. In: M.V. Conlin and T. Baum (eds.) *Island Tourism: Management Principles and Practice*, Chichester: Wiley International, pp. 217–229.

Carlile, L. (2000) 'Niche or mass market? The regional context of tourism in Palau', *The Contemporary Pacific*, 12(2): 415–436.

Cassidy, F. and L. Brown (2010) 'Determinants of small Pacific Island tourism: A Vanuatu study', *Asia Pacific Journal of Tourism Research*, 15(2): 143–153.

Cheer, J., K. Reeves and J.H. Laing (2013) 'Tourism and traditional culture: Land diving in Vanuatu', *Annals of Tourism Research*, 43(4): 435–455.

Cheng, D. (2016) 'Countering Chinese inroads into Micronesia', *Issue Brief, The Heritage Foundation*, 27 October. Available online at http://thf-reports.s3.amazonaws.com/2016/IB4618.pdf (accessed 24 February 2019).

Chung, J. (2018) 'Paulauan airline stops flights to China', *Taipei Times*. Available online at https://www.taipeitimes.com/News/front/archives/2018/07/19/2003696936 (accessed 20 July 2020), 29 July.

Collison, F.M. and D. Spears (2010) 'Marketing cultural and heritage tourism: The Marshall Islands', *International Journal of Culture, Tourism and Hospitality Research*, 4(2): 130–142.

Commonwealth Network (2019) 'Accommodation in Tuvalu', *Nexus Commonwealth Network*. Available online at http://www.commonwealthofnations.org/sectors-tuvalu/travel/accommodation (accessed 9 September 2019).

Connell, J. (2006) 'Nauru: The first failed Pacific state?', *The Round Table*, 95(383): 47–63.

Countryeconomy.com (2020) 'Federated States of Micronesia GDP – Gross domestic product'. Available online at https://countryeconomy.com/gdp/micronesia (accessed 29 January 2021).

de Burlo, C. (1996) 'Cultural resistance and ethnic tourism on South Pentecost, Vanuatu'. In: R. Butler and T. Hinch (eds.) *Tourism and Indigenous Peoples*, Wallingford: CAB International, pp. 255–276.

de Burlo, C. (2003) 'Tourism, conservation and the cultural environment in rural Vanuatu'. In: D. Harrison (ed.) *Pacific Island Tourism*, New York: Cognizant Communication Corporation, pp. 69–81.

Evans, D. (2017) 'Hard work: Youth employment programming in Honiara, Solomon Islands', *SSGM Discussion Paper 2016/7*, Canberra, Australia: State, Society and Governance in Melanesia Program (SSGM), Australian National University. Available online at https://dpa.bellschool.anu.edu.au/sites/default/files/publications/attachments/2016-10/dp_2016_7_evans_online.pdf (accessed 20 July 2020).

Everett, H., D. Simpson and S. Wayne (2018) 'Tourism as a driver of growth in the Pacific: A pathway to growth and prosperity for Pacific Island countries', *Issues in Pacific Development*, No. 2 June, Manila, Philippines: Asian Development Bank.

Fair, H. (2018) 'Three stories of Noah: Navigating religious climate change narratives in the Pacific Island region', *Geo: Geography and Environment*, 5(2). Online first: doi.org/10.1002/geo2.68.

Fitzpatrick, S.M. and V. N.Kanai (2001) 'An applied approach to archaeology in Palau', *Cultural Resource Management*, 24(1): 41–43.

Fraenkel, J. (2017) 'Bringing the politics back in', *Inside Story*, 6th July. Available online at https://insidestory.org.au/bringing-the-politics-back-in/ (accessed 5 January 2020)

Friberg, E., K. Schaefer and L. Holen (2006) 'US economic assistance to two Micronesian nations: Aid impact, dependency and migration', *Asia Pacific Viewpoint*, 47(1): 123–133.

Gibson, D. (2012) 'The cultural challenges faced by indigenous-owned small medium tourism enterprises (SMTES) in Fiji', *The Journal of Pacific Studies*, 32(2): 103–126.

Godfrey, D. (2019) 'Vanuatu aims to quadruple visitor numbers in next decade', *RNZ* (Radio New Zealand), 11 March. Available online at https://www.rnz.co.nz/international/programmes/datelinepacific/audio/2018685832/vanuatu-aims-to-quadruple-visitor-numbers-in-next-decade (accessed 10 July 2020).

Government of the Cook Islands (2018) *Fiscal Strategy and Economic Update*, Extract from *Budget Book 1*: 2018/19, Rarotonga: Ministry of Finance and Economic Management, September.

Government of Kiribati (2013) *Kiribati Islands*, 9th–14th September, Novotel, Nadi, Fiji: *Pacific Regional Tourism Capacity Building Programme* (PRTCBP).

Haccius, J. (2011) 'The interaction of modern and custom land tenure systems in Vanuatu', *Discussion Paper 2011/1*, Australian National University, School of International, Political and Strategic Studies.

Hajkowicz, S. and P. Okotai (2005) *An Economic Evaluation of Watershed Pollution in Rarotonga, the Cook Islands*, Brisbane, Australia: CSIRO Sustainable Ecosystems.

Hamdi, R. (2019) 'Palau campaign stands tough on environment despite hit on tourism', *Skift*, 4 January. Available online at https://skift.com/2019/01/04/palau-campaign-stands-tough-on-environment-despite-hit-to-tourism/ (accessed 5 March 2019).

Harrison, D. (1997) 'Globalization and tourism: Some themes from Fiji'. In: M. Oppermann (ed.), *Pacific Rim Tourism*, Wallingford: CAB International, pp. 167–183.

Harrison, D. and B. Prasad (2013) 'The contribution of tourism to the development of Fiji and other Pacific Island countries'. In: C. Tisdell (ed.) *Handbook of Tourism Economics: Analysis, New Applications and Case Studies*, Singapore: World Scientific, pp. 741–761.

Harrison, D. and S. Pratt (2015) 'Tourism in Pacific Island countries: Current issues and future challenges'. In: S. Pratt and D. Harrison (eds.) *Tourism in Pacific Islands*, Abingdon: Routledge, pp. 3–21.

Harrison, D. and R. Sharpley (2018) 'Introduction: Mass tourism in a small world'. In: D. Harrison and R. Sharpley (eds.) *Mass Tourism in a Small World*, Wallingford: CAB International, pp. 1–14.

Harvey, D. (2020) 'Anti-capitalist politics in the time of COVID-19', *Jacobin Magazine*, 20 March. Retrieved from https://jacobinmag.com/2020/03/david-harvey-coronavirus-political-economy-disruptions (accessed 20 August 2020).

Hermann, E. and W. Kempf (2017) 'Climate change and the imagining of migration: Emerging discourses on Kiribati's land purchase in Fiji', *The Contemporary Pacific*, 29(2): 231–263.

Hezel, F.X. (2017) *On Your Mark, Get Set… (Tourism's Take-off in Micronesia)*, Honolulu, Hawai'i: East-West Centre.

HKEJ (Hong Kong Economic Journal) (2015) 'Palau to halve charter flights from China to stem influx', 17 March. Available online at https://www.ejinsight.com/eji/article/id/1008181/20150317-palau-to-halve-charter-flights-from-china-to-stem-influx (accessed 10 July 2020).

Kendall, D. (2009) 'Doomed island', *Alternatives Journal*, 35(1): 34–37.

Khor, H.E., R.P. Kronenberg and P. Tumbarello (2016) *Resilience and Growth in the Small States of the Pacific*, Washington District of Columbia: International Monetary Fund.

Loop Pacific (2018) 'PNG-owned Taumeasina Island Resort tops Samoa award', 12 January. Available online at https://www.loopsamoa.com/business/png-owned-taumeasina-island-resort-tops-samoa-award-72787 (accessed 5 March 2019).

Ministry of Tourism (2009) 'Overview of tourism in Tonga', *Presentation to Tonga Update*, Nuku'alofa, Tonga: University of the South Pacific.

Morrison, C. (2012) 'Impacts of tourism on threatened species in the Pacific region: A review', *Pacific Conservation Biology*, 18(4): 227–239.

Movono, A., S. Pratt and D. Harrison (2015) 'Adapting and reacting to tourism development: A tale of two villages on Fiji's coral Coast'. In: S. Pratt and D. Harrison (eds.) *Tourism in Pacific Islands: Current Issues and Future Challenges*, Abingdon: Routledge, pp. 101–117.

Nero, K.I., F.B. Murray and M.L. Burton (2000) 'The meanings of work in contemporary Palau: Policy implications of globalization in the Pacific', *The Contemporary Pacific*, 12(2): 319–348.

Orams, M. (2013) 'Economic activity derived from whale-based tourism in Vava'u, Tonga', *Coastal Management*, 41(6): 481–500.

Pemberton, C. and G. Kapak (2003) *Tourism Investment Guide in the Pacific Islands*, Sydney, Australia: The Pacific Islands Trade and Investment Commission. Available online at http://www.pitic.org.au/pdfdocs/publications/publi-13914-200601.pdf (accessed 8 January 2007)

Perry, N. (2019) 'China comes to Tonga', *The Diplomat*, 10 July. Available online at https://thediplomat.com/2019/07/china-comes-to-tonga/ (accessed 5 January 2020).

Pratt, S., S. McCabe and A. Movono (2016) 'Gross happiness of a "tourism" village in Fiji', *Journal of Destination Marketing and Management*, 5(1): 26–35.

Raicola, V. (2009) 'Uproar over Fiji hotelier's deal', *Sunday Times* (Fiji), 10 May.

Robinson, P. and J. Connell (2008) '"Everything is truthful here": custom village tourism in Tanna Vanuatu'. In: J. Connell and B. Rugendyke (eds.) *Tourism at the Grassroots: Villagers and Visitors in the Asia-Pacific*, Abingdon, Oxon: Routledge, pp. 77–97.

Roy, E.A. (2017) 'Cook Islands faces its "worst case scenario", being granted developed country status', *The Guardian*, 8 October.

Samoa Tourism Authority (2014) *Samoa Tourism Sector Plan 2014–2019*, Apia: STA.

Sanggaran, J-P. and D. Zion (2016) 'Is Australia engaged in torturing asylum seekers? A cautionary tale for Europe', *Journal of Medical Ethics*, 42(7): 420–423.

Scheyvens, R. (2006) 'Sun, sand, and beach fale: Benefiting from backpackers - The Samoan Way', *Tourism Recreation Research*, 31(3): 75–86.

Scheyvens, R. and M. Russell (2013) *Sharing the Riches of Tourism in Vanuatu*, School of People, Environment and Planning, Massey University.

Sharpley, R. and D. Harrison (2018) 'Conclusion: Mass tourism in the future'. In: R. Sharpley and D. Harrison (eds.) *Mass Tourism in a Small World*, London: Routledge, pp. 232–240.

Sofield, T.H.B. (2003) *Empowerment for Sustainable Tourism Development*, Oxford (UK): Pergamon.

SPTO (South Pacific Tourism Organisation) (2018) *Annual Review of Visitor Arrivals in Pacific Island Countries, 2017*, Suva, Fiji: SPTO.

SPTO (South Pacific Tourism Organisation) (2019) *Annual Visitor Arrivals Report: 2018*, Suva, Fiji: SPTO.

SPTO (Pacific Tourism Organisation) (2021) *Annual Visitor Arrivals Snapshot: 2020*, Suva, Fiji: SPTO.

Stefanova, M. (2008) 'The price of tourism: Land alienation in Vanuatu', *Justice for the Poor*, 2(1): 1–4.

Syme-Buchanan, F. (2019) 'Call for Cook Islands to diversify its economy', *Radio New Zealand*, 21st January. Available online at https://www.rnz.co.nz/international/pacific-news/380530/call-for-cook-islands-to-diversify-its-economy (accessed 5 March 2019).

Taufatofua, R.G. and S. Craig-Smith (2010) 'The socio-cultural impacts of visiting friends and relatives on hosts: A Samoan study', *WIT Transactions in Ecology and the Environment*, 130: 89–100.

Taylor, J.E. (2001) 'Tourism to the Cook Islands: Retrospective and prospective', *Cornell Hotel and Restaurant Administration Quarterly*, 42(2): 70–81.

Tisdell, C. (2014) 'The MIRAB model of small island economies in the Pacific and their security issues: *Revised version', Working Paper No. 58*, March, University of Queensland, Social Economics, Policy and Development.

Trau, A.M. (2012) 'Beyond pro-poor tourism: (re)interpreting tourism-based approaches to poverty alleviation in Vanuatu', *Tourism Planning and Development*, 9(2): 149–164.

United Nations (2019) 'The Micronesia challenge'. Available online at https://sustainabledevelopment.un.org/partnership/?p=2502%20 (accessed 29 September 2020).

UNWTO (United Nations World Tourism Organization) (2019a) *Compendium of Tourism Statistics*: 2013–2017, Madrid, Spain: UNWTO.

UNWTO (United Nations World Tourism Organization) (2019b) *International Tourism Highlights, 2019 edition*, Madrid, Spain: UNWTO.

UNWTO (United Nations World Tourism Organization) (2021) *International Tourism Highlights, 2020 edition*, Madrid, Spain: UNWTO.

U.S. Department of State (2019) 'U.S. relations with Marshall Islands', 9 December. Available online at https://www.state.gov/u-s-relations-with-marshall-islands (accessed 16 September 2020).

Wabnitz, C.C.C., A.M. Cisneros-Montemayor, Q. Hanich and Y. Ota (2018) 'Ecotourism, climate change and reef fish consumption in Palau: Benefits, trade-offs and adaption strategies', *Marine Policy*, 88: 323–332.

Wallace, C. (2018) 'Success of humanitarian aid to Tuvalu slowly brings sustainability', *Borgen Project*, 4 February. Available online at https://borgenproject.org/humanitarian-aid-to-Tuvalu (accessed 31 March 2019).

Watson, R. (2018) 'Tourism development in Niue and the impact of New Zealand's aid', *MA Thesis*, New Zealand: Department of Geography, University of Otago.

World Bank (2019) 'World development indicators'. Available online at https://data.worldbank.org/data-catalog/world-development-indicators (accessed 13 March 2019).

WTO (World Tourism Organization) (2001) *Compendium of Tourism Statistics*, Madrid, Spain: WTO.

Worldometers (2019) 'Countries in the world by population'. Available online at https://www.worldometers.info/world-population/population-by-country (accessed 19 September, 2019).

Zhang, D. (2015) 'China – New Zealand – Cook Islands triangular aid project on water supply'. *SSGM Discussion Paper 2016/7*, Canberra, Australia: State, Society and Governance in Melanesia Program (SSGM), Australian National University. Available online at https://bellschool.anu.edu.au/sites/default/files/publications/attachments/2016-10/dp_2016_7_evans_online.pdf (accessed 12 May).

8

DECIPHERING TOURISM'S AUSPICIOUS AND INAUSPICIOUS RELATIONSHIPS WITH FOOD AND AGRICULTURE IN PACIFIC ISLAND STATES

Identifying problems and solutions

Daniel Ka Leong Chong and Marcus L. Stephenson

Introduction

Although not all tourists wish to be adventurous in terms of consuming local food products (Sengel et al. 2015), local fare is arguably part of the authentic tourist desire and sustainable tourism experience (Sims 2009). The ability of a country or region to produce local food (and beverage) for tourist consumption can help strengthen a destination's tourism image and identity (Lin et al. 2011). Furthermore, according to the United Nations World Tourism Organization (UNWTO 2015: n.p.): 'Tourism can spur agricultural productivity by promoting the production, use and sale of local produce in tourist destinations and its full integration in the tourism value chain'. The production and consumption of local produce is crucial to local communities and small island states which are economically dependent on tourism development, especially if strategic alliances between the agriculture and tourism sectors are fully established (Telfer 2008). This chapter initially provides an overview of the challenges concerning the production of local food for tourism consumption within the context of small Pacific Island states. The work will then present two case illustrations concerning Fiji and the Solomon Islands, especially to help validate and extend our understanding of the prevailing challenges and also to explore specific ways forward for each country to proactively create stronger intersections between agriculture, food and tourism. Finally, the chapter highlights the impacts of the COVID-19 pandemic on food production and how communities and businesses adapted to the change. The work provides recommendations for Pacific Island states to advance ways to ensure that inter-sectoral linkages between tourism and food/agriculture are, indeed, auspicious.

DOI: 10.4324/9780429019968-9

Challenges faced by Pacific Island states in the production of local food for tourist consumption

There are prevailing and often inter-connected challenges which ensure that the relationship been local food production and tourism consumption remains problematic in Pacific Island states. The notable challenges concern inconsistent food supply channels and logistical hindrances, globalization and the adverse transformation of local and authentic dietary habits, urbanization and rural migration, limited structural intersections between tourism and agriculture, food security and safety concerns, climate change and natural disasters (Barnett 2011; Cauchi et al. 2019). Cyclone Harold, for instance, struck Vanuatu in April 2020, damaging at least 60% of croplands and causing losses to livestock and stored food (FAO 2020). Honey production was also impacted as the cyclone destroyed the bee feeding sources of pollen producing flowers (TVET 2020). Climatic conditions have damaged honey production elsewhere in the Pacific Islands, for instance, flash flooding in Samoa in 2018 impacted the livelihoods of beekeepers, in some cases, halving the number of beehives (Samoa Observer 2018). Natural disasters, therefore, effect food security, which is a major concern in Pacific Island states. A preliminary assessment of the Hunga Tonga-Hunga Ha'apai volcano eruption and tsunami in Tonga in January 2022 indicated that 80% of crops in communities in 'Eua, Ha'apai and Tongatapu were affected (Food Safety News 2022). Climate change is also an ongoing hinderance, as increasing sea levels threaten the yield of crop production and agriculture in the Marshall Islands (Ahlgren et al. 2014).

Some Pacific Island states struggle to grow adequate crops, as only the hardiest of plants survive (McIver et al. 2014). Limitations in the production of fresh produce naturally restrict the tourism (and hospitality and restaurant) industry from being able to consistently offer tourists local food options and choices. This is apparent in countries such as Kiribati, for which one Trip Advisor user advised other visitors not to expect five-star meals (Trip advisor user 2019). Nonetheless, local experiences of food consumption can impact tourist satisfaction (Björk and Kauppinen-Räisänen 2017), and some Pacific Island destinations are sensitive to this fact. Samoa, for instance, is described as a 'food lover's paradise' (Spacifica Travel 2021). One immediate problem in the growth of adequate crops concerns the lack of fresh water for growing vegetables. In Kiribati, freshwater for agriculture in the form of ponds are endangered by sea incursions destroying taro plant pits (Aretaake 2019). Nonetheless, the Pacific Islands are quite diverse, and some islands do have scope for agricultural advancement and integration with the tourism industry, especially in islands which have fertile soil. Pohnpei in the Federated States of Micronesia produces a diversity of indigenous crops, including 131 breadfruit varieties, 55 banana varieties and 24 giant swamp taro varieties (Corsi et al. 2008: 310). Importantly, there are economic and cultural reasons for ensuring that island governments continue with the agricultural advancement of indigenous crops. The task, however, is to ensure there are clear structural connections to the tourism and hospitality industries.

Another challenge impeding the ability of Pacific Islands to develop their food and agricultural sectors relates to changes in local diets and, hence, food demand, as islanders are becoming disconnected from their culinary roots. Internal migration from rural to urban areas, especially in search of employment in tourism, hospitality and the service sectors, has encouraged communities to move away from traditional farming and rural life (Magbalot-Fernandez and Umar 2019), thereby affecting agricultural production. In the Solomon Islands, 'rapid urbanization' and 'population drift' impact traditional food habits, and this can be seen on the island of Guadalcanal, the main tourist hub, where Western food consumption is popular (Bottcher et al. 2021: 2). The popularity of fast food in Nauru has led to the island being named as the 'fattest nation in the world' (Independent 2011). Despite the perceived socio-political status attached to being obese in some Pacific Islands, as personified by religious and political elites in Samoa, serious health concerns pertaining to high levels of type-2 diabetes mellitus and metabolic syndrome prevail (Lameka 2020). Diabetes in the Cook Islands impacts 26.8% of the country (Tu'akoi et al. 2020) and is a notable concern for the government, who have been involved in a diabetes health awareness campaign (Figure 8.1).

Declining health standards and the increase of non-communicable diseases are partly attributed to the consumption of inferior food imports, as the local culinary scene is dominated by such imported products as rice and flour and canned butter, fish and meat. In Pohnpei, there has been a prominent shift away from 'locally-grown vitamin-, mineral- and fibre-rich carbohydrate staples' (Corsi et al. 2008: 310). In the Marshall Islands, an estimated 90% of the food is imported (Ahlgren et al. 2014: 72) and an estimated 50% of fresh produce, including dairy, meat and seafood is imported by hotels and restaurants in Vanuatu (Connell and Taulealo 2021:108). Arguably, 'dietary colonialism' is partly responsible for the transformation of food habits, as major food-export countries such as Australia, New Zealand and the United States (US) have extensively exported such high-fat products as corn beef, chicken backs, mutton-flaps and turkey tails to the Pacific Islands (Hughes and Lawrence 2005: 299).

In some islands, such as Nauru, an over-focus on food imports can be partly associated with the failure of community gardening initiatives (McLennan 2017). Integration of agriculture with the tourist industry is popularly seen as valuable in encouraging agricultural development and poverty reduction, but in the context of Pacific Island states, local agricultural produce is often dealt with in low volumes and, thus, its supply to hotels and restaurants is not constant and reliable (Connell and Cruz 2019). Moreover, there is often a lack of tourist awareness of local food produce and cuisine. Singh et al.'s (2015) study on tourism and agriculture in Niue indicated that there is a missed opportunity to promote local food to tourists, especially as major tourist websites often lack detailed information on local food and how it can be accessed at show days and market days.

The following two main sections provide case exemplars of the Pacific Islands, notably Fiji and the Solomon Islands. The cases are concerned with critically addressing tourism's inter-con-

Figure 8.1 A billboard located outside of the hospital in Aitutaki (Cook Islands), representing a diabetes health awareness campaign. Source: M. Stephenson, May 2017.

nections, disconnections and potential connections with the food/agriculture sectors, drawing further on the challenges that these two islands face together with positively identifying ways forward to resolving ongoing concerns.

Fiji – case scenario

Food, beverages and the role of tourism

Fijian food has customarily been healthy (Movono et al. 2018), with fish being a significant component of the diet (Singh et al. 2016). Fiji's cuisine also embraces traditional Indian cooking, as the Indian Fijian community represents the second largest ethnic population in the country. Local cuisine has also been influenced by cooking methods and flavours from Southeast Asia and China (Dignan et al. 2004). Chinese food is also prevalent in local cuisine, as it is historically rooted to Chinese immigration to Fiji from the late 1870s when settlers came as cooks, market gardeners and small traders, impacting mainstream culinary habits and traditions (Ng Kumlin Ali 2002). However, in terms of traditional and indigenous food and food preparation in Fiji, the 'lovo' (a pit or earth oven) has maintained a central role. This involves an underground oven created by producing a hole in the ground and lining it with coconut husks, wood and stones. On top of the heated stones, meats and vegetables are wrapped in banana or palm leaves to cook (Figure 8.2). As this process captivates tourists, hotels and resorts organize and regularly market their 'lovo nights' (Latte 2020). Also, 'kokoda', which can be made of raw mahi-mahi fish infused with coconut cream, lime, onions and tomatoes (Singh et al. 2016), is an iconic dish in Fiji, and there have been attempts to market this culinary delight to tourists (Lifespice 2020) (Figure 8.3).

The most celebrated local beverage in Fiji is 'kava', known locally as 'yaqona' (or 'grog'), which is the root of a shrubby pepper plant whose crushed roots produce an entheogenic drink with sedative, anaesthetic and euphoriant properties. From the 1980s manufacturers helped to

Figure 8.2 Lovo preparation in Suva (Fiji). Source: Kosita Butratana, November 2016.

Figure 8.3 Dish of kokoda. Source: bonchan (www.shutterstock.com/image-photo/kokoda-coconut-milk-ceviche-fijian-cuisine-1798875904).

commodify the plant material for multiple purposes, including dietary supplements and natural medicine (Pollock 2009). Kava is an important authentic product which has become part of the 'tourist gaze' (Urry 1990) and a 'must do' activity for tourists. Popular travel sites, hotels, village tour operators and travel companies promote the kava drinking ritual as an essential component of the tourism experience, linked to Indigenous culture and society (Adams 2020). This ceremonial ritual, known as Sevusevu, significantly symbolizes the high value placed on 'sociality' and 'hospitality' in Fijian society (Brison 2001:309). Sevusevu involves individuals sharing the same kava bowl as the kava is passed around a circle of people. This ritual is central to Fijian social life and performed on a range of occasions from daily life events to such events as births, marriages, funerals and welcoming guests (Figure 8.4), though kava consumption was originally restricted to chiefs, priests and high-ranking men (Shaver 2015), whereas now kava is internationalized and made available to anyone who wishes to purchase the product. Fiji gained approximately $FJ30.7 million (US$14.59 million) from kava export earnings in 2018, with the US being the largest customer, receiving 148,000 kg, New Zealand receiving 80,000 kg and Hawaii receiving 13,000 kg (Tora 2020). Fiji Kava Limited, also known as Taki Mai, is a major player in kava distribution and also commercializes kava shots mixed with such local flavours as banana, coconut, chocolate banana, guava and pineapple (Singh 2014).

Although niche kava production has the potential to stimulate local economies, foreign companies have seized opportunities to benefit from this market. Fiji Kava Limited, for instance, is the largest operating kava production company domicile in Australia. Other reputable beverage companies that have developed from localized products and sources in Fiji also have foreign ownership: FIJI Water (bottled mineral water company owned by an American company); and Fiji Bitter, Fiji Gold, Vonu Pure Lager, Fiji Premium and Bounty Rum (produced by Paradise Beverages Fiji which is owned by Coca-Cola Amatil) (Drinks Insight Network 2017). The sustainability of these business eco-systems is not particularly conducive to a localized and community-driven approach to the development and expansion of natural and local products. This situation is not isolated to Fiji alone. Samoan Vailima beer, for instance, though brewed locally is also owned by Coco-Cola Amatil (Mex 2015).

Figure 8.4 A kava bowl ('tanoa fai'ava') at Fiji Marriott Resort Momi Bay, Fiji. Source: Kosita Butratana, November 2019.

However, despite such products being foreign-owned, their localized association to a specific place strengthens the destination image as food (and beverage) consumption is perceived as cultural resources (Zain et al. 2018). There are natural benefits to a country to market and develop local products, as exemplified in the case of FIJI Water, which is a natural artesian water founded in 1996 under the name Natural Waters of Viti Limited, though now owned by Justin Vineyards and Winery of Paso Robles, California (Inside FMCG 2017). FIJI Water is the premium water of choice for many luxury hotels. FIJI Water has marketed its water as 'untouched' and where 'every drop is green', and 'unspoiled by the contaminated air of the 21st century' (Jones et al. 2017: 113). The bottles narrate how the water is naturally formed with minerals contributing to its soft taste, making a relatively simple commodity 'linked to an "exotic" location and sold to elite consumers as a form of cultural capital' (Connell 2006: 342). Crucially, FIJI Water helps the country market itself as tourism destination, through which it has 'strategically sold the image of Fiji as an environmental paradise' (Connell 2006: 345). In a subtle yet sophisticated way, the product personifies the naturalness of Fiji to tourists and potential tourists and is a product that could be viewed as a primary destination brand component. FIJI Water capitalizes on Fiji's 'exotic' nature to differentiate its product in a competitive global market. Yet Jones et al. (2017: 119) are critical of the company, emphasizing that it utilizes 'tourism stereotypes of Fiji' and plays on the company's 'clichéd tropical theme', despite the real location of the water plant being far from the tropical rainforests but located in the Fijian highlands. Moreover, Jones et al. (2017) implicates the company for over utilizing fossil fuels for bottling, packaging and transportation, questioning the extent to which the local community fully benefits. Although the company's ownership structure does raise serious issues concerning the wider benefits to Fijian society and the economy, from a tourism perspective Fiji Water arguably contributes to the destination personality of Fiji and enables image building.

Challenges of integrating local food production into the hotel and tourism industry

The reliance on international food imports has an impact on the economic (and cultural) fabric of Fijian society. Around $FJ74.4 million (US$36.4 million) was spent on the procurement of fresh produce in 2017, and 52% of this amount was spent on imported items (Turaga 2018). There is, thus, potential to create significant demand for local produce, boosting farm incomes for those who deliver to restaurants and resorts. Nevertheless, the type and standard of the hotel as well as the location often has a bearing on the extent to which local produce in Fiji can or cannot be utilized. Hotel purchasing of local products is significant in areas where farming is popular, such as in Lautoka, Nadi, the Mamanuca Islands and the Yasawa Islands, whereas in Denarau, local purchasing is not as significant given that it is a more separate resort enclave (IFC 2018). The structure of multi-national based resorts in terms of food requirements means that they are not always conducive to partnering with local food producers and farmers (Scheyvens and Laeis 2021). As these writers indicate, the need for large-scale hotels to pursue corporate policies on food safety can limit the utilization of local produce, irrespective of any socio-economic or environmental benefits. Accordingly, Denarau is home to high quality hotels where food safety requirements are stricter and where food preferences are more particular, and, thus, there is more demand for items such as imported mushrooms, lettuce varieties, coloured capsicums, broccoli and celery (IFC 2018: 3).

There are five key barriers to increasing local production for Fiji's hotel sector: (1) lack of networking between key decision-makers in hotels and local producers/suppliers; (2) inconsistent supply (particularly fruits, vegetables, seafood and dairy products); (3) seasonality of available local produce (particularly fruits and vegetables); (4) poor quality of products (particularly meat, seafood and dairy products); and (5) lack of food safety standards for meat and seafood (IFC 2018: 8–9). Nonetheless, local chefs have arguably been socialized into the production of European cuisine to appease tourists, a traditional trend that has been popular since the establishment of tourism in Fiji from the 1950s, which also implies a lack of appreciation of Fijian food (Laeis et al. 2019). Berno (2011) notes that local products are mainly utilized in the hotel industry to produce dishes for staff meals and for specific tourist events (e.g., 'lovo nights'), also utilizing fruits and vegetables for breakfasts and for accompanying Western-based meals. She further emphasizes the challenges faced in utilizing local produce, such as poor infrastructure and limited access to technology and capital for farmers, incompatibility of some local dishes with the taste sensibilities of tourists, over-fishing of local stocks and poor perception of Fijian foods amongst tourists. The United Nations Food and Agriculture Organization estimates that there are approximately 30,000 subsistence fishers in Fiji, where over-fishing reached a crisis situation when communities had to fish to survive during the first year of the COVID-19 pandemic (Chanel and Singh 2021).

Innovation and human resources

There are opportunities for hotels and restaurants to develop Fijian menus and for encouraging tourists to have a real insight into the local influences of food within the country. Several resorts have marketed the use of fresh, locally grown organic produce, helping to create a socially responsible perception that the hospitality product prioritizes local sectoral linkages. Musket Cove Island Resort and Marina in the island of Malolo Lailai has a long-established organic farm supplying its restaurants with various fruits and vegetables (Fiji Sun 2010). Although the inter-linking of local cuisine and tourism for the benefit of both tourists and rural communities appears to be a logical step forward, many developing countries find the process difficult to implement (Rogerson 2012), especially given the constant threat of such dominant forces as the globalization and Westernization of food and culinary habits, which obviously influence food demand and consumption patterns.

While progress has been made and emphasis has been placed on improving the capacity building of skilled workers, the tourism industry continues to struggle with securing workers who have the necessary industry practices and standards (Cheer et al. 2018). Furthermore, in the past five decades, small developing countries have been among the most disadvantaged countries in terms of technological innovation (Shah and Biden 2021). Fiji's over-reliance on food imports has the potential to undermine the motivation of individuals and communities to take seriously the need for innovation and entrepreneurial ideas in the food and beverage sectors, whether in terms of technological or non-technological skills. Fiji should, thus, continue to expand its hospitality and culinary training through a range of educational platforms, including providing business training opportunities to engage in small- to medium-scale food entrepreneurship. There may well be scope for the government to formally advance their support to food micro-businesses targeted at the tourism and hospitality sector where this informal food sector should be valued, as it can provide an important 'safety net' for members of poor communities with the potential to become entrepreneurial in their ventures (see Chong and Stephenson 2020), leveraging localized food and agricultural resources. There are opportunities for such businesses to trade directly with the hospitality and restaurant sectors as well as tourist consumers themselves, and this could include agritourism initiatives, as explained in the next case illustration of the Solomon Islands. These initiatives could help to stimulate direct sectoral linkages between the food/agricultural and tourism sectors, thereby creating strong auspicious relationships.

The Solomon Islands – case scenario

Local food and beverage products

Although the tourism industry in the Solomon Islands is far from the level of institutional maturity that Fiji has experienced (see Chapters 10 and 11), with only around 28,900 tourist arrivals at the end of 2019 (World Bank 2021), tourists do have access to international restaurants in Gizo (the capital of the Western Province) and Honiara (capital city of Solomon Islands located on the island of Guadalcanal) (iExplore n.d.). Nonetheless, like all Pacific Island states, each country has its own unique dishes, and, in the case of the Solomon Islands, the country's dish is 'poi' (fermented taro roots). This unofficial national dish is highlighted on the destination guide of the website of Solomon Airlines (Solomon Airlines n.d.). In terms of local beverages, the country hosts Solomon Breweries Limited, established in 1993 as the first brewery in the Solomon Islands selling SB (Special Brew) and SolBrew Larger beer, pre-mixed spirits and soft drinks, which eventually came under the management of the Dutch multi-national company, Heineken, in 2014 (Solomon Times 2014). Like Fiji, the country has been capable of incubating local products and developing local resources, but there is often a point at which external/overseas interests comes to the forefront of business advancement.

Although the Solomon Islands does not have as many popularized local food products and culinary dishes commonly associated with the tourism industry as compared to Fiji, coffee production is emerging as a notable agricultural and industrial development. Thus, a large number of coffee plants have been established in the country since the 1980s, and an increasing amount of Arabica has been grown in the highlands of Guadalcanal (Bunaa 2018). There are various established coffee associations operating in the country, for instance, the farms of the Labuhila Coffee Farmers Association (LCFA), founded in 1978, are a family-owned business; King Solomon Coffee involves more than 1,000 coffee farmers from 40 villages; and Varivao Holdings Limited, which began buying and selling Solomon Gold Coffee from 1993 and has since grown to become one of the Solomon Islands' leading coffee buyers and processors (Bunna 2018).

Some of the islands are popularly associated with specific food items, notably the island of Makira, which produces many varieties of banana. Prior to the pandemic, the island held the annual banana festival which emphasized the cultural dynamics of banana farming (Jayawardena et al. 2020). Surprisingly, the island has not significantly diversified the use of bananas, such as utilizing unripe bananas and producing banana flour to bake bread (Atkin 2021). Such forms of product diversification would help to strengthen the sectoral links to tourism, given the novelty and scope of particular products.

Agritourism initiatives and food tours

Solomon Islands' Ministry of Agriculture and Livestock indicated an interest in inter-linking the agriculture and tourism sectors through the establishment of an agritourism policy (Steinmetz 2018), involving tourists interacting with primary production or staying in local farm establishments (i.e., farm stays) for the purpose of consuming (and learning about) local produce (Roberts and Hall 2001). Hence, agritourism can create local business opportunities and promote local products and services. Malkanthi and Routry (2011) note that farmers can use agritourism and farm stay tourism as a strategy to contribute to the agricultural development of their farms and utilize such forms of tourism to promote sustainable rural development.

There are productive opportunities for food tours to be part of the country's tourism scene. One inbound tour operator, for instance, Travel Solomons, launched a 'chocolate tour' in Honiara which illustrates to tourists the process of chocolate production and consumption through visiting a cocoa plantation, processing factory and chocolate shop (SPTO n.d.). Given that there is a growing export market in chocolate, the internationalization of cocoa products will help to profile the country to potential tourists. Other food tours that already have been established as part of the tourism experience are fishing tours and tours of food markets (Fishing Booker 2021). Nonetheless, there has been little activity in terms of developing tours of coconut plantations, despite coconuts representing one of country's largest export products and having socio-cultural value for Solomon Islanders. These tours could also incorporate tours of villages where coconut oil is produced, looking at the key stages of the production process as well as the role of women facilitating this process. Hence, there are substantial opportunities to incorporate food tours and related educational events into the tourist agenda, which could have an eco-tourism theme. For instance, the village of Baniata on the coast of Rendova in the Western Province is renowned for producing the indigenous ngali nuts that are certified as organic (Organic Without Boundaries 2020). Consequently, there are endless opportunities for a strong auspicious relationship to develop between the food/agriculture and tourism in the Solomon Islands, but the approach taken at national level is rather fragmented despite a realization that strong intersections between the sectors are needed.

Capacity building

Moreover, there is a real need for a talented workforce in the food service industry (Takahashi 2019), especially in terms of developing local resources and creating ways to link these resources to the tourism and hospitality industry. Programmes that support chefs to develop and promote the exchange of skills and best practices across the country should be encouraged. This approach could then cultivate more professional chefs who could promote local food and cuisine, working closely with farmers to improve the quality of food needed by hotels and restaurants. Also, as the Fiji case illustrated, there is a need for capacity building in the informal and self-employed sector through ensuring that business skills and related training extends to the field of entrepreneurship.

Capacity building also involves dealing with food security concerns which are rife in the Solomon Islands (Hardy et al. 2013). One challenge involves accessing fresh, nutritious and healthy products. In the country's fishing industry, around 94% of fresh tuna is exported to foreign markets leaving only 6% of tuna for the local market, which is often small in size, poorer quality and old (Cleasby et al. 2014). Therefore, the objective in capacity building is to work toward easing the impediments to accessing local resources, which, in this case, may well mean revisiting trade and fishery policies. Government efforts in strengthening national capacity for food control can help to safeguard food safety in the islands, leading to economic and agricultural development (including fisheries). As the country has multiple outer islands, transporting food from the main island to the outer islands presents a significant logistical challenge as the quality, safety and perishability of the items may be compromised on arrival (FAO 2021). This limits both tourism development in the outer islands and the types of tourism development that can be pursued.

From the standpoint of food tourism, agritourism can create local business opportunities and promote local products and services through the development of farm stay tourism. Furthermore, there are numerous opportunities for food tours to be integrated into the country's tourism scene. Capacity building, however, is essential for the advancement of food development and ensuring a synergistic and auspicious relationship with the tourism industry.

Rethinking the intersections between agriculture, food and tourism in a post-pandemic context

The COVID-19 pandemic encouraged Pacific Island inhabitants to reconnect with agriculture, subsistence living and rural livelihoods. The Tongan government allocated TOP3.2 million (US$1.36 million) to the ministries overseeing fisheries and agriculture to increase local food production, and the Tuvalu government encouraged the development of home gardens through providing seedlings to residents (Sherzad 2020). In Samoa, 200 displaced hotel workers from Savai'i participated in waged work to rehabilitate the island's coconut and cocoa industries (Samoa Global News 2021). The revitalization of gardens and local planting initiatives could stimulate micro-businesses that could serve the hospitality industry once full recovery occurs. The pandemic also brought to light the importance of improving food commodity storage, processing and distribution, which are all crucial for food security and for the consumption of fresh foods, which can be a challenge in the outer islands (Snowdon et al. 2011). Some Pacific Island states responded to the COVID-19 movement control orders by leveraging technology and utilizing existing skill sets to seek alternative trade channels, and e-commerce revived bartering systems in a digital format and through social media platforms. For instance, the Facebook group 'barter for better Fiji', formed in April 2020, was a community response to enable non-cash product exchanges during the lockdowns. Similarly, in the Solomon Islands, local produce (e.g., herbs) that was previously sold to hotels was sold via Facebook (Davila and Wilkes 2020). In Fiji, one local supplier of pineapples for 45–50 hotels redirected his produce to local markets, where the farm gate price per pineapple reduced from an average price of FJ$1.50 (US$0.71) in the pre-COVID-19 period to FJ$0.75 during the pandemic (US$0.35) (Iese et al. 2021).

As the pandemic encouraged individuals and communities to appreciate the rural and agricultural environment more, a current progressive rethink is required concerning how agriculture and tourism can intersect to the advantage of all interest groups, especially both producers and consumers. There is a pressing need to look beyond import dependency toward the formation of stronger inter-sectoral linkages between tourism and agriculture (and fisheries), with the outcome of providing economically and culturally sustainable alternatives to an import-led approach. It is inferred that import dependency represents a challenge to the appreciation of traditional foodways and food heritages, especially given the broader influences of urbanism, colonialism, migration and globalization (Berno 2020). Progressive initiatives may well be required to challenge import dependency. Vanuatu's Torba province, for instance, wishes to take advantage of its agricultural land and rich natural resources, where local leaders encourage tourism bungalows to serve guests only locally grown organic food (Roy 2017).

Conclusion and research implications

This chapter has demonstrated that the intersections between the agriculture/food and tourism sectors in Pacific Island states are generally inauspicious, especially as a range of prevailing challenges exist: food security concerns, seasonal demands from tourists, urbanization and Westernization, climate change and natural disasters, food import dependence, poor agricultural infrastructure and limited production capacity, inefficient transport and logistical infrastructures, limitations in marketing and the need for skilled workers and skills-based training. It is essential, therefore, that Pacific Island states embrace more progressive capacity building initiatives, work toward installing stronger food security directives, and institute more localized forms of food and beverage production and commercialization.

The two case illustrations did indicate that, despite the challenges, there is certainly potential in strengthening the inter-sectoral linkages. The Solomon Islands case scenario demonstrated the potential for agritourism, food tours and farm-to-table tourism initiatives, which could form part of the tourism planning strategies and destination marketing portfolios of the Pacific Island states. The Fiji case scenario demonstrated how particular products can represent the symbolic and cultural capital of a destination, though product innovation in developing countries can be vulnerable to foreign ownership, in which business and community decisions may not always satisfy local and national interests.

Nonetheless, local food business models and national support are critical in advancing auspicious relationships between the tourism, food and agricultural sectors. Local ownership structures not only help to maintain income generation and regeneration in the community but can also create vibrant communities and foster an entrepreneurial spirit. The pandemic reconnected people to rural life and the land from which new reinvigorated community approaches to farming could be advanced. Consequently, it is necessary to preserve cultural and community values around food, through conserving the culture, history and tradition associated with local food production and consumption. This approach can help inspire and satisfy the needs and desires of tourists though, at the same time, strengthen the nation's ethnic, cultural and national identities. It would be enlightening to conduct more research on sectoral linkages between tourism and agriculture/food in Samoa, especially in terms of recognizing the existence of localized forms of tourism and hospitality as well as local ownership structures of particular accommodation units (see Chapter 22). There is certainly a continual need for dialogue between the public and private sector, especially in light of the application of stronger and more robust tourism and agriculture linkages in the post-COVID redevelopment stage of the tourism industry. Continual investment in infrastructure and community development is vital for strengthening local agricultural production and facilitating the tourism and food-services industries. It is also necessary for creating a consistent and robust food supply chain.

Over a decade ago, McGregor et al. (2009) emphasized the crucial importance of strengthening land productivity to improve food security. Therefore, more research concerning genetic diversity is crucial in ensuring that farmers can manage risks concerning food security. The key is to utilize localized food and traditional practices and adapt them in ways that attract and accommodate global tourism demands, as well as embrace new innovative procedures and technological advancements. There is an urgent need for research across all Pacific Island states to look at how agricultural systems can be developed in relation to the needs of the hospitality and tourism industries, focusing on how sectoral linkages can empower local communities and secure local livelihoods.

References

Adams, T. (2020) 'Drinking the kava: A visit to the real Fiji', *GoNomad Travel, December 18*. Available online at https://www.gonomad.com/2907-drinking-the-kava-a-visit-to-the-real-fiji (assessed 3 January 2020).

Ahlgren, I., S. Yamada and A. Wong (2014) 'Rising oceans, climate change, food aid, and human rights in the Marshall Islands', *Health and Human Rights Journal*, 16(1): 69–80.

Atkin, G. (2021) 'Makira should regret not diversifying banana', *Solomon Star*, 17 February.

Aretaake, R. (2019) 'Traditional Kiribati beliefs about environmental issues and its impacts on rural and urban communities', *Disaster Prevention and Management*, 28(1): 25–32.

Barnett, J. (2011) 'Dangerous climate change in the Pacific Islands: Food production and food security', *Regional Environmental Change*, 11(S1): 229–237.

Berno, T. (2011) 'Sustainability on a plate: Linking agriculture and food in the Fiji Islands tourism industry'. In: R.M. Torres and J.H. Momsen (eds.) *Tourism and Agriculture: New Geographies of Production and Rural Restructuring*, London, UK: Routledge, pp. 87–103.

Berno, T. (2020) 'Linking food biodiversity and food traditions to food tourism in small island developing states (SIDS)'. In: D. Hunter, T. Borelli and E. Gee (eds.) *Biodiversity, Food and Nutrition: A New Agenda for Sustainable Food Systems*, London, UK: Routledge, pp. 236–254.

Björk, P. and H. Kauppinen-Räisänen (2017) 'Interested in eating and drinking? How food affects travel satisfaction and the overall holiday experience', *Scandinavian Journal of Hospitality and Tourism*, 17(1): 9–26.

Bottcher, C., S.J.R. Underhill, J. Aliakbari and S.J. Burkhart (2021) 'Food access and availability in Auki, Solomon Islands', *Journal of Hunger and Environmental Nutrition*, 16(6): 751–769.

Brison, K.J. (2001) 'Constructing identity through ceremonial language in rural Fiji', *Ethnology*, 40(4): 309–327.

Bunaa (2018) 'Solomon Island – Coffee cultivation in 14.379 km distance' *Bunna News*, 15 December. Available online at https://bunaa.de/en/solomon-islands/ (assessed 20 December 2021).

Cauchi, J.P., I. Correa-Velez and H. Bambrick (2019) 'Climate change, food security and health in Kiribati: A narrative review of the literature', *Global Health Action*, 12(1) DOI: 10.1080/16549716.2019.1603683.

Chanel, S. and S. Singh (2021) 'Fishing in Fiji under strain', *Earth Journalism Network*, 3 February. Available Online at https://earthjournalism.net/stories/fishing-in-fiji-under-strain (accessed 27 March 2022).

Cheer, J.M., S. Pratt, D. Tolkach, A. Bailey, S. Taumoepeau and A. Movono (2018) 'Tourism in Pacific Island countries: A status quo round-up', *Asia and the Pacific Policy Studies*, 5(3): 442–461.

Chong, K.L. and M.L. Stephenson (2020) 'Deciphering food hawkerpreneurship: Challenges and success factors in franchising street food businesses in Malaysia', *Tourism and Hospitality Research*, 20(4): 493–509.

Cleasby, N., A.M. Schwarz, M. Phillips, C. Paul, J. Pant, J. Oeta, T. Pickering, A. Meloty, M. Laumani and M. Kori (2014) 'The socio-economic context for improving food security through landbased aquaculture in Solomon Islands: A peri-urban case study', *Marine Policy*, 45: 89–97.

Connell, J. and P. Cruz (2019) Vanuatu, *Pacific Economic Monitor, May* 15–17.

Connell, J. and T. Taulealo (2021) 'Island tourism and COVID-19 in Vanuatu and Samoa: An unfolding crisis', *Small States & Territories*, 4(1): 105–124.

Corsi, A., L. Englberger, R. Flores, A. Lorens and M.H. Fitzgerald (2008) 'A participatory assessment of dietary patterns and food behavior in Pohnpei, Federated States of Micronesia', *Asia Pacific Journal of Clinical Nutrition*, 17(2): 309–316.

Davila, F. and B. Wilkes (2020) 'COVID-19 and food systems in Pacific island countries'. In: L. Robins, S. Crimp, M. van Wensveen, R.G. Alders, R.M. Bourke, J. Butler, M. Cosijn, F. Davila, A. Lal, J.F. McCarthy, A. McWilliam, A.S.M. Palo, N. Thomson, P. Warr and M. Webb (eds.) *COVID-19 and Food Systems in the Indo-Pacific: An Assessment of Vulnerabilities, Impacts and Opportunities for Action*, ACIAR Technical Report 96, Australian Centre for International Australian Research, Australian Government, pp. 94–126. Available online at https://www.aciar.gov.au/publication/covid-19-and-food-systems (assessed 21 December 2021)

Dignan, C., B. Burlingame, S. Kumar and W. Aalbersberg (2004) *The Pacific Islands Food Composition Tables*, Rome: Food and Agriculture Organization of the United Nations.

Drinks Insight Network (2017) 'Paradise beverages launches new larger brand Vonu Export', *VerMEXdict Media Limited*, 2 November. Available online at https://www.drinks-insight-network.com/news/news-paradise-beverages-launches-new-lager-brand-vonu-export-5962930/ (accessed 19 August 2022).

FAO (Food and Agricultural Organization of the United Nations) (2020) 'The Republic of Vanuatu: Tropical Cyclone Harold caused widespread damage to the agriculture sector', 8 May, Italy. Available online at https://reliefweb.int/report/vanuatu/republic-vanuatu-tropical-cyclone-harold-caused-widespread-damage-agriculture-sector (accessed 19 August).

FAO (Food and Agricultural Organization of the United Nations) (2021) *Implementation of the Global Strategy in Samoa*, FAO Regional Office for Asia and the Pacific. Available online at https://www.fao.org/asiapacific/perspectives/agricultural-statistics/global-strategy/results-in-the-region/samoa/en/ (accessed 20 December 2021).

Fiji Sun (2010) 'Musket Cove on the move', 12 April. Available online at https://fijisun.com.fj/2010/04/12/musket-cove-on-the-move/ (28 April 2022).

Fishing Booker (2021) 'Solomon: 46 fishing charters available', *FishingBooker.com*. Available online at https://fishingbooker.com/charters/search/us/MD?search_location=solomons (accessed 17 August 2022).

Food Safety News (2022) 'United States contributes to Tonga aid after natural disaster', 9 March. Available online at https://www.foodsafetynews.com/2022/03/united-states-contributes-to-tonga-aid-after-natural-disaster/ (accessed 28 April 2022).

Hardy, P-Y., C. Béné, L. Doyen and A.M. Schwarz (2013) 'Food security versus environment conservation: A case study of Solomon Islands' small-scale fisheries', *Environmental Development*, 8: 38–56.

Hughes, R.G. and M. Lawrence (2005) 'Globalisation, food and health in Pacific Island countries', *Asia Pacific Journal of Clinical Nutrition*, 14(4): 298–306.

iExplore (n.d.) 'Solomon Islands – Food and restaurants'. Available online at https://www.iexplore.com/articles/travel-guides/australia-and-south-pacific/solomon-islands/food-and-restaurants (accessed 28 April 2022).

IFC (International Finance Corporation) (2018) *From the Farm to the Tourist's Table – A Study of Fresh Produce Demand from Fiji's Hotels and Resorts*, July, World Bank Group. Available online at https://www.ifc.org/wps/wcm/connect/a31b8a6c-70ee-4a94-9b75-e2e3750c2f25/From+the+Farm+to+the+Tourists+Table+Final+Report.pdf?MOD=AJPERES&CVID=mlVfcmM (accessed 5 January 2021).

Independent (2011) 'How tiny Nauru became world's fattest nation', *The Independent*, 4 February. Available online at https://www.independent.co.uk/life-style/health-and-families/health-news/how-tiny-nauru-became-worlds-fattest-nation-2203835.html (accessed 20 January 2021).

Inside FMCG (2017) 'Fiji Water expands in hotels', *Inside FMCG*, 21 March. Available online at https://insidefmcg.com.au/2017/03/21/fiji-water-expands-in-hotels/ (accessed 20 January 2021).

Jayawardena, N.S., J. Boe, A.B. Rohoia and P. Sharma (2020) 'Effective marketing strategies for promoting SMEs in Pacific Island countries: A systematic literature review and a future research agenda', *Griffith University-South Pacific Central Banks Joint Policy Research Working Paper Series,* No. 18, Queensland, Australia: Griffith University.

Jones, C., W.E. Murray and J. Overton (2017) 'FIJI Water, water everywhere: Global brands and democratic and social injustice', *Asia Pacific Viewpoint*, 58(1): 112–123.

Laeis, Gabriel C.M., R.A. Scheyvens and C. Morris (2019) 'Cuisine: A new concept for analysing tourism-agriculture linkages?', *Journal of Tourism and Cultural Change*, 18(6): 643–658.

Lameka, V. (2020) 'Obesity in Samoa: Culture, history and dietary priorities', *Journal of Samoa Studies*, 10: 25–39.

Latte (2020) 'Fiji's Yasawa Island unveils buyout package', *Luxury Australian Travel Trade E-News,* 20 August. Available online at https://latteluxurynews.com/2020/08/17/fijis-yasawa-island-unveils-buyout-package/ (accessed 6 January 2021).

Iese, V., M. Wairiu, G.M. Hickey, D. Ugalde, D.H. Salili, Jr., J. Walenenea, T. Tabe, M. Keremama, C. Teva, O. Navunicagi, J. Fesaitu, R. Tigona, D. Krishna, H. Sachan, N. Unwin, C. Guell, E. Haynes, F. Veisa, L. Vaike, Z. Bird, M. Ha'apio, N. Roko, S. Patolo, A.R. Dean, S. Kiran, P. Tikai, J. Tuiloma, S. Halavatau, J. Francis and A.C. Ward (2021) 'Impacts of COVID-19 on agriculture and food systems in Pacific Island countries (PICs): Evidence from communities in Fiji and Solomon Islands', *Agricultural Systems*, 190. DOI: 10.1016/j.agsy.2021.103099.

Lifespice (2020) 'Discovering the food of Fiji', 6 February. Available online at http://lifespice.in/the-good-life/discovering-the-food-of-fiji/ (accessed 5 January 2021).

Lin, Y.C., T.E. Pearson and L.A. Cai (2011) 'Food as a form of destination identity: A tourism destination brand perspective', *Tourism and Hospitality Research*, 11(1): 30–48.

Magbalot-Fernandez, A. and M. Umar (2019) 'A review on root crops processing for food security and health', *Journal of South Pacific Agriculture*, 21: 26–33.

Malkanthi, S.H. and J.K. Routry (2011) 'Potential for agritourism development: Evidence from Sri Lanka', *Journal of Agricultural Sciences*, 6(1): 45–58.

McGregor, A., M. Manley, S. Tubuna, R. Deo and M. Bourke (2009) 'Pacific island food security: Situation, challenges and opportunities', *Pacific Economic Bulletin*, 24(2): 24–42.

McIver, L., A. Woodward, S. Davies, T. Tibwe and S. Iddings (2014) 'Assessment of the health impacts of climate change in Kiribati', *International Journal of Environmental Research and Public Health*, 11(5): 5224–5240.

McLennan, A.K. (2017) 'Local food, imported food, and the failures of community gardening initiatives in Nauru'. In: M. Wilson (ed.) *Postcolonialism, Indigeneity and Struggles for Food Sovereignty*, London: Routledge, pp. 145–163.

MEX (2015) 'Paradise beverages Fiji case study', *MEX Maintenance Software, 4 November.* Available online at https://www.mex.com.au/Blog/Article/paradise-beverages-fiji-case-study (accessed 19 August 2022).

Movono, A., H. Dahles and S. Becken (2018) 'Fijian culture and the environment: A focus on the ecological and social interconnectedness of tourism development', *Journal of Sustainable Tourism*, 26(3): 451–469.

Ng Kumlin Ali, B. (2002) *Chinese in Fiji*, Suva: Institute of Pacific Studies, University of the South Pacific, Fiji: Suva.

Organic Without boundaries (2020) 'Ngali nuts- from the forest floor to the food store', 3 March. Available online at https://www.organicwithoutboundaries.bio/2020/03/03/ngali-nuts/ (accessed 28 April 2022).

Pollock, N.J. (2009) 'Sustainability of the Kava trade', *The Contemporary Pacific*, 21(2): 265–297.

Roberts, L. and D. Hall (eds.) (2001) *Rural Tourism and Recreation: Principles to Practice*, United Kingdom: CABI Publishing.

Rogerson, C.M. (2012) 'Strengthening agriculture-tourism linkages in the developing world: Opportunities, barriers and current initiatives', *African Journal of Agricultural Research*, 7(4) 616–623.

Roy, E.A. (2017) 'South Pacific islands ban western junk food and go organic', *The Guardian*, 3 February. Available online at https://www.theguardian.com/world/2017/feb/03/south-pacific-islands-vanuatu-torba-ban-western-junk-food-health-organic (accessed 21 December 2021).

Samoa Global News (2021) *Unemployed Hotel Workers in Asau Benefit from Replanting Scheme*, 8 February. Available online at https://samoaglobalnews.com/unemployed-hotel-workers-in-asau-benefit-from-replanting-scheme/ (accessed 20 December 2021).

Samoa Observer (2018) 'Honey bee "must be preserved"'. Available online at https://www.samoaobserver.ws/category/samoa/13813 (accessed 20 December 2021).

Scheyvens, R. and G. Laeis (2021) 'Linkages between tourist resorts, local food production and the sustainable development goals', *Tourism Geographies*, 23(4): 787–809.

Sengel, T., A. Karagoz, G. Cetin, F.I. Dincer, S.M. Ertugral and M. Balık (2015) 'Tourists' approach to local food', *Procedia -Social and Behavioral Sciences*, 195: 429–437.

Shah, K. (2021) 'Science, technology and innovation for sustainability in small island states: Key implications for policy and practice', *IPL Policy Brief Series*, Issue 1, 28 February. Available online at https://papers.ssrn.com/sol3/papers.cfm?abstract_id=3798220 (accessed 24 March 2022).

Shaver, J.H. (2015) 'The evolution of stratification in Fijian ritual participation', *Religion, Brain and Behavior*, 5(2): 101–117.

Sherzad, S. (2020) 'Impacts of COVID-19 on the Food Systems in the Pacific Small Island Developing States (PSIDS) and a Look into the PSIDS Responses', FAO (Food and Agriculture Organization of the United Nations) Publication. Available online at https://www.fao.org/uploads/pics/COVID-19_impacts_on_food_systems_in_PICs_CRFS_.pdf (accessed 19 August 2022).

Sims, R. (2009) 'Food, place and authenticity: Local food and the sustainable tourism experience', *Journal of Sustainable Tourism*, 17(3): 321–336.

Singh, S. (2014) 'Taki Mai makers named 2014', *CPG Editor's Choice Award*, 16 October. Available online at https://www.mailife.com.fj/taki-mai-makers-named-2014-cpg-editors-choice-award/ (accessed 27 April 2022).

Singh, G., K.K. Devi, R. Naz and K.S. Fam (2016) 'An empirical investigation into the cultural antecedents of food consumption: Study of Fijian consumers', *Amity Global Business Review*, 11: 15–27.

Singh, E., S. Milne and J. Hull (2015) 'Linkages between tourism and agriculture: Stakeholder perspectives and online marketing and promotion on the Island of Niue in the South Pacific'. In: S. Pratt and D. Harrison (eds.) *Tourism in Pacific Islands: Current Issues and Future Challenges*, Abingdon: Routledge, pp. 189–204.

Snowdon, W., M. Moodie, J. Schultz and B. Swinburn (2011) 'Modelling of potential food policy interventions in Fiji and Tonga and their impacts on noncommunicable disease mortality', *Food Policy*, 36(5): 597–605.

Solomon Airlines (n.d.) 'Dining and food in the Solomon Islands'. Available online at https://www.flysolomons.com/destination-guide/food (accessed 27 April 2022).

Solomon Times (2014) 'SolBrew launches "6 pack" in provinces', *Solomon Times Online*, 11 June. Available online at https://www.solomontimes.com/news/solbrew-launches-6-pack-in-provinces/8187 (accessed 12 January 2021).

Spacifica Travel (2021) 'Samoa – A food lover's paradise', *spacificatravel.com*. Available Online at https://spacificatravel.com/blog/samoa-a-food-lovers-paradise (accessed 22 December 2021).

SPTO (South Pacific Tourism Organisation) (n.d.) 'One for chocolate lovers', *Pacific Tourism Organisation*. Available online at https://southpacificislands.travel/one-for-chocolate-lovers/ (accessed 18 August 2022).

Steinmetz, J.T. (2018) 'Agriculture, cuisine and tourism: a winning combination in Solomon Islands', *eturbonews*, 9 February. Available online at https://eturbonews.com/177519/agriculture-cuisine-tourism-winning-combination-solomon-islands/ (accessed 8 January 2020).

Takahashi, K. (2019) 'Tourism demand and migration nexus in Small Island developing States (SIDS): Applying the tourism demand model in the Pacific region', *Island Studies Journal*, 14(1): 163–174.

Telfer, D.J. (2000) 'Tastes of Niagara: Building strategic alliances between tourism and agriculture', *International Journal of Hospitality and Tourism Administration*, 1(1): 71–88.

TVET (Technical and Vocational Education and Training) (2020) 'Training for Sanma beekeepers', 17 August. Available online at http://www.vanuatutvet.org.vu/news-events/news/training-for-sanma-beekeepers (accessed 8 February 2022).

Tora, T. (2020) 'The great kava boom: how Fiji's beloved psychoactive brew is going global', 4 February, *The Guardian*. Available online at https://www.theguardian.com/world/2020/feb/05/the-great-kava-boom-how-fijis-beloved-psychoactive-brew-is-going-global (accessed 29 January 2021).

Trip Advisor user (2019) 'Don't expect a 5-star room or meals', *Tripadvisor*, 13 October. Available Online at https://www.tripadvisor.com.my/ShowUserReviews-g779183-d2515061-r717803813-The_Villages-Kiritimati Line_Islands.html (accessed 23 January 2021).

Tu'akoi, S., N. Tamarua-Herman, K. Tairea, M.H. Vickers, Y.Y.M. Aung and J.L. Bay (2020) 'Supporting Cook Island communities to access DOHaD evidence', *Journal of Developmental Origins of Health and Disease*, 11(6): 564–572.

Turaga, S. (2018) 'Fiji's main tourism areas spent $74.4 million on the procurement of fresh produce in 2017', *Fiji Village*, 22 August. Available online at https://fijivillage.com/news/Fijis-main-tourism-areas-spent-744-million-on-the-procurement-of-fresh-produce-in-2017-rsk925 (accessed 20 January 2021).

UNWTO (United Nations World Tourism Organization) (2015) *Tourism and the Sustainable Development Goals,* Madrid, Spain. Available online at https://www.e-unwto.org/doi/pdf/10.18111/9789284417254 (accessed 17 October 2019).

Urry, J. (1990) *The Tourist Gaze: Leisure and Travel in Contemporary Societies*, London: Sage.

World Bank (2021) 'International tourism, number of arrivals – Solomon Islands', World Bank Group. Available on at https://data.worldbank.org/indicator/ST.INT.ARVL?locations=SB (accessed 19 December 2021).

Zain, N.A.M., M.S.M. Zahari and M.H.M. Hanafiah (2018) 'Food and tourism destination image: Moderating effect of local food consumption', *E-Review of Tourism Research*, 15(1): 21–36.

PART II

Tourism and island states in Melanesia

9

AUSTRALIAN TRAVEL WRITING AND NOTIONS OF SAVAGERY IN MELANESIA

Nicholas Halter

Introduction

Like their European and American counterparts, Australian travellers were fascinated with the possibility of encountering the 'savage' in the Pacific Islands. Australian knowledge of the islands in the early twentieth century was significantly influenced by the discourse of the savage, a persistent and dominant literary trope since the first European explorer accounts appeared. The dualism of the 'noble' and 'ignoble' savage became a popular image in Australian travel writing, and one that was distorted, conflated, and contested over time. Australians were exposed to the savage in multiple forms, a convenient and ambiguous figure used in children's literature, newspaper reports, missionary and government propaganda, film, and photography. It also featured prominently in tourism materials and travel accounts to entertain, educate, and justify colonial rule to tourists and travellers alike. In Melanesia, accounts of supposed savagery reinforced and legitimized political discourse at home in the late nineteenth and early twentieth centuries, which called for Australian sub-imperialism in the region. The work of professional authors blurred the line between travel writing and political promotion. This chapter will, thus, explore the history of European representations of Melanesia and consider the ways in which Australian tourists confirmed or modified popular ideas of race and savagery. The travel accounts of three authors, Beatrice Grimshaw in Fiji, Jack McLaren in the Solomon Islands, and Colin Simpson in Vanuatu, show how the growth of tourism in the region prompted a shift in Australians' perceptions of the Melanesian savage during the twentieth century.

Melanesian misconceptions

The arbitrary boundaries that Europeans have used to demarcate the islands of Melanesia and Australia have often overlooked the ancient connections between them. Lower sea levels allowed peoples to migrate from Southeast Asia in canoes or rafts because the distances between islands were much less than the present. Archaeological evidence documents the migrations of these Archaic Papuans between 60,000 and 35,000 years ago, a first wave of movement through the New Guinea peninsula and spreading throughout the Bismarck and Solomon archipelagos (Lal and Fortune 2000: 54). The colonization of the western Pacific

DOI: 10.4324/9780429019968-11

continued approximately 3,500 years ago when a second Austronesian wave of migration introduced a new cultural complex from Asia and spread further into present-day Vanuatu, New Caledonia, and Fiji (Bellwood et al. 2006: 7). Though Melanesia's ancient origins have often featured as a tourist attraction, its people have been portrayed as primitive and static. As a result, their traditional mobility and vast regional connections, including those with Australia, have largely been ignored.

Though the Spanish were the first to reach the South Pacific in the sixteenth century (notable for the discovery of the Solomon Islands by Alvaro de Mendaña in 1568), it was French botanist and explorer Dumont d'Urville who is credited with the division of the Pacific into 'Melanesia', 'Polynesia', and 'Micronesia'. Pacific historians have since shown that these classifications were built upon pre-existing European ideas about slavery and physical differences already in existence (Tcherkezoff 2003). Yet, the category continues to be used today to describe the islands of Fiji, the Solomon Islands, New Caledonia, Vanuatu, and Papua New Guinea. The category 'Melanesia' has been especially problematic because it is the only one of the three island groups named after the skin colour of its inhabitants, literally meaning 'the dark islands'. The other two categories described the geography of the islands. Subsequently this mapping became racialized and 'verified' by science in the nineteenth and twentieth centuries. Pseudo-scientific theories based in evolutionary theory supposed that Islanders could be situated within a racial hierarchy in which primitive races not only exhibited physical differences but also differences of an emotional and psychological nature. These theories placed Melanesians on the same scale as Australian Aborigines, as the 'Oceanic Negroes' were compared to Africans and Asiatics (Kabutaulaka 2015: 197). The absence of a centralized authority, as was found in Polynesia, was perceived to imply inferiority, so Melanesians and Australian Aboriginals were considered more primitive than the kingdoms and chiefdoms in the eastern Pacific.

In contrast to Melanesia, Polynesia was romanticized in European literature and art as a natural paradise which was home to the 'noble savage', distinguished by his/her innocence, primitivism and a simple life close to nature (Howe 1977; Smith 1985). Subsequent tourist imagery in the twentieth century repeated tropes of a sexual and feminine Polynesia, an image which subtly justified the colonization of islands by masculine European powers (Jolly 1997). Melanesia was a much more challenging ground for colonial forces to penetrate by comparison. Challenges posed by geographical isolation, disease, and local political organization frustrated colonial attempts to control the Melanesian islands. This meant that some islands, particularly Papua New Guinea and the highland areas, encountered colonial travellers and tourists as late as the 1880s and 1890s.

Developing out of Christian mission outreach in the Pacific in the early 1800s, the 'ignoble savage' was a stereotype used to present Islanders negatively. Emphasizing their nakedness, savage dances, warring, and idol worship, missionaries used the ignoble savage to justify their conversion efforts (Smith 1985: 317). By the twentieth century, tourists were following in the footsteps of many of the pioneering missionaries, and early tourist accounts of Melanesia were informed by this Christian narrative. Today, Christianity remains a powerful force in the Pacific and a prominent feature of tourism promotions. Over time, Christian representations of the ignoble savage softened, and it is important to note that the distinction between noble and ignoble savage was not always clear cut (see Weir 2013). Steamship companies often represented both aspects at the same time, emphasizing both the dangerous and exotic aspects of Melanesia to attract customers but also reassuring them it was a safe destination. This ambiguity was a key feature of subsequent tourism promotions, and lack of knowledge about Melanesia ensured it remained a mysterious destination for many Australians.

Australian tourists in the Pacific

In her pioneering study of the history of Melanesian tourism, Douglas (1996) divided travellers according to three particular groups. The first group were the 'allocentrics', who were the early explorers attracted by the isolation of the islands, the difficulty of access, and the absence of published material. The second group were the 'excursionists', residing less than 24 hours in the country and were associated with the rise of ocean cruising in the interwar period. The third group were the 'pilgrims', travelling to the Islands after World War II to visit battlefield sites. Though these generalizations are useful for thinking about the broader patterns in tourism in Melanesia, in reality, travellers did not fit neatly into these groups. There were adventurers, traders, missionaries, colonial officials, scientific researchers, professional travel writers, photographers, and other artists who possessed different motivations for visiting Melanesia and whose points of origin shaped the ways they viewed the region. It is for this reason that an analysis of Australian travel writing is important to appreciate how Australians understood their neighbours in nuanced and contextual ways.

Prior to Federation in 1901, the Australian continent consisted of a series of British colonies with different motives that shaped their relationships with the Pacific (see Lake 2015; Thompson 1980). Melanesia was viewed as a convenient source of labour and a region that produced copra, gold, rubber, and sugar for companies based in colonial Australia. The proximity of the eastern seaboard to the Pacific region meant that colonial Australia became a springboard or staging point for competing commercial and Christian expeditions into Melanesia. Queensland was concerned with preventing German expansion to the north while also developing its sugar industry using Islander labour, recruited through the controversial practice of 'blackbirding'. The first tourist venture in Melanesia was a shooting and fishing expedition in 1884 offered by Sydney-based Burns, Philp & Company (commonly known as 'BP'). Travellers could travel aboard the *Elsea* to Port Moresby and return for 25 pounds (*Sydney Morning Herald* 1884). Like many of these early allocentric encounters, contact with Indigenous peoples was by default rather than design. BP would later earn the nickname 'Bloody Pirates' for its monopoly of Melanesian shipping and trade, reflecting the close ties between commerce and tourism in the late nineteenth and early twentieth centuries. Colonial Australian travellers, like renowned fiction writer Louis Becke, conformed to English adventure tropes which pitted courageous white men against supposedly savage people (Day 1967). Becke's travels around the Pacific at the end of the nineteenth century gave his accounts legitimacy and contributed to the popularity of his texts as educational and informative (Halter 2016: 380).

Initially used to refer to a wild and untamed forest, it was not until the sixteenth century that the term 'savage' was used to describe a wild person. By the 1880s, descriptions of the savage environment and people in the Pacific Islands tended to the formulaic and predictable, and travellers increasingly reproduced stereotypical descriptions. This was a process of confirming one's expectations, validating and authenticating one's travel, and often a case of writing for a commercial market demanding adventure and excitement. In Australia in the nineteenth and twentieth centuries, the word 'savage' was commonly used to describe Aborigines and Pacific Islanders, but it was also applied to convicts, drunkards, politicians, and vagrants as well as people living in remote rural areas (Russell 2010: 3). The context in which it was used was important. For the new nation of Australia at the turn of the twentieth century, Melanesian savagery served to justify its strict racial immigration policies, known collectively as the 'White Australia Policy', as well as defend the nation's sub-imperial desires in the region. For example, in the territories of Papua and New Guinea, which fell under Australian mandate from 1918 to 1975, 'savage' was interpreted in the context of Australia's successes or failures as a colonial power.

Over time, 'savage' became an ambiguous, versatile and value-laden term used interchangeably with words such as 'cannibal', 'headhunter', 'primitive', and 'native' to describe Melanesia (see Halter 2021). In their work on nineteenth century notions of savagery in colonial Australia, Anderson and Perrin (2008: 148) criticize an oversimplification of the colonial encounter by drawing attention to 'the discursive failure to represent or construct Australia's Indigenous peoples in the terms that had been applied to other peoples in other colonial contexts'. A similar process can be observed in travel writing about the Pacific Islands produced by Australians in the twentieth century, as Dixon (1995; 2001; 2007) highlights the diverse forms of representation of the Pacific which reflected the shifting literary and political landscapes within the nation.

'From Fiji to the Cannibal Islands'

Early tourism ventures in Melanesia were enabled by commerce, and tourism products promoted the benefits of foreign enterprise in the region. This was no better demonstrated by Beatrice Grimshaw, an Irish sportswriter who became a famous travel writer in the Pacific and was later claimed by Australia as one of their own. Grimshaw was born in 1870 in Ireland. She initially worked as a sports journalist and for various shipping companies until 1904 when she travelled to the Pacific funded by government and company commissions to write tourist publicity for the Cook Islands, Niue, Samoa, Tonga, and New Zealand and on the prospects for settlers in Fiji. These became the subject matter of three books she published in Europe which launched her into fame: *In the Strange South Seas* (Grimshaw 1907b), *From Fiji to the Cannibal Islands* (Grimshaw 1907a) and her first Pacific fiction, *Vaiti of the Islands* (Grimshaw 1908). When she returned to the Pacific to report on the territory of Papua in 1907, she ended up staying for 27 years and concentrated on her fiction writing. Most were escapist, outdoor romances with a Pacific setting, and one was produced into the movie *The Adorable Outcast* (1928). In 1934, she left Papua, visiting Fiji, Samoa, and Tonga one last time before retiring in Bathurst, New South Wales in 1936. She died in 1953 having published four travel books, 24 novels, ten volumes of short stories, and countless articles for newspapers and magazines (Evans 1993; Gardner 1977; Laracy 2013).

Travel writers like Grimshaw were a useful tool employed by Australian investors to promote their economic interests in the region. Her 1907 travelogue *From Fiji to the Cannibal Islands* focused exclusively on Melanesia and mostly on Fiji and its potential for settlement and commercial enterprise. In fact, it was published in New York under the title *Fiji and Its Possibilities.* Fiji was 'the richest of all the rich Pacific archipelagos' she claimed and 'were not one-hundredth part developed' (Grimshaw 1907a: 31). Her writing style blended personal narrative with commercial advice, and she did not disguise these ulterior motives from her readers. In her other travelogue, which documented her Polynesian adventures, *In the Strange South Seas* (1907b: 51), she wrote: 'To find out, as far as possible, what were the prospects for settlers in some of the principal Pacific groups, was the main object of my journey to the Islands. It had always seemed to me that the practical side of Pacific life received singularly little attention, in most books of travel'.

Grimshaw spent more than six months in Fiji at a time when the tourism industry was booming. When she arrived by boat in Suva harbour, she observed 'the pavement is dotted with tourists – British, American, Colonial – armed with guide-books and cameras' (Grimshaw 1907a: 24). Colonized by the British in 1884, Fiji became a central hub for trade and transport in the Pacific, which attracted significant investment and tourist interest from Australia. It was also located on the edge of the Melanesia–Polynesia boundary, so, according to Grimshaw (1907a:

7), it was not considered as savage as the 'wicked cannibal group' in the West, but it did not have 'the nameless dreamy charm of the Eastern Islands'.

After arriving in Suva, Grimshaw then toured plantations on the two main islands of Viti Levu and Vanua Levu. She ignored warnings from Suva residents that an inland trip was 'too rough' and travelled by horse with three Fijian guides. Grimshaw was unusual in that she was one of only 15 Australian women who wrote about their Pacific travels between 1880 and 1941. She used this to her advantage by styling herself in a pioneering role as a white woman going where few had gone before. On her travels, Grimshaw (1907a: 31) remarked on the 'satisfactory tropical climate' and, on several occasions, declared the land unclaimed and untouched: 'there were tens of thousands of acres all over the islands unused and unoccupied; white settlers and planters seldom or never came to try their luck'. In reality, this narrative ignored Indigenous claims to land and should be read in the political context of the time in which the British-owned Colonial Sugar Refinery (CSR) rapidly acquired land for sugar plantations. Staffed by Australian overseers, CSR's growth meant that Fiji was becoming an 'Australian economic colony' in 1900 according to Roger Thompson (1998: 19). Despite promoting the economic potential of Fiji, Grimshaw could not resist referring to its supposedly savage past. At the beginning of her book, she warned her readers that the 'Fijian was the most determined cannibal known to savage history' (1907a: 10). However, she never witnessed such savage behaviour first hand, imagining 'the not-so-long-ago days of the death-drum, the strangling noose, and the "bokolo" (human body) served up smoking hot, with savoury herbs for the sunset meal' (1907a: 72). Although cannibalism was practiced in pre-colonial Fiji, by the 1900s, the practice had been abolished by missionaries and much of the Indigenous population were Christian. However, the growth in tourism in the early 1900s brought a flood of tourists whose accounts show a fascination with finding evidence of cannibalism, be it human bones, cannibal forks, or even dubious 'ex-cannibals'. As Grimshaw (1907a: 330) observed:

> It is very gratifying, from a moral point of view, to see the clean, tidy, school-attending, prosaically peaceful folk that have replaced the original savage; but to the traveller, original savages are a good deal more interesting.

This explains the persistence of the savage trope in travel accounts, though, over time, the idea of the savage Fijian was sanitized and commercialized for foreign tourists.

'My Odyssey'

Post-World War I tourism in the Pacific continued to expand as improvements in transportation made travel more accessible to a wider range of Australians (Halter 2017; Douglas and Douglas 2004). Like Fiji, the Solomon Islands was a British colony with significant numbers of Australian settlers and investment, though cruise ship visits were less frequent and brought smaller numbers of 'excursionists'. One of them was Melbournian author Jack McLaren (1884–1954), who described his experiences working and travelling in Papua and New Guinea, the Solomon Islands, and Fiji in his 1923 autobiography *My Odyssey* (McLaren 1923). Moving to London in 1925 to pursue his writing career, he established a reputation as a writer of the Pacific and worked as a radio broadcaster for the British Broadcasting Corporation. For many excursionists of the inter-war period, cruising the Pacific was popular for escaping the cold winters and drudgery of daily life. McLaren searched for the 'Real Wild' (1923: 11) because he regarded civilization 'with scorn' (1923: 66). He was critical of city life because 'people seemed unreal-artificial like' and were 'denied the spice of existence' (1923: 66). These ideas were encouraged by guidebooks and travel magazines which grew in number during the 1920s and '30s as shipping companies expanded and promoted their routes. Presented as informative and educational, many of these magazines stressed the benefits of economic enterprise in the Pacific and depicted the white trader living amongst Melanesian savages as a tourist attraction (Halter 2016). Magazines such as *BP Magazine* (1928–1942) and *Walkabout* (1934–1978) featured similar stereotypical illustrations which compared the traditional canoes with luxury steamships, and the Melanesian was often portrayed in a masculine and warlike form (see Johnston and Rolls 2016; Kuttainen, Liebich and Galletly 2018).

The British protectorate of the Solomon Islands was a profitable colony in the early 1900s, with many Australian plantations harvesting copra for the British-owned Lever Pacific Plantations (Moore 2019). The main port at Tulagi was not as developed as the port in Fiji, so the island-hopping schedule of shipping services meant travellers had more opportunities to visit plantations and farmlands. Jack McLaren was typical of many travellers at the time who continued to repeat the racialized stereotypes when describing the Solomon Islands. One such feature that McLaren (1923: 68) was fearful of was their supposed unpredictable behaviour, stating:

> I could never be sure of them, for they were possessed of instincts at which I could only vaguely guess and over which they had no control. At all times they were liable to give expression to certain queer impulses which were their age-old heritage, and causelessly murder the stranger in their midst – to regret it deeply afterwards no doubt.

Another perceived danger to foreign visitors was disease. McLaren (1923: 244) had suffered malaria and yaws on different occasions during his travels, leading him to conclude that disease was far more dangerous than 'the barbarity, treachery and cunning of the natives'. Though McLaren retained a colonial view of Islanders and used colonialist language to describe the Solomon Islands, he did consider that there were 'degrees of savagery'. Face-to-face encounters in the islands forced McLaren to question his presumption of Melanesian barbarity. He admitted that Solomon Islanders were 'rather good chaps' who had been 'maligned at least in part', and recognized their basic equality:

> And I, watching, concluded that in the matters of superstition, tradition and, above all, keen sensibility to scorn the man of the Palaeolithic Wild and the man of the Civilized Wild were brothers all the while.
>
> *(1923: 256)*

This response became more common in the inter-war period and later as increasing numbers of travellers visited the once-remote Melanesian group and, in the face of humanity, realized 'savagery' was a dubious label.

'The New Hebrides: A Tragicomedy'

World War II had devastating effects on the islands of Melanesia and its inhabitants, particularly in the Solomon Islands and Papua New Guinea (Kwai 2017). Yet, the construction of airfields by Japanese and Allied forces during the war enabled an expansion of air transportation and tourism from the 1950s onward. Many of these air travellers were part of the 'pilgrim' wave to visit war sites, but there was also growth in visitors travelling on package tours. One of these travellers was Australian travel writer Colin Simpson, who arrived by the Qantas flying boat in the New Hebrides (present-day Vanuatu) in 1950. Edwin Colin

Simpson (1908–1983) was born in Sydney and began working as a journalist during the Depression with Sydney's *Daily Guardian, Daily Telegraph and Sunday* papers. In 1945, he worked for three years for the Australian Broadcasting Commission, writing radio documentaries, mostly for a travel series called 'Australian Walkabout'. This sparked an interest in Aboriginal anthropology which resulted in his first book *Adam in Ochre* (Simpson 1951). It gained international popularity and was shortly followed by two books about New Guinea, *Adam with Arrows* (Simpson 1953) and *Adam in Plumes* (Simpson 1954). In 1955, he published *Islands of Men: A Six-part Book about Life in Melanesia* which included a chapter titled 'The New Hebrides: A Tragicomedy'.

Discovered and named by British explorer James Cook, the New Hebrides was divided between French and British colonial rule from 1887 until their independence in 1970. The condominium arrangement resulted in a dual colonial administrative system which created some confusion. British and French traders competed under different laws, British and French missionaries competed for converts, and two education systems operated in English and French. Like many travellers before him, Simpson (1955) was critical of the condominium. When he arrived in Efate in 1950 to write a radio documentary for the Australian Broadcasting Corporation, he recalled that it looked the same as when he had first visited in 1934 (1955: 85). Simpson's account is also typical in that it focused overwhelmingly on the details of the colonial administration at the expense of any real consideration of Indigenous peoples.

In the *Australian Dictionary of Biography*, White (2012, n.p.) notes:

> Seeking international sales, he rejected the strident Australianness of his main travel writer competitor, Frank Clune, and instead presented a self-consciously sophisticated persona, indicated by an interest in the arts and being frank about sex. He nevertheless defended the ordinary tourist against anti-tourist 'snobs'.

Though Simpson's writing style was distinct from some of his predecessors, he reinforced many of the popular stereotypes of the pre-war decades. Drawing on the work of historical and anthropological experts for legitimacy, Simpson (1955: 93–94) reminds the readers 'the people are very dark skinned', naked, polygamous, male dominant, possess a 'primitive culture' and engage in 'fairly constant warfare, of the guerrilla kind, with a life-for-a-life as tribal law'. The chapter is arranged in the style of a play with 'Acts' that mainly focus on the deeds of foreign traders, blackbirders, and missionaries who all fell victim to New Hebridean savagery in the past. The 'missionary martyrs' as he termed them 'bore the brunt of savagery', and he observed that the 'native attitude' was still 'very anti-white' in parts (Simpson 1955: 106–108). Simpson attempts to give more detailed ethnographic descriptions of Indigenous culture, but the aspects he describes (of land diving, homosexuality, and cargo cults) are tokenistic and serve to accentuate the exoticism of the islands. It is likely that Simpson was aware of some of the complexities of colonial rule in the New Hebrides. He recalls meeting Australian zoologist A.J. Marshall, whom he admired and quoted, but rather than publish Marshall's sympathetic comments about the New Hebrideans he emphasized past cannibal rituals and reprinted a photograph of a dead warrior's body. In reality, cannibalistic practices would not have been likely in the New Hebrides in the 1950s, and, indeed, Simpson (1955: 117–120) admitted it 'has now not much significance'. Whether Simpson was actually fascinated with Melanesian cannibalism or merely used it as a ploy to sell copies of his book is unclear. Nonetheless, his books were widely read both at home and internationally and perpetuated an image of the New Hebrides as dangerous and savage.

Contemporary trends and representation

Australian routes have changed significantly in the past 50 years, and a larger proportion of Australians have visited the Pacific Islands as tourists. Although the COVID-19 pandemic significantly slowed down tourist visitation to the region in 2020, with virtual stagnation from April of that year, in 2018, Australia represented the highest share of tourist arrivals in the Pacific, at 28.9% (South Pacific Tourism Organisation 2019: 21). Of those Australian tourists recorded, the most popular country was Fiji, visited by 367,000 Australian tourists in the Pacific in 2019 (Australian Government 2019a). Though successive Australian governments have stated that Melanesia is its closest and most important neighbour, misperceptions about the region have persisted in public discourse and debate. The term 'Arc of Instability' used in Australian foreign policy circles in the 2000s to describe a number of island governments that were in political strife (or perceived to be at risk) continued to perpetuate inaccurate notions of Melanesian culture and society as volatile, dangerous, and inferior (Kabutaulaka 2015: 202). Travel patterns also reveal that Asia is still foremost in Australian minds (Australian Government 2019b). In order to improve relations between Australian travellers and their Pacific hosts, a better understanding of the historical context which has shaped the ways in which Australians have imagined and represented their Melanesian neighbours for much of the twentieth century is needed.

Australian travel accounts are also part of a broader global corpus that presented stereotypical and racial descriptions of Melanesian savagery throughout the twentieth century, both in written travelogues and in magazines and promotional images. A longitudinal study of the *National Geographic Magazine* by Lutz and Collins (1993: 145) identified the dominance of masculine and primitive imagery of Melanesia which suggests that racial tropes and symbols are persistent. Despite the growth of the tourism industry bringing more visitors to the Melanesian islands and increasing public awareness of their diversity and culture, publishing has continued to satisfy a demand for sensationalized and essentialized representations of Pacific Islanders. Paul Theroux's (1992) bestseller, *The Happy Isles of Oceania,* has since been criticized for its scathing descriptions of Melanesians (see for example, Kabutaulaka 2015, Lyons 1994, Va'ai 2005). Subsequent travel writers like J. Maarten Troost continue to encourage savage stereotypes with titles like *The Sex Lives of Cannibals* (Troost 2004) and *Getting Stoned with Savages: A Trip Through the Islands of Fiji and Vanuatu* (Troost 2006).

Limited research has been conducted on how Pacific Islanders have responded to these skewed representations or how national tourism bodies have attempted to counter the narrative. A survey of contemporary tourism websites of the nations of Fiji, the Solomon Islands, and Vanuatu shows that some imagery has changed since the 1950s. Though images of decorated or armed Melanesian men persist, many destinations now emphasize the 'sun, surf and sand' as a key attraction, and there is greater emphasis on cultural diversity and natural wonders. The largest tourist destination, Fiji, continues to promote its 'friendly' nature in contrast to preconceptions of savagery. Nonetheless, the history of cannibalism remains a popular tourist attraction in the Pacific, and, in some cases, Pacific Islanders have taken advantage of this foreign fascination. For some national tourism organizations, cannibalism is a clever marketing ploy for foreign markets. For example, of the '10 Reasons Why You Must Visit Solomon Islands' listed by Tourism Solomons (2018), Number 1 is 'cultural/multicultural' and the explanation begins: 'Journey through an era where headhunters collected trophy skulls'. A search of the website for Vanuatu Tourism (2019) delivered 25 hits on the word 'cannibal'. There is also appropriation of the cannibal trope at the local level. In Fiji, for example, locals sell cannibal forks at souvenir shops in the same fashion as they did in the 1890s. There are also guided tours offered that are specifically marketed to the macabre, such as the 'Cannibal Cave Tour' of Naihehe Cave in

Fiji (Banivanua Mar 2016) or the Amelbati Cannibal Site Tour in Malekula, Vanuatu (Vanuatu Tourism 2019). Not surprisingly, there are attempts within Pacific Island countries to resist and challenge Western notions of savagery or cannibalism. The Fiji ecotourism company Talanoa Treks (2019), for example, actively challenges historical misconceptions by offering guided treks to remote villages, some of which have been involved in historic cases of murder or cannibalism. By providing opportunities for tourists to meet local villages, Talanoa Treks provides a more complex and unique understanding of Pacific history.

Conclusion and research implications

This chapter has presented a broad overview of the ways in which the trope of the savage has been perpetuated and adapted by travellers and the tourism industry since the late nineteenth century. Stereotypes of Melanesian savagery have been stubbornly persistent since they were first invented by Europeans in the 1800s, and a study of the travel writing produced by Australian travellers since the 1880s shows that visitors to the Pacific Islands were exposed to various representations of the savage which shaped their expectations of, and engagement with, the islands. The works of three professional authors, Beatrice Grimshaw, Jack McLaren, and Colin Simpson point to subtle changes in the way that Melanesian savagery was represented over time. The potential for settlement and economic development in the Pacific was a major factor at the turn of the century, and Melanesians were portrayed as inferior subjects to legitimize colonial exploitation. Yet, by the 1930s, more frequent contact encouraged a more complex portrayal of the savage, as travellers questioned familiar generalizations and acknowledged distinct island cultures and histories. Racial distinctions had been further undermined in later decades by these travellers, despite the tourism and publishing industries continuing to promote ideas of exotic savagery for their own commercial gain.

The contemporary implications of this historical review suggest that Australians, and Westerners in general, continue to fall back on familiar stereotypes when visiting the Pacific Islands. The tourism industry today continues to adapt and reinterpret the idea of the savage for their own purposes, and, in many cases, the figure is a convenient marketing ploy. An understanding of the historical contexts which shaped the Pacific Islands shows that each island has a distinct experience of colonialism and tourism which shapes the ways in which the savage trope has been applied by individual travellers. Rather than replicating generalizations, contemporary tourism ventures could provide a more localized and historically informed approach to tourism that complicates notions of Melanesian savagery and locates it within the colonial contexts that created it.

Future research in this area should be multidisciplinary and connect tourism studies with the discipline of history. Few scholars have taken a historical approach to Pacific tourism that seeks to understand the colonial roots of travel in the Pacific and the ways in which history continues to shape the contemporary tourism industry. Cultural and heritage tourism is an emerging and understudied subject in the Pacific, and little is known about how Pacific Island countries and local tourism operators negotiate and present history to visitors. Future critical discussions in this area must problematize the colonial narratives of the past and look at successful cases in the tourism industry in which ideas of Melanesian savagery are negotiated and challenged. For the national governments of the Pacific Island states, many of whom have celebrated or will soon celebrate 50 years since independence, efforts to challenge Western stereotypes are making progress. The challenges of 'basic infrastructure, fragile governments, violent inter-clan rivalry, poor international media exposure, rampant malaria and administrative indifference' (Douglas 2004: 47) are eroding as Pacific Island countries recognize the importance of tourism to expand and strengthen their economies and societies. More work needs to be done to understand how the heritage and history of the Pacific Islands can be presented in a culturally sensitive way that challenges exaggerated myths of past violence and recognizes the ways in which Indigenous people have responded to and resisted foreign influence.

References

Anderson, K. and C. Perrin (2008) 'Beyond savagery: The limits of Australian "Aboriginalism"', *Cultural Studies Review*, 14(2): 147–169.

Australian Government (2019a) 'Fiji country brief', Department of Foreign Affairs and Trade. Available online at https://www.dfat.gov.au/geo/fiji/fiji-country-brief#:~:text=Overview,Suva%20is%20Fiji's%20capital (accessed 20 November 2019).

Australian Government (2019b) 'Australian international travellers: The places we go', Department of Foreign Affairs and Trade. Available online at https://www.dfat.gov.au/sites/default/files/australian-international-travellers-the-places-we-go.pdf (accessed 20 November 2019).

Banivanua Mar, T. (2016) 'Performing cannibalism in the South Seas'. In: K. Alexeyeff and J. Taylor (eds.) *Touring Pacific Cultures*, Canberra: ANU Press, pp. 323–331.

Bellwood, P., J.J. Fox and D. Tryon (2006) 'The Austronesians in history: Common origins and diverse transformations'. In: P. Bellwood, J.J. Fox and D. Tryon (eds.) *The Austronesians: Historical and Comparative Perspectives*, Canberra: ANUE Press, pp.1–16.

BP Magazine (1928–1942) Sydney, Burns: Philp & Co.

Day, A.G. (1967) *Louis Becke*, Melbourne: Hill of Content.

Dixon, R. (1995) *Writing the Colonial Adventure: Race, Gender, and Nation in Anglo-Australian Popular Fiction, 1875–1914*, Cambridge: Cambridge University Press.

Dixon, R. (2001) *Prosthetic Gods: Travel, Representation, and Colonial Governance*, St Lucia: University of Queensland Press and API Network.

Dixon, R. (2007) 'What was travel writing? Frank Hurley and the Media contexts of early twentieth-century Australian travel writing', *Studies in Travel Writing*, 11(1): 59–81.

Douglas, N. (1996) *They Came for Savages: 100 Years of Tourism in Melanesia*, Lismore: Southern Cross University Press.

Douglas, N. (2004) 'Towards a history of tourism in Solomon Islands', *The Journal of Pacific Studies*, 26(1&2): 29–50.

Douglas, N. and N. Douglas (2004) *The Cruise Experience: Global and Regional Issues in Cruising*, Frenchs Forest: Pearson Education.

Evans, J. (1993) 'Feminism, postcolonialism and Beatrice Grimshaw: A case study of a white woman in imperialism', *M.A. Thesis*, La Trobe University.

Gardner, S.J. (1977) 'For love and money: Early writings of Beatrice Grimshaw, colonial Papua's woman of letters', *New Literature Review*, 1: 10–20.

Grimshaw, B. (1907a) *From Fiji to the Cannibal Islands*, London: Eveleigh Nash.

Grimshaw, B. (1907b) *In the Strange South Seas*, London: Hutchinson & Co.

Grimshaw, B. (1908) *Vaiti of the Islands*, New York: A. Wessels Co.

Halter, N. (2016) 'Searching for the Land of the Golden Cocoa-Nut: Australian travel writing about commercial enterprise in the Pacific Islands', *The Journal of Pacific History*, 51(4): 375–391.

Halter, N. (2017) 'Ambivalent mobilities in the Pacific: "savagery" and "civilisation" in the Australian interwar imaginary', *Transfers*, 7(1): 34–51.

Halter, N. (2021) *Australian Travellers in the South Seas*, Canberra: ANU Press.

Howe, K.R. (1977) 'The fate of the savage in Pacific historiography', *New Zealand Journal of History*, 11(2): 137–154.

Johnston, A. and M. Rolls (2016) *Travelling Home, Walkabout Magazine and Mid-Twentieth Century Australia*, London: Anthem Press.

Jolly, M. (1997) 'From point Venus to Bali Ha'i: eroticism and exoticism in representations of the Pacific'. In: L. Manderson and M. Jolly (eds.) *Sites of Desire, Economies of Pleasure: Sexualities in Asia and the Pacific*, Chicago: University of Chicago Press, pp. 99–122.

Kabutaulaka, T. (2015) 'Re-presenting Melanesia: ignoble savages and Melanesian alter-natives', *The Contemporary Pacific*, *27*(1): 110–145.

Kwai, A. (2017) *Solomon Islanders in World War II: An Indigenous Perspective*, Canberra: ANU Press.

Kuttainen, V., S. Liebich and S. Galletly (2018) *The Transported Imagination: Australian Interwar Magazines and the Geographical Imaginaries of Colonial Modernity*, New York: Cambria Press.

Lake, M. (2015) 'The Australian dream of an island Empire: Race, reputation and resistance', *Australian Historical Studies*, 46(3): 410–424.

Lal, B.V. and K. Fortune (eds.) (2000) *The Pacific Islands: An Encyclopedia*, Honolulu: University of Hawai'i Press.

Laracy, H. (2013) 'Beatrice Grimshaw (1870–1953): Pride and prejudice in Papua'. In: H. Laracy (ed.), *Watriama and Co: Further Pacific Islands Portraits*, Canberra: ANU Press, pp. 141–169.

Lutz, C. and J.L. Collins (1993) *Reading National Geographic,* Chicago: University of Chicago Press.

Lyons, P. (1994) 'Review of the Happy Isles of Oceania: Paddling the Pacific, by P. Theroux', *Manoa*, 6(2): 262–265.

McLaren, J. (1923) *My Odyssey*, London: Ernest Benn.

Moore, C. (2019) *Tulagi: Pacific Outpost of British Empire*, Canberra: ANU Press

Russell, P. (2010) *Savage or Civilised?: Manners in Colonial Australia*, Sydney: University of New South Wales Press.

Simpson, C. (1951) *Adam in Ochre: Inside Aboriginal Australia*, Sydney: Angus & Robertson.

Simpson, C. (1953) *Adam with Arrows: Inside New Guinea*, Sydney: Angus & Robertson.

Simpson, C. (1954) *Adam in Plumes*, Sydney: Angus & Robertson.

Simpson, C. (1955) *Islands of Men: A Six-Part Book about Life in Melanesia*, Sydney: Halstead Press.

Smith, B. (1985) *European Vision and the South Pacific*, Sydney: Harper & Row.

South Pacific Tourism Organisation (2019) *Annual Visitor Arrivals Report 2018* Suva: SPTO.

Sydney Morning Herald (1884) 'Clipper yacht Elsea, for New Guinea', 29 September. Available online at http://nla.gov.au/nla.news-article28369242 (accessed 10 June 2019).

Talanoa Treks (2019) Offical website, Suva, Fiji. Available online at https://talanoa-treks-fiji.com/ (accessed 22 November 2019).

Tcherkezoff, S. (2003) 'A long and unfortunate voyage towards the "invention" of the Melanesia/Polynesia Distinction 1595–1832', Translated from French by Isabel Ollivier, *The Journal of Pacific History*, 38(2): 175–196.

Theroux, P. (1992) *The Happy Isles of Oceania: Paddling the Pacific*, London: Penguin Books.

Thompson, R.C. (1980) *Australian Imperialism in the Pacific: The Expansionist Era, 1820–1920*, Carlton: Melbourne University Press.

Thompson, R.C. (1998) *Australia and the Pacific Islands*, Melbourne: Australian Scholarly Publishing.

Tourism Solomons (2018) *'10 reasons why you must visit Solomon Islands'*. Available online at https://www.visitsolomons.com.sb/10-reasons-why-you-must-visit-solomon-islands/ (accessed 22 November 2019).

Troost, J.M. (2006) *Getting Stoned with Savages: A Trip Through the Islands of Fiji and Vanuatu*, New York: Broadway Books.

Troost, J.M. (2004) *The Sex Lives of Cannibals*, New York: Broadway Books.

Va'ai, S. (2005) 'Pacific utopias and national identities in the twenty-first century', *PORTAL: Journal of Multidisciplinary International Studies*, 2(2): 1–23.

Vanuatu tourism (2019) 'Amelbati cannibal site tour'. Available online at https://www.vanuatu.travel/en/amelbati-cannibal-site-tour (accessed 22 November 2019).

Walkabout (1934–1978) Melbourne: Australian National Travel Association.

Weir, C. (2013) '"Deeply interested in these children whom you have not seen": The Protestant Sunday School View of the Pacific, 1900–1940', *The Journal of Pacific History*, 48(1): 43–62.

White, R. (2012) 'Simpson, Edwin Colin (1908–1983)', *Australian Dictionary of Biography,* Canberra: National Centre of Biography, Australian National University. Available online at http://adb.anu.edu.au/biography/simpson-edwin-colin-15926/text27127 (accessed 13 July 2019).

10

TOURISM DEVELOPMENT IN THE SOLOMON ISLANDS

Identifying the concerns and challenges

Farah Atiqah Mohamad Noor, Jefferson Patovaki, and Marcus L. Stephenson

Introduction

The Solomon Islands is made up of 996 islands in the southwest Pacific Ocean and lies approximately 1,900 kilometres northeast of Australia, with a total landmass of around 27,500 square kilometres (Leal Filho et al. 2020). Until the early 2000s, the economy of the Solomon Islands was focused significantly on the logging industry. However, driven by an influx of foreign investment since the 1980s, the country shifted its attention toward tourism development. Nonetheless, the country has experienced economic and political challenges as well as the recent COVID-19 pandemic, impacting the inbound tourism market and the economy as a whole. The chapter initially looks at the development agenda of the Solomon Islands and then the evolution of the tourism industry, for which attention is drawn to such tourism development challenges as past and re-emerging ethnic and political conflicts, limited investment in the accommodation sector, and other interrelated challenges. The discussion recognizes the country's need to pursue tourism as a development directive, especially to diversify away from a logging industry that has been impactful on the environment. The chapter indicates that, although there is a need to pursue tourism development, it must be done carefully to ensure that the local community is represented and directly benefit. This has not been the case for those communities situated in the United Nations Educational, Scientific and Cultural Organization (UNESCO) Word Heritage Site of East Rennell, as the subsequent discussion explains. The work then highlights a region that has potential to become UNESCO recognized, Marovo Lagoon, especially if key stakeholders support the site and the vested interests of the local communities living in the area are seriously considered. The chapter then draws attention to the socio-cultural and economic concerns pertaining to tourism development, notably land disputes and child labour. The discussion finally explains the impacts of the COVID-19 pandemic on the tourism industry, implicating ways forward for tourism development.

The Solomon Islands – background context

The islands were frequented by European explorers from the late 1500s, with a notable visit from the Spanish explorer Alvaro de Mendaña in 1568, apparently naming them the Islands

DOI: 10.4324/9780429019968-12

of Solomon (Douglas 2004). The Solomon Islands became a British protectorate in 1896 and remained committed to the Allies during the conflict with Japan in World War II, with self-government being achieved in 1976 and full independence in 1978 (Treloar and Hall 2005: 253). Honiara, located on the northwestern coast of Guadalcanal (Figure 10.1), rose as the new capital city of the independent Solomon Islands, though it had become the capital city of the British Protectorate of the Solomon Islands in 1952. The British administration encouraged churches to fund and operate health clinics and schools in the main islands, especially in Guadalcanal, lifting the administration of such responsibilities (Cassells 2019). The country, thus, witnessed uneven development, especially in the islands located further north and south of Guadalcanal, and particularly in terms of healthcare, transportation, and education (McDougall and Kere 2011). Around 74% of the population (approximately 720,000) lived in rural areas in 2020, which is seen as a challenge for infrastructure development and service delivery (DFAT 2020).

The country's main source of income concerns fishing and logging and also export markets to Australia and China for such plantation crops as cacao, copra, and palm oil (Leal Filho et al. 2020). However, the rural population relies heavily on small-scale fisheries and subsistence agriculture as their source of food and income (Allen and Dinnen 2018). The economy of the Solomon Islands has been significantly impacted by the COVID-19 pandemic, for which gross domestic product (GDP) growth has been estimated to have declined to -4.3% in 2020 due to a slowdown in commodity exports, domestic activity, and tourism (IMF 2021). Arguably, the key development objective should be focus on the advancement of sustainable industries, unlike the logging industry, which has been a source of political tension in the Solomon Islands. Excessive logging activities occurred from the 1980s to the mid-1990s, with the Solomon Islands becoming the world's sixth-largest tropical log exporter at that time (Hviding and Bayliss-Smith 2000). However, the damage was significant, as logging contributed to a substantial reduction of forest

Figure 10.1 Map of the Solomon Islands, indicating the capital city of Honiara, Guadalcanal. Source: Produced by Peter Hermes Furian (www.shutterstock.com/image-vector/solomon-islands-political-map-capital-honiara-457239373).

resources (Piemonte and Fabregas 2020). Moreover, Dyer (2017:2) emphasizes that the logging industry 'has been marked by environmental destruction, political malfeasance, financial trickery, unethical practices by foreign logging companies and a failure of the state to regulate or control the logging industry'.

The evolution of tourism – developments and challenges

James Burns and Robert Philp, under the company name of Burns Philip, established shipping connections between Australia and Solomon Islands (and the New Hebrides) in 1894, pioneered cruise trips and transported tourists (Douglas 2004). Douglas notes that, by the 1930s, the arrival of the steamships in the Solomon Islands took place every six weeks from Sydney. Megapode Airways was established in 1963 and became Solomon Islands Airways (SolAir), which was purchased by the government in 1984, and from 1989 new routes were established between the Solomon Islands and such countries as Fiji, New Zealand, Papua New Guinea (PNG), Tonga, and Vanuatu (Panakera 2007). Nonetheless, even by the turn of the twenty-first century, tourism development was described as 'negligible', though the service industry was identified as having 'untapped potential' (Hou 2001: 15). However, the country does have some recognized tourism attractions, such as beaches and marine areas with unique reef fish, marine life, and sunken vessels from World War II (Treloar and Hall 2004).

Tourism development was weakened as a consequence of ethnic tensions between the people of Malaita and Guadalcanal during the late 1990s until the early 2000s, especially in the capital city of Honiara, with frequent physical altercations taking place between both island communities (Iroga 2008). Internal political conflicts and a rise in street violence not only impacted opportunities for the tourism industry to advance and for the destination to have a sustainable image as a peaceful destination, but it also affected economic and infrastructural development vital for tourism. Accordingly, ethnic conflict led to widespread destruction of transport infrastructure, schools, water supplies, and sanitation systems, estimated to cost over US$250 million (Hou 2001:18). The aftermath resulted in widespread displacement and the return of approximately 20,000 islanders to their provinces of origin (Foukona and Timmer 2016). In 2003, the Australian and New Zealand governments assisted the Solomon Islands through a nation-building project known as the Regional Assistance Mission to Solomon Islands (RAMSI), though it has been argued that RAMSI's 14-year tenure in the Solomon Islands had limited impact in terms of re-building the social and economic policies in the island nation, with rural communities continuing to be politically and economically marginalized (Allen and Dinnen 2018).

In November 2021, unsettled tensions re-emerged in Honiara when Solomon Islanders rioted in the Chinatown district where they set fire to businesses owned by ethnic-Chinese, calling for the prime minister to resign (Sora 2021). According to the Central Bank of Solomon Islands, the riots impacted the economy to the value of SB$534 million (US$66 million) (Ligaiula 2021). From a Malaitan perspective, there were concerns over the government switching allegiance from Taiwan to China in 2019, despite Malaita continuing to maintain political alliances with (and receive economic aid from) Taiwan (Sora 2021). Ironically, the changing geopolitical circumstances of the Solomon Islands may have great potential in attracting a lucrative market segment of Chinese tourists, despite the fact that the Solomon Islands only received around 2,000 Chinese (mainly business) visitors annually, prior to the COVID-19 pandemic (Lee 2019). For the Solomon Islands, in 2020 China represented 34.4% of imports and 64.4% of exports and, thus, this country's growing influence on the Solomon Islands' development agenda cannot be overlooked (Beattie 2022). However, although the changing geopolitical circumstances of the Solomon Islands may have great potential in attracting a lucrative market segment, they have also provoked the re-emergence of political volatility that is not conducive to any paradisiacal image associated with the Pacific Islands (see Schellhorn and Perkins 2004).

The geographical accessibility of the Solomon Islands could be of strategic importance for future tourism growth. The country is accessible for the Australian market, as the flight time from Brisbane to Honiara is only around 3 hours and 15 minutes. In 2010, the Solomon Islands received 20,500 international tourists, peaking at 28,900 in 2019 (World Data.info 2021). This upward trend was partly influenced by the hosting of various international conferences and events. Tourism growth encouraged the government in 2019 to announce an annual target of 60,000 tourist arrivals by 2025, along with recognition of the potential increase of the Chinese market given the country's politico-economic re-alignment to China (Solomon Times 2019). In terms of developing the leisure tourism segment, the country has focused on niche adventure-based tourism (e.g., diving and nature-based tourism) (Tovmasyan 2016). Unique marine life and pristine coral reef ecosystems are a major draw for diving and snorkelling tourism. One of the marine tourist spots is known as the 'Iron Bottom Sound', associated with remnants from World War II and unique marine geography (Tourism Solomons 2019). While Australians and New Zealanders visit the Solomon Islands because of its proximity and tropical climate, Japanese and Americans have far more historic connections to the country through World War II nostalgia (Panakera 2007).

Given the realization of the unsustainable nature of the logging industry and its impact on the environment (Wairiu 2007), it would be beneficial to ensure a smooth transition toward tourism development as a way to strengthen economic growth. Accordingly, the government proactively launched the National Development Strategy 2016–2035, emphasizing the importance of constructing new hotels, improving hotelier services, advancing the conservation of marine resources, promoting cultural heritage, rehabilitating domestic and international airports, improving marketing, and upgraded training institutions and hospitality courses (Solomon Islands Government 2016). Limited accommodation is a significant barrier to the advancement of the tourism industry. Prior to the COVID-19 pandemic, Trip Advisor (2019) identified 80 accommodation properties in the country with only 19 classified as hotels. The hotels are generally of moderate quality and mostly located in Honiara and Guadalcanal, with guesthouses and dive resorts located in Gizo in the Western Province. Hotels in Honiara cater to business travellers who access the city through Honiara International Airport, which is the primary international airport in the country and is located 8 kilometres from the capital. The ability to attract foreign investment in the hotel industry is identified as a major challenge because of the absence of internationally branded and luxury hotels (Bakker 2019).

For international tourists, air travel to Honiara has been possible from Australia, Fiji, New Zealand, PNG, and Vanuatu. Until the COVID-19 pandemic, such airlines as Air Niugini, Air Vanuatu, Fiji Airways, Solomon Airlines, and Virgin Australia were providing scheduled international flights into Honiara (Tourism Solomons 2018). In 2019, Solomon Airlines launched a direct flight route from Brisbane to the re-opened Munda Airport on the island of New Georgia in the Western Province, which resulted in a rise in international visitation to Munda, described as one of the world's most desirable dive destinations (Travel Weekly 2019). In order to accommodate more international flights as the country's second main airport, Munda Airport endured a large-scale refurbishment initiative that was facilitated by the Solomon Islands Roads and Aviation Project, a broad initiative involving the upgrading of Malatia's road network and improving Honiara International Airport (Solomon Star 2019). With continued infrastructure upgrades to the airports, there may well be significant opportunities for new markets to develop from such countries as Japan and the United States (US) (Bakker 2019). Improved air accessibility is fundamental to reducing flight costs to the Solomon Islands, helping the country to be more competitive and attract more international tourists.

World recognition for natural and cultural heritage: addressing ongoing problems

Two globally significant resources with potential to attract tourists that are fraught with unresolved problems are East Rennell, located in the southern portion of Rennell Island (south of Guadalcanal), and Marovo Lagoon, located in New Georgia Islands surrounded by Vangunu Island and Nggatokae Island (northwest of Guadalcanal).

East Rennell was included in the World Heritage List in 1998 because of its natural environment and was the first UNESCO World Heritage Site (WHS) in the independent Pacific Island states; it was later added to the List of World Heritage in Danger in 2013 (Kiddle 2020). The island is an important site for studying biogeography and hosting many endemic plant and wildlife species, along with the unique flora of Lake Tengano (Huo et al. 2021). East Rennell faces challenges due to invasive species, marine resource exploitation, climate change, ongoing logging, and inefficient legal protection (Kiddle 2020). Villagers from Hutuna, Tegano, Niupani, and Tevaitahe in East Rendell have not been convinced of the economic benefits of the WHS, voicing their concerns over limited quantifiable outcomes, including government failure to stimulate tourism development and lack of development of transportation infrastructure (Wate 2019). Wate (2019) notes that it is crucial to develop sustainable livelihoods for the site's customary owners, which would need to be fostered through appropriate stakeholder consultation. There were concerns that biocultural values, customary land tenure, environmental knowledge, and traditional resource use practices were not represented nor recognized internationally, and initial hopes to develop ecotourism through the UNESCO designation did not materialize (Smith 2011).

Marovo Lagoon is an ecologically diverse environment supported by a chain of volcanic islands which host biodiverse rainforests (Hviding 2005). The lagoon is the world's largest double barrier enclosed lagoon system (UNESCO 2008). Marovo Lagoon supports high biodiversity and a rich cultural heritage connecting 50 coastal villages which are inhabited by a total population of 13,000 (Hviding 2012). It drew significant attention in the 1990s from such non-government organizations (NGOs) as Greenpeace, New Zealand Aid, and the World Wildlife Fund in the pursuance of ecologically sustainable development projects (Hviding 2006). Hviding (2006) indicates that there was limited success due to the lack of synergy between the environmentalists and landowners, as the former did not appreciate the importance of the resource rights and tenure patterns of local communities, especially as Marovo's customary land and sea tenure system concerns the interests of kinship groups who have ancestral ownership of designated sections of the land and/or the lagoon and the barrier reef. It was further noted that the challenges associated with Marovo Lagoon becoming a UNESCO WHS, despite being proposed, related to overreliance on logging companies and lack of 'alternative incomes to rival timber royalties from logging companies' (Hviding 2006: 83). Designating the Marovo Lagoon as a UNESCO WHS could be socio-economically beneficial for the Indigenous community through engaging in entrepreneurship activities and advantageous for the Western Provincial Government through procuring taxes and local business revenues from increased activities (Patovaki 2020). Islanders could also benefit from any subsequent influx of tourists through infrastructural improvements.

One significant way forward for Marovo Lagoon to become a WHS would be to generate local awareness concerning the lagoon's outstanding universal values, especially low-land rainforests and marine biodiversity. This could facilitate local support for the project. It is, thus, important that the relevant authorities (e.g., the Environment and Conservation Division of the Ministry of Environment, Climate Change, Disaster Management and Meteorology; the Culture Division of the Ministry of Culture and Tourism; Ministry of Fisheries and Marine Resources; and the Western Provincial Government) proactively and collectively encourage communities to take pride and have greater responsibility for the environment and culture of the lagoon. Increased international awareness could help to promote eco-friendly forms of sustainable tourism that benefit local communities, together with systematic local consultation and public accountability. However, international organizations and NGOs should initially comprehend the socio-cultural and political fabric associated with the Marovo Lagoon before determining the right course of action. It is crucial that internal stakeholders first work toward a clear consensus in pursuing constructive ways forward. The national government, provincial government, and the house of chiefs should establish a joint tripartite working team to look seriously into the case of the lagoon becoming a WHS (Patovaki 2020). The resources for such a venture would need to be secured, which is a challenge as the Solomon Islands government has not had the human and financial resources available to adequately assess whether a particular site has the potential to become a WHS (Smith 2011).

The two case studies illustrated above largely draw attention to the social and cultural aspects of development that are important to consider in developing and maintaining sites of global significance. Accordingly, there are other concerns to now consider that can impact tourism development's synergistic role with local culture, society, and the economy.

Socio-cultural and economic concerns

Tourism development prospects in the Solomon Islands are tempered by cautionary concerns regarding the harmful impacts that land disputes can have on tribal culture and traditional ways of life. Land disputes often manifest situations whereby tribal communities have been disinherited or isolated from decision-making concerning customary land (Allen 2011). Although tribal communities may be cautious of tourism development, there may well be equal access to financial benefits and opportunities for more decision-making at the community level, especially if community forms of tourism are developed. However, decision-making regarding customary land has been increasingly decentralized from tribal communities, creating conflict between such communities and international firms. In the case of Anuha Island, for instance, one resort was burned to the ground, as it was alleged that an international firm had not fully considered the interests and local traditions of customary land (Diedrich and Aswani 2016). Tradition is a significant marker of local identity and any form of modification of a local community's territory can lead to a widespread local conflict, which can be an ongoing challenge for tourism development (Yang et al. 2013). In early 2022, the Solomon Islands had to provisionally suspend flights between Honiara and Marau because of a continuing dispute between local landowners, despite this route encouraging domestic tourism and supporting local tourism businesses (Iroga 2022).

Diedrich and Aswani (2016: n.p.) imply that education and capacity building are necessary components of tourism development in the Solomon Islands, as they are concerned with 'raising awareness about foreign cultural differences' and able to provide local communities with 'the tools they need to absorb social change and still maintain important cultural attributes'. Their study indicates that those Solomon Islanders who experienced more secure economic circumstances felt more prepared to adapt to social change. As tourism development in the Solomon Islands is not as advanced as in Fiji and Vanuatu (see Chapter 7), the country is in a suitable position to work toward ensuring that the socio-cultural impacts of tourism development are minimalized.

One of the main economic vulnerabilities that can delegitimize the benefits of tourism development is child labour. The Solomon Islands became a member of the International Labour Organization (ILO) in 1984 and ratified the Worst Forms of Child Labour Convention in 1999, thereafter sanctioning the Minimum Age Convention, 1973 (SINSO, SIMoHMS and SPC 2017: 292). Despite these legislative manoeuvres, ILO assessed the nature of child labour in the Solomon Islands, concluding that it has the 'worst forms of child labour' (2017: 292), in which more than 3 out of 5 children between the ages of 5 and 11 have been involved in child labour activities (2017: 293). This survey report emphasizes that children, especially those with limited access to quality education, have been forced to enter the labour market by their own families and employers, where they have worked in dangerous and exploitative environments.

Although the common types of economic sectors that children work in have not been fully classified and, despite limited statistical evidence of child labour in the tourism industry, children may still enter the tourism workforce to help their parents financially. Businesses, including family-operated enterprises, can take advantage of the loose labour policies in the country by offering low compensation rates to employees. Therefore, child labour in the tourism industry, whether in terms of children working in boating services, guesthouses, diving lodges and other small tourism businesses, could be a significant social concern (Allen and Dinnen 2018). The government needs to promote awareness of the main triggers for child labour in the Solomon Islands, instituting suitable preventative strategies. For example, pro-poor tourism activities could be developed to assist local families to seek productive employment and avoid reliance on children's additional earnings so that tourism can be utilized directly to alleviate poverty (Jiang et al. 2011). Moreover, the improvement of accessibility to education may ultimately improve the literacy rate and reduce the number of child laborers in the country (Watson-Gegeo and Gegeo 1992). Another preventative method by the tourism authority would be to introduce a 'no child labour' code or guideline by partnering with law enforcement authorities and trade associations to curb the number of working children (Hagedoorn 2013).

COVID-19 impacts on the tourism industry

As a consequence of the COVID-19 pandemic, there were various challenges faced by the tourism and hospitality industry due to the immediate decline of international and domestic tourists from March 2020. There was a mass retrenchment of employees in several public and private industries. The hotel sector was amongst the sectors facing economic pressure. Although the hotel sector retrenched one-third of its staff, some public sector support for the tourism industry transpired. For instance, Honiara City Council worked on ways to help support locally owned tourism and hospitality operators as well as assist local craftsmen and artists in subsidizing their fees for business licenses (Patovaki et al. 2020). In May 2020, the government announced an Economic Stimulus Package (ESP) of SI$309 million (currently US$38.2 million), with SI$90 million (US$11.1 million) being allocated to such infrastructural projects as roads, wharves, bridges and airports, and 40% of the ESP being directed to farmers, fisheries, forestry, and tourism (Connell and Cruz 2020: 22).

It is crucial for the Solomon Islands to try to increase domestic tourism activities, especially as international tourist numbers dropped to an all-time low of 4,400 in 2020 (World Data.info 2021). The wider social message concerns the substance of Solomon Islander unity, implicated in the central theme 'lumi tugeda', which means 'you and me together'. The initiative involved around 47 tourism operators offering travel options and discounted holidays in 12 destinations in the country (Solomon Airlines 2021). This effort has been focused on the development of a local travel bubble initiative established by the national government and Tourism Solomons, the country's official destination marketing organization. The initiative has been encouraging locals to experience a vacation at a special rate that is designed to revamp the domestic tourism market and sustain local tourism operators that have been affected by a lack of international mobility. However, for long-term recovery the Asian Development Bank (ADB) emphasized that investment in human capital and infrastructure is crucial in strengthening the country's resources (ADB 2021b).

Around 8,491 people contracted the COVID-19 virus in the Solomon Islands by 13 March 2022, and there has been internal political vocalization emphasizing that the country is spending SBD100 million (US$12.3 million) in constructing stadiums for the hosting of the 2023 Pacific Games rather than investing in COVID-19 prevention (Houston 2022). However, the country has had some past destination visibility in terms of promoting festivals and events, evident in the hosting of the quadrennial Festival of Pacific Arts 2012, involving more than 3,000 regional and international attendees and several thousand Solomon Islanders (Shennan 2013). The country arguably needs to re-establish a tourism industry, particularly as a projected 60%–70% of tourism businesses have ended operations because of tourism immobility since the outbreak of the pandemic (Solomon Islands Government 2022).

Conclusion and research implications

This chapter asserts that despite a steady surge forward in tourism in the immediate years prior to the COVID-19 pandemic, the evolution of tourism development in the Solomon Islands has not been intensive, and there has been a lack of infrastructural capacity to develop a robust tourism industry. The Solomon Islands should seize the opportunity to develop tourism in a more sustainable way, learning from the mistakes associated with the unsustainable practices of the logging industry. The socio-cultural and economic challenges in the Solomon Islands require public and private stakeholders to improve the tourism viability of the island nation. Importantly, community-based tourism ought to be encouraged, in which community members are a primary stakeholder in decision-making and income allocation, especially to ensure that tourism is economically and socio-culturally sustainable.

Significant market research is necessary to tap into the potential of the Solomon Islands as an international tourism destination and to assist the country in recovering from the pandemic, which has, no doubt, placed economic challenges on the country. In terms of product development, heritage tourism could encourage social sustainability through considering those target markets interested in 'war tourism'. The branding of the Solomon Islands as a 'war tourism' or 'memorial tourism' destination would be pertinent through benchmarking existing destinations with proven success in advancing such forms of tourism. Gibson et al. (2021: 5) asserted that battlefield sites in the Solomon Islands could provide 'comparative advantage over other regional destinations'. Accordingly, the government of the Solomon Islands could devise a plan to upgrade the tourist facilities at its post-war tourism sites. Furthermore, the incorporation of storytelling narratives concerning war nostalgia and attractions could create interest for tourists and potential investors (Christou et.al. 2018).

Given that the country has been a source of conflict, as in the war years and in terms of more recent ethnic and political disputes, it could well be important to critically examine how tourism in the Solomon Islands can be developed as a 'force for good' (Bianchi and Stephenson 2014: 174), helping to strengthen 'civil society' (2014:191) and be 'mobilized as a tool for justice and reconciliation' (2014: 206). Future researchers could also identify feasible approaches to develop special interest tourism, such as pro-poor tourism and community-based tourism, especially as these forms of tourism could strategically fit with a culturally sustainable approach to tourism development. The important role of tourism in terms of poverty alleviation has been usefully evaluated in the Melanesia countries of Fiji (Scheyvens and Russel 2012) and Vanuatu (Trau 2012); however, more critical research is required on how specific types of tourism in the Solomon Islands can function effectively to alleviate poverty, which could be purposeful in the context of a post-COVID-19 recovery plan.

References

Allen, M. (2011) 'The political economy of logging in the Solomon Islands'. In: R. Duncan (ed.) *The Political Economy of Economic Reform in the Pacific*, Manila, Philippines: Asian Development Bank, pp. 277–301.

Allen, M.G. and S. Dinnen (eds.) (2018) *Statebuilding and State Formation in the Western Pacific: Solomon Islands in Transition?* London: Routledge.

ADB (Asian Development Bank) (2021) 'Investment in infrastructure, human capital can help Solomon Islands' recovery from COVID-19', 29 April, Philippines, Manila: Asian Development Bank. Available online at https://www.adb.org/news/investment-infrastructure-human-capital-can-help-solomon-islands-recovery-covid-19-adb#:~:text=HONIARA%2C%20SOLOMON%20ISLANDS%20(29%20April,a%20new%20report%20from%20the (accessed 15 August 2022).

Bakker, M. (2019) 'A conceptual framework for identifying the binding constraints to tourism-driven inclusive growth', *Tourism Planning and Development*, 16(5): 575–590.

Beattie, E. (2022) 'Paradise lost? China, the Solomons and the battle of the Pacific', *Nikkei Asia*, Nikkei INC. Available online at https://asia.nikkei.com/Spotlight/Asia-Insight/Paradise-lost-China-the-Solomons-and-the-battle-for-the-Pacific (16 August 2022).

Bianchi, R. and M. Stephenson (2014) *Tourism and Citizenship: Rights, Freedoms and Responsibilities in the Global Order*, London: Routledge.

Cassells, R. (2019) 'Engaging with churches to address development-related challenges in Solomon Islands', *Sites: A Journal of Social Anthropology and Cultural Studies*, 16(1): 109–134.

Christou, P., A. Farmaki and G. Evangelou (2018) 'Nurturing nostalgia?: A response from rural tourism stakeholders', *Tourism Management*, 69: 42–51.

Connell, J. and P. Cruz (2020) 'Supporting a sustainable recovery in Solomon Islands', *Pacific Economic Monitor*, pp. 22–24, Manila, Philippines: Asian Development Bank, December. Available online at https://www.adb.org/sites/default/files/publication/662406/pem-december-2020.pdf (accessed 4 April 2022).

DFAT (Department of Foreign Affairs and Trade) (2020) 'Market insights: Connecting Australian business to the world', Australian Government, pp. 1–4. Available online at https://www.dfat.gov.au/sites/default/files/solomon-islands-market-insights-2021.pdf (accessed 31 March 2022).

Diedrich, A. and S. Aswani (2016) 'Exploring the potential impacts of tourism development on social and ecological change in the Solomon Islands', *Ambio*, 45(7): 808–818.

Douglas, N. (2004) 'Towards a history of tourism in Solomon Islands', *The Journal of Pacific Studies*, 26(1–2): 29–49.

Dyer, M. (2017) 'Eating money: Narratives of equality on customary land in the context of natural resource extraction in the Solomon Islands', *The Australian Journal of Anthropology*, 28(1): 88–103.

Foukona, J.D. and J. Timmer (2016) 'The culture of agreement making in Solomon Islands', *Oceania*, 86(2): 116–131.

Gibson, D., E. Yai and S. Pratt (2022) 'Journeying into the past to discover the potential for WWII dark tourism in the Solomon Islands', *Current Issues in Tourism*, 25(14): 2285–2302.

Solomon Islands Government (2022) 'Tourism sector gears up for border re-opening', 24 March, Office of the Prime Minister & Cabinet: Government Communication Unit. Available online at https://solomons.gov.sb/tourism-sector-gears-up-for-border-re-opening/ (accessed 30 March 2022).

Hagedoorn, E. (2013) 'Child labour and tourism: how travel companies can reduce child labour in tourism destinations', *ICRT Occasional Paper* (OP26), May, International Centre for Responsible Tourism. Available online at https://respect.international/wp-content/uploads/2017/10/Child-Labour-and-Tourism.pdf (accessed 31 March 2022).

Hou, R.N. (2001) 'The Solomon Islands economy: Recent developments and the impact of ethnic tensions', *Pacific Economic Bulletin*, 17(2): 15–32.

Huo, S., M. Wang, G Chen, H. Shu, H. and R. Yang (2021) 'Monitoring and assessment of endangered UNESCO World Heritage Sites using space technology: A case study of east Rennell, Solomon Islands', *Heritage Science*, 9(101): 1–13.

Houston, M. (2022) 'Solomon Islands government confident of holding 2023 Pacific Games despite COVID-19', *Inside the Games*, 13 March. Available online at https://www.insidethegames.biz/articles/1120501/solomon-islands-2023-pacific-games-covid (accessed 17 March 2022).

Hviding, E. and T. Bayliss-Smith (2000) *Islands of Rainforest: Agroforestry, Logging and Eco-tourism in Solomon Islands*, Hampshire: Ashgate.

Hviding, E. (2005) *Reef and Rainforest: An Environmental Encyclopedia of Marovo Lagoon, Solomon Islands*, Knowledges of Nature 1 (second edition), United Nations Educational, Scientific and Cultural Organization: Paris. Available online at file:///C:/Users/courtsq/Downloads/138643qaa.pdf (16 August 2022).

Hviding, E. (2006) 'Knowing and managing biodiversity in the Pacific Islands: Challenges of environmentalism in Marovo Lagoon', *International Social Science Journal*, 58(187): 69–85.

Hviding, E. (2012) 'Sciences in the plural: The UNESCO Environmental Encyclopedia of Marovo Lagon, Solomon Islands'. *The Journal of Pacific Studies*, 32(2): 128–143.

IMF (International Monetary Fund) (2021) 'IMF staff concludes visit to Solomon Islands', 5 March. Available online at https://www.imf.org/en/News/Articles/2021/03/05/pr2162-solomon-islands-imf-staff-concludes-visit-to-solomon-islands (accessed 7 May 2022).

Iroga, R. (2008) 'Local media's role in peace building in post-conflict Solomon Islands'. In: E. Papoutsaki and U.S. Harris (eds.) *South Pacific Islands Communication: Regional Perspectives, Local Issues*, Singapore: Nanyang Technological University & Asia Pacific Information Communication Centre, pp.152–174.

Iroga, R. (2022) 'Airlines suspends services between Honiara and Marau due to land dispute', *SBM (Solomon Business Magazine) Online,* 7 January. Available online at https://sbm.sb/airlines-suspends-services-between-honiara-marau-due-to-land-dispute/ (accessed 16 August 2022).

Jiang, M., T. DeLacy, N.P. Mkiramweni and D. Harrison (2011) 'Some evidence for tourism alleviating poverty', *Annals of Tourism Research*, 38(3): 1181–1184.

Kiddle, G.L. (2020) 'Achieving the desired state of conservation for east Rennell, Solomon Islands: Progress, opportunities and challenges', *Asia and the Pacific Policy Studies*, 7(3): 262–277.

Leal Filho, W., M. Otoara Ha'apio, J.M. Lütz and C. Li (2020) 'Climate change adaptation as a development challenge to small Island states: A case study from the Solomon Islands', *Environmental Science and Policy*, 107: 179–187.

Lee, R.A.J. (2019) 'China poised to be a key growth driver for Pacific Islands' tourism', *TTG Asia*, 25 November. Available online at https://www.ttgasia.com/2019/11/25/china-poised-to-be-a-key-growth-driver-for-pacific-islands-tourism/ (accessed 17 March 2022).

Ligaiula, P. (2021) 'Solomon Islands riot cost US$66 million', *Pacific Island News Association*, 6 December. Available online at https://pina.com.fj/2021/12/06/solomon-islands-riot-cost-us66million/ (1 April 2022).

McDougall, D. and J. Kere (2011) 'Christianity, custom, and law: Conflict and peacemaking in the postconflict Solomon Islands'. In: M.J. Brigg and R. Bleiker (eds.) *Mediating Across Difference*, Honolulu, Hawai'i: University of Hawaii Press, pp. 141–162.

Panakera, C. (2007) 'World War II and tourism development in Solomon Islands'. In: C. Ryan (ed.), *Battlefield Tourism*: History, *Place and Interpretation*, Oxford: Elsevier: 125–141.

Patovaki, J. (2020) 'Challenges of maintaining a World Heritage Site in an emerging tourism destination: A case study of Morovo Lagoon in the Solomon Islands', *Master Thesis*, Suva, Fiji: University of the South Pacific.

Patovaki, J., B. Belo and H. Sore (2020) *Impact of State of Public Emergency on Locally Owned Tourism Operators within Honiara City During the COVID-19 Pandemic*, 6 December, Honiara, Solomon Islands: Tourism and Culture Division, Honiara City Council.

Piemonte, C. and A. Fabregas (2020) 'Solomon Islands transition finance country diagnostic: Preparing for graduation from Least Developed Country (LDC) status', *OECD Development Co-Operation Working Papers*, No 86, Paris: OECD Publishing.

Schellhorn, M. and H.C. Perkins (2004) 'The stuff of which dreams are made: Representations of the South Sea in German-language tourist brochures', *Current Issues in Tourism*, 7(2): 95–133.

Scheyvens, R., M. Russell (2012) 'Tourism and poverty alleviation in Fiji: Comparing the impacts of small- and large-scale tourism enterprises', *Journal of Sustainable Tourism*, 20(3): 417–436.

Shennan, J. (2013) 'Festival of Pacific arts', *DANZ Quarterly: New Zealand Dance*, 30: 2–3.

SINSO (Solomon Islands National Statistical Office), Solomon Islands Ministry of Health and Medical Services (SIMoHMS) and SPC (Pacific community) (2017), *Solomon Islands Demographic and Health Survey 2015: Final Report*, June. Prepared for publication at SPC's Noumea Headquarters, New Caledonia: Noumea Cedex.

Smith, A. (2011) 'East Rennell World Heritage Site: Misunderstandings, inconsistencies and opportunities in the implementation of the World Heritage Convention in the Pacific Islands', *International Journal of Heritage Studies*, 17(6): 592–607.

Solomon Airlines (2021) '"Let's keep our country moving" as national as national tourism effort lumi Tugeda Holidays expands', 6 July. Available online at https://www.flysolomons.com/about-us/news/general/lets-keep-our-country-moving-iumi-tugeda-holidays-expansion-july-2021 (accessed 9 September 2021).

Solomon Islands Government (2016) *National Development Strategy 2016–2035: Improving the Social and Economic Livelihoods of all Solomon Islanders*, April, Honiara, Solomon Islands: Ministry of Development Planning and Aid Coordination. Available online at https://www.adb.org/sites/default/files/linked-documents/cobp-sol-2017-2019-ld-01.pdf (accessed 16 March 2022).

Solomon Star (2019) '$433 infrastructural improvement project launched', 6 September. Available online https://www.solomonstarnews.com/433m-infrastructure-improvement-project-launched/ (accessed 5 September 2021).

Solomon Times (2019) 'Tourism Solomons hopeful of more Chinese tourists', *Solomon Times Online*, 4 November. Available online at https://www.solomontimes.com/news/tourism-solomons-hopeful-of-more-chinese-tourists/9440 (accessed 30 March 2022).

Sora, M. (2021) 'Australia's early intervention can help Solomon Island but the roots of the conflict run deep', *The Guardian*, 26 November. Available online at https://www.theguardian.com/commentisfree/2021/nov/26/australias-early-intervention-can-help-solomon-islands-but-the-roots-of-the-conflict-run-deep (accessed 16 August 2022).

Tourism Solomons (2018) 'Solomon Islands: How to get there'. Available online at https://www.visitsolomons.com.sb/plan-your-adventure/how-to-get-here/. Available online at https://www.visitsolomons.com.sb/plan-your-adventure/how-to-get-here/ (accessed 15 January 2019).

Tourism Solomons (2019) 'Solomon Islands tourism looks to attract 60,000 visitors annually by 2025', 7 February. Available online at https://www.visitsolomons.com.sb/solomon-islands-tourism-looks-to-attract-60000-visitors-annually-by-2025/ (accessed 14 September 2019).

Tovmasyan, G. (2016) 'Tourism development trends in the world', *European Journal of Economic Studies*, 3: 429–434.

Trau, A.M. (2012) 'Beyond pro-poor tourism: (Re)interpreting tourism-based approaches to poverty alleviation in Vanuatu', *Tourism Planning and Development*, 9(2): 149–164.

Travel Weekly (2019) 'Solomon Airlines launches Brisbane-Munda direct flight', 2 April. Available online at https://www.travelweekly.com.au/article/airline-wrap-low-cost-airline-launches-lufthansas-new-service-fresh-links-to-solomon-islands-more/ (accessed 12 September 2021).

Treloar, P. and C.M. Hall (2005) 'Tourism in the Pacific islands'. In: C. Cooper and C.M. Hall (eds.) *Oceania: A Tourism Handbook*, Clevedon: Channel View Publications, pp. 173–294.

Trip Advisor (2019) 'Solomon Island hotels'. Available online at https://www.tripadvisor.com.my/Hotels-g294139-Solomon_Islands-Hotels.html (accessed 1 October 2019).

UNESCO (United Nations Educational, Scientific and Cultural Organization) (2008) *Marovo – Tetepare Complex*, UNESCO World Heritage Convention. Available Online at https://whc.unesco.org/en/tentativelists/5414/ (accessed 20 September 2019).

Wairiu, M. (2007) 'History of the forestry industry in Solomon Islands: The case of Guadalcanal', *Journal of Pacific History*, 42(2): 233–246.

Wate, L. (2019) 'World Heritage Site program criticised', *Environment, Media: Solomon Islands*, 22 May. Available online at https://environment.islesmedia.net/world-heritage-site-program-criticized/ (accessed 29 March 2022).

Watson-Gegeo, K.A. and D.W. Gegeo (1992) 'Schooling, knowledge, and power: Social transformation in the Solomon Islands', *Anthropology and Education Quarterly*, 23(1): 10–29.

WorldData.info (2021) 'Tourism on the Solomon Islands'. Available online at https://www.worlddata.info/oceania/solomon-islands/tourism.php (30 March 2022).

Yang, J., C. Ryan and L. Zhang (2013) 'Social conflict in communities impacted by tourism', *Tourism Management*, 35: 82–93.

11

TOURISM POLICY AND PLANNING IN FIJI

A critique

Apisalome Movono and Marcus L. Stephenson

Introduction

Tourism has long been acknowledged as a tool by which the developmental ambitions of small islands states can be operationalized (Movono and Becken 2018; Prasad 2014). Recently, more attention is being given to tourism because of its potential to accomplish the United Nations Sustainable Development Goals at the localized community level (Scheyvens 2018). In Fiji and in other parts of the Pacific, tourism projects an ideal economic alternative, appealing to the deficiencies typical of Pacific Island states (Movono et al. 2015; Prasad 2014). However, there remains much debate, as the body of literature about Pacific Island tourism and its relevant policies remain fragmented. Although tourism policy and its examination has gained momentum amongst scholars in developing countries, very few studies have yet fully explored this area in the context of a specific tourism-dependent country in the Pacific. The examination of tourism-related policies becomes an important point for debate, as these are now usually contextualized as part of a sustainable development meta-discourse. The impacts of tourism policies are complex in nature and are foremost in influencing tourism development and its impacts at all levels within a country. As such, it is essential to explore how tourism policy development evolved in a country such as Fiji, especially to better understand how tourism policies are implemented and the extent to which they are actionable.

This study will provide a contextual background to the study area by outlining the history of tourism policy development and the implementation challenges in the Republic of Fiji. Ultimately, this chapter will argue that tourism policy and planning frameworks must be encouraged through a holistic, multi-national and regional level context, especially to ensure that national tourism policies are in line with the United Nations Sustainable Development Goals (UNSDG). This chapter will first offer some background information on Fiji and its tourism industry. The work will then provide an overview of tourism policy and its interconnected role to planning and will look closely at policy objectives, issues, concerns and implementation in the context of the Republic of Fiji. It will outline the historical development of tourism policies in Fiji, discuss specific challenges and propose the use of binding multi-national agreements as a means to encourage sustainable tourism policy development (and implementation) as a pathway to attaining the UNSDGs. Finally, the chapter will discuss the impacts of the COVID-19 pandemic on

DOI: 10.4324/9780429019968-13

the tourism industry, indicating how tourism policies and related planning directives will need to be revisited in light of the economic drawbacks of being overly dependent on tourism. This enquiry is an extension of ongoing research by the first author of this chapter, which initially began in 2009 with postgraduate research then leading to an ethnographic-based PhD thesis, focusing on tourism-dependent communities and community resilience in Fiji (Movono 2017).

Tourism in Fiji - overview

The country is located approximately 2,000 kilometres north of Auckland and 5,000 kilometres southwest of Honolulu (Rao 2002). Fiji has a population of some 884,887 people spread across its 14 provinces (Figure 11.1), over a quarter of whom are concentrated on the main island of Viti Levu (Fijian Government 2018). The Fiji Islands became part of the British Crown Colony in 1874, and it was not until 1974 that Fiji became politically independent from Britain (Foster and Macdonald 2021). Although sugar was the principal industry in Fiji in terms of its contribution to gross domestic product (GDP), foreign exchange earnings and employment, this industry was exceeded by tourism in the late 1990s, as it became the largest exchange earner (Rao 2002).

Tourism formally began in Fiji's capital city Suva in the early twentieth century with the establishment of the White Settlers League, a body representing European settlers who promoted Fiji to passengers disembarking from ships that crossed the Pacific. The White Settlers League and its charter became the first tourism-related policy in Fiji, endorsed by the colonial government and established in 1923, setting up the Suva Tourism Board and later the Fiji Publicity Board in 1925 (Scott 1970). The Publicity Board was replaced by the Fiji Visitor

Figure 11.1 Map of Fiji, indicating the provinces. Source: Khaiinauylovinns (www.shutterstock.com/image-vector/fiji-map-vector-illustration-on-white-2144071061).

Bureau, formed under the Fiji Commission and Visitors Bureau Act of 1969 (Prasad and Tisdell 2006), later becoming Tourism Fiji, the principal statutory body responsible for promoting, marketing and coordinating tourism. Despite its early beginnings, tourism only started to develop in Fiji after World War II, prompted by the rise of disposable incomes in Australia and New Zealand (Fiji's main sources markets) and by developments in hotel infrastructure, transportation and specific government policies that laid the foundation for a robust tourism industry (Movono et al. 2015; Movono et al. 2018). The Belt Collins Report in 1973 became the Fijian government's first significant attempt to plan for (and anticipate) tourism development (Belt, Collins and Associates Ltd 1973). The report was timely given the emerging increase in tourism visitation, in which the Fijian tourism industry grew significantly from the late 1960s and early 1970s, with tourists increasing from 40,000 visitors in 1973 to 186,000 visitors in 1983 (Treloar and Hall 2005). The government saw tourism as an integral and favourable element of the nation's economy, and this vision was embodied within the 1985 National Development Plan which perceived tourism as making a constructive contribution to the economy (Kanemasu 2015).

Fiji recorded the highest level of visitor arrivals in 2019, standing at 894,389 and representing an increase of 2.8% from the previous year in which there were 367,020 visitors from Australia and 205,998 from New Zealand, with 656,249 visitors travelling to the country for holiday purposes, 92,026 visiting friends and relatives, 29,882 travelling for business purposes, and 116,232 for other reasons (Singh 2020). Tourism represented nearly 40% of Fiji's GDP, employing around 150,000 people though by mid-April 2020 around 95% of the country's flights were grounded due to border closures and travel restrictions, along with the closure of 279 hotels and resorts and 25,000 job losses (Chanel 2020). The pandemic brought to light the extent to which Fiji was economically dependent on tourism, especially as a consequence of a range of pro-poor tourism policies which encouraged significant tourism development.

Tourism policy and planning - overview

Tourism policy and planning are value-laden and typically political activities, despite their economic importance (Hall and Jenkins 1995). The literature on tourism policy from the 1960s onward is not only fragmented but also is often nestled under more prominent areas such as tourism planning and development. It is only after the 1980s that sustainability and, more recently, climate change emerged as influential issues woven into public policy (Brendehaug et al. 2017; Hall and Jenkins 1995). However, for public policy to be effective, then, there needs to be a clear planning framework in place. Williams and Lew (1998: 125) emphasize that it is planning that has a key role in ensuring tourism development is sustainable, especially if it is to be a central mechanism for 'conserving scarce or important resources'. The many policy development frameworks that exist are described as having either reactive and proactive approaches, which define the tourism policy setting process (Hall and Jenkins 1995). As a result, most policies relating to tourism can be grouped into three broad categories: (1) policies designed to encourage tourism, (2) policies which maximize revenue from tourism and (3) those which aim to control tourism impacts. The literature also indicates that government policies implemented in developing countries are more economically motivated, with countries setting agendas in which tourism is regulated to maximize economic benefits for the country and relevant stakeholders (Kerr 2003). The literature also implies that, despite international benchmarking, each country is responsible for setting its own tourism policy agenda based on its own priorities (Cheer et al. 2017; Hall and Jenkins 1995). As such, it becomes relevant to question how policies are designed and for what purpose.

Tourism-specific policies have many benefits which allow a country to control its growth and influence how the industry should behave. Having sound integrated policies are beneficial for providing a platform for planning and absorbing shocks, and managing the benefits and non-benefits that stem from such a complex industry. Nonetheless, Kerr (2003) notes that often there is a lack of interest and political will to establish and implement policies that govern tourism-based activities. National priorities for developing countries such as Fiji have been influenced by the global and competitive tourism marketplace, in which tourism has been marketized and seen as a tool for economic rather than social development (Hall 1996, 2011). As such, tourism policies may lack the holistic attention required to encourage internal, local and multi-sectoral linkages to increase opportunities for poverty alleviation and local empowerment in the tourism production system (Adiyia et al. 2015).

Another feature of tourism policy implementation involves the lack of attention to long-term implementation, which is common in countries, particularly in the Pacific, where national interests are determined through political priorities and election cycles (Hall 1996; Harrison and Pratt 2010). The continuity and constant monitoring of tourism policies is a grey area in which very little discussion is held and not enough questions are asked about the long-term sustainability of tourism policy. In recent years, a high-level sustainability agenda has prevailed, in which more countries have become inspired to incorporate strategies for sustainable development within tourism policy (Brendehaug et al. 2017; Scheveyns 2018). This has prompted countries to include sustainable measures within strategic development plans, though often without proper reporting mechanisms to monitor the effectiveness of such policies at regular intervals. This is particularly true in Fiji where there is often very little information about the effectiveness or status of the implementation of its strategic tourism plans. Furthermore, there is a lack of empirical studies which focus on critically evaluating government policies that govern tourism, thereby indicating another gap in the literature. This study, thus, attempts to shed light on a pathway forward to ensure that a cohesive policy regime can exist for a Pacific Island state.

Tourism policy and planning in Fiji

Since the development of Fiji's pioneering tourism infrastructure in the 1960s, the country has developed specific tourism-based policies that can be broadly divided into policies designed to entice tourism development, manage tourism's impacts and maximize government revenue. In tourism's early years in Fiji, the first tourism specific policies developed and ratified were those designed to encourage investors, many of whom were convinced by the rapid increase in arrival numbers to help advance the tourism industry (Goundar 1983). Specific policies were first implemented by the Native Land Trust Board (NLTB) (now iTaukei Land Trust Board in 2011), which is the administrator of all native land in Fiji, to create new lease titles that could cater to tourism development (Goundar 1983; Scott 1970). NLTB's primary function was to negotiate fair lease payments on behalf of landowners, supporting major tourism projects and plans for the development of tourism infrastructure (Scheyvens and Russell 2012). A former retired tourism minister states:

> Being part of negotiations were not only interesting but we had a duty to landowners to properly interpret tourism's potential and promise…a task that we did not take lightly.
>
> *(pers. comm. May 2019)*

The 1970s heralded a new wave of optimism in Fiji, where the potential of sugar was fully realized, and it was time to venture into tourism. The period of optimism in the evolving stages of tourism development is best described by Doxey (1976) as being a time of extreme euphoria, in which there is significant optimism concerning the foreseeable benefits of tourism. Landowners were encouraged to lease their lands in return for preferential employment agreements and the promise of lucrative land lease payments (Goundar 1983; Scott 1970). Nonetheless, not all Fijians were ecstatic about tourism development. In the late 1960s and 1970s, the chief minister and the first prime minister of Fiji, Ratu Kamisese Mara, was critical of tourism development and concerned with the limitations of local ownership, exploitation of local artisans and foreign tour operators' lack of cultural awareness (Kanemasu 2015). Eventually the leasing of customary land to foreign citizens and companies prevailed, especially in terms of private investment in tourism ventures and tourism resorts (Britton 1983).

Another significant tourism policy designed to entice investors and tourists was the 1958 Hotel Aid Ordinance Act, which focused on tax concessions on certain goods and investments and represented a positive step forward in the provision of assistance and incentives for hotel development (Goundar 1983). This includes subsidies, cash grants, investment allowances and tax incentives given to developers in return for their investment in Fiji. Simultaneously, the government focused on infrastructural assistance such as transport, roads and electricity in tourism areas while also establishing a zero-duty policy on certain items such as electronics and other items to entice visitors to choose Fiji as a shopping destination (Belt, Collins and Associates Ltd. 1973; Goundar 1983). This multi-pronged and incentivized approach to tax management kick-started the Fijian tourism industry and, according to a former government planner, 'contributed to the building of a strong platform for Fiji to enter the tourism scene' (pers. comm. June 2019). However, over-focusing on incentivizing tourism development and infrastructural development echoes the planning tradition of 'boosterism'. This is a popular planning approach that emerged from the 1960s and predicated on the importance of capital injection to advance development, though it has been criticized for not equitably involving hosts communities in the decision-making process (Sharpley 2008).

Nonetheless, since tourism's establishment in Fiji, the industry remained on a steady path of growth and became successful as a revenue earner, prompting the government to invest in continued tourism development plans but also explore the development of policies to protect its golden goose. The Belt Collins Report (1973) and its predecessors became the central framework for how tourism was to develop and how public infrastructure investments were to be made. This came at a critical moment in the history of Fiji because this development plan encouraged future tourism growth. This initiative gave way to the government's adoption of a policy strategy entitled 'ecotourism and village-based tourism', produced by the Ministry of Tourism and Transport in 1999 (Bricker 2002) (see Chapter 15). As Bricker (2002: 235) emphasizes, this policy recognized that ecotourism should complement conventional tourism but that conservation should be more of a primary objective, where village tourism should be 'centrally available' but 'regularly monitored' and linked to the improvement of the quality of village life, and in which the government clarifies rights of ownership to avoid tourism development conflicts. In fact, King's (1997) study of the Mamanuca Islands in Fiji found that, although resorts do have an intense commercial focus, there is significant opportunities for tourists and local Fijian communities to closely interact. Nevertheless, the policy concerning ecotourism development not only recognizes the need for alternative forms of tourism but also acknowledges the shortcomings of tourism both at the community and micro level as well as the leakages and revenues that are lost through the large multi-national companies (Britton 1983).

In addition to pursuing resort-based tourism, Fiji also encouraged backpacker establishments and villages to operate ecotourism products involving Indigenous Fijians. This gave rise to projects such as the Tribe Wanted Project (Keene 2005) and a wide disbursement of small Indigenous owned ecoresorts such as the Wayalailai Ecohaven Resort, which provided direct economic and social benefits to local communities (Gibson 2015). In 2005, the Environment Management Act (EMA) was enacted, the first legislation to implement the precautionary principle encouraging developments to undertake Environmental Impact Assessments (EIAs) (Finau et al. 2013). This was the most significant policy implemented in Fiji to date, seeking to control and limit the ecological impacts of tourism. Prior to the EMA, the Fiji government endorsed environmental regulations to deal with the negative effects of economic development on environmental degradation, such as the Land Conservation and Improvement Act of 1953 and the Forestry Act of 1963, but directives were not effective, as the agencies accountable for implementing such legislation were not proactive (Finau et al. 2013). Nonetheless, whilst EIAs are part of a system for managing the impacts of tourism, they 'do not effectively hold anyone to account for the impacts of development nor do they build the foundations for ongoing environmental monitoring' (Goodwin 2011: 221). Controlling and limiting the impacts may well be conducive but only within the context of avoiding high-impact and conventional tourism activities, particularly as they influence biodiversity loss and threaten Fiji's mangroves, estuaries and reef and foreshore ecosystems (Tyllianakis et al. 2019).

It was not until the post-2006 coup that more aggressive policies were implemented as a means to boost government revenue. This began with the introduction of the Surfing Areas Decree 2010 (President of the Republic of Fiji 2010), Super Yacht Decree 2010 (Fijian Government 2010) and the Fiji Casino (Operator) Decree 2012 (Fijian Government 2014), moves which indicated the diversification of Fiji's tourist market beyond the traditional honeymoon and family markets. There is little data to support the worth of the two latter decrees, implying the value in sustaining and growing Fiji's image as a family destination. The introduction of the 'Service Turnover Tax' (STT) in 2012 and the 'environmental levy' in 2015, marked the introduction of policies to increase tourism's potential as a means of earning financial gains for the government (Fiji Times 2015). Nonetheless, the environmental levy broadened out to the 'Environment and Climate Adaptation Levy' (ECAL) on prescribed services, items and income for which is utilized to protect the natural environment (Fiji Ministry of Economy 2018). Nonetheless, as a consequence of the COVID-19 pandemic and its effect on tourism and industry, the Fiji government announced that the 6% SST has been eliminated and ECAL will decrease to 5% from 10% (Narayan 2020). One tourism operator raised concerns relating to such levies, stating:

> Since 2007 there have been many government policies that specifically target tourism, which means that we have to transfer costs to consumers and this affects our business, which is mainly nature-based and small scale.
>
> *(pers. comm. May 2019)*

The challenge is that tourism has been Fiji's largest economic contributor, and because of governance regimes, the country has become more dependent than ever on tourism for economic benefits and tax revenue. A former CEO of the Fiji Visitors Bureau emphasizes that 'the challenges of implementing the tourism development plans and policies of the government can be attributed to political instability, quick turnover of government and ministry staff, and the lack of accountability in reporting and policy review' (pers. comm. May 2019). Another former official of the Ministry of Industry, Trade and Tourism stresses:

> Progress in implementing policies and tourism development plans were greatly hindered by the unstable political situation in 1987 and later in 2000 and 2006… changes in governments following the coups brought many new priorities which were meant

> to please the electorate yet strayed attention away from implementing, monitoring and adequately streamlining government policy on tourism.
>
> *(pers. comm. May 2019)*

There is a notable absence of clear timelines and reporting procedures for the tourism development plans. As a former high-ranking civil servant claims:

> The development plans were more to show that plans were being prepared and that we were doing something, however, little attention was given to the mechanisms involved in its thorough implementation…, which is compounded by the fragmented nature of tourism governance laws and policies
>
> *(pers. comm. June 2019)*

Tourism policy and planning frameworks must be adopted at a multi-national and regional level first, especially to ensure that national tourism policies are in line with the UNSDGs so that longevity and sustainability of policy implementation are comprehensively achieved in Fiji (Scheyvens 2018) and that localized policies and plans in Fiji are legitimized by a process of alignment to globally accredited frameworks. Nonetheless, Fiji has been trying to prioritize climate change concerns, as indicated in SDG13 which emphasizes the importance of taking urgent action to tackle climate change and its impacts. Climate change concerns are a potential threat to tourism development in Fiji and noticeable through such impacts as coral bleaching, coastal deterioration and beach erosion (Moreno and Becken 2009). However, although climate change is recognized at the industry and government level in Fiji, integrating climate change into sectoral policies, there is still a prevailing lack of awareness across the tourism sector and a reluctance to work toward most favourable forms of climate change adaption (Jiang et al. 2012). These authors note:

> Climate change risks are threatening the sustainable growth of the tourism industry and therefore, tourism-specific adaptation strategies need to be developed to assist the sector in protecting and growing local livelihoods in Fiji. The Fijian government will need to collaborate with the private sector, communities and other stakeholders to address those adaption gaps.
>
> *(2012: 257)*

Singh et al. (2021) indicate that, irrespective of various local regulations (e.g., Environmental Levy, Land Use Policy, Code of Logging, and the Sustainable Development Bill) and endorsement of worldwide conventions (e.g., International Convention on Biological Diversity and United Nations Convention on the Law of the Sea), serious environmental challenges and impacts continue to unfold in Fiji. They further assert that there is a need for the government to ensure more enforcement and guidelines for environmental institutions and environmental evaluation and monitoring, as well as encouraging more affirmative 'corporate social responsibility' (CSR) strategies for marine and coastal protection and restoration from the private sector. Moreover, a CSR approach for the hospitality industry in Fiji should arguably pursue a 'development first' approach, which involves such directives as 'human development and community well-being', 'reducing vulnerability of the poorest', 'consideration of gender norms and inequalities', 'distribution of benefits', 'valuing cultural capital', 'utilization of local knowledge' and 'accountability to local communities' (Hughes and Scheyvens 2016: 476).

The problems associated with policy implementation are complex. In Fiji, the concerns are threefold: the first and most obvious refers to shifts in government policies with changes in political leadership and a change of perspective toward tourism development; the second concerns the lack of a holistic approach to tourism policy agenda setting; and the third relates to the clear need for a well-aligned tourism policy agenda – together with a genuine and united commitment to see through the implementation in attainment of the UNSDGs. As such, this chapter broadly implies that the option of adopting binding multi-national agreements for the regional cooperation in the attainment of the UNSDGs should be explored first. This will provide the political will and reflect a communal effort in ensuring that tourism is utilized as a tool with which the goals of the UNSDGs can be realized (Scheyvens 2018; Scheyvens and Movono 2018).

Nonetheless, the government did establish the Fijian Development Plan 2017–2021 ('Fijian Tourism 2021'), which recognized the need for improvements in sustainable tourism development through more environmental compliance, enforcement and monitoring of conditioned approvals of EIAs, and more formalized guidelines to cultural conservation (Ministry of Industry, Trade and Tourism 2017). This latter recommendation challenges the way in which tourism can commodify culture, which can often happen in tourism destinations (see Greenwood 1989). Fijian Tourism 2021 actually highlighted the importance of investing in 'Fijian Made' products (Ministry of Industry, Trade and Tourism 2017:35), which can be seen as a form of cultural preservation through reinforcing the importance of 'authenticity' (see MacCannell 1999).

Tourism development, the COVID-19 pandemic and post-pandemic recovery

The problem with tourism policy is not being able to predict the various impeding crisis scenarios that can impact the tourism industry and render current policies less relevant. In addition to focusing on the sustainable development of the industry, 'Fijian Tourism 2021' also established such key objectives as encouraging domestic and foreign investment, infrastructural development and investment, and enhancing a risk management agenda – with the intention of targeting a diverse range of niche forms of tourism to help Fiji achieve 930,000 visitors by 2021 (Ministry of Industry, Trade and Tourism 2017). However, the outbreak of the COVID-19 pandemic in March 2020 ensured that these projections were no longer feasible, especially as Fiji's borders were closed to international tourism and travel – with devastating consequences for Fijian nationals. Accordingly, 'micro, small and medium enterprises' in the tourism sector lost seven times more income than those in the non-tourism sector, and by April 2020, tourism revenues declined by 59% compared with April 2019 (IFC 2020). Around 93% of 279 members of the Fiji Hotel and Tourism Association (FHTA) closed down their businesses (Fijian Trade Union Congress 2020). Although Fiji managed to control the numbers of COVID-19 transmissions to low numbers in 2020, the situation deteriorated throughout the latter half of 2021, with confirmed cases increasing from 72 in mid-April to more than 52,000 by the end of October 2021 (Wainiqolo and Del Castillo 2021).

Nonetheless, as the country experienced undertourism and controlled movements through travel bubbles, it is purposeful to reassess and revisit tourism policies and related planning directives in light of lessons learnt in being over-dependent on tourism, as was the case prior to the outbreak of the pandemic. Fiji's over-reliance on tourism development was facilitated by pro-tourism policies. Therefore, there is a need to ensure that policies, supported by clear planning directives, can integrate tourism more with the other sectors of the economy (see Chapter 8). However, the government continues to foster a pro-tourism policy approach, and this was acknowledged in the recent Fiji Investment Act of 2021, prearranged by Fiji's Ministry of Commerce, Trade, Tourism and Transport to strengthen protection guarantees for foreign investors; an act that replaced the Foreign Investment Act of 1999 (UNCTD 2021). Post-pandemic recovery planning should arguably look at ways in which policy can encourage people to diversify their livelihoods in tourism-related communities, which should not be an arduous task, as many unemployed Fijians during the height of the pandemic returned to their villages and re-learnt traditional skills and embraced subsistence living in order to survive (Movono et al. 2022).

Since 2020, the government placed significant emphasis on advancing domestic tourism and encouraging businesses to restart operations. However, the reduction of ECAL (as indicated above) is arguably a 'risky decision', as it destabilizes climate change goals (UNSDG 13) (Becken and Loehr 2022). Therefore, future planning directives and policies need to be careful not to prioritize economic development at the expense of both the ecological environment and the socio-cultural environment. Nonetheless, an emphasis on domestic tourism and transport policies could continue to help diversify the tourism market dominated by international tourism and travel. Appropriately, Hutchison et al. (2021: n.p.) assert, 'for too long tourism has been focused on economic growth and meeting tourist demands and not on what is sustainable for people and places'. These authors emphasize that it is crucial that tourism development, planning and policy in Fiji are aligned to the needs and interests of its citizens and national development.

Conclusion and research implications

This chapter has exemplified that tourism policy in Fiji is fraught with many cultural, political and socio-economic complexities, often affecting (and intersecting with) policy setting and implementation. However, it would be useful for Pacific Island states, like Fiji, to adopt a multi-pronged approach to the policy setting, particularly where the UNSDGs are utilized as a critical focus by which tourism policy, planning and implementation are aligned and encouraged using a regional, multi-government agreement to encourage long-term implementation (Scheyvens 2018). Furthermore, a more holistic approach to the tourism policy setting is necessary, one that integrates with other sectors and builds resilience within society. The UNSDGs offer hope to tackle socio-economic and environmental vulnerabilities in destinations, providing a policy framework for a balanced approach to tourism development for a sustainable future.

The work encourages more empirical-based enquiries to take place, focusing on the degree of effectiveness of the governance of tourism development and government policies. Therefore, the intention would be to thoroughly deconstruct the policies themselves, especially the political ideologies that underpin policy formulation and policy application. Also, clear case assessments concerning the monitoring of tourism and tourism-related policies and degrees of effectiveness of implementation would be useful, along with empirical assessments of the effectiveness of key stakeholders involved in the policy making and implementation process. Crucially, it is pertinent to consider what policies are now necessary to ensure that tourism adopts a restorative function in light of the impending post-COVID era, ensuring there is a clear symbiotic relationship between tourism development and sustainable development; one which is sensitive to the needs of Fijian nationals.

References

Adiyia, B., A. Stoffelen, B. Jennes, D. Vanneste and W.M. Ahebwa (2015) 'Analysing governance in tourism value chains to reshape the tourist bubble in developing countries: The case of cultural tourism in Uganda', *Journal of Ecotourism*, 14(2–3): 113–129.

Becken, S. and J. Loehr (2022) 'Asia–Pacific tourism futures emerging from COVID-19 recovery responses and implications for sustainability', *The Journal of Tourism Futures*, 8:1. https://doi.org/10.1108/JTF-05-2021-0131 (accessed 27 February 2022).

Belt, Collins and Associates Ltd (1973) *Tourism Development Program for Fiji*, Washington: United Nations Development Program/International Bank for Reconstruction and Development.

Brendehaug, E., C. Aall and R. Dodds (2017) 'Environmental policy integration as a strategy for sustainable tourism planning: Issues in implementation', *Journal of Sustainable Tourism*, 25(9): 1257–1247.

Bricker, K. (2002) 'Ecotourism development in the rural highlands of Fiji'. In: D. Harrison (ed.), *Tourism and the Less Developed World: Issues and Case Studies*, Oxford: CABI Publishing, pp. 235–249.

Britton, S.S.G. (1983) *Tourism and Underdevelopment in Fiji*, Canberra: Australian National University Press.

Chanel, S. (2020) 'It's catastrophic: Fiji's colossal tourism', *The Guardian*, 15 April. Available online at https://www.theguardian.com/world/2020/apr/16/its-catastrophic-fijis-colossal-tourism-sector-devastated-by-coronavirus (accessed 20 February 2021).

Cheer, J.M., S. Pratt, D. Tolkach, A. Bailey, S. Taoumoepeau and A. Movono (2017) 'Tourism in Pacific island countries: A status quo round up', *Asia and the Pacific Policy Studies*, 5(3): 442–461.

Doxey, G.V. (1976) 'When enough's enough, the natives are restless in old Niagara', *Heritage Canada*, 9(2): 26–29.

Fijian Government (2010) 'Super Yacht Charter Decree endorsed', 31 March, Suva, Fiji. Available online at https://www.fiji.gov.fj/Media-Centre/News/Super-Yacht-Charter-Decree-endorsed (accessed 12 April 2022).

Fijian Government (2014) 'Constitution of the Republic of Fiji (Section 92(3))', *Government of Fiji Gazette Supplement*, No. 22. 25 September. Available online at http://macbio-pacific.info/wp-content/uploads/2017/08/LN-43-Ministerial-AssignmentsFiji.pdf (accessed 12 April 2022).

Fijian Government (2018) 'Fiji Bureau of Statistics releases 2017 census results', 10 January. Available online at https://www.fiji.gov.fj/Media-Centre/News/Fiji-Bureau-of-Statistics-Releases-2017-Census-Res (accessed 12 April 2022).

Fijian Trades Union Congress (2020) *Impact of COVID-19 on Employment & Business: In-crisis Rapid Assessment – 13 May–19 June 2020 (Volume 1)*, 29 August. Available online at https://www.ilo.org/wcmsp5/groups/public/---asia/---ro-bangkok/---ilo-suva/documents/publication/wcms_754703.pdf (accessed 27 February 2022).

Fiji Ministry of Economy (2018) 'ECAL in action: How your environment and climate adaptation levy is building a better, stronger Fiji', *Bulletin*, 1 June, Fiji Climate Change and National Designated Authority (NDA) Portal, Suva, Fiji. Available online at https://fijiclimatechangeportal.gov.fj/publication/ecal-in-action-how-your-environment-and-climate-adaptation-levy-is-building-a-better-stronger-fiji-bulletin-01-2018-june-2018-environment-and-climate-adaptation-levy/ (accessed 20 August 2020).

Fiji Times (2015) 'Tax collection', 12 November. Available online at https://www.fijitimes.com/tax-collection/ (accessed 20 February 2021).

Finau, G., J. Samuwai, M. Rotuivaqali, C. Kuma, L. Kanaenabogi and T. Veituna (2013) 'The impact of the Environment Management Act (EMA) on the accountability of companies in Fiji', *Journal of Modern Accounting and Auditing*, 9(9): 1216–1234.

Foster, S. and B.K. Macdonald (2021) 'Fiji', *Encyclopedia Britannica*, 10 March. Available online at https://www.britannica.com/place/Fiji-republic-Pacific-Ocean (accessed 27 February 2022).

FVB (Fiji Visitor Bureau) (2005). *Annual Report*, Suva: FVB.

Gibson, D. (2015) 'Community-based tourism in Fiji: A case study of Wayalailai Ecohaven Resort, Yasawa island group'. In: S. Pratt and D. Harrison (eds.) *Tourism in Pacific Islands: Current Issues and Future Challenges*, Abingdon: Routledge, pp. 142–157.

Goodwin, H. (2011) *Taking Responsibility for Tourism*, Woodeaton, Oxford: Goodfellow Publishers Limited.

Goundar, R. (1983) 'Government policies of tourism in Fiji and the development towards this industry', *Unpublished Research Paper*, Suva, Fiji: University of the South Pacific.

Greenwood, D.J. (1989) 'Culture by the pound: An anthropological perspective on tourism as cultural commodification'. In: V.L. Smith (ed.), *Hosts and Guests: The Anthropology of Tourism*, 2nd edition, Philadelphia: Pennsylvania Press, pp. 171–185.

Hall, M. (1996) 'Political effects of tourism in the Pacific'. In: M. Hall and S. Page (eds.) *Tourism in the Pacific: Issues and Cases*, London: International Thompson Business Press, pp. 91–108.

Hall, C.M. (2011) 'A typology of governance and its implications for tourism policy analysis', *Journal of Sustainable Tourism*, 19(4–5): 437–457.

Hall, C.M. and J. Jenkins (1995) *Tourism and Public Policy*, London: Routledge.

Harrison, D. and S. Pratt (2010) 'Political change and tourism'. In: R. Butler and W. Santikul (eds.) *Tourism and Political Change*, Oxford: Goodfellow Publishers Ltd, pp. 160–172.

Hughes, E. and R. Scheyvens (2016) 'Corporate social responsibility in tourism post-2015: A development first approach', *Tourism Geographies*, 18(5): 469–482.

Hutchison, B., A. Movono and R. Scheyvens (2021) 'Resetting tourism post-COVID-19: Why Indigenous peoples must be central to the conversation', *Tourism Recreation Research*, https://doi.org/10.1080/02508281.2021.1905343.

IFC (International Finance Corporation) (2020) 'Fiji COVID-19 business survey: Tourism focus – impacts, responses and recommendations', *World Bank Group*, July. Available online at https://www.ifc.org/wps/wcm/connect/4fc358f9-5b07-4580-a28c-8d24bfaf9c63/Fiji+ COVID-19+Business+Survey+Results+-+Tourism+Focus+Final.pdf?MOD=AJPERES&CVID=ndnpJrE (accessed 11 April 2022).

Jiang, M., E. Wong, L.M. Klint, T. DeLacy and D. Dominey-Howes (2012) 'Tourism adaptation to climate change–analysing the policy environment of Fiji', *International Journal of Tourism Policy*, *4*(3): 238–260.

Kanemasu, Y. (2015) 'Fiji tourism half a century on: Tracing the trajectory of local responses'. In: S. Pratt and D. Harrison (eds.) *Tourism in Pacific Islands: Current Issues and Future Challenges*, Abingdon: Routledge, pp. 63–84.

Keene, B. (2005) *Tribe Wanted: My Adventures on Paradise or Bust*, London: Ebury Press.

Kerr, W.R. (2003) *Tourism Public Policy and the Strategic Management of the Failure*, London: Pergamon.

King, B.E.M. (1997) *Creating Island Resorts*, London: Routledge.

MacCannell, D. (1999) *The Tourist: A New Theory of the Leisure Class*, Berkeley: University of California Press (first published by 1976 by Schocken Books Inc.).

Ministry of Industry, Trade and Tourism (2017) 'Fijian tourism 2021. ministry for industry, trade and tourism'. Available online at mitt.gov.fj/wp-content/uploads/2019/04/FT2021.pdf (accessed 6 April 2022).

Moreno, A. and S. Becken (2009) 'A climate change vulnerability assessment methodology for coastal tourism', *Journal of Sustainable Tourism*, 17(4): 473–488.

Movono, A. (2017) 'Na Irevurevu: Exploits, resilience, and tourism development in Vatuolalai Village, Coral Coast, Fiji', *Unpublished PhD Thesis*, Queensland, Australia: Griffith University.

Movono, A. and S. Becken (2018) 'Solesolevaki as social capital: A tale of a village, two tribes, and a resort in Fiji', *Asia Pacific Journal of Tourism Research*, 23(2): 146–157.

Movono, A., H. Dahles and S. Becken (2018) 'Fijian culture and the environment: A focus on the ecological and social interconnectedness of tourism development', *Journal of Sustainable Tourism*, *26*(3): 451–469.

Movono, A., D. Harrison and S. Pratt (2015) 'Adapting and reacting to tourism development: A tale of two villages on Fiji's coral coast'. In: S. Pratt and D. Harrison (eds.) *Tourism in Pacific Islands: Current Issues and Future Challenges*, Abingdon: Routledge, pp. 100–114.

Movono, A., R. Scheyvens and S. Auckram (2022) 'Silver linings around dark clouds: Tourism, Covid-19 and return to traditional values, villages and the vanua', *Asia Pacific Viewpoint*. https://doi.org/10.1111/apv.12340.

Narayan, V. (2020) 'Major changes announced in the 2020/2021 national budget', 17 June. Available online at https://www.fijivillage.com/feature/National-Budget-2020-2021-Announcement-fx485r/ (accessed 20 January 2022).

Prasad, B.C. (2014) 'Why Fiji is not the "Mauritius" of the Pacific? Lessons for small island nations in the Pacific', *International Journal of Social Economics*, 41(6): 467–481.

Prasad, B.C. and C. Tisdell (2006) *Economic Performance and Sustainable Development: A Case Study of Fiji Islands*, New York: Nova Science Publishers.

President of the Republic of Fiji (2010) *'Regulation of surfing areas decree 2010: Decree no. 35 of 2010'*, Suva: Republic of Fiji Islands.

Rao, M. (2002) 'Challenges and issues for tourism in the South Pacific island states: The case of the Fiji Islands', *Tourism Economics*, 8(4): 401–429.

Scheyvens, R. (2018) 'Linking tourism to the sustainable development goals: A geographical perspective', *Tourism Geographies*, 20(2): 341–342.

Scheyvens, R. and A. Movono (2018) 'Development and change: Reflections on tourism in the South Pacific', *Development Bulletin* (No. 80), Development Studies Network, Canberra, Australia: Australian National University, pp. 134–139.

Scheyvens, R. and M. Russell (2012) 'Tourism and poverty alleviation in Fiji: Comparing the impacts of small- and large-scale tourism enterprises', *Journal of Sustainable Tourism*, 20(3): 417–436.

Scott, R.J. (1970) 'The development of tourism in Fiji since 1923', *Transactions and Proceedings of the Fiji Society*, 12: 40–50.

Sharpley, R. (2008) 'Planning for tourism: The case of Dubai', *Tourism and Hospitality: Planning and Development*, 5(1): 13–30.

Singh, I. (2020) 'Fiji sets record in visitor arrivals', *FBC News*, 22 January. Available online at https://www.fbcnews.com.fj/business/fiji-sets-record-in-visitor-arrivals/ (accessed 8 August 2020).

Singh, S., J.A. Bhat, S. Shah and N.A. Pala (2021) 'Coastal resource management and tourism development in Fiji Islands: A conservation challenge', *Environment, Development and Sustainability*, 23(3): 3009–3027.

Treloar, P. and C.M. Hall (2005) 'Tourism in the Pacific islands'. In: C. Cooper and C.M. Hall (eds.) *Oceania: A Tourism Handbook*, Clevedon: Channel View Publications, pp. 173–294.

Tyllianakis, E., G. Grilli, D. Gibson, S. Ferrini, H. Conejo-Watt and T. Luisetti (2019) 'Policy options to achieve culturally-aware and environmentally-sustainable tourism in Fiji', *Marine Pollution Bulletin*, 148: 107–115.

UNCTD (United Nations Conference on Tourism and Development) (2021) 'Fiji adopts a new investment act', *Investment Policy Hub*, Geneva, Switzerland. Available online at https://investmentpolicy.unctad.org/investment-policy-monitor/measures/3717/fiji-adopts-a-new-investment-act (accessed 14 April 2021).

Wainiqolo, I. and C. Del Castillo (2021) 'Reducing gender inequality for a sustainable recovery in Fiji', *Pacific Economic Monitor*, December, Asian Development Bank, Mandaluyong City, Philippines. Available online at https://www.adb.org/sites/default/files/publication/757271/pem-december-2021.pdf (27 February 2022).

Williams, S. and A.A. Lew (1998) *Tourism Geography: Critical Understanding of Place, Space and Experience*, 3rd edition, London: Routledge.

12

GENDER EMPOWERMENT IN TOURISM DEVELOPMENT

Female bungalow hosts in Vanuatu

Nicole Orsua, Joseph M. Cheer, and Madelene Blaer

Introduction

Before the COVID-19 pandemic, more than half of employees in the global tourism industry were women, though there were notable gender gaps in wages, senior management positions and political representation (UNWTO 2019). As Alarcòn and Cole (2019) argue, there can be no sustainability in tourism without gender equality and, thus, this proposition is examined in this chapter. The work poses a fundamental question: 'To what extent do experiences of female bungalow hosts imply progress toward gender empowerment in Vanuatu?' To understand the experiences of the ni-Vanuatu (Indigenous Melanesians native to Vanuatu) bungalow hosts, this chapter will first explore the cultural and historical impacts on gender equality in Vanuatu and the role sustainable tourism has in its development. The second part of the chapter will be an analysis of their experiences through utilizing four dimensions of empowerment developed by Scheyvens and Lagisa (1998) which Scheyvens then applied to ecotourism (1999, 2010). The conclusion discusses the identified barriers to the development of empowerment for female entrepreneurs, recognizing how the implementation of gender-specific policies can help foster gender empowerment in the Pacific.

Vanuatu is a lower-middle-income economy situated in the Melanesia division of the South Pacific (World Bank 2020). It has a curious historical past as a condominium of both France and Great Britain, giving it distinctive Francophone-Anglophone characteristics. Although there are scores of dialects across the country's scattered archipelago, Bislama, a creolized version of English is universally spoken. Over the past two decades, tourism has proved to be a significant economic driver in the South Pacific with international arrivals increasing by around 50% between 2005 and 2015, and Vanuatu is no exception (Everett et al. 2018). 2019 was a record year for Vanuatu, as international air traffic increased by 18.3%, leading to a total of 219,000 arrivals (Air Vanuatu 2020). Vanuatu's tourism growth halted in 2020 due to the COVID-19 pandemic (DFAT 2020a). In the same year, Cyclone Harold struck Vanuatu's central and upper islands (DFAT 2020b). The closure of borders due to the pandemic led to loss of employment and income for more than 2,000 employees in the formal economy, mostly within the hospitality and construction industries (Tugeta 2020). The economic loss in the private sector between the period of March to June 2020 was estimated at VUV$7.58 billion (US$70 million) (Naupa 2020).

DOI: 10.4324/9780429019968-14

While there is an increase in research surrounding the intersection between tourism and gender empowerment, there is a general lack of understanding, interpretation and implementation of gender equality into the policies and practices of global tourism (Alarcòn and Cole 2019). Nevertheless, gender-sensitive tourism policies at a national level can give rise to economic empowerment for women in tourism (UNWTO 2019). Expressions of gender inequalities in Vanuatu vary and are most evident in data relating to domestic violence, marrying age for women, education attainment, (under)representation in employment, poor land ownership rights and constraints to accessing money and bank accounts. For every 100 males attending primary and secondary school there is a 10% decrease in female attendance, a gap which widens significantly at the tertiary level (Naidu 2010). Men are employed at a 65% higher rate than women in the formal economy, whereas women are 50% more likely than men to be working in unpaid family labour (VNSO 2016).

The Melanesian economy can be categorized into three categories: formal, informal and non-cash based (Carnegie et al. 2012). The formal economy is defined as engaging in waged or salaried-based work for a registered business that produces goods or services. In the tourism sector, this includes restaurants and cafes, accommodation providers, tour operators and souvenir stores. The informal economy includes paid domestic work and the creation of goods and services for non-registered businesses and does not include lodging of taxes or licensing fees. The non-cash economy refers to unpaid work such as community work, family and household duties, and church and community service.

Gender in tourism

Kinnaird and Hall's (1994) gender analysis represents one of the foremost forays addressing the issue of gender from a tourism development perspective. Margaret Swain's (1995) seminal publication followed, arguing that gender analysis needed to be brought into mainstream tourism research as an agent of change, especially as it was not significantly utilized in destination planning. Women spend more hours on unpaid care and domestic work than men (UN 2020), an imbalance that continues when a woman enters formal or informal economies. This imbalance is referred to as 'double burden' (IWDA 2016) and is a common theme in gender-based enquiries in tourism (Alarcòn and Cole 2019; Suarez 2018). Moreover, the identification of the 'third shift' exposes inequity whereby a woman's first shift is the caretaking role, second is household responsibilities, and the third comprises formal employment (Morgan and Winkler 2020).

Sustainable tourism development in Vanuatu

Sustainability in tourism exists on a spectrum with interpretations between opposing ends of 'micro-sustainability' and 'macro-sustainability' (Stoddard et al. 2012). The 'micro' approach to sustainability concerns a business-like structure, whereas the 'macro' approach incorporates a more holistic approach. Gender equality and empowerment are arguably aligned with the 'macro' approach and contextualized within a diverse set of societal goals. Vanuatu's commitment to sustainability is cemented by the *National Sustainable Development Plan 2016–2030* (DSPPAC 2016) and the Vanuatu Sustainable Tourism Policy (VSTP) 2019–2030 (VDOT 2019). The overall vision of VSTP is 'to protect and celebrate Vanuatu's unique environment, culture, *kastom* (customs and cultures) and people through sustainable and responsible tourism' (2019: 6). Such initiatives are vital to Vanuatu's economy and daily livelihoods because of heavy reliance on tourism (Cheer and Peel 2011).

Female entrepreneurs in tourism

The tourism industry typically consists of small to medium enterprises (SMEs), highlighting the importance of considering the place of entrepreneurship within tourism development. Coincidentally, there remains a significant gender gap when it comes to female entrepreneurship (GEDI 2014). While employment opportunities for women in tourism are increasingly evident, gender equality barriers persist, limiting advancement of female empowerment in tourism (UNWTO 2020). Barriers to entry for female entrepreneurs are often expressed through gender bias laws, anachronistic cultural norms and lack of access to collateral (UNWTO 2019). More broadly, the accommodation sector has been a gateway for women to enter tourism as the responsibilities of operating a guesthouse are compatible with traditional female work, which includes cleaning and cooking. Micro-businesses such as bungalows and homestays are praised for their positive impacts to rural areas by stimulating regional dispersion by driving visitors away from city centre hotels (Scheyvens 1999, 2010), potentially leading to poverty reduction (Scheyvens and Russell 2012) and stimulating community development (Zapalska and Brozik 2017).

Movono and Dahles (2017) highlight that, while tourism can create opportunities for female empowerment through entrepreneurship, there tends to be a lack of empirical evidence of instances in which empowerment was fully achieved. A sense of empowerment can be explained as the ability to break through traditional gendered norms, thereby increasing power and decision-making within households and communities. Gaining empowerment does not need to be limited to an individual experience and can also be achieved through the creation of female-centric networks and associations (Vujko et al. 2016). These authors note that women usually consolidate resources to create opportunities in which they can nurture each other's skills and create supportive environments for economic and social empowerment.

Kabeer (1999) divides empowerment into three categories: (1) resources, which are not only material goods but also include social and human relationship gains that assist with the ability to make decisions for oneself and one's family; (2) agency, or the capacity to drive decision-making is explained by individuals having the opportunity to not only define personal goals but also to work toward achievement, even in the face of adversity; and (3) achievement occurs when resources were combined with agency, and one was able to make strategic choices about one's life. Therefore, resources, agency, and achievement are the building blocks for empowerment. However, it is important to acknowledge that empowerment is multifaceted and evolves with societal progression (Tucker and Boonabaana 2012). Scheyvens and Lagisa (1998) classify four key dimensions of female empowerment: economic, psychological, social and political. Here, we apply these four dimensions of empowerment to help identify barriers that exist for female bungalow entrepreneurs in Vanuatu.

A qualitative method of enquiry

A qualitative research methodology approach was applied to analyse emotions, perceptions and experiences of ni-Vanuatu female bungalow entrepreneurs. Phenomenology was selected, as it is grounded in the human experience and is ideal to describe the experiences of the local hosts (Pernecky and Jamal 2010). The application of feminist phenomenology adds 'conceptual richness to gender theories' by putting women at the centre of the study (Gardiner 2018: 291) and highlights the unique experiences of female bungalow hosts in what is a largely patriarchal society. To ensure academic rigor 'bracketing' was utilized, a process by which the researcher recognizes and actively attempts to set aside their own inherent biases and preconceived notions that affect the analysis. However, hermeneutic phenomenology acknowledges that the researcher cannot be separated from the world they are studying (Larkin et al. 2006), and feminist phenomenology is reflexive by nature (Bensemann 2010). To reconcile these opposing views regarding the proper location of the researcher's reflections, Creswell (2013) notes that researchers utilizing bracketing while embracing their experiences through descriptive data often apply bracketing before analysing the experiences of the study's respondents. Two types of descriptive data are applied here: 'textual description' for the experiences of respondents and 'structural' data when referring to the lead researcher's participant observations of the environment in which the interviews took place.

Five semi-structured interviews were employed in which discussion topics were predetermined and a conversational style allowed for diving deeper into participant responses (Galletta and Cross 2013). Each interview lasted for approximately one hour. Purposive sampling was employed to align with time and practical constraints and led the researcher to develop close links with tourism sector stakeholders before the commencement of fieldwork, which was supported by the Vanuatu Department of Tourism. The small sample (Table 12.1) suited the exploratory nature of this study while confirming the dearth of female entrepreneurship in the sector. The data analysis was guided by 'interpretive phenomenological analysis' (IPA), an inductive approach that utilizes concept mapping to uncover themes in the participant's responses. From these overarching themes, subthemes are then uncovered.

Findings and analysis

Despite the utilization of a small sample size, demographic patterns did emerge (Table 12.2). None of the hosts had an education level higher than secondary school, and this finding was

Table 12.1 Bungalow ownership in Shefa province

Island	*Total number of listings*	*Total number owned and operated by men*	*Total number owned and operated by couples*	*Total number owned and operated by women*
Efate	11	5	4	2
Nguna	6	3	2	2
Pele	9	5	3	1
Total	26	13	9	5

Source: Authors' own compilation derived from the study.

Table 12.2 Demographic background of the study's respondents

Respondent	*Age*	*Gender*	*Island of residence*	*Level of education*	*Relationship status*	*Number of children*	*Years of operation*	*Timeframe for bungalow construction*
R1	55–64	Female	Efate	Secondary	Married	Two	6 to 9	Three months
R2	35–44	Female	Nguna	Secondary	Married	Four	10 or more	Three years
R3	55–64	Female	Nguna	Primary	Married	Five	3 to 5	Six years
R4	35–44	Female	Nguna	Secondary	Married	Four	10 or more	Not Applicable
R5	25–34	Female	Pele	Primary	Married	Three	6 to 9	Three years

Source: Authors' own compilation derived from the study.

consistent with data concerning gender-based participation in Vanuatu's education sector (Naidu 2010) (as highlighted earlier). All the women were married with more than two children, indicating that they had domestic duties in addition to their bungalow responsibilities. The participants ranged in age from their 20s to 60s, and only one had prior work experience in the formal economy.

Respondent 1 (R1)

Located on the main island of Efate, R1 lives in a northern village next to the boat launch for Ngnua and Pele. She has two bungalows, the first of which was built in 2012 as a homestay experience and attached to her home. The second is a freestanding unit completed in 2016 that incorporates modern and traditional designs (Figure 12.1). In addition to the bungalows, she opened a small shop located in her village.

Respondent 2 (R2)

Located in Nguna, R2 does not have any experience working in hospitality or tourism prior to building her first bungalow more than a decade ago. She started with a building that had two bedrooms, a kitchen and a service area. Over time, she added four standalone structures and the final building was for the toilet and shower. Her previous work experience was as an entrepreneur, as she opened a small village shop in 2015. R2's commitment to her enterprise is apparent by the fact that she engages in the vast majority of the work, despite being married and having four children.

Figure 12.1 Bungalow belonging to Respondent 1 (R1). Source: N. Orsua, February 2020.

Respondent 3 (R3)

R3's bungalow was a long building with two bedrooms and a large common area that opened out onto a large porch overlooking the water. Her husband and his two brothers established the enterprise, though he was the only surviving brother. R3 and the wives of the deceased brothers oversaw daily operations and were an example of a self-help group (see Vujko et al. 2016), involving women sharing their resources, time and responsibilities. In the start-up process, the three brothers would meet to make business decisions, which included their wives. At the time of the interview, the three women and one surviving brother met to make decisions about the operation. They utilized an overseas-based administrator who paid their salaries based on the work contributed, which came from the money that was deposited into the bungalow's bank account.

Respondent 4 (R4)

R4's situation was unique compared to the other participants. She was the elected manager of a village-owned bungalow enterprise, another example of a self-help group. The four bungalow units were owned and under collective control by women villagers, an arrangement that was compared to an association. Every two years a manager is elected and a committee of five women was formed to facilitate bungalow operations. In addition to managing bungalows, R4 worked at a local primary school overseeing the food programme. The bungalows were not a significant part of life in the village, they were 'just there'. This response indicated that, though the women of the village had identified the opportunity for having bungalows as an economic stimulus for the community, there was no emphasis on maximizing profitability through consistent occupancy. The association had a bank account for the bungalows into which proceeds were deposited. Annually, an appointed treasurer paid the participating women their share of the profits.

Respondent 5 (R5)

The village of Worearu on the island of Pele had five bungalow establishments, with the first erected by R5's father. After watching her father run his bungalow, she suggested the idea to her husband, who then built the first structure in 2016 (Figure 12.2). R5 was the youngest participant and had three children, the youngest was a toddler who she parents alone when her husband works overseas. He was away during the interview and, upon his return, she had planned to expand her business by adding another bungalow. Unlike R1 and R2, who received unpaid assistance from family members, R5 employed family members. As a result, the bungalow supported both her immediate and extended family.

The collection of resources

All participants are members of the informal and non-cash based economies, having little experience in the formal economy prior to building their bungalows. R1 cleaned for expatriates and sold produce at the roadside markets. R3 also sold produce on the roadside. R4 worked for an expatriate in 2012 as a housekeeper on a private island estate before becoming the head chef of this establishment. R2, R3 and R5 spoke of the importance of financial resources to fund the construction of the bungalows. R2 and R5 received financial and emotional support from family members. R2 and her family decided to start the bungalow enterprise in 2007, taking three years to complete. The delay in completing the first bungalow, especially the bathroom

Figure 12.2 Bedroom belonging to the bungalow of Respondent 5 (R5). Source: N. Orsua February 2020.

and kitchen, was because they 'didn't have enough money'. To fund construction, R2 utilized the proceeds from her small village shop and her father's boat business. However, the income from these ventures was not regular and 'when the money stopped flow[ing], we stop. When [the money was] back, we build'. R3 sold fruit, vegetables and nuts at the markets in Port Vila to help fund the establishment of her family's bungalow, which took six years to complete. In contrast, R5 relied on her husband's earnings from working overseas, opening her bungalow after three years of intermittent construction.

Agency: the opportunity to define one's goals, even in the face of adversity

R2 indicated that the decision to open a bungalow was made with her family, specifically her parents who assisted in collecting the financial resources to build the infrastructure. The inclusion of family members in the decision-making process did not diminish R2's agency because it became clear she was the primary labourer. R4 worked in a self-help group, though the village women all decided to join the association and take on the responsibility of managing the bungalow when elected. It is important to note that if a woman leverages self-help groups it does not diminish her decision-making capacity and, therefore, does not discount her agency (Vujko et al. 2016). While R5's decision-making process to become an entrepreneur included her husband, his involvement does not diminish her autonomy, as she was the driving force behind the business plan. R3 was the only participant whose goal was not to own her own bungalow, as it was her husband and his brothers that decided to start the business.

The economic dimensions of female empowerment

Empowerment often starts with an increase in resources (Kabeer 1999), which can lead to ending women's economic dependency on men and creating economic empowerment (Movono and Dahles 2017) as well as freedom of financial choice (Moswete and Lacey 2015). The experiences of the ni-Vanuatu respondents suggest that the hosts did not achieve the indicators of Scheyvens's (1999, 2012) nor Moswete and Lacey's (2015) definition of economic independence. However, all the hosts (except for R3) achieved the ability to make financial decisions regarding their businesses. When addressing how revenue earned from the bungalows was utilized, two clear categories emerged: (1) to reinvest (and help to maintain) the business and (2) to support family members. While hosts may not have achieved economic independence at home, R1, R2 and R5 had the autonomy to make financial decisions regarding the needs of the bungalow business. R4 was part of a collective decision-making process, as the association members voted on property updates and maintenance schedules. Therefore, in terms of the instances indicated, it was clear that respondents were active participants in business decision-making.

When it comes to family expenses, R1, R2 and R3 all indicated that they utilized money earned on basic family necessities (shelter, food and clothing), needs (education and utilities) and medical expenses. It was not discerned whether the respondents had equal decision-making power as their spouses when it came to household financial decisions. R1, R2 and R5 indicated that earnings from their bungalow also went to support their extended family. R2 and R5 had children in their households with no access to childcare assistance while working. The burden of childcare would be classified as part of the Melanesian non-paid economy, reinforcing the traditional role of women in the household. Lack of childcare is an indicator of economic disempowerment, as it can be a barrier for women to enter the paid workforce. However, the idea of paid childcare could be framed as a Western preconceived notion of empowerment and was bracketed by the researcher. In village communities, it is more common to have unpaid childcare assistance and, therefore, may not be an indicator of disempowerment in the context of Vanuatu.

When asked about challenges experienced, R1 replied: 'sometime[s] no money'. She explained that there were times when she did not have the money required to purchase the ingredients for the meals included in her overnight rates. In these instances, she had to borrow ('took credit') from her small village shop. When asked if she had any plans on expanding her business her response was: 'yes, but money talks', indicating that lack of funds was keeping her from building a kitchen. R4 identified money as a challenge through disagreement amongst the association members regarding the bungalow's nightly rate. The bungalows were not consistently occupied, and when they were, occupants were mostly from non-governmental organization (NGOs) or governmental agencies requiring discounted rates. The association members were not always pleased with the discounted rates that R4 negotiated, indicating a different perspective to her revenue management strategies.

The psychological dimension of female empowerment

Psychological empowerment is defined as dignity, self-esteem/pride and self-worth (Scheyvens and Lagisa 1998). The participants were asked questions that address self-image and feelings associated with operating their bungalow enterprise to uncover psychological impacts. The findings concluded that the hosts exhibited dimensions of all indicators and definitions of psychological empowerment. When asked about challenges she experienced as a host, R2 spoke about negative perceptions from members of her village. When asked how she overcame the negativity, she stated: 'One thing about business, you have to be a good person at all times. If I know that person is talking about [me], I do positive. I do everything good'. What she was describing was a sense of dignity that was developed as a result of being an entrepreneur, a facet of psychological empowerment identified by Scheyvens (1999, 2012). Her dignity was expressed through her understanding that she must retain a sense of composure when faced with criticism from her community. R2 articulates the criticism against her as an entrepreneur from the other women in her village, expressing:

> My friends say they are saying negative stories, saying I am proud. Sometimes I focus on the bungalow and no time for village work, and they say I forget about them. Mamas get together every week, but when there are guests, I don't know… [They] think I forget about village needs.

In communities where female entrepreneurship was limited, negative reactions from fellow female villagers toward the pioneering female entrepreneur were observed in which reluctance to accept a pioneering female entrepreneur leads to the female villagers perpetuating traditional gender roles rather than accepting new dimensions (Movono and Dahles 2017). Pioneer female entrepreneurs are the societal outliers that become agents of change, increasing the social capital in their community for women, even if they are not aware of their social capital gains (Diaz-Carrion 2018). For this study, the definition of social capital concerns 'social relations that generate benefits for women' (2018: 109). Diaz-Carrion (2018) also notes that this lack of awareness creates conflicting opinions between what is deemed appropriate behaviour for women based on traditional roles, compared to behaviour influenced by social progress and the creation of new gender roles. R2 articulated experiencing cultural constraints when asked about how she believes members in her village felt about her being a female entrepreneur. It became clear why R2 did not want to call herself 'proud' as there appeared to be a negative perception of her entrepreneurship activities in her village.

Respondents were also asked how owning their own business makes them feel. R2 responded with:

> I am not proud, but happy that I am running a bungalow business. Not proud [because] of so many challenges [that I] have over[come], some, but not all…Am happy and glad [to] run own business and can do it on my own. Because they depend on [me] for everything.

In this context, the word 'they' refers to her family. Her answer displayed self-reliance, higher self-esteem and faith in her own abilities. R2 overcame numerous obstacles to gain the levels of achievements she has experienced. The respondents were asked if owning their own bungalow caused a change in their self-identity or self-image. In response, R5 answered: 'Yes, I learn a lot from how I run the business, [because] I don't know anything before, and I faced challenges and overcome. At first Shefa [Travel, a division of The Department of Tourism] helped a lot. [I am] not really experienced but going on well now'. While displaying an increase in self-esteem and self-reliance, R5 stopped short of allowing herself to be proud of her hard work, as did R2.

According to a Shefa Travel representative, pride in ni-Vanuatu culture tends to have a negative connotation, as it implies that individuals are 'thinking highly of themselves… they will not use (pride), but the term happy covers the term pride as well' (pers. comm. February 2020). Being proud of their businesses was apparent through the appearance of their accommodation establishments, which were clean, tidy, bright and colourful, showcasing shells and flowers as decorations. R5 drew a connection between her self-worth and supporting her family, stating: 'I feel glad [because] it is very helpful for me and my family. Guests give money to develop [the] needs of family and for kids to go to school and get [an] education'. Her experience supported the theory that female entrepreneurs are more likely to re-invest their earnings into their children's education (Kirkwood 2009). R5 was able to support her children and extended family, enabling her to feel a sense of self-worth and pride.

The social dimension of female empowerment

The United Nations World Tourism Organization (UNWTO 2019) argues that gender discriminatory laws and cultural constraints are challenges for female entrepreneurs in tourism. In Vanuatu, there is a saying that 'women work, men talk' (Bronwen 2002: 12), indicating that it is not uncommon for women to be left with household and caretaking responsibilities. The hosts received support from daughters-in-law, sisters-in-law and parents but made no mention of support from husbands. The exclusion of assistance from male members of the family may have indicated the existence of the double burden effect, which is the imbalance of responsibility of the non-cash economy. All respondents were asked about their household responsibilities and if any changes to those responsibilities have been made since opening their business. These questions were asked to identify whether respondents had experienced the 'double burden' effect, which was alluded to earlier.

R1 continued all her normal household chores when there were guests, which she felt was manageable with support from her daughter and daughter-in-law. R2 indicated that the 'bungalow is very demanding, sometimes all day. Not every time, but now with help from Mom can go home some'. In addition to her responsibilities at the bungalow, she still conducted household chores such as 'small shopping sometimes...some small cleaning. Parents run [the] shop [and] parents help a lot'. R5 responded:

> Husband travels a lot so [I] take on a lot of responsibility. So, it's good to have family help and they all receive money to help support. We work together to set up for guest and it's not all for my sake, it benefits all.

When asked about the affects the bungalow had on the family's financial situation, especially in the context of her extended family, R5 commented that the bungalow made family members 'happy' as they receive a 'small amount of money'. There was no mention of daily tasks that their husbands were responsible for, or assisted with, during any of the interviews.

R3 expressed that 'village life is a busy life' in which 'there [is] a lot expected in [the] village and at home'. As noted in the methods section, there were a limited number of women in the Shefa Province who operated their own bungalow businesses. This finding suggests that bungalow ownership does not yet seem to be widely accepted as women's work. This was not surprising, as there were cultural constraints imposed upon the participants by community members. R2's experience with negative sentiments within her village was similar to that of the female-shared accommodation entrepreneur in Fiji who also experienced jealousy from her female villagers (Movono and Dahles 2017). As Movono and Dahles (2017) explain, this participant was able to overcome the barrier of social jealousy and eventually gain a higher social status (which led to political empowerment) by encouraging other female villagers to become entrepreneurs themselves. At the time of the interview, however, R2 had not overcome such social barriers.

It became evident during the interviews that women had responsibilities within the village in addition to their family and household needs. Building upon the third shift theory (Morgan and Winkler 2020), it can be argued that women experience four shifts when becoming an entrepreneur. Their first and second shifts were at home, their third shift being the responsibilities to their village, and their fourth shift their bungalows. When they became entrepreneurs, the needs of their guests superseded those of the community. Even R4, who was an elected manager, felt the tension between managing and community responsibilities. The women in her village met bi-weekly to talk about life and the church and to sew, cook and weave. However, if there were guests then R4 and committee members could not attend, postponing the gathering. Only R5 did not feel any animosity in her village regarding her bungalow business. Her experience may have been because her village had a higher concentration of bungalows than the other villages, and, therefore, it was more socially acceptable to be an entrepreneur. The village's acceptance of accommodation entrepreneurs was then extended to R5, showcasing that women can gain the social capital necessary to challenge existing gender roles in Vanuatu.

Conclusions and research implications

This chapter explored the pivotal role that female empowerment plays in sustainable tourism development by examining the lived experience of ni-Vanuatu women entrepreneurs in the shared accommodation sector. All respondents experienced challenges securing financial resources to build their bungalows and were often subject to disruption when funds ran out. Four faced jealousy or a lack of understanding from their fellow female villagers when their businesses would interfere with the amount of time available for village activities. There was no indication that the respondents' husbands assisted with the day-to-day operations, with women receiving support from their parents, in-laws and children. Respondents also had successes and achieved indicators of psychological empowerment: 'dignity', 'self-worthiness', 'self-reliance' and 'pride'. They contributed to the financial improvement of their own nuclear family and their extended families. While economic independence and social empowerment have not yet been fully achieved, it is evident that hosts are in the process of working toward empowerment. In the social sphere hosts have gained social capital, a key ingredient in challenging gender norms and a building block for social empowerment. While not all women will use their acquired social capital to challenge norms, some will, which could lead to empowerment in the political arena. Political empowerment is acknowledged as one of the more insurmountable challenges, as it requires overcoming social biases toward cultural gender roles (Cole 2018a) and is an area that could be explored in further research. The lack of political empowerment achieved by the respondents suggests that traditional ni-Vanuatu gender roles must be changed before ensuring gender equality and empowerment in Vanuatu's tourism industry (Cole 2018b).

The Vanuatu Department of Tourism recently released the Vanuatu Sustainable Tourism Strategy 2021–2025 (VDOT 2021), identifying four themes for recovery: well-being, resilience, diversification and sustainability. Although there is no discussion in the strategy addressing the cultural and social barriers impacting the entrepreneurial advancements and experiences of women, the prime minister of Vanuatu launched an updated National Gender Equality Policy 2020–2030 in August 2021 (Daily Post 2021). The policy included strategic areas addressing women's economic empowerment as well as climate and disaster resilience and acknowledging a whole governmental approach.

Future research recommendations include a longitudinal study across all provinces of Vanuatu, as this research provides a snapshot in time from which comparative analysis can be made. A longitudinal study would create the opportunity to observe whether the social capital gains uncovered in this research would lead to social and political empowerment. Further research is required to address the impacts of the COVID-19 global pandemic on women entrepreneurs in tourism and their levels of empowerment. Early research suggests that the pandemic will cause a regression on gender equality and empowerment in vulnerable countries (Park and Inocencio

2020). Empowerment itself is ascertained through the building of social capital and is, therefore, an important component of community resilience. For South Pacific Indigenous communities to build resilience and withstand external shocks that impact the global tourism industry, implementation of strategic gender policies is vital. There is a need to educate men and women on the uneven distribution of household responsibilities as well as showcase the rewards that female entrepreneurs and their families can receive from running successful businesses. Programmes that encourage the use of self-help groups to combat jealousy and create a collective voice for lobbying for gender issues at both a political and village level would be valuable. As showcased in this chapter, stakeholder engagement is one tool by which the assessment of Indigenous female empowerment can be developed through the implementation of gender specific tourism policies. The four dimensions of empowerment (Scheyvens 1999, 2010) can assist with the analysis of a community's degree of female empowerment, especially by uncovering areas in which empowerment and disempowerment exist. However, this research shows that the inclusion of capacity building through social capital is an important building block to achieve female empowerment. Disempowerment arises when challenges and barriers that hinder advancement through the empowerment process are present. Through the identification of barriers to the empowerment of women, the development of strategic policies can occur, thus, paving the way toward the achievement of empowerment across all levels – economic, psychological, social and political.

References

Air Vanuatu (2020) 'Air Vanuatu achieves a record year in 2019', 26 February. Available online at https://www.airvanuatu.com/news/air-vanuatu-achieves-a-record-year-in-2019 (accessed 11 March 2020).

Alarcòn, D. M. and S. Cole (2019) 'No sustainability for tourism without gender equality', *Journal of Sustainable Tourism*, 27(7): 903–919.

Bensemann, J. (2010) 'Allowing women's voices to be heard in tourism research: Competing paradigms of method'. In: M. Hall (ed.) *Fieldwork in Tourism: Methods, Issues and Reflections*, London: Routledge, pp. 151–167.

Bownwen, D. (2002) 'Christian citizens: Women and negotiations of modernity in Vanuatu', *The Contemporary Pacific*, 14(1): 1–38.

Carnegie, M., C. Rowland, K. Gibson, K. McKinnon, J. Crawford and C. Slatter (2012) *Gender and Economy in Melanesian Communities: A Manual of Indicators and Tools to Track Change*, November, University of Western Sydney, Macquarie University and International Women's Development Agency.

Cheer, J. and V. Peel (2011) 'The tourism-foreign aid nexus in Vanuatu: Future directions', *Tourism Planning and Development*, 8(3): 253–264.

Cole, S. (2018a) 'Conclusions: Beyond empowerment'. In: S. Cole (ed.), *Gender Equality and Tourism*, Wallingford: CAB International, pp. 132–141.

Cole, S. (2018b) 'Introduction: Gender equality and tourism - beyond empowerment'. In: S. Cole (ed.), *Gender Equality and Tourism*, Wallingford: CAB International, pp. 1–11.

Creswell, J. W. (2013) *Qualitative Inquiry and Research Design: Choosing among Five Approaches*, Third Edition, Thousand Oaks: SAGE Publications, Inc.

Daily Post (2021) 'Prime Minster launches national gender equality policy', 28 August. Available online at https://www.dailypost.vu/news/prime-minster-launches-national-gender-equality-policy/article_8aa98d48-4e13-53db-95f3-7ced4fb04e08.html (accessed 27 December 2021).

DFAT (Department of Foreign Affairs and Trade) (2020a) 'Tropical cyclone Harold', *Crisis Hub*, DFAT, Australian Government. Available online at https://www.dfat.gov.au/crisis-hub/Pages/tropical-cyclone-harold (accessed 27 December 2021).

DFAT (Department of Foreign Affairs and Trade) (2020b) 'Vanuatu COVID-19 development response plan', DFAT, Australian Government. Available online at https://www.dfat.gov.au/sites/default/files/covid-response-plan-vanuatu.pdf (accessed 6 May 2022).

DSPPAC (Department of Strategic Policy, Planning and Aid Coordination) (2016) *Vanuatu 2030: The People's Plan. National Sustainable Development Plan 2016–2030*, Port Vila, Efate, Sanma Province, Vanuatu: DSPPAC. Available online at https://www.gov.vu/index.php/resources/vanuatu-2030 (accessed 6 May 2022).

Diaz-Carrion, I. A. (2018) 'Tourism entrepreneurship and gender in the global South: The Mexican experience'. In: S. Cole (ed.), *Gender Equality and Tourism, Beyond Empowerment*, Oxfordshire: CAB International, pp. 108–117.

Everett, H., D. Simpson and S. Wayne (2018) 'Tourism as a driver of growth in the Pacific', Manila, Philippines: Asian Development Bank. Available online at https://www.adb.org/sites/default/files/publication/430171/tourism-growth-pacific.pdf (accessed 18 February 2020).

Galletta, A. and W. Cross (2013) *Mastering the Semi-Structured Interview and Beyond: From Research Design to Analysis and Publication*, New York City: New York University Press.

Gardiner, R. A. (2018) 'Hannah and her sisters: Theorizing gender and leadership through the lens of feminist phenomenology', *Leadership*, 4(3): 291–306.

Global Entrepreneurship Development Institute (GEDI) (2014) *Gender and Economic Development: The Puzzle*, GEDI. Available online at http://thegedi.org/gender-and-economic-development-the-puzzle/ (accessed 8 November 2019).

IWDA (International Women's Development Agency) (2016) *The Double Burden: The Impact of Economic Empowerment Initiatives on Women's Workload*, December, Melbourne: IWDA. Available online at https://iwda.org.au/resource/the-double-burden-the-impact-of-economic-empowerment-initiatives-on-womens-workload/ (accessed 8 March 2020).

Kabeer, N. (1999) 'Resources, agency, achievements: Reflections on the measurement of women's empowerment', *Development and Change*, 30(3): 435–464.

Kinnaird, V. and D. Hall (eds.) (1994) *Tourism: A Gender Analysis*, Chichester: Wiley.

Kirkwood, J. (2009) 'Motivational factors in a push-pull theory of entrepreneurship', *Gender in Management*, 24(5): 346–364.

Larkin, M., S. Watts and E. Clifton (2006) 'Giving voice and making sense in interpretative phenomenological analysis', *Qualitative Research in Psychology*, 3(2): 102–120.

Morgan, M. S. and R. L. Winkler (2020) 'The third shift? Gender and empowerment in a women's ecotourism cooperative', *Rural Sociological Society*, 85(1): 137–164.

Moswete, N. and G. Lacey (2015) '"Women cannot lead": Empowering women through cultural tourism in Botswana', *Journal of Sustainable Tourism*, 23(4): 600–617.

Movono, A. and H. Dahles (2017) 'Female empowerment and tourism: A focus on businesses in a Fijian village', *Asian Pacific Journal of Tourism Research*, 22(6): 681–692.

Naidu, S. (2010) 'Right to education for all children in Vanuatu – Are girls getting an equal opportunity to education compared to boys?' *Law, Social Justice & Global Development Journal*, 16. Available online at https://warwick.ac.uk/fac/soc/law/elj/lgd/2010_2/naidu/ (accessed 6 May 2022).

Naupa, A. (2020) 'A tale of two sectors: Women leaders bridging the formal and informal sectors during Vanuatu's COVID crisis', 28 July, Griffith Asia Insights, Griffith Asia Institute, Griffith University. Available online at https://blogs.griffith.edu.au/asiainsights/a-tale-of-two-sectors-women-leaders-bridging-the-formal-and-informal-sectors-during-vanuatus-covid-crisis/ (accessed 02 February 2021).

Park, C-Y. and A. M. Inocencio (2020) 'COVID-19 is no excuse to regress on gender equality', *ADB Briefs*, Asia Development Bank, November. Available online at https://dx.doi.org/10.22617/BRF200317-2 (accessed 6 May 2022).

Pernecky, T. and T. Jamal (2010) '(Hermeneutic) phenomenology in tourism studies', *Annals of Tourism Research*, 37(4): 1055–1075.

Scheyvens, R. (1999) 'Ecotourism and the empowerment of local communities', *Tourism Management*, 20(2): 245–249.

Scheyvens, R. (2010) 'Promoting women's empowerment through involvement in ecotourism: Experiences from the third world', *Journal of Sustainable Tourism*, 8(3): 232–249.

Scheyvens, R. and L. Lagisa (1998) 'Women, disempowerment and resistance: An analysis of logging and mining activities in the Pacific', *Singapore Journal of Tropical Geography*, 19(1): 51–70.

Scheyvens, R. and M. Russell (2012) 'Tourism and poverty alleviation in Fiji: Comparing the impacts of small - and large - scale tourism enterprises', *Journal of Sustainable Tourism*, 20(3): 417–436.

Suárez, P.V. (2018) 'Tourism as empowerment: Women artisan's experiences in Central Mexico'. In: S. Cole (ed.) *Gender Equality and Tourism: Beyond Empowerment*, Wallingford: CAB International, pp. 46–54.

Swain, M. B. (1995) 'Gender in tourism', *Annals of Tourism Research*, 22(2): 247–266.

Stoddard, J. E., C. E. Pollard and M. R. Evans (2012) 'The triple bottom line: A framework for sustainable tourism development', *International Journal of Hospitality and Tourism Administration*, 13(3): 233–258.

Tucker, H. and B. Boonabaana (2012) 'A critical analysis of tourism, gender and poverty reduction', *Journal of Sustainable Tourism*, 20(3): 437–455.

Tugeta, Y. E. (2020, July) *Vanuatu Recovery Strategy 2020–2030: TC Harold & COVID-19*, July, Port Vila, Vanuatu: Government of Vanuatu. Available online at https://reliefweb.int/report/vanuatu/vanuatu-recovery-strategy-2020-2023-tc-harold-covid-19-vanuatu-july-2020 (accessed 16 February 2021).

United Nations (UN) (2020) 'Sustainable Development Goal 5: Achieve gender equality and empower all women and girls', Department of Economic and Social Affairs, United Nations. Available online at https://sustainabledevelopment.un.org/sdg5 (accessed 5 April 2020).

UNWTO (United Nations World Tourism Organization) (2019) *Global Report on Women in Tourism*, Second Edition, Madrid: UNWTO. Available online at https://www.e-unwto.org/doi/book/10.18111/9789284420384 (accessed 1 December 2019).

UNWTO (United Nations World Tourism Organization) (2020) 'Our focus: Sustainable development', Madrid: UNWTO. Available online at https://www.unwto.org/sustainable-development (accessed 11 April 2020).

VDOT (Vanuatu Department of Tourism) (2019) *Sustainable Tourism Policy 2019–2030*. Available online at https://tourism.gov.vu/images/DoT-Documents/Plans/Vanuatu_Sustainable_Tourism_Strategy_LR.pdf (accessed 6 May 2022).

VDOT (Vanuatu Department of Tourism) (2021) *Vanuatu Sustainable Tourism Strategy 2021–2025*. Available online at https://tourism.gov.vu/images/DoT-Documents/Plans/Vanuatu_Sustainable_Tourism_Strategy_LR.pdf (accessed 27 August 2022).

VNSO (Vanuatu National Statistics Office) (2016) *Shefa Province: Mini Census 2016, Key Facts*, Ministry of Finance and Economic Management. Available online at https://vnso.gov.vu/index.php/en/census-and-surveys/census/2016-mini-census#shefaa-fact-sheets (accessed 6 May 2022).

Vujko, A., T. N. Tretiakova, M. D. Petrovic, M. Radovanovic, T. Gajic and D. Vukovic (2016) 'Women's empowerment through self-employment in tourism', *Annals of Tourism Research*, 76: 328–330.

World Bank (2020) 'Data: World bank country and lending groups'. Available online at https://datahelpdesk.worldbank.org/knowledgebase/articles/906519-world-bank-country-and-lending-groups (accessed 21 February 2021).

Zapalska, A. and D. Brozik (2017) 'Māori female entrepreneurship in the tourism industry', *Tourism: An International Interdisciplinary Journal*, 65(2): 156–172.

13

TOURISM, MICRO-ENTREPRENEURSHIP, AND HANDICRAFTS IN THE SOLOMON ISLANDS

Alexander Trupp

Introduction

Despite a rich potential for natural and cultural tourist attractions, international tourist arrivals in the Solomon Islands have remained low for the past two decades. Before the COVID-19 pandemic, in 2019, only 29,000 tourist arrivals and more than 60 cruise ship calls were registered (UNWTO 2020). Reasons for this relatively slow tourism development are manifold but include internal political and ethnic conflicts, limited accessibility, and comparably high prices for accommodation and transport. The capital city Honiara represents the main point of entry for international flights and cruise ships. Honiara's main attractions for domestic and international visitors are its various handicrafts and artworks at local markets, the museum shop, and stores in the city. These include wood and stone carvings, shell products, and different types of weavings. Handicrafts mainly refer to products with practical or decorative functions, made by hand or with simple tools, made of local material, and requiring specific artisan skills (Grobar 2019). Souvenirs, on the contrary, are usually mass-manufactured items that are produced, distributed, and consumed with few emotional attachments (Graburn 1976; Swanson and Timothy 2012). This distinction may become blurry if products are handmade and/or based on local materials but simultaneously produced and sold on a large scale.

Tourism micro-entrepreneurship can offer prospects of self-determination, empowerment, and economic betterment. As economic actors in travel destinations, handicraft micro-entrepreneurs can contribute to the production and consumption of tourist places from which tourists, micro-businesses, and the destination can all benefit (Hall and Rath 2007). Souvenir producers and vendors can carve out their own niches in the tourism industry by becoming entrepreneurial and commodifying some of their cultural features, such as producing weavings or wood carvings (Kumar et al. 2022; Movono and Dahles 2017; Trupp 2017). However, in the Pacific Island region, positive multiplier and trickle-down effects for small-scale businesses and vendors are often limited by the 'all-inclusive' resorts or cruise ships, which offer most tourist-oriented products and services within their confined premises. Moreover, the emergence of shopping malls that draw arriving tourists away from small souvenir stands as well as the appearance of imported and mass-manufactured products, lead to economic leakages and loss

DOI: 10.4324/9780429019968-15

of cultural capital in terms of arts and crafts skills or regional identity (Azarya 2004; Cole 2007; Lacher and Nepal 2011). Scholars argue that the role of Indigenous peoples in the souvenir/craft sector and the value chain of tourism ought to be aligned to sustainable development principles in which crafts are mainly made and sold locally and represent local cultures and identities (Saarinen 2016).

This chapter, thus, assesses the current development and status quo of Honiara's handicraft businesses. Based on an inventory of craft businesses and interviews with different stakeholders, the author analyzed the current product range of Honiara's handicraft businesses and its associated socio-cultural impacts and economic benefits for local micro-enterprises. Data collection for this article took place in the capital city Honiara over a two-month period prior to the COVID-19 pandemic. The research deploys semi-structured interviews with souvenir and handicraft micro-entrepreneurs, producers, shop vendors, and stakeholders such as a representative of the Ministry of Culture and Tourism and the president of the Solomon Islands Artist Association. Interviewed micro-entrepreneurs may take on several roles, such as business owners, managers, producers, and vendors. Moreover, observations, including field notes, were conducted at the various handicraft markets and shops in Honiara.

Souvenirs and handicrafts

Tourists, souvenir micro-entrepreneurs, and their products are connected to a global–local relationship characterized by social, cultural, and economic dynamics (Cave et al. 2013). For example, global fashion or design trends can impact the pattern and style of handicrafts in tourism destinations at the local level. Souvenirs and handicrafts are an important part of the tourist experience and represent a broad spectrum of the material culture of tourism (Husa 2020). Handicrafts, mass-manufactured items, and traditional travel mementos such as books, clothing, jewellery, antiques, food, and toys fulfil functions of showing (off) travel evidence, providing gifts for family, friends, or colleagues, and symbolically representing memories (Cohen 2000; Swanson and Timothy 2012; Wilkins 2011). Souvenir and handicraft businesses can lead to economic development by contributing to local ownership, local employment, and income generation (Saarinen 2016). Grobar's (2019) study suggests that at least 10% of the labour force in many developing countries where tourism plays a significant role are employed in the handicraft sector. Abisuga-Oyekunle and Fillis (2017) show that half of the interviewed crafters in South Africa view their business as profitable. According to the International Trade Center (ITC 2012:6), the average tourist visiting developing countries spends around US$20 to US$80 on crafts and souvenirs, making handicraft businesses the primary 'pro-poor income earners in the tourism value chain'. Earlier studies in Vanuatu show that the souvenir and craft sector may only supplement existing income rather than creating new full-time jobs (de Burlo 1996; Milne 1991). In this way, handicraft production may complement other activities such as farming or fishing and be perceived as a household diversification strategy. Nevertheless, souvenirs in remote areas are often imported from other countries, greatly increasing leakages out of the local economy and hampering economic development (Lacher and Nepal 2011).

Cruise ship tourism has increased in the Pacific region but has also been criticized for its enclavic nature (Cheer 2017). Recent research regarding souvenir purchasing in Fiji shows that cruise ship visitors buy less than stayover tourists (Kumar et al. 2022). Nevertheless, Douglas and Douglas argue that handicraft markets on cruise ship days constitute one of the few areas in which small local businesses gain direct access to the cash expenditure of tourists, particularly in remote and more isolated island communities (Douglas and Douglas 2004). Furthermore, the '"souvenirization" and "touristification" of material culture' (Husa 2020:279) always involves the process of cultural commodification, which leads to changes in the meaning of cultural products (Cohen 1988). While culture may become meaningless, such processes are also evaluated positively as tourism may bring economic benefits, human and cultural capital, or a sense of pride (Cole 2007). Silverman's (2013) research on tourism development in a Sepik River community in Papua New Guinea shows that, although the wood carver's motivation to produce tourist art is 'singularly economic' (2013: 239), their earned money has instrumental and representational qualities. Therefore, economic motivation does not necessarily erode symbolic meaning. Other scholars agree that commodification incorporates handicraft producers into the new economy but criticize that such economic integration simultaneously keeps them culturally at the margins of society (Azarya 2004).

Even though souvenirs constitute a vital part of the tourism experience, academic studies on such topics in the Pacific Island region are limited. Literature in the regional context focuses on marine products and assesses the relationship between tourism and the pearl-shell industry in Fiji and the broader South Pacific region (Chand et al. 2015). A recent study by Taylor (2016) explores how touristic images of ethnicity become miniaturized in the production of souvenirs, and Morgado (2003) analyses how the Hawaiian shirt as a commodity for the tourist market has transformed 'from kitsch to chic'. Research on cruise ship tourists in the Pacific further shows that the authenticity of handicrafts does not play any significant role in cruise passengers' purchasing decisions (Douglas and Douglas 2004), especially as these products are rather seen as gifts or memory holders. More research concerning Pacific crafts has been carried out about Pacific diaspora communities in New Zealand and their work within formal and informal cultural economies (Cave 2009; Cave and Buda 2013; Thode-Arora 2012).

Tourism and handicraft development in the Solomon Islands

The current tourism context

The Solomon Islands belong to the least visited destinations worldwide (UNWTO 2020). Factors that have impeded tourism development in the Solomon Islands are manifold: political instability and ethnic tensions; limited accessibility; high airfares and travel costs; health issues, such as malaria; lack of a well-developed tourist infrastructure and amenities; natural disasters; and disputes over land (Diedrich and Aswani 2016; Gay 2009). Despite the country's low visitor volume, in 2019, the total contribution of travel and tourism accrued to 9.3% of the gross domestic product (GDP) and 8.4% to overall employment (WTTC 2021). According to the International Visitor Survey (IVS) data from 2019, the majority of tourists to the Solomon Islands can be classified under business/conference visitors (46%), followed by holidaymakers and vacationers (26%), and visiting friends and relatives (14%) (NZTRI n.d.). The most important source markets have been Australia, New Zealand, the United States (US), Papua New Guinea, Fiji, and the Asian market, particularly China and Japan (NZTRI n.d.; SIVB 2016). The vast majority of tourists visit the province of Guadalcanal, where the capital city Honiara is located (NZTRI n.d.).

The Solomon Islands National Tourism Development Strategy (2015–2019) acknowledged that the country underperformed in tourism sector growth and highlighted the need for small-scale tourism based on niche markets and a focus on cruise ship tourism (SPTO 2015). Moreover, the country's *Micro, Small and Medium Enterprise (SME) Policy and Strategy* identified the tourism and hospitality industry as a priority sector with significant growth potential (Solomon Islands Government, n.d.). In the Solomon Islands, micro-enterprises are defined as businesses with a maximum number of five full-time equivalent employees, an annual turnover of less than SBD0.3 million (currently US$36,250), and a net capital investment of less than SBD0.5 million (US$60,410) (Solomon Islands Government, n.d.). The development of handicraft and souvenir micro-businesses can play an important role in this context.

Development of crafts

Before Solomon Islanders transitioned from subsistence cultivation to a cash-oriented lifestyle, carving and weaving served important secular and religious functions (Horoi 1980). Warfare and hunting required spears, warclubs, bows, and arrows. Body ornaments were used to present oneself, wooden figures or stone images were placed at traditional shrines, and shell and feather products served as a medium of exchange similar to banknotes or coin money. Such Solomon Islands artefacts and ornaments have attracted Western visitors since the arrival of the European expeditions in the sixteenth and seventeenth centuries. In the decades and centuries to come, thousands of objects from the Solomon Islands were bought, exchanged, or taken by seamen, traders, colonial officers, missionaries, researchers, collectors, and tourists and have been displayed in museums around the globe (Burt 2009; Specht and Bolton 2005). One of the interview respondents from Western Province also recalled stories from his grandfather, who had sold wooden bowls and carvings to foreigners who came by sailing boats 100 years ago (Interview 6, male woodcarver and vendor, Honiara).

In the 1950s, commercial airlines settled in Honiara, and international tourism subsequently grew, though at relatively low levels (Douglas 1997). Simultaneously, urbanization and rural–urban migration accelerated, and small-scale industrialization took off (Horoi 1980). This was also the time when tourist-oriented crafts in the Solomon Islands appeared, based on carving (both wood and stone), weaving (not textiles but mainly mats, bags, or hats based on pandanus or coconut leaves), in-lay work with mother of pearl shells, and shell money (Austin 2011; Guo 2007; Horoi 1980). According to interview respondents, only two main handicraft sale venues existed in Honiara until the late twentieth century (Interview 7, female handicraft business owner, Honiara). In the past 15–20 years, more shops, markets, and products have emerged. Artisanal and business associations such as the Solomon Islands Arts Association (founded in 1991) and the Solomon Islands Women in Business Association (SIWIBA, founded in 2004) assisted in strengthening the links between arts, handicrafts, and tourism by providing promotional materials for tourists, organizing events, and facilitating workshops and training.

The types of handicrafts available in the Solomon Islands vary by region. Shell money comes from the island of Malaita (Figure 13.1), red feather money from the eastern islands of Santa Cruz, wooden carvings mostly derive from Western Province and Makira, many of the fine patterns of pandanus shoulder bags are produced by people from the Polynesian settlements (e.g., Rennell and Bellona), and many of the woven baskets originate from the island of Guadalcanal, where the capital city of Honiara is located (Austin 2011). The following section provides a brief overview of Honiara's existing handicraft businesses and products.

Current handicraft businesses and products

The author identified 13 shops, businesses, and markets that are directly geared at selling handicrafts and souvenirs to international (and partly also domestic) visitors. Generally, shops and markets in Honiara place a strong emphasis on regionally and locally (within the Solomon Islands) made and sourced products. All surveyed businesses offer necklaces, jewellery, or shell

Figure 13.1 Shell money production performance for tourists in the Langalanga Lagoon, Malaita, Solomon Islands. Source: Kosita Butratana, June 2017.

money made of seeds, shells, feathers, and tusks. The second most popular product category concerns regionally made woven products such as bags, baskets, mats or bowls, and different types of wood carvings, including figures, animals, small boats, and tableware. Paintings from Solomon Islands artists have also become increasingly popular in the tourism market (see Table 13.1). The handicrafts sector in Honiara can be categorized into three main types of businesses.

The first category consists of vending stalls at *handicraft, arts, or souvenir markets* (S2, S3, S4, S11). These are designated marketplaces bringing together consumers with producers/artists or vendors and retailers. Such markets constitute relatively controlled and managed sites that also reflect the interest of governmental organizations and/or their supporting associations. While the cruise ship market (S11) operates only on cruise ship days, other markets operate daily. The number of craft stalls and present vendors depends on seasonal and weather conditions. Micro-entrepreneurs at these market stalls often have several business functions, including craft production, retailing, and business management.

The second category consists of *souvenir and handicraft shops* scattered around town (S1; S5–S10). While shops S1 and S5–S7 strongly focus on regional handicrafts and products, shops S9–S11 also offer imported clothing (often with locally branded destination images) or mass-manufactured items representative of the destination, such as key chains, magnets, or mugs. These shops are mainly retailers, and some of them are tied to a more prominent institution such as a foreign-owned hotel or the national museum.

The third category relates to a fluctuating number of *mobile or footpath vendors*. They carry their products in a bag and often display their items near or in front of the major hotels, markets, or the museum (S12 and S13). They may change their location, approach tourists actively while walking through the city, and may not have a license for selling. Most of these vendors sell wood carvings, paintings, necklaces, and earrings.

Table 13.1 Range of handicraft and souvenir shops and products in Honiara

13 Shops/Markets (S)		*14 Product Categories (PC)*	
S1	Museum Handicraft Shop (PC 1–6, 12, 14)	PC1	Postcards or photographs about the region
S2	Arts Village – open-air crafts market (PC 2–4, 6, 10, 13)	PC2	Paintings from local Solomon Islands (SI) artists
S3	National Art Gallery (PC 2–4, 6, 9, 13, 14)	PC3	Regional (SI) woven products such as bags, baskets, mats, or bowls
S4	Honiara Central Market (PC 6–8)	PC4	Regional (SI) carvings and wooden items such as small figures, animals, boats, or tableware; carved bamboo and coconut products
S5	Hotel Souvenir Shop (PC 1, 3, 4, 6–10, 12, 13)	PC5	Tribal art such as original or reproduced masks or sculptures from Indigenous groups of SI
S6	Nautilus Books and Gifts - Shop (PC 1-14)	PC6	Regional (SI) items (necklaces, jewellery, shell money) made of seeds, shells, feathers, and tusks
S7	King Solomon Arts and Craft Centre – Shop (PC 2–4, 6, 8, 10, 12)	PC7	Regional (SI) culinary products such as coffee, chocolate, pepper, coconut oil, vanilla, and kava
S8	Island Artifacts – Shop (PC 2–4, 6, 10, 13, 14)	PC8	Regional (SI) hygiene, health, or cosmetic products such as soaps, creams, and oil
S9	Beaufut Island Souvenirs Shop (PC 1, 3, 4, 6, 9, 10, 13)	PC9	Other items representative of the destination, such as key chains, fridge magnets, and mugs
S10	DJ Graphics – shop focusing on clothing and shirts (PC 1–4, 6–10, 12, 13)	PC10	Caps, T-shirts, or other clothing branded with the destination, hotel, and attraction
S11	Cruise ship market with a fluctuating number of stalls (PC 1–4, 6, 9, 10)	PC11	Non-regional arts and crafts such as paintings, toys, and ornaments
S12	Footpath vendors – fluctuating numbers (PC 2–4, 6, 9, 10)	PC12	Books or DVDs about the country
S13	Mobile vendors – fluctuating numbers (PC 2–4, 6, 14)	PC13	Other
		PC14	Regional (SI) stone carvings

Source: Author's own compilation based on fieldwork.

Supply perspectives

The commodification of material culture

While some mass-manufactured and imported products have entered the craft and souvenir landscape of Honiara, the Solomon Islands have kept a relatively strong focus on locally crafted products and artworks. The localness of Honiara's craft products is also reinforced by existing market regulations, associations, and even hotel shops which discourage and prohibit the sales of imported items (Interviews 3, 5, 8). Honiara's crafts and souvenirs, thus, differ from those in other Melanesian countries, such as Vanuatu, where many products are being imported. One of the interviewed entrepreneurs visited Vanuatu and highlighted her astonishment about the lack of local crafts and artworks at main markets and shops in the capital city of Port Vila.

The commodification of Indigenous arts and crafts has also taken place in the Solomon Islands. Items that were initially used to fulfil utilitarian needs or served religious purposes have been transformed to meet the needs and interests of international travellers. This development

Figure 13.2 Shell money products at Honiara market. Source: A. Trupp, June 2017.

was first noted by Horoi (1980), who stated that carving in the Solomon Islands has lost its traditional function and purpose because it is mainly done to earn cash income. The *Nguzunguzu*, for example, is a carved wooden figurehead that traditionally affixed canoes. Such carvings are now available as standalone crafts and souvenirs in all shapes and sizes. Shell money in the form of strings of polished shell beads has been used as a bridal gift and as local currency to exchange goods in Malaita (Burt 2009). Over the years, modified necklaces and bracelets have become popular souvenirs and fashionable items for tourists and Solomon Islanders. Interviews with craft entrepreneurs and artists indicate that the commodification of arts and material culture has spread geographically and has gained broader acceptance. For example, shell money has served as a bridal gift and a local currency but is now widely sold at shops and markets across town and has become a popular souvenir for tourists (Figure 13.2). One of the respondents stated that such change of functions and meanings of products also lead to local resentment and criticism.

> Before, yes, but now everyone is keen on income. It used to be big business before already with the cruise boats in Langalanga. But they did not expand to urban areas. Now, they produce and sell here in Honiara, too. I think one should not display the shell money in a hotel, you should do it in a proper place.
>
> *(Interview 9, female shop owner, Honiara)*

Respondents' views on the commodification of culture also show that they understand culture as a dynamic and interactive process. Tourists and entrepreneurs are tangled in a global–local relationship (Cave et al. 2013). Such interdependencies between tourists, hosts, and source destinations are also recognized by respondents. As one local artist and carver argues:

> Things are changing and we have to be adaptive, and so I think, I go for it. Selling this kind of artwork [...] I mean, we are adjusting to it [...]. If I sell my artwork [...]

> people from other parts of the world will see it, and they ask about it. They would say something about this piece from the Solomon Islands. This exchange is taking place. That's my opinion.
>
> *(Interview 2, male carver/artist Honiara)*

Other vendors also explained that they adapt their products according to tourists' requirements. In some cases, tourists tell handicraft producers how to design their products or ask specifically for 'made-to-order products'. Tourists give ideas for a particular carving, the design for a necklace, or the colours and shape of a weaving product. Moreover, many craft stalls are now selling *Bilum*, a string handmade bag originating from Papua New Guinea. This is the result of the exchange between craft production techniques between female handicraft producing groups of the Solomon Islands and Papua New Guinea over the past few years. Indeed, several of the interviewed micro-entrepreneurs gained experience abroad, i.e., through workshops or participation in cultural or art fairs.

As shown above, most interviewed micro-entrepreneurs and artists acknowledge processes of cultural change and commodification as realities that they feel relatively confident to deal with. However, some micro-entrepreneurs and public stakeholders expressed concerns about issues relating to 'indigenous cultural and intellectual property rights' (ICIPR). ICIPR relates to the heritage of Indigenous people and can include material cultural expressions such as shell money or particular carving or weaving products tied to a specific Indigenous group or territory. Breaching of copyright or misappropriation of Indigenous arts and crafts occurs when non-indigenous artists create and market them as Indigenous, or when mass-produced souvenirs by non-indigenous people copy cultural symbols or display inauthentic designs (Janke and Sentina 2018). The importance of ICIPR is increasingly acknowledged by stakeholders of the Indigenous arts and cultural industries. Nevertheless, its development and implementation in the Solomon Islands context is still in an embryonic state and this was clearly evoked in the following narrative:

> The issues of intellectual property come in. The issues of cultural rights also come in. [...] Traditional Knowledge and Expressions of Culture Bill. In short, we say TKEC. That particular bill focuses on the traditional rights of the owners, of cultural products, cultural resources, and of course, traditional knowledge. We are just at the beginning of this (Interview 4, Ministry of Culture and Tourism, Honiara).

While this section has focused on perspectives of cultural change and commodification in the context of handicrafts and souvenir businesses, the following discussion will outline the economic benefits for Honiara's micro-entrepreneurs.

Economic impacts of handicraft sales

The Solomon Islands are classified as a low-income country by the World Bank and have a world ranking of 151 out of 189 countries and territories on the global human development index for the year 2019 (UNDP 2020). The minimum wage was US$0.49 per hour during the time of fieldwork, but the government doubled this amount, which was effective as of August 2019. However, the pandemic led to an increase in unemployment, especially in the capital city Honiara (NDI 2021). Large parts of the population, especially in the outer islands, depend on agriculture, livestock, fishing, and forestry, while the country imports oil and many manufactured products. Considering the country's low-wage structure and the high costs for many (imported) products, self-employment in the Solomon Islands may offer a relatively secure economic basis (Ongoa 2017).

Interviews with micro-entrepreneurs at different markets in Honiara indicate that they deem their earnings as quite satisfactory. Income, however, is not stable and fluctuates strongly according to tourist numbers, location, weather, and cruise ship arrivals. Monthly income can, thus, vary between US$150 and US$2,000. Such fluctuations are also indicated in the two interview narratives below.

> My income depends on the customers every day and what they want. Sometimes it's good, sometimes no income a day. [...]. Last month, I was at the market for two weeks. [...]. I roughly made 800SBD (US$95). [...]. That is medium good income for me.
>
> *(Interview 1, female retailer/craft producer, Honiara)*

> I mean, in terms of income, what all of us artists here receive is, to me, quite satisfactory. It helps most of our needs, so I would say good money, but I can't really put it in figures because, how do I say, it's not the same every time. [...] Sometimes, in a week, I can make around 5,000 SBD (US$600).
>
> *(Interview 2, male carver/artist, Honiara)*

The interviewed artist quoted above can display and sell his paintings and carvings in the newly constructed Solomon Islands National Art Gallery. The Ministry of Culture and Tourism formally governs this indoor exhibition and marketplace, while the Solomon Islands Artist Association has also played a significant role in developing the idea and concept of the gallery. Adjacent to the National Art Gallery is the open-air space of the Arts Village, where micro-entrepreneurs set up their vending stalls and tents. Sales operations are often interrupted by heavy rain, which respondents identified as one of their main challenges to operating their business (Figure 13.3). Nonetheless, a new roofed crafts market centre was constructed to cater for the exhibition and sales of Solomon Islands handicrafts (The Island Sun 2018).

Because Honiara receives a rather low number of international tourists, the impact of day visitors in the form of cruise ship tourists is significant for many respondents. On cruise ship days, many of the market vendors and city shops set up stalls and tents at the temporary cruise ship market. This market is located at the wharf next to the landing for cruises and is managed by the Solomon Islands Visitor Bureau. In the interview quote below, one of the interview respondents explains the economic difference between cruise ship and non-cruise ship days.

Respondent: If there is a market at the wharf, hopefully, I get 3,000SBD (US$354) for one day when the tourists come.

Interviewer: Wow, that's quite a difference.

Respondent: Yeah, very different. When I make a low income on such days, I get 1,000SBD (US$118) just for one day.

(Interview 1, female retailer/craft producer, Honiara)

Another micro-entrepreneur based at the Arts Village market enthusiastically stated: 'Every day I dream about cruise ships because this means money inflow [laughing]' (Interview 9, female retailer/craft producer, Honiara). Vendors and micro-entrepreneurs receive information about cruise ship arrivals through such associations as the Solomon Islands Artist Association and SIWIBA. While most of the interviewed vendors and micro-businesses clearly stated that they benefit economically from cruise visitors, vendors from one shop in town stated that they hardly

Figure 13.3 Challenges concerning the outdoor handicraft market in Honiara. Source: A. Trupp, June 2017.

profit from the influx of cruise passengers. They view the increasing competition from market sellers and mobile vendors as their main concern.

COVID-19 pandemic and its impact on the handicraft industry

Handicraft micro-businesses in the Solomon Islands have mainly been concentrated in the capital city Honiara which, until recently, hosted the only international airport in the country. With the opening of Munda International Airport in Western Province in April 2019, it was anticipated that international tourist arrivals would increase, and handicraft and souvenir businesses would expand. However, international immobility as a consequence of the COVID-19 virus significantly impacted the tourism industry's prospects. In 2020, the total contribution of travel and tourism dropped by 67.5% (WTTC 2021), and international borders remained closed for most types of visitors throughout 2021. In addition, at the early stages of the pandemic, the Honiara City Council suspended the sale of non-essential goods, including products and handicrafts such as clothes, jewellery, and accessories (Gwao 2020). Furthermore, the government encouraged people in Honiara to return to their home villages and communities, which can provide alternative livelihoods and help to reduce the virus spreading across crowded places such as markets (Nanau and Labu-Nanau 2021).

Consequently, the demand and supply of handicraft products have been significantly disrupted, and an environment of uncertainty among small businesses remains, despite the fact that the region is currently attempting to open up to tourism. A survey of market vendors conducted by Market for Change in June 2020 showed that the pandemic strongly impacted vendors' economic gains, with more than 90% experiencing a substantial decrease in income (UN Women 2021). In response to the COVID-19 pandemic, the Solomon Islands launched the so-called Economic Stimulus Package, including rental relief packages for small and medium businesses. In a recent National Democratic Institute (NDI 2021) survey, 61% of respondents rated the government's financial assistance during the crisis as positive. However, there has also been concern that there were no specific considerations for the protection and support of vulnerable groups such as women, the elderly, and marginalized communities (Nanau and Labu-Nanau 2021).

Globally, there is a COVID-19 driven trend toward digital forms of communication, marketing, and online sales. However, for many small handicraft businesses and vendors in the Solomon Islands, access to the internet and lack of e-commerce training represents challenges that need to be addressed in future initiatives led by associations and public sector bodies.

Conclusion and research implications

Based on an inventory of handicraft shops and interviews with different stakeholders, this chapter assessed the development of Honiara's handicraft businesses from a socio-cultural and economic perspective. The Solomon Islands feature rich natural and cultural resources, including long-established traditions of woodcarving, weaving, and production of shell money. Handicrafts and artwork are strongly linked with the tourism industry as visitors often meet different cultures and return home with gifts or souvenirs that reflect that experience (Wilkins 2011). Moreover, the unique features of the country's arts and crafts could help strengthen the promotional efforts of the destination.

The production, display, and sales of arts and crafts at markets and shops throughout Honiara represent a visible and vital part of cultural heritage tourism. The crafts and arts exhibited and sold in Honiara primarily consist of handmade and locally sourced and produced items representing different regions of the country. This sets the Solomon Islands apart from many other destinations in the Pacific, where mass-manufactured and often imported souvenirs dominate. However, many of the products have been adopted in design, size, and meaning to meet the demand of the tourist market. While objects such as shell money or wooden carvings initially fulfilled utilitarian needs or served religious purposes, they have been transformed to meet the needs and interests of domestic and international travellers. Hence, products become culturally adapted or even partially acculturated as a consequence of the influence of tourism, touristification, and commodification. While commodification may transform cultural goods into commodities that serve the needs and interests of tourists, this process also helps to keep traditions of weaving and carving alive. It can also lead to the advancement, international representation, and diversification of arts and crafts. Interviewed artists also understand culture and its related products, such as paintings and other artworks, as dynamic and subject to international exchanges and globalization.

Processes of the commodification of crafts and arts can clash with concerns about Indigenous cultural and intellectual property rights. The Solomon Islands are culturally and ethnically diverse and protecting its cultural heritage against the misappropriation of Indigenous arts and crafts remains a challenge. Implementing the utilization of 'geographical indication' (GI) can promote cultural products by helping consumers differentiate between local and imported products (Grobar 2019). The GI indicates that the handicraft 'belongs' to a particular place of origin as the examples of the Chanderi fabric in India or the Parma ham in Italy illustrate (Grobar 2019). Indigenous communities, however, will need a more holistic approach to protect their knowledge and cultural expression, including deeper multi-stakeholder consultation and legislative change (Janke and Sentina 2018).

From an economic perspective, interviewed micro-entrepreneurs were relatively satisfied with their income from selling crafts and arts prior to the pandemic. Market vendors highlighted that they could multiply their sales on cruise ship days. Given the low number of international tourist arrivals, cruise tourists often make a big difference in the earnings of micro-businesses.

However, such sales are not consistent because cruise calls are infrequent and last for only one day (IFC 2016). Moreover, not all interviewed entrepreneurs can benefit from cruise ship passengers who usually buy from the stalls at the temporarily erected cruise ship market or the centrally located handicraft market. Businesses that cannot join these locations are disadvantaged. More research is needed to assess the long-term economic and socio-cultural impacts of cruise ship tourism on local businesses. This approach should consider the perspectives of the different types of micro-entrepreneurs identified in this chapter, as well as perspectives from cruise ship tourists and the cruise industry.

References

Abisuga-Oyekunle, O. A. and I. R. Fillis (2017) 'The role of handicraft micro-enterprises as a catalyst for youth employment', *Creative Industries Journal*, 10(1): 59–74.

Austin, R. (2011) *Handicrafts of the Solomon Islands*, Noumea: Secretariat of the Pacific Community.

Azarya, V. (2004) 'Globalization and international tourism in developing countries: Marginality as a commercial commodity', *Current Sociology*, 52(6): 949–967.

Burt, B. (2009) *Body Ornaments of Malaita, Solomon Islands*, Honolulu: University of Hawai'i Press.

Cave, J. (2009) 'Embedded identity: Pacific islanders, cultural economies, and migrant tourism product', *Tourism, Culture and Communication*, 9(1–2): 65–77.

Cave, J., T. Baum and L. Jolliffe (2013) 'Theorizing tourism and souvenirs, glocal perspectives on the margins'. In J. Cave, T. Baum and L. Jolliffe (eds.) *Tourism and Souvenirs: Glocal Perspectives from the Margins*, Bristol: Channel View Publications, pp. 1–25.

Cave, J. and D. Buda (2013) 'Souvenirs as transactions in place and identity: Perspectives from Aotearoa New Zealand'. In: J. Cave, T. Baum and L. Jolliffe (eds.) *Tourism and Souvenirs: Glocal Perspectives from the Margins*, Bristol: Channel View Publications, pp. 98–118.

Chand, A., S. Naidu, P. C. Southgate and T. Simos (2015) 'The relationship between tourism, the pearl and mother of pearl shell jewellery industries in Fiji'. In: S. Pratt and D. Harrison (eds.) *Tourism in Pacific Islands: Current Issues and Future Challenges*, Abingdon: Routledge, pp. 148–164.

Cheer, J. M. (2017) 'Cruise tourism in a remote small island-high yield and low impact?' In: R. Dowling and E. Cowan (eds.) *Cruise Ship Tourism*, 2nd edition, Wallingford: CABI, pp. 408–423.

Cohen, E. (1988) 'Authenticity and commoditization in tourism', *Annals of Tourism Research*, 15(3): 371–386.

Cohen, E. (2000) *The Commercialized Crafts of Thailand: Hilltribes and Lowland Villages*, Richmond: Curzon Press.

Cole, S. (2007) 'Beyond authenticity and commodification', *Annals of Tourism Research*, 34(4): 943–960.

de Burlo, C. R. (1996) 'Vanuatu'. In: C. M. Hall and S. J. Page (eds.) *Tourism in the Pacific: Issues and Cases*, London: International Thomson Business Press, pp. 235–255.

Diedrich, A. and S. Aswani (2016) 'Exploring the potential impacts of tourism development on social and ecological change in the Solomon Islands', *Ambio*, 45(7): 808–818.

Douglas, N. (1997) 'Applying the life cycle model to Melanesia', *Annals of Tourism Research*, 24(1): 1–22.

Douglas, N. and N. Douglas (2004) 'Cruise ship passenger spending patterns in Pacific island ports', *International Journal of Tourism Research*, 6(4): 251–261.

Gay, D. (2009) *Solomon Islands: Diagnostic Trade Integration Study*, Port Vila: Blue Planet Media.

Graburn, N. H. H. (1976) *Ethnic and Tourist Arts: Cultural Expressions from the Fourth World*, Berkeley, Los Angeles: University of California Press.

Grobar, L. M. (2019) 'Policies to promote employment and preserve cultural heritage in the handicraft sector', *International Journal of Cultural Policy*, 25(4): 515–527.

Guo, P. (2007) '"Making money": objects, productions, and performances of shell money manufacture in Langalanga, Solomon Islands'. In: M. H. H. Hsiao (ed.) *New Frontiers of Southeast Asia and Pacific Studies*, Taipei: Center for Asia-Pacific Area Studies, Academia Sinica, pp. 211–240.

Gwao, J. (2020) 'Suspension of sale of non-essential goods at central market questioned', *Solomon Times*. Available at https://www.solomontimes.com/news/suspension-of-sale-of-nonessential-goods-at-central-market-questioned/9706 (accessed 3 January 2022).

Hall, M. C. and J. Rath (2007) 'Tourism, migration and place advantage in the global cultural economy'. In: J. Rath (ed.), *Tourism, Ethnic Diversity and the City*, Abingdon: Routledge, pp. 1–24.

Horoi, S. R. (1980) 'Tourism and Solomon handicrafts'. In: The Institute of Pacific Studies (ed.), *Pacific Tourism: As Islanders See It*, Suva: The Institute of Pacific Studies, USP, pp. 111–114.

Husa, L. C. (2020) 'The "souvenirization" and "touristification" of material culture in Thailand–mutual constructions of "otherness" in the tourism and souvenir industries', *Journal of Heritage Tourism*, 15(3): 279–293.

IFC (International Finance Corporation) (June 2016) *Assessment of the Economic Impact of Cruise Tourism in Papua New Guinea & Solomon Islands*, Washington: International Finance Corporation .

ITC (International Trade Centre) (2012) 'Inclusive tourism: Linking the handicraft sector to tourism markets', *International Trade Centre Website*. Available online at https://doksi.net/en/get.php?lid=23672 (accessed10 May 2020).

Janke, T. and M. Sentina (2018) *Indigenous Knowledge: Issues for Protection and Management*, IP Australia: Commonwealth of Australia.

Kumar, N., A. Trupp and S. Pratt (2022) '"Linking tourists" and micro-entrepreneurs' perceptions of souvenirs: The case of Fiji', *Asia Pacific Journal of Tourism Research*, 27(1): 1–14.

Lacher, R. G. and S. K. Nepal (2011) 'The economic impact of souvenir sales in peripheral areas a case study from Northern Thailand', *Tourism Recreation Research*, 36(1): 27–37.

Milne, S. (1991) 'Tourism and economic development in Vanuatu', *Singapore Journal of Tropical Geography*, 11(1): 13–26.

Morgado, M. A. (2003) 'From kitsch to chic: The transformation of Hawaiian shirt aesthetics', *Clothing and Textiles Research Journal*, 21(2): 75–88.

Movono, A. and H. Dahles (2017) 'Female empowerment and tourism: A focus on businesses in a Fijian village', *Asia Pacific Journal of Tourism Research*, 22(6): 681–692.

Nanau, G. L. and M. Labu-Nanau (2021) 'The Solomon Islands' social policy response to Covid-19: Between Wantok and economic stimulus package', *CRC 1342 Covid-19 Social Policy Response Series No 18*, University of Bremen.

NDI (National Democratic Institute) (2021) 'Solomon Islands road to recovery from the COVID-19 pandemic', August. Available online at https://www.ndi.org/sites/ default/files/NDI%20Solomon%20 Islands%20Public%20Opinion%20Research%20Report.pdf (accessed 3 January 2022).

NZTRI (New Zealand Tourism Research Institute) (n.d.) 'Solomon Islands International Visitor Survey: January to December 2019'. Available online at https //www.nztri.org.nz/sites/all/themes /nztribootstrap/ufy/nztriassets/ptdireports/solomons/Solomon-Islands-IVS-Jan-Dec-2019-rs.pdf (accessed 3 January 2022).

Ongoa, E. G. (2017) 'Understanding entrepreneurial competencies of women entrepreneurs in Solomon Islands', *Master Thesis*, Suva, Fiji: The University of the South Pacific.

Saarinen, J. (2016) 'Cultural tourism and the role of crafts in Southern Africa: The case of craft markets in Windhoek, Namibia', *Tourism: An International Interdisciplinary Journal*, 64(4): 409–418.

Silverman, E. K. (2013) 'After cannibal tours: Cargoism and marginality in a post-touristic Sepik River society', *The Contemporary Pacific*, 25(2): 221–257.

SIVB (Solomon Islands Visitor Bureau) (2016) *International Visitor Survey 2016*, Honiara: Solomon Islands Visitors Bureau.

Solomon Islands Government (n.d.) 'SMEs policy and strategy', Ministry of Commerce, Industry, Labour and Immigration website. Available online at https://www.commerce.gov.sb/component/edocman/39 -smes-policy-and-strategy.html (accessed 29 October 2019).

Specht, J. and L. Bolton (2005) 'Pacific islands' artifact collections: The UNESCO inventory project', *Journal of Museum Ethnography*, 17: 58–74.

SPTO (South Pacific Tourism Organisation) (2015) *The Solomon Islands National Tourism Development Strategy 2015–2019*, South Pacific Tourism Organisation Website. Available online at http://macbio-pacific.info /wp-content/uploads/2017/08/National-Tourism-Strategy-2015.pdf (accessed 29 October 2019).

Swanson, K. K. and D. J. Timothy (2012) 'Souvenirs: Icons of meaning, commercialization and commoditization', *Tourism Management*, 33(3): 489–499.

Taylor, J. (2016) 'Pikinini in paradise: Photography, souvenirs and the "child native" in tourism'. In: K. Alexeyeff and J. Taylor (eds.) *Touring, Pacific Cultures*, Canberra: ANU Press, pp. 361–378.

The Island Sun (2018) 'New crafts market centre to open in Honiara', *The Island Sun Daily News*, 1 November. Available online at http://theislandsun.com.sb/new-crafts-market-centre-to-open-in-honiara/ (accessed 29 October 2019).

Thode-Arora, H. (2012) '"Tablemats helped Niue to survive", Flechterinnen von der Insel Niue in Auckland/Neuseeland'. In: M. Dabringer and A. Trupp (eds.) *Wirtschaften mit Migrationshintergrund. Zur soziokulturellen Bedeutung, ethnischer Ökonomien in urbanen Räumen*, Vienna Studienverlag, pp. 65–79 .

Trupp, A. (2017) *Migration, Micro-Business and Tourism in Thailand: Highlanders in the City*, London: Routledge.
UNDP (United Nations Development Programme) (2020) 'Briefing note for countries on the 2020 human development report. Solomon Islands'. Available online at http://hdr.undp.org/sites/default/files/Country-Profiles/SLB.pdf (accessed 15 December 2021).
UN Women (2021) 'Markets for change continues COVID-19 response framework'. Available online at https://asiapacific.unwomen.org/en/news-and-events/stories/2021/03/markets-for-change-continues-covid-19-response-work (accessed 3 January 2022).
UNWTO (United Nations World Tourism Organization) (2020) *UNWTO Tourism Highlights,* Madrid: UNWTO. Available online at https://www.e-unwto.org/doi/book/10.18111/9789284422456 (accessed 8 December 2021).
Wilkins, H. (2011) 'Souvenirs: What and why we buy', *Journal of Travel Research*, 50(3): 239–247.
WTTC (World Travel and Tourism Council) (2021) 'Solomon Islands: 2021 annual research: Key highlights', London: World Travel and Tourism Council. Available online at https://wttc.org/Research/Economic-Impact/moduleId/704/itemId/203/controller/DownloadRequest/action/QuickDownload (accessed 8 December 2021).

14

COMMUNITY MANAGEMENT OF CULTURAL TOURISM AT A WORLD HERITAGE SITE

Intersections of the 'local' and 'global' at Chief Roi Mata's Domain, Vanuatu

Adam M. Trau and Chris Ballard

Introduction

In common with most Indigenous communities in Vanuatu and other independent Pacific Island states, the villages of Lelepa Island and Mangaliliu (known collectively as the 'Lelema' community) are engaged in a constant renegotiation of local livelihoods under the changing conditions of a globalizing world. Village life is typified by new and complicated blends of individual, family and community responsibilities, obligations and aspirations; not least of which is ensuring that there is enough customary land and subsistence agricultural output derived from their landholdings to provide basic shelter and nutrition. It is within this context that the Lelema community is attempting to navigate the challenges and take advantage of opportunities relating to the World Heritage status of their landscape.

The Lelema community owns and manages the traditional sites associated with the life and death of Chief Roi Mata. Along with much of the surrounding cultural landscape, these sites and the stories associated with them constitute Chief Roi Mata's Domain (CRMD), Vanuatu's first United Nations Educational, Scientific and Cultural Organization (UNESCO) World Heritage Site (Figures 14.1 and 14.2). In oral traditions widely recounted throughout central Vanuatu, Roi Mata is said to have lived at the village of Mangaas, at the mouth of Havannah Harbour on mainland Efate. During a period of intense warfare across Efate, Roi Mata summoned representatives from across the island to a feast at Mangaas, at which he ended the warring by distributing totemic identities to the different matrilineal social groups, binding them together. As a much older man, while feasting on nearby Lelepa Island, Roi Mata fell ill and drew his last breath in the large chamber cave of Fels. Fearful of his extraordinary powers, the community decided not to bury him at Mangaas but, instead, to remove a small residential settlement from the nearshore island of Retoka (also known as Hat, Artok or Eretoka Island) and bury him there. As many as 300 individuals, including members of his family and court, are said to have been buried alive around his grave to accompany him to the afterworld. Retoka Island and his former residence at

DOI: 10.4324/9780429019968-16

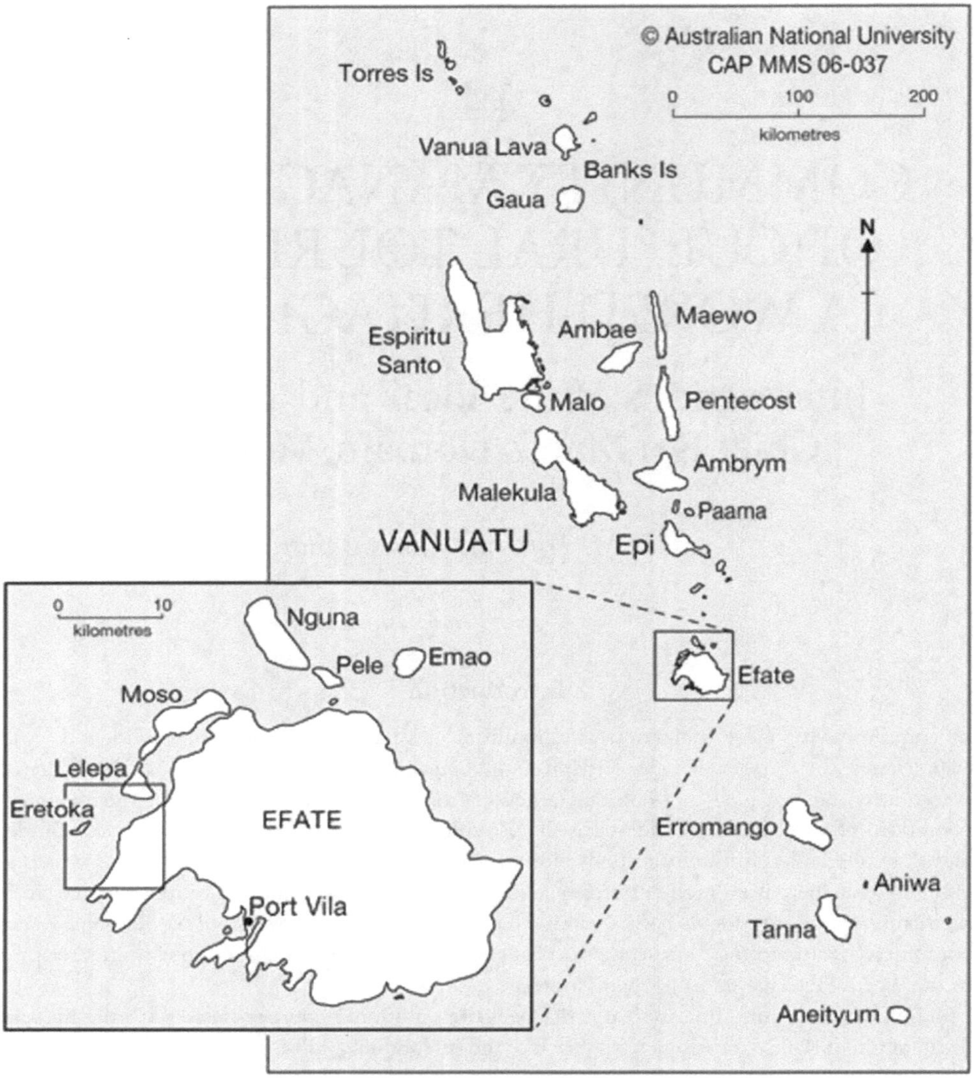

Figure 14.1 Map ofVanuatu locating Chief Roi Mata's Domain in northwest Efate. Source: Republic of Vanuatu (2007: 11).

Mangaas were then placed under traditional *tapu* prohibitions, and neither locale has been resettled since (Republic ofVanuatu 2007). Mangaas and the burial site at Retoka were excavated in 1967 by archaeologist José Garanger (1972) and the Retoka excavation, which covered only a portion of the burial yard. This excavation uncovered the remains of more than 50 individuals arranged around the central figure of an old man, seemingly confirming the oral traditions. Both Roi Mata's burial and the apparently simultaneous abandonment of his Mangaas settlement have since been archaeologically dated to about 1600 CE (Bedford et al. 1998).

The World Heritage property of CRMD consists of a triangle encompassing the three main cultural sites (Mangaas, Fels Cave and Retoka Island) and the stretch of sea between them, delin-

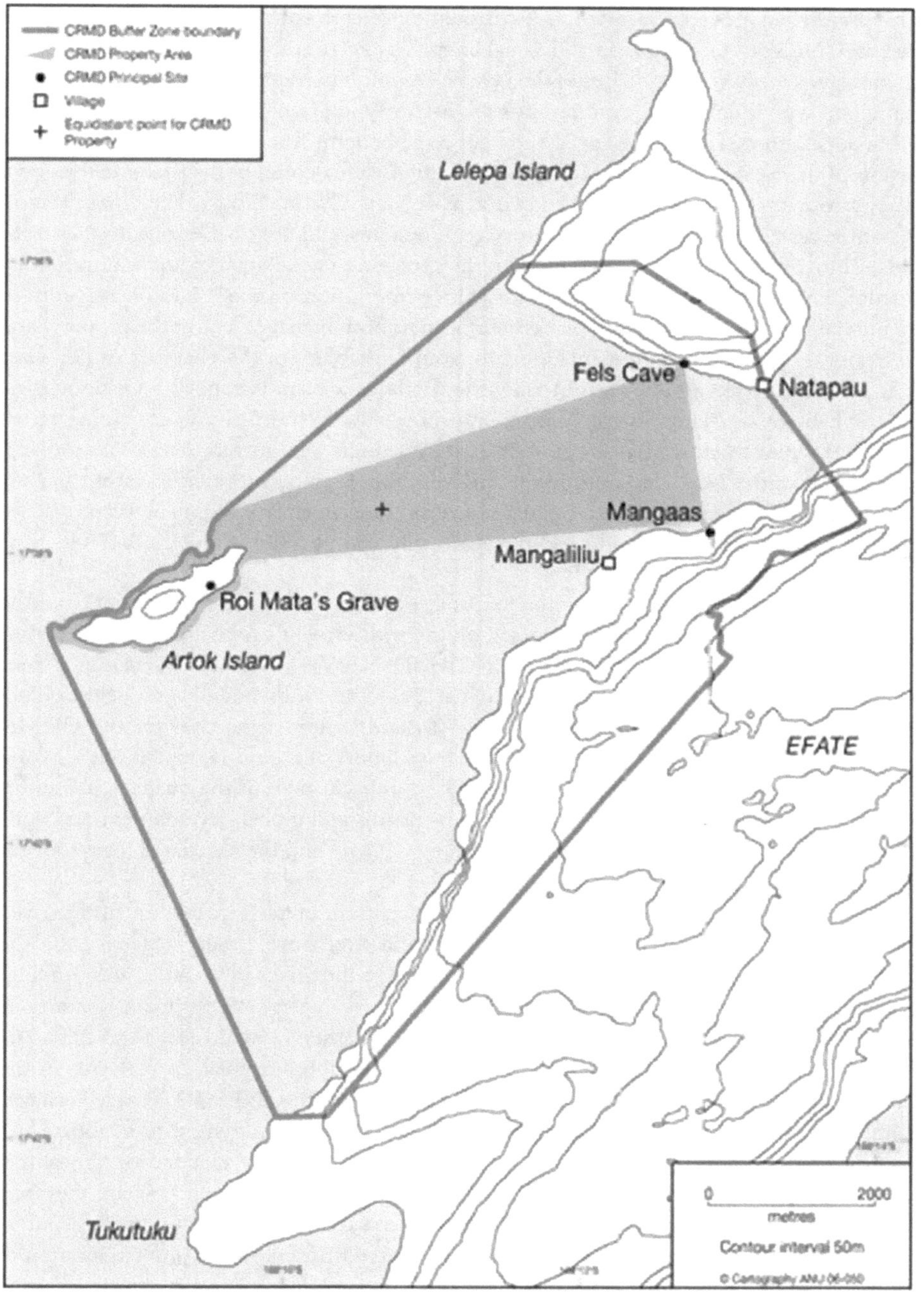

Figure 14.2 Boundaries of the World Heritage property and buffer zone of Chief Roi Mata's Domain. Source: Republic of Vanuatu (2007: 8).

eated by the shaded area in Figure 14.2. A surrounding buffer zone marked by the grey line, required by UNESCO to provide an additional layer of protection to the World Heritage property, encompasses the two main villages of the Lelema community, Mangaliliu (Efate) and Natapau (Lelepa), along with much of the community's productive land (Trau et al. 2014; Wilson 2006).

Endeavouring to maximize the social and economic benefits from inclusion of a substantial proportion of their land on the World Heritage list, the Lelema community established its own community tourism business in 2005, known as Roi Mata Cultural Tours. This formalization brings with it certain reporting and regulatory requirements, and the goal of this business is to provide exclusive access to commercial tours of the sites and to generate jobs and training along with income and dividends for the community. Since inscription on the World Heritage list in 2008, a number of significant challenges have surfaced to both management of the property and the success of the tourism business, including: ongoing instability of the chamber of Fels cave, which is periodically closed due to roof falls after earthquakes; the Category 5 Cyclone Pam, which struck Efate in March 2015, constraining international tourism and damaging much of the CRMD infrastructure (Ballard et al. 2020; Department of Tourism 2015); sales of land, including much of the area lying outside the CRMD buffer zone to foreign investors, including Caribbean Cruises, which aims to build a resort on the eastern half of Lelepa Island (Jacobs 2019; McDonnell 2015); and the effects both nationally and globally of the COVID-19 pandemic (Naupa et al. 2021).

This chapter provides a critical examination of the heritage and tourism experience of the Lelema community. The observations are based on the involvement of the chapter's first author (Trau), initially as a development volunteer at CRMD in 2008 (Trau 2013), then as a doctoral researcher (Trau 2012a; 2012b; 2015; Trau et al. 2014), and subsequently as a development worker across Vanuatu; and of the second author (Ballard), a lead researcher for the CRMD nomination from 2004 to 2008, providing continuing support of the management of the World Heritage Site since its inscription. The authors seek to unpack some of the nuances and ambiguities at the intersections between local customary norms and global development and conservation agendas, as these are expressed in relation to CRMD and the associated local tourism enterprise, Roi Mata Cultural Tours.

The chapter's goal is to move beyond prevailing conceptions of heritage and tourism derived from international discourses and, instead, to foreground alternative, local perspectives on the success and failure of community-led tourism and heritage initiatives in Vanuatu, with findings that are potentially relevant for other small island states. Several questions pertinent to community conservation and development are addressed in this chapter. How do ideas and initiatives associated with development and conservation that are in global circulation play out in the Lelema community and in the context of their engagement with UNESCO World Heritage? What are some of the global and local drivers informing community attempts to negotiate and interpret these externally derived ideas and concepts? What forms of blended or 'glocalized' tourism and heritage management emerge from this negotiated process, and how are they best supported, complemented and enhanced? Further context on the Lelema community and its UNESCO World Heritage property of CRMD is presented first, followed by an exploration of the inadequacy of conventional local–global binaries to capture the more nuanced 'glocal' realities of community-based tourism. Our central case study of Roi Mata Cultural Tours illustrates how local management, led by the Lelema World Heritage Committee, meets the challenge of conforming with and being judged against a mixture of both local and international development and conservation norms. We conclude with a call for further support for local management bodies confronting similar challenges and indicate a number of possible future research directions for the field.

The Lelema community and World Heritage

The UNESCO World Heritage Convention, launched in 1972, is the most prominent global mechanism for the identification, protection, presentation and transmission of cultural and natural heritage that is considered to have 'outstanding universal value' for future generations. Initially focused largely on monumental sites in Europe and parts of Asia, World Heritage has been deliberately extended to other regions and other forms of site since 1994 through the 'Global Strategy for a Representative, Balanced and Credible World Heritage List' and, in 1992, the creation of a new category of 'cultural landscape' designed to accommodate alternative definitions of 'monumental' and to integrate local cultural understandings of heritage significance (Taylor and Lennon 2011). Vanuatu ratified the World Heritage Convention in 2002 but moved quickly to consider its options for nomination, announcing a Tentative List of possible sites in 2004. Awareness at both a national and local level of the significance of the Roi Mata stories and the archaeological finds of the 1960s led the Vanuatu Cultural Centre (Vanuatu Kaljoral Senta or VKS), in collaboration with the Lelema community, to propose CRMD, in 2004, as an obvious candidate for Vanuatu's first World Heritage nomination. The VKS then invited several Australian heritage researchers and volunteers to provide technical assistance to the Lelema community and, between 2005 and 2008, this collaborative team produced a nomination dossier and plan of management (Republic of Vanuatu 2007; Wilson et al. 2011). In July 2008, two Lelepa Islanders, Douglas Kalotiti, as Chair of Lelema's World Heritage and Tourism Committee, and Donald Kalpokas, as Vanuatu's Ambassador to the United Nations, successfully made the case for the inscription of CRMD as a cultural landscape to the 32nd Meeting of the World Heritage Committee in Quebec City, Canada.

The management of CRMD takes place primarily at the local level through the Lelema World Heritage Committee. As of 2022, the committee consists of six individuals from Mangaliliu and Lelepa, including two men and four women. Together they work on behalf of the three landowning chiefs of the World Heritage property as well as the Lelema Council of Chiefs and the community more generally. The committee's responsibilities include managing the community-owned World Heritage tour, Roi Mata Cultural Tours, and ensuring that the modest profits generated by the tour are shared and distributed as widely as possible throughout the community. The development of the tour was a core component of the World Heritage nomination bid upon the insistence of chiefs and the community. The committee also oversees an increasingly diverse range of small project areas associated with CRMD World Heritage, including a pilot ranger programme, bungalow development, craft revitalization, and water security.

The Constitution of the Republic of Vanuatu vests the ownership of all land with 'the indigenous custom owners and their descendants' (Government of Vanuatu 1980, Article 73). The World Heritage property and surrounding buffer zone of CRMD are collectively held under customary tenure by the Lelema community in the villages of Mangaliliu and Lelepa Island. Mangaliliu, on the mainland of Efate, is connected by road to the nation's capital, Port Vila, a 20-minute drive away, while Lelepa must be accessed by boat. The total population of the Lelema community is approximately 700 in just over 80 households. In the face of intensifying pressures that are both local and global in origin, Lelema villagers are trying to mobilize the World Heritage listing of CRMD to further develop tourism as a benefit stream while also strengthening local heritage conservation.

Moving beyond local–global binaries

Given the strength and pace of globalization and the exploitative development and detrimental cultural changes playing out within ni-Vanuatu society (Slatter 2006; Timothy 2019;

Wittersheim 2011), it is tempting to portray the Lelema community and similar communities elsewhere as largely powerless in the face of modernity, relinquishing culture and tradition and replacing them with the modern. This concern is evident in the founding philosophy of the UNESCO World Heritage Convention, which proposes to protect universally important cultural and natural heritage sites from the destructive, unwieldy and homogenizing forces of globalization writ large (Hall 2011; Huxley 1947). It is evident that globalization is further heightening inequalities of income and opportunity within and between many local communities such as Lelema, but this chapter also argues that local responses and reactions within polities such as Vanuatu are more imprecise, convoluted and even contradictory than is implied by much of globalization theory and that this is particularly true when that theory is applied to the fields of heritage and tourism.

Globalization in Vanuatu is neither top-down nor bottom-up but, rather, a continuous process operating simultaneously in both directions, which is highly characteristic of what Robertson (1995) terms 'glocalization' – the paradox which sees globalization largely experienced and interpreted at the local level, while those localities are themselves produced through globalized contexts. The case study of CRMD reinforces the need to move beyond thinking about globalization in the analysis of tourism as exclusively top-down and driven purely by external forces; equally, we can no longer view processes of localization at the community level as straightforward, universally consensual and informed solely by local agendas and concerns. The Lelema community uses its authority over customary land to actively select and adapt both global and local principles that relate to heritage conservation and tourism development in order to give shape to its own distinct vision of World Heritage or 'Wol Eritej', as expressed locally. Within the Lelema community, written and signed formal agreements are accorded little value, weight or respect. In the context of local sovereignty over customary land, signatures on paper mean very little. The protection of CRMD World Heritage is reliant instead on the verbal commitments and social practices and principles (in other words, the *kastom* or tradition) of the customary landowners, both individually and collectively.

Acknowledging Lelema villagers as critical and powerful agents, blending the useful aspects of both globalization and localization through a process referred to here as glocalization, we aim to draw attention to new conceptual understandings of heritage conservation and tourism development. How might we begin to re-think, re-fashion and re-vision the complex global–local intersections of conservation and development both for CRMD and, more broadly, for Vanuatu's future? The Lelema community has consistently demonstrated that any global force that impedes (or is seen to even affect) local level sovereignty over customary land is at best re-shaped according to local needs or beliefs and, at worst, expelled or obstructed to the point where it is no longer viable. Most regularly, this takes the form of roadblocks using various means such as chains, logs or signs, but it can also include groups of afflicted traditional landowners travelling to the location that represents or is at the centre of the dispute and voicing their demands directly. In the context of Vanuatu customary land tenure, these actions are not just instances of protest and resistance but profound, embodied articulations of local power and authority.

CRMD World Heritage and its associated tourism initiatives are characterized by their glocalized nature. However, this also presents a challenge as it is then subject to multiple benchmarks of success at the global and local level. The glocal form of heritage conservation at CRMD is evaluated by the codified international system of UNESCO World Heritage operating through the Vanuatu state, while it is also valued locally for its conformity with the complex and ever-changing pluralities of everyday life within the Lelema community. Similarly, the results of glocal tourism development at CRMD are assessed by international tourists operating within an international market economy, while also being judged by the Lelema community in relation to locally valued measures of poverty reduction. These tensions are especially evident in the community's experience of developing its local tourism business, Roi Mata Cultural Tours.

Roi Mata Cultural Tours

Community benefit through development of a commercial tourism operation was an essential requirement of the planning process for CRMD from the outset, at the insistence of the land-owning chiefs and communities of Lelepa and Mangaliliu (Wilson et al. 2012). Tourism significantly contributes to the gross domestic product (GDP) and to local economies in Vanuatu, and does so to a greater degree than in most other Pacific Island countries (Cheer 2017). Lelema community members have extensive experience in the tourism industry both locally and in the capital Port Vila, and tourism was thus an obvious focus for economic benefit from World Heritage inscription. A community-owned tour operator was an integral component of the World Heritage nomination for CRMD (Republic of Vanuatu 2007), and Roi Mata Cultural Tours was launched in 2006. At the outset, Lelema community members articulated their vision for the Roi Mata Cultural Tour:

> It [the tourism project] is a dream that each person gathered here shares, just like we share our Church, our school and much more. It will make us have more respect for our own cultural strengths. This gathering brings every household in the community together, and we will make sure that we pass this dream on to the upcoming generations.
>
> *(Mangaliliu Tourism Vision Statement, quoted in Greig 2006: 6)*

> The Tourism project that goes ahead in the Lelepa region must benefit the community and each household within the community. It must be a sustainable (long-term) project. We want to make sure that we look after all of our cultural and natural places, as well as all of our sacred (tabu) places.
>
> *(Lelepa Island Tourism Vision Statement, quoted in Greig 2006: 6)*

From a local perspective, there have been two key grounds for tour business success: 1) the training and employment of as many individuals and households from the community as possible and 2) the investment of profits into community health and education services. Training for tourism at CRMD has had the additional benefit of preparing younger Lelema women and men for the hospitality industry elsewhere in the local Havannah Harbour area or more widely in Efate. As part of the full-day Roi Mata Cultural Tour, tourists are guided to all three sites within the World Heritage property for approximately US$100 – slightly above the price of similar tours available elsewhere in the region. The Lelema World Heritage Committee is responsible for all financial and operational aspects of the tour business as well as ensuring that community members are involved in every aspect of the tour. Locally owned transport is preferred whenever available, notably mini-vans and boats with a maximum capacity of 12 persons. The committee has also decided to set local wages approximately 20% higher than the income that villagers might derive from comparable work in Port Vila. Tour guides, for instance, receive approximately US$15 per tour. Each tour caters to an average of between five and six tourists, which allows the committee to cover costs and generate a small profit.

Prior to the 2008 inscription of CRMD, tours were run irregularly, with only a handful of pre-arranged package tours occurring during the first two years of operation. Immediately after inscription, between July and December 2008, 31 tours were undertaken with a total of 323 tourists. Since then, tour numbers have waxed and waned, creeping up during the peak tourist season (June–July) and then dropping to only a handful over the summer months (December–February). However, the tourism business needs to satisfy both the tour desks of foreign-owned hotels and resorts, from which more than 90% of all international tourists hail, and the powerful expectations of wealth redistribution within the Lelema communities. This tension highlights the difficulties associated with developing a glocalized business model such as the community-owned Roi Mata Cultural Tours, which are subjected to the contrasting expectations of both international visitors and the local community.

A number of operational and functional aspects to Roi Mata Cultural Tours are indicative of the nuances and contradictions of the glocalization process. From local tour guide wages to donations for community services, the Lelema World Heritage Committee is continuously attempting to balance conservation and development outcomes, a challenge which is simultaneously subject to both internal and external demands and pressures. Roi Mata Cultural Tours aims to generate revenue for individual community members while investing profits derived from the business in communal goods and services. Lelema community conceptions and measures of development success through Roi Mata Cultural Tours, such as paying for the school fees of all community children, are just as important locally as the provision of full-time employment and increases in personal and household income levels.

The Lelema World Heritage Committee is acutely aware that community emphasis on the equitable distribution of tour benefits in terms of both income and profits reduces the efficacy of the tour and the capacity for the tour to remain competitive, especially against foreign competitors who are not bound by local cultural concepts or rules. Instead of investing profits from the tour back into the business, by improving infrastructure or marketing materials for example, profits are channelled toward community projects such as stocking the village aid post or sponsoring cultural ceremonies and community events. Prior to the COVID-19 pandemic and the closure of the country's international borders, the tour also employed a large rotating labour force of more than 100 villagers (tour guides, cultural dancers, craft makers and caterers) which is well beyond the operational needs of the business. However, this approach has been necessary to meet community expectations of the tour and maintain the tour's social license within the community. This keeps operational costs high, creates overly complex accounting for the committee and often results in an inconsistent tour product. Nonetheless, distributing profits within the community, employing the maximum number of villagers and paying relatively high wages are all measures judged locally to be more important than maximizing operational and cost efficiency.

For the World Heritage Site of CRMD and Roi Mata Cultural Tours, the ongoing glocalizing process of balancing local and global demands is by no means a guarantee of overall success or sustainability. More than fourteen years after the inscription of CRMD on the World Heritage list in July 2008, the possibility of significantly increasing profit margins remains as challenging as ever. While Roi Mata Cultural Tours has generally managed to fund the expenses of the Lelema World Heritage Committee, including transport, stationery and a nominal sitting allowance, even this becomes difficult during the low tourism season. The key indicators of poverty reduction for the Lelema communities – increase in incomes for individuals and households and the provision of services such as schooling and health care for all community members – are still elusive (at least to the imagined levels of most community members) despite the extraordinary drive and commitment of committee members.

Since 2015, a series of additional challenges threatened the viability of the tourism business at CRMD. Cyclone Pam struck Vanuatu in March 2015, devastating infrastructure and livelihoods, as well as the national economy, and massively impacting the tourism industry in particular (Department of Tourism 2015; Dornan and Cain 2019). At CRMD, much of the tour infrastructure and equipment was destroyed, but damage to the historical sites was superficial, reflecting traditional familiarity with cyclones and their careful location (Ballard et al. 2020). Tourism had almost recovered, both locally and nationally, when the twin calamities of a second major natural hazard, Cyclone Harold, and the COVID-19 pandemic struck Vanuatu early in 2020. The impact of COVID-19 on the livelihoods of the Lelema community has been profound, with the effective cessation of tours along with loss of jobs and food supply contracts not just at CRMD but also in other areas of the tourism industry (Naupa et al. 2021).

Tourist demand was relatively constant prior to being curtailed by the COVID-19 border closures from March 2020. Tour records show that in 2019 there were 221 tourists to the site, dropping dramatically to 31 tourists in 2020 and just 13 in 2021. The tour operation continues to function, but the possibility of growing the business to the point where it might be considered an unequivocal success by most Lelema community members is now even more elusive (see Trau 2012a for earlier discussion of these same challenges). Rather like the numerous challenges to glocal heritage conservation (Trau 2012b; Trau et al. 2014; Wilson et al. 2011), glocal economic development at CRMD often struggles to be judged a success when measured from the perspectives of either the foreign-owned hotels or tour agents, which tend to regard the tour as a 'niche' interest, especially for cruise ship tourists. Even at pre-pandemic levels, Lelema community members were frustrated at the low numbers of tourists on the Roi Mata Cultural Tour compared to other tours in the area which focus on generic activities such as fishing, snorkelling or scuba diving. A 2019 internal review by the Lelema World Heritage Committee proposed diversifying the tour's offerings, featuring these activities with wider appeal against a backdrop of local culture and World Heritage (Ballard, fieldnotes: November 2019).

Climate change presents an inexorable and existential threat both to the physical landscape of CRMD, much of which lies close to the present sea level, and to the wellbeing of the community that sustains the outstanding universal value of the site and its significance (Gravari-Barbas 2020; Loehr 2020). However, a much more immediate danger from cruise ship tourism threatens to overwhelm much that the community has achieved through its engagement with World Heritage and the local tourism business. Cruise ship passengers have never contributed substantially to tour numbers at CRMD, in part because Indigenous heritage is not generally a focus for mass cruise tourism but also because of the physical distance between CRMD and the main landing location for short-stay cruise passengers in Port Vila (see Cheer 2017 and Watt and Bremner 2020 on cruise tourism's contribution to income in Vanuatu more widely).

Roi Mata Cultural Tours has allowed the Lelema community to present and promote its particular vision of its own history, culture and landscape. Characterization of the tour as 'dark tourism' (Cheer et al. 2015) fails to understand how the community has curated its heritage, deliberately underplaying the sensationalist potential of the burial site on Artok Island in its focus on the peace-making power of Chief Roi Mata. In 2019, Royal Caribbean acquired a long-term lease to the eastern half of Lelepa Island and announced plans to construct a major tourist resort there – the second in its 'Perfect Day Collection' (Royal Caribbean International (n.d.)), after the Coco Cay resort in the Bahamas (Jacobs 2019). The plans promoted for Lelepa by Royal Caribbean are unlikely to respect local preferences for the representation of their history and culture given the vision of the lead architect on the resort project: 'think of Indiana Jones, with a bit of Mad Max mixed in' (Jacobs 2019). Ultimately, control over these competing visions rests on control over land titles. McDonnell (2018) indicates how earlier exposure to mass tourism – in the form of the Survivor Vanuatu series shot in and around what would become the CRMD World Heritage Site – led directly to an increase in land sales in the area, and further loss of customary lands has also been linked to the use of tourism as a tool for disaster recovery in Vanuatu (Neef 2021). This problematic nexus between tourism and loss of traditional lands – and thus loss of control over livelihoods – will continue to present the most significant challenge to the quest for equitable forms of tourism and development at CRMD.

Glocalization on the periphery

Local benchmarks for success of the tour business within the traditional or subsistence economy are largely invisible to (and undervalued by) contemporary mainstream economic and development indicators (Tanguay 2015). Local cultural factors, such as wealth redistribution within the community or familial and communal ritual obligations, are commonly perceived by the 'development establishment' as barriers to economic development (Crewe and Harrison 2005; Woolcock and Davern 2015). When glocalized community businesses are measured against global development indicators that focus on the individual, the broader enterprise of businesses such as Roi Mata Cultural Tours is only partially represented. Local cultural measures of success, such as community solidary and wealth redistribution, are rendered invisible to international development discourse and largely ignored. If the benefits of glocalized business models and heritage management systems are to be realized by local or Indigenous communities, then greater support in the form of technical, budgetary and regulatory assistance will be needed. The support of national and transnational actors at CRMD, such as the Vanuatu Cultural Centre, volunteers, researchers and others, is beneficial in augmenting local agency and promoting alternate grassroots development and conservation narratives. Such an approach is comparable to what Appadurai (2000: 3) refers to in his seminal text as 'grassroots globalization', or 'globalization from below'. He contends that non-government organizations and transnational advocacy networks 'are the crucibles and institutional instruments of most serious efforts to globalise from below' (2005: 15).

Notably absent at CRMD, however, is input from the national government of Vanuatu, and this is particularly striking given its role as signatory to the World Heritage Convention. Other than the critical support of the Vanuatu Cultural Centre, there has been no financial investment in CRMD or its committee and only limited recognition of its pioneering role as an Indigenous tourism business (Stepwise Heritage and Tourism 2013). In the case of CRMD, an approach to heritage conservation and tourism development that emphasizes grassroots perspectives and promotes local cultural reconfigurations through an ongoing process of glocalization is imperative. Local Lelema community conservation and development strategies will require constant readjustment and recalibration well into the future. Indeed, the shape and scope of this truly 'glocal' entity that is CRMD World Heritage is still emerging and becoming apparent, more than fourteen years since inscription.

World Heritage and other conservation projects that position community benefit and community management at their centre must contend with competing visions and interests from within the community. Communities are naturally heterogeneous entities composed of interests that are often divergent. Therefore, there are a range of perspectives within the Lelema community on what successful development and conservation through UNESCO World Heritage can or should look like. Community consensus on CRMD is thus an ongoing process of negotiation; as Bushell and Staiff (2012: 247) observe, '"Tourism", "community" and "heritage" are not neat entities … Rather each term is a reductive abstraction of what is very messy, porous, relational, contested and in motion'. The on-ground reality of community-based conservation is that there will always be fluctuating proportions of supporters and opponents for a project within the community (Read 2011). Even with a tourism enterprise that addresses both local and global benchmarks of success, there is still a limit to which World Heritage can be viewed as useful in the everyday lives of the more than 700 community members of Lelema. Other types of developments that are more detrimental and less sustainable, including the sale or long-term leasing of traditional lands both within and beyond the CRMD buffer zone, will negatively affect the outstanding universal value of the World Heritage property.

Conclusion and research implications

Delivering a consistently high-quality heritage tourism product for the competitive tourist market, while meeting both the conservation requirements of UNESCO World Heritage status and the shifting needs, desires and expectations of community members, presents a complex and evolving challenge for local managers. Roi Mata Cultural Tours and other tourism development opportunities that the Lelema community might seek to develop, leveraging from the World Heritage status of their landscape, are critical not only for revenue generation to support site maintenance, especially in the absence of government budget support, but also for sustaining community support for CRMD. The CRMD World Heritage property and Roi Mata Cultural Tours rely on local community members, the most publicly prominent of whom are the chiefs and site landowners, who subscribe to the vision that this heritage and tourism initiative will reap more respect for customs and more benefits than other opportunities on offer. The critical agents of conservation at CRMD will always be members of the Lelema community, and particularly those individuals who choose to serve on the Lelema World Heritage Committee.

There is a pressing need for greater international and national recognition and support for local or Indigenous groups or committees, such as the Lelema World Heritage Committee, with responsibilities for meeting local, national and international conservation and development expectations or requirements. At CRMD, the role of the committee is central to everything, articulating and mediating those desires and demands on the local, national and international scales, but the capacity of the committee's members, both individually and collectively, is finite and subject to expiry over time. Such committees need to be adequately supported – financially, technically and legislatively on both national and international levels. Well-equipped local management plays the most important role in sustaining the levels of local commitment and enthusiasm necessary for community-based conservation to remain a potent partner in the development process. The various national and international stakeholders of UNESCO World Heritage – broadly including the World Heritage Centre, the World Heritage Committee and the Advisory Bodies as well as state parties and their relevant government authorities – need to think strategically and pragmatically in terms of the ways in which they can collaboratively support these local management efforts in Vanuatu and other small island states.

The chapter has sought to shed some light on the critical bottom-up forces at play within the World Heritage areas such as CRMD in Vanuatu. However, there is still much to be done to understand how tourism development in association with UNESCO World Heritage should be designed, implemented and measured. The need for further research which focuses on the specific conditions, factors and influences that lead to successful community-based heritage conservation and poverty alleviation is both essential and urgent. Key questions that invite further research include: a more detailed exploration of the webs of decision-making and information exchange that extend between international bodies such as UNESCO and local managers and communities, a better understanding of the way a generalized desire within the community for heritage conservation and transmission articulates with local economic and development priorities and needs and the scope for improving the integration of locally owned and operated tourism enterprises within national and industry bodies and strategies.

Other lines of enquiry could include investigating the socio-cultural and economic value of heritage tourism locally for both host communities and domestic tourists but also, as COVID-19 border restrictions ease, inbound tourists arriving from both air and sea. This is especially pertinent for heritage tourism businesses such as Roi Mata Cultural Tours which essentially operate within two systems – the local community system of benefit redistribution and the global capitalist system of open and competitive markets. In this context, how can the gap between foreign and locally owned and operated accommodation and tour providers be reduced? For Vanuatu and many other small island states, the marketability of heritage and culture is considerably lower than that of sandy beaches, reef snorkelling and – in Vanuatu's case – abundant duty-free shopping. Further understanding how this value is ascribed within the tourist industry – and how it can be potentially shifted and increased in the niche sectors of heritage and cultural tourism – would be highly relevant especially for those businesses that are locally owned and operated.

References

Appadurai, A. (2000) 'Grassroots globalization and the research imagination', *Public Culture*, 12(1): 1–19.

Ballard, C., M. Wilson, Y. Nojima, R. Matanik and R. Shing (2020) 'Disaster as opportunity? Cyclone pam and the transmission of cultural heritage', *Anthropological Forum*, 30(1/2): 90–107.

Bedford, S., M. Spriggs, M. Wilson and R. Regenvanu (1998) 'The Australian National University- National Museum of Vanuatu Archaeological Project 1994–7: A preliminary report on the establishment of cultural sequences and rock art research', *Asian Perspectives*, 37: 165–193.

Bushell, R. and R. Staiff (2012) 'Rethinking relationships: World heritage, communities and tourism'. In: P. Daly and T. Winter (eds.) *Routledge Handbook of Heritage in Asia*, London: Routledge, pp. 247–265.

Cheer, J. M. (2017) 'Cruise tourism in a remote small island – high yield and low impact?' In: R. Dowling and C. Weeden (eds.) *Cruise Ship Tourism*, 2nd edition, Wallingford, Boston: CABI, pp. 408–423.

Cheer, J. M., K. J. Reeves and J. H. Laing (2015) 'Debunking Pacific utopias: Chief Roi Mata's Domain and the re-imagining of people and place in Vanuatu'. In: S. Pratt and D. Harrison (eds.) *Tourism in Pacific Islands: Current Issues and Future Challenges*, Abingdon: Routledge, pp. 109–122.

Crewe, E. and E. Harrison (2005) 'Seeing culture as a barrier'. In: M. Edelman and A. Haugerud (eds.) *The Anthropology of Development and Globalisation: From Classical Political Economy to Contemporary Neoliberalism*, Oxford: Blackwell, pp. 232–234.

Department of Tourism (2015) *Tourism Assessment Report, Tropical Cyclone Pam*, Port Vila: Department of Tourism.

Dornan, M. and T. Newton Cain (2019) 'Vanuatu and cyclone pam: An update on fiscal, economic, and development impacts', *Pacific Economic Monitor*, May 30-35, Manila, Philippines: Asian Development Bank. Available online at https://www.adb.org/sites/default/files/publication/498406/pem-may-2019.pdf (accessed 16 August 2022).

Garanger, J. (1972) 'Archéologie des Nouvelles-Hébrides: Contribution à la connaissance des îles du centre', *Publications de la Société des Océanistes No. 30*, Paris: ORSTOM.

Government of Vanuatu (1980) 'Constitution of the Republic of Vanuatu'. Available online at https://www.gov.vu/images/legislation/constitution-en.pdf (accessed 27 March 2022).

Gravari-Barbas, M. (ed.) (2020) 'Climate change, World Heritage and tourism / changement climatique, patrimoine mondial et tourisme', *10e séminaire de la Chaire UNESCO et du réseau UNIWIN-UNESCO 'Culture, Tourisme, Développement'*, Paris: UNESCO. Available online at https://chaire-unesco-culture-tourisme.pantheonsorbonne.fr /sites/default/ files/inline-files/Actes_Seminaire_UNESCO_2019_0.pdf (accessed 9 February 2022).

Greig, C. (2006) *Cultural Tourism Strategy for Chief Roi Mata's Domain*, Port Vila: Vanuatu Cultural Centre and World Heritage and Tourism Committee for the Lelepa Region.

Hall, M. (2011) 'Introduction: Towards World Heritage'. In: M. Hall (ed.) *Towards World Heritage: International Origins of the Preservation Movement 1870–1930*, Surrey and Burlington: Ashgate Publishing Ltd, pp. 1–22.

Huxley, J. (1947) *UNESCO: Its Purpose Land and its Philosophy*, London: Preparatory Commission of UNESCO, The Frederick Printing Co.

Jacobs, R. (2019) 'Revealed: We discover Royal Caribbean's Lelepa Island design', *Cruise Passenger*, 8 November. Available online at https://cruisepassenger.com.au/news/exclusive-how-royal-caribbeans-private-island-will-chill-and-thrill/ (accessed 11 February 2022).

Loehr, J. (2020) 'The Vanuatu tourism adaptation system: A holistic approach to reducing climate risk', *Journal of Sustainable Tourism*, 28(4): 515–534.

McDonnell, S. (2015) '"The land will eat you": Land and sorcery in North Efate, Vanuatu'. In: M. Forsyth and R. Eves (eds.) *Talking it Through: Responses to Sorcery and Witchcraft Beliefs and Practices in Melanesia*, Canberra: Australian National University Press, pp. 137–160.

McDonnell, S. (2018) 'Selling "sites of desire"', *The Contemporary Pacific*, 30(2): 413–436.

Naupa, A., S. Mecartney, L. Pechan and N. Howlett (2021) 'An industry in crisis: How Vanuatu's tourism sector is seeking economic recovery'. In: Y. Campbell and J. Connell (eds.) *COVID in the Islands: A Comparative Perspective on the Caribbean and the Pacific*, Singapore: Palgrave Macmillan, pp. 231–252.

Neef, A. (2021) 'The contentious role of tourism in disaster response and recovery in Vanuatu', *Frontiers in Earth Science*, 9: 771345.

Read, J. L. (2011) *The Last Wild Island: Saving Tetepare*, Kensington Park: Page Digital Publishing Group.

Republic of Vanuatu (2007) *Chief Roi Mata's Domain: Nomination by the Republic of Vanuatu for Inscription on the World Heritage List*, Port Vila: Vanuatu Cultural Centre.

Robertson, R. (1995) 'Glocalization: Time–space and homogeneity–heterogeneity'. In: M. Featherstone and S. L. R. Robertson (eds.) *Global Modernities*, London: Sage Publications, pp. 25–44.

Royal Caribbean International (n.d.) 'Perfect day at Lelepa, webpage'. Available online at https://www.royalcaribbean.com/aus/en/lelepa-cruises (accessed 11 February 2022).

Slatter, C. (2006) *The con/dominion of Vanuatu? Paying the Price of Investment and Land Liberalisation - A Case Study of Vanuatu's Tourism Industry*, Auckland: OXFAM New Zealand.

Stepwise Heritage and Tourism (2013) *Strengthening World Heritage and protected area governance in Vanuatu*, Report for AusAID Pacific Public Sector Linkage Program, Canberra: Stepwise, Heritage and Tourism.

Tanguay, J. (2015) 'Alternative indicators of wellbeing for Melanesia: Cultural values driving public policy'. In: L. MacDowall, M. Badham, E. Blomkamp and K. Dunphy (eds.) *Making Culture Count: New Directions in Cultural Policy Research*, London: Palgrave Macmillan, pp. 162–172.

Taylor, K. and J. Lennon (2011) 'Cultural landscapes: A bridge between culture and nature?', *International Journal of Heritage Studies*, 17(6): 537–554.

Timothy, D. J. (2019) 'Introduction to the Handbook of Globalisation and Tourism'. In: D. J. Timothy (ed.) *Handbook of Globalisation and Tourism*, Cheltenham: Edward Elgar Publishing Limited, pp. 2–11.

Trau, A. M. (2012a) 'Beyond pro-poor tourism: (Re)interpreting tourism-based approaches to poverty alleviation in Vanuatu', *Tourism Planning and Development*, 9(2): 149–164.

Trau, A. M. (2012b) 'The glocalisation of World Heritage at Chief Roi Mata's Domain, Vanuatu', *Historic Environment*, 24(3): 4–11.

Trau, A. M. (2013) 'World Heritage at Chief Roi Mata's Domain: The global-local nexus of community heritage conservation and tourism development in Vanuatu', *Unpublished PhD*, New South Wales, Australia: University of Western Sydney.

Trau, A. M. (2015) 'Challenges and dilemmas of international development volunteering: A case study from Vanuatu', *Development in Practice*, 25(1): 29–41.

Trau, A., C. Ballard and M. Wilson (2014) 'Bafa zon: Localising World Heritage at Chief Roi Mata's Domain, Vanuatu', *International Journal of Heritage Studies*, 20(1): 86–103.

Watt, G. and H. Brenner (2020) 'Cruise tourism in Vanuatu: Impacts and issues', *the Council for Australasian Tourism and Hospitality Education 2020 Conference*, pp. 344–349.

Wilson, M. (2006) *Plan of Management for Chief Roi Mata's Domain*, Port Vila: Vanuatu Cultural Centre and World Heritage and Tourism Committee for the Lelepa Region.

Wilson, M., C. Ballard and D. Kalotiti (2011) 'Chief Roi Mata's Domain: Challenges for a World Heritage property in Vanuatu', *Historic Environment*, 23: 5–11.

Wilson, M., C. Ballard, R. Matanik and T. Warry (2012) 'Community as the first C: Conservation and development through tourism at Chief Roi Mata's Domain, Vanuatu'. In: A. Smith (ed.) *World Heritage in a Sea of Islands: Pacific 2009 Programme*, World Heritage Papers 34, Paris: UNESCO, pp. 68–73.

Wittersheim, E. (2011) 'Paradise for sale: The sweet illusions of economic growth in Vanuatu', *Journal de la Société des Océanistes*, 133: 323–332.

Woolcock, G. and M. Davern (2015) 'Creative accounts: Reimagining culture and wellbeing by tapping into the global movement to redefine progress'. In: L. MacDowall, M. Badham, E. Blomkamp and K. Dunphy (eds.) *Making Culture Count: New Directions in Cultural Policy Research*, London: Palgrave Macmillan, pp. 129–144.

15

AN ASSESSMENT OF THE SOCIAL MEDIA MARKETING PERFORMANCE OF FIJI'S HOTEL INDUSTRY BEFORE AND DURING COVID-19

Karishma Sharma and Alexander Trupp

Introduction

The spread of the coronavirus (COVID-19) and large-scale travel bans have continued to cause significant damage to the global tourism and hospitality industry (Jiang and Wen 2020). Prior to COVID-19, tourism's total contribution towards Fiji's gross domestic product (GDP) was approximately 46% for the year 2019 (Lockington 2020). However, due to the closure of international borders from March 2020, visitor arrivals in Fiji declined by 83.6% in 2020 compared to 2019. Tourism earnings for the September 2020 quarter declined by 99.5% compared to the September quarter of 2019 (Fiji Bureau of Statistics 2020). Hotels are especially susceptible to reduced tourism alongside decreased economic activity (Hoisington 2020). During and after such crises, tourists have increasing concerns related to health and overall safety (Zhan et al. 2022). Such concerns and changes in consumer demand need to be addressed by hoteliers as part of their information and marketing campaigns (Jiang and Wen 2020).

To overcome the problems caused by the COVID-19 pandemic, hotels in Fiji are using 'social media' (SM) to diversify their products and target markets. 'Social media marketing' (SMM) has grown to be one of the most influential marketing and communication techniques for hotels. Hotels use SM to engage and collaborate with customers and learn about their expectations to create innovative product ideas (Li et al. 2021). The 'Love our Locals' marketing campaign, initiated by both the Ministry of Commerce, Trade, Tourism and Transport (MCTTT) and Fiji's government marketing body Tourism Fiji, allowed the hotel industry to operate with a 55% average occupancy rate from June until September 2020; while historically off-peak seasons recorded a 30% average occupancy rate prior to the pandemic (Turaga 2020). This is primarily due to a domestic tourist market staying at hotels for an average stay of two nights (Krishant 2020). Recent research conducted by Sami (2021) suggests that domestic and external shocks to the Fijian tourism industry have only temporary effects to room occupancy rates.

DOI: 10.4324/9780429019968-17

Against this background of the current health crisis and the shifting tourism markets in Fiji, this study assesses Fiji's hotel industry's SMM performance before and during the COVID-19 pandemic. Existing research examines the implications of adopting digital marketing solutions in the service industry (Buhalis and Jun 2011; Chan and Guillet 2011), though limited research concerns SMM in Pacific Island states. Like other small island developing states, Fiji is characterized by its small size and scale, geographical isolation, limited resources in information and communications technologies, and internet connectivity (Mate et al. 2019). Existing studies on SMM in the hospitality industry show that, whilst many businesses are aware of the significance of using digital solutions, SM adoption was hindered by human, financial, geographical, and training issues (Jones et al. 2015; Roult et al. 2016). Therefore, this study seeks to address and examine SMM performance within the context of an island state in the Pacific. The 'digital marketing framework' (DMF) was developed to assess SMM performance for hotels. Models for digital marketing solutions, utilized by Kierzkowski et al. (1996) and Pani and Moharana (2013), were modified and adapted to create a DMF suitable for assessing SMM. There are many similarities between SM and other interactive technologies, including building relations with customers and learning from customer feedback. Pani and Moharana's (2013) study indicates benefits deriving from the implementation of the DMF, notably companies being able to deliver information more effectively, learn about user preferences, build relationships, and maintain customer loyalty.

This chapter aims to assess the hotel industry's SMM performance through the application of the DMF in the context of a resort in Fiji. This study deploys a netnography of the resort's SM platforms and is further supported with two semi-structured interviews with the resort's marketing personnel. The research took place in two stages. Stage 1 was prior to the COVID-19 pandemic (December 2017 to December 2018), and Stage 2 was during the COVID-19 pandemic (March 2020 to July 2020).

Social media marketing in tourism and hospitality

SM has substantially impacted the tourism and hospitality industry (Xiang and Gretzel 2010). The increased usage of SM for marketing purposes allows tourism and hospitality businesses to reach a wider audience and disseminate information more quickly (Sajid 2016). Tourist decision-making behaviour has been influenced significantly by the tools and information available on SM (Buhalis and Jun 2011). These functions and features include online bookings, online payments, online shopping, virtual tours, and planning travel itineraries (Fotis et al. 2012). Tourists can use SM to express their own opinions and evaluate and review the products and services purchased, thereby enabling electronic word-of-mouth promotion (Chan and Guillet 2011). Most tourism-related products and services are not well differentiated. Therefore, purchasing tourism products requires higher customer involvement and engagement (Lepkowska-White and Parsons 2019). Tourists use SM sites such as TripAdvisor and Facebook to look for more comprehensive information about the destination to minimize financial risks and uncertainty about the quality of the services and safety. Tourism companies need to utilize SM efficiently to better respond to customer requests and preferences (Mate et al. 2019).

Adapted digital marketing framework

The abilities of new marketing mediums imply the necessity of new marketing models. Accordingly, Kierzkowski et al. (1996) provide a conceptual framework for digital marketing which features five essential elements: (1) 'attract', (2) 'engage', (3) 'retain', (4) 'learn', and (5) 'relate back to users'. Similarly, Pani and Moharana (2013) developed a framework concerning another five broad categories: (1) 'reach', (2) 'volume', (3) 'engagement', (4) 'influence', and (5) 'share of voice', which is used to calculate the impact that SM presence has on the return on investment. The authors argue that the fundamental marketing aspects and technical characteristics of digital marketing cannot be viewed in isolation. Therefore, elements from existing frameworks are used to derive the theoretical framework for this study. The proposed framework adopts elements from Kierzkowski et al's (1996) original digital marketing framework and incorporates Pani and Moharana's (2013) SM key performance indicators (KPI), including reach, volume, and influence. Our framework identifies key areas in the marketing process where SM is likely to have a significant impact and uncovers SMM issues that directly or indirectly impact the resort.

The adapted framework (Figure 15.1) consists of eight interacting attributes: (1) 'attract' – there are various visual and technical ways to attract customers to SM platforms; (2) 'reach' – the spread of an SM conversation can be measured, helping the company to understand the spread of the content and the size of the audience for the message; (3) 'volume' – provides insight into the size of the conversation about a brand, company, or campaign; (4) 'influence' – the group of users does not necessarily have to be large but, rather, influential; (5) 'engage' – in order to build and maintain influence, companies need to identify and attract users that *engage* with the company and subsequently engage on its behalf (Peters et al. 2013); (6) 'retain' – once the customer has been attracted and engaged, marketers will need to *retain* users; (7) 'learning' – studying consumers' demographics, attitudes and behaviours; and (8) 'relate back' – there is a need to 'relate back' to the customer to provide customized interactions, where 'marketers have the opportunity to deliver a personalized service or communication about the availability of a personalized service' (Matikiti et al. 2016: 5).

There is a need to measure the content that is provided by the organizations and how people interact with the content (Attribute 3: volume), especially in regards to how numbers of use interactions change over time (Pani and Moharana 2013); and how influential the organization is

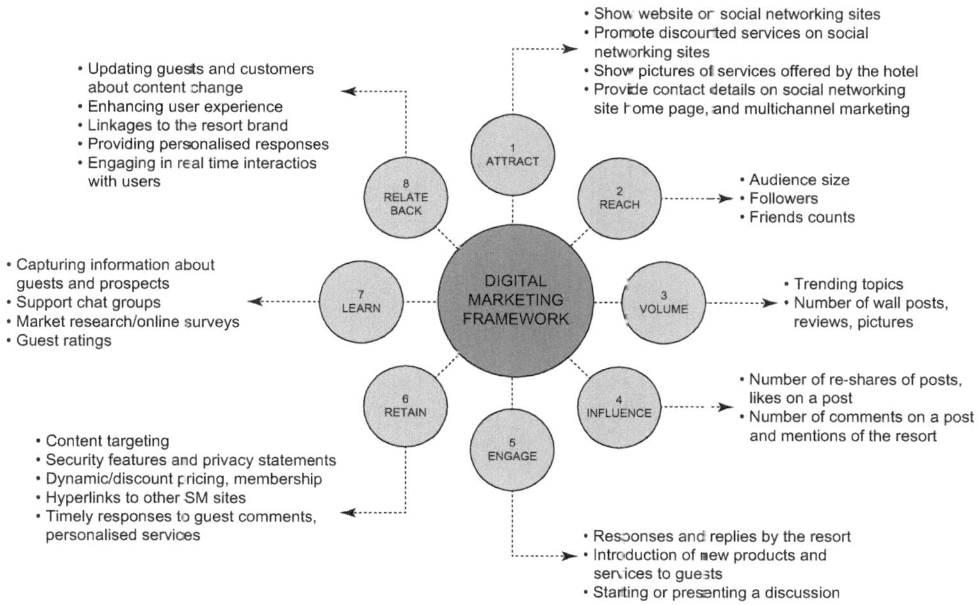

Figure 15.1 Digital marketing framework. Source: Adapted from Kierzkowski et al. (1996) and Pani and Moharana (2013).

(Attribute 4) in encouraging action among their followers. Marketers should 'engage' (Attribute 5) with users' interests and participation to achieve some form of interaction. Engagement ratio and duration are two important activities that organizations monitor (Pani and Moharana 2013). Attribute 6, 'retain', involves marketers building relationships with customers by maintaining contact with them (Kierzkowski et al. 1996). This reduces the chance of customers switching to competitor sites. Information for Attribute 7, 'learn', may come in email communications and opinions shared on SM, or information gathered in surveys and questionnaires, or registration processes. Additionally, marketers can get feedback from customers by using digital media to test their products or services (Kiani 1998).

Research methods

The research used a netnography of SM platforms of a five-star resort located on Denarau Island, located on the western side ofViti Levu. It is a popular tourist destination and home to numerous high-end properties, especially among Australians and New Zealanders (Dave 2019) and, thus, represents a typical upmarket enclave resort in the South Pacific. While only a single resort has been analysed, this study includes an SMM assessment at two points in time: Stage 1, prior to COVID-19 (December 2017–December 2018); and Stage 2, during COVID-19 (March–July 2020). The research collected a large proportion of its data from SM platforms such a TripAdvisor, Facebook, and Instagram. Kozinets (2002: 12) states that 'online communities form or manifest cultures, the learned beliefs, values and customs that serve to order, guide and direct the behaviour of a particular society or group'. Netnographic research involves searching for and analyzing relevant computer-generated data to address identified research questions. A researcher, for example, interested in travel motivations of tourists visiting Fiji can generate data using search engines, such as Google, through strategic keyword searches. Researchers then use their discretion to include or exclude portions of generated data. The data may be quantitative or qualitative. In a South Pacific context, Mate et al. (2019) used netnography to analyze managerial hotel responses on TripAdvisor, which was useful for generating a strategy to respond to negative SM comments and improving service recovery.

Netnographic data from Stage 1 was supported with two semi-structured interviews designed to capture the interviewee's thoughts and ideas about the hotel's SMM performance. The interviews were conducted with the director of sales and the marketing executive manager of the resort in June 2018. Unfortunately, the researchers were unable to conduct semi-structured interviews during the second stage of the research because the sales and marketing personnel did not respond to requests for interviews.

DMF findings

The following section introduces the findings of the eight attributes of the proposed DMF framework (also see Table 15.1). The findings include the SMM assessments from Stage 1 and Stage 2 of the study.

Attract

During Stage 1, promotion was the most common activity which took up most of the advertising efforts. New products or additional benefits in the resort's food and beverage outlets were often promoted, followed by images of families and staff at the property. Other popular images included beach and poolside scenery. According to the resort's director of sales, SMM helped

Table 15.1 DMF attributes before and during COVID-19

	Prior to COVID-19 (December 2017–December 2018)	*During COVID-19 (March–July 2020)*
Attract	• Promotions about new products; additional benefits included as part of the resort package • Pictures: Food and beverage (96%); families (93%); staff (86%); property-beach and pool (79%); events (64%) • Search engine optimisation (SEO) marketing	• Discounted packages on accommodation (5), food and beverage (11), entertainment (4), and events (1). • Pictures: Food and beverage (18%); families (16%); staff (21%); property-beach/ pool/ facilities (31%); promotions (14%) • Video was used twice in the five-month period.
Reach	• Facebook @ the resort: 26,390 followers: 24,874 likes; 4.4/5 rating • Instagram @ the resort: 5,064 followers	• Facebook @ the resort: 37,873 followers; 37,165 likes; 4.8/5 rating • Instagram @ the resort: 8,154 followers
Volume	• Facebook @ the resort: 285 photos; 10 videos • Instagram @ the resort: 102 photos on Instagram	• Facebook @ the resort: 135 photos; 12 videos • Instagram @ the resort: 29 photos on Instagram
Influence	*Facebook Corporate Page @ the resort:* • Average frequency of update of one post: every two days; total shares: 76; average shares per posts: less than one • Total comments: 630; average comments per post: 1 comment; total reactions: 6,192; average reactions: 516 *Instagram @ the resort* • Total shares: every two days; total comments: 520; average comment per post: 5 • Total reactions: 5,100; average reactions: 100	*Facebook Corporate Page @ the resort* • Average frequency of update of one post: every two days; total shares: 535; average shares per posts: one in four days • Total comments: 2514; average comments per post: 16 comments; total reactions: 3267; average reactions: 516 *Instagram @ the resort* • Total comments: 313; average comment per post: 10 comments • Total reactions: 3267; average reactions: 112
Engage	• 660 TripAdvisor reviews; response rate: 100% to positive and 83% to negative reviews • Guest complaints related to: staff empathy (42%); reliability (28%); resort amenities and other tangible factors (15%); availability of rooms (15%) • Service recovery strategies included: (1) apologizing to the guest and (2) justifying the complaint	• 86 TripAdvisor reviews; response rate: 83% (overall) • Guest complaints (as recorded) related to: Resort amenities and other tangible factors (5%) • Minimal engagement on TripAdvisor (March–May) • Inconsistent content delivery on SM platforms • Service recovery strategies included: (1) apologizing to the guest and (2) justifying the complaint
Retain	• TripAdvisor, Facebook, Instagram, and YouTube featured on resort website • Hyperlinks integrated to access resort website	• Additional strategies to interact with users such as #membermoments
Learn	• Rewards loyalty programme to learn about guest preferences • Additional strategies used by the resort to learn about the guests is from gathering feedback from complaints from TripAdvisor	• Using SM to understand the guests'/users' preferred levels of engagement and engagement channels
Relate Back	• Personalized SMM marketing by responding with personalized messages on Facebook, Instagram, and TripAdvisor • The resort also tries to update their users and guests with recent news related to the resort	• Inconsistent responses with personalized messages • Acknowledging COVID-19 pandemic in 16 TripAdvisor reviews • Uploading posts that celebrate special holidays such as Easter and Mothers' Day and paying tribute to Anzac Day

Source: Authors' own compilation derived from the research study.

build the hotel image and promote the sister hotels. During Stage 2, promotional strategies were utilized to attract guests to the resort. In accordance with the national 'Love our Locals' marketing campaign, the resort offered discounted packages on accommodation, food and beverage, entertainment, and events. The resort also continued to upload pictures to attract guests. However, more emphasis was placed on the resort property and facilities, followed by the staff, food and beverage, and then family pictures. Videos used focused mostly on the resort properties, particularly the pool area and the beach. Previous studies show that changing consumer demands and travel trends (e.g., shortened length of stay) have changed the importance placed on resort amenities and activities (Brey et al. 2011). Different amenities, such as spa offerings and the hotel pool, have become increasingly important for extending the length of stay. The resort consistently used images in Stage 1 and Stage 2 that highlight local/Fijian cultural attractions, special events, and festivals celebrated such as 'Christmas on the Islands' and the friendliness of the resort's staff to attract guests. This coincides with Dwyer and Kim (2010) suggestions that human resources (employees and locals) play a significant role in a destination's ability to compete.

Reach, volume, and influence

Reach, volume, and influence are the next elements of the DMF. The resort utilizes SMM to achieve brand awareness, generate online traffic, and encourage users to follow their SM platforms, in particular Facebook and Instagram. The resort reported that their current marketing strategies are effectively based on traffic generated and results projected on social networking sites. According to the marketing executive (Interview, 20 June 2018):

> People are responding much more to digital marketing rather than the offline marketing and that is how we have allocated our marketing budgets. There is FJD$500 dollars allocated to social media marketing in a month. That's a lot of money for social media. But that's a lot of boosts for Facebook.

The resort also uses tools related to communication and understanding how customers interact with web content. The director of sales (Interview, 3 June 2018) states:

> The tools that we consider in a marketing plan are Facebook Insights, a social media management system which helps us a lot with posting, analytics, and monthly reports for engagement. The e-marketing manager reports the statistics during our weekly meetings – which are paid for.

In 2018, the resort had a larger following on Facebook (24,827 followers) than on Instagram (5,064 followers). During Stage 1, the resort made one to two posts every second day on average on Facebook and Instagram. The engagement levels between Facebook and Instagram varied, even though posts were updated or new content was uploaded on the same days. Users shared at least one post per day on Facebook but were more engaged on Instagram, averaging five comments per post as opposed to the average of one comment per post on Facebook. During Stage 2, the resort's SM followers increased, reaching 26,390 via Facebook and 8,154 via Instagram. However, the resort's content updates were inconsistent. For example, no content had been uploaded onto Instagram from June to July 2020. Generally, users were more engaged on Facebook, averaging 16 comments per post compared to the 10 comments per post on Instagram. This may be due to the features built into Instagram, which allow viewers to only react and comment on posts. Whereas with Facebook, it may be easier and faster to press a like button, or sharing a post as opposed to writing a comment.

Engage

Marketers should arguably engage users' interests and participation to achieve some form of interaction or a transaction. SM is used to create communities that facilitate engagement between online users and the resort. According to the marketing executive (Interview, 20 June 2018):

> A review site such as TripAdvisor gathers all the reviews coming in for the hotel, the comments going on social media for our hotel, and pictures. This comes to a single screen, a single monitor where we can of course reply to all these reviews in Facebook. Also, we have the possibility of communicating messages automatically to all the social media platforms. So, we're just doing it once.

The findings reveal that the users engage differently across different websites. For example, on TripAdvisor, users mainly post the reviews about their experiences, whereas on Facebook and Instagram they discuss and interact with each other. There were 660 TripAdvisor reviews analyzed during Stage 1. Over this period, 35 negative reviews were identified, accounting for 5% of the total reviews. This means that the overall rating for the resort was positive, although 8% of the reviews were rated as average. The resort also responded to all the positive reviews and 83% of the negative reviews. Most of the complaints were about lack of staff empathy (42%), followed by the reliability of service (28%), availability of rooms (15%), and resort amenities (15%). During Stage 2, 86 TripAdvisor reviews were analyzed. Out of these, 44% were posted by local tourists. Over this time period, the overall rating for the resort was positive, with only 2% accounting for both negative and average reviews. The resort did not respond to 7% of all reviews. There were also inconsistencies with the response time from management, who appeared to be inactive on TripAdvisor since mid-March to mid-May 2020. Also, 2% of the TripAdvisor comments were guest complaints concerning the resort amenities and lack of staff empathy.

Handling guest complaints on SM was another activity performed by the resort to engage guests. Davidow (2000) found that repurchase intent, customer satisfaction, and word of mouth (WOM) marketing are positively correlated to the company's willingness to accept and fix a problem. Therefore, hotel managers need to own up to mistakes, respond to complaints, and make corrections. The two strategies used by the resort when addressing guest complaints were 'apologizing to the guest' and 'justifying the complaint'. The resort did not 'excuse' the complaint or 'offer a refund or discount' to the guest. In most cases, the perceived service outcome was justified and accompanied by an apology, as demonstrated in the following manager's account:

> On behalf of all of us, I must sincerely apologise to you and your daughter for the tape fiasco. So, it was definitely mind boggling for me that one of our staff would deny your daughter a piece of tape. Upon further investigation this appears to be a case of miscommunication. The staff member involved misunderstood the request completely and deeply regrets that this soured your experience with us.

In both assessment stages, management understood that complaints often arose due to miscommunication and the guest's subjective perspective/experience. Lee and Hu (2004) posit that most complaints made to hotel staff are usually about service failures, equipment failures, and guest dissatisfaction. Guests can easily forgive the hotel when recovery actions are implemented (e.g., correcting the problem, sending out follow-up letters, and upgrading guests).

Retain

Retaining the guests involves building relationships with the customer by maintaining contact and providing easy access to information. To ensure that users would re-visit the resort's digital platforms, the resort ensured accessibility to all their digital platforms (SM and website). The primary website is used to accept online bookings and communicate with their customers and use the website as a gateway to the resort's SM platforms. The director of sales (Interview, 3 June 2018) explained:

> The website is adapted to all screens, computer tablets, mobiles. But at the same time, the guests use our websites to reach our social networking platforms to communicate with us for various issues or become informed for our updates.

Other ways to retain users include preparing activities or discussions relating to the resort's core business, such as uploading posts that request the user to interact. An activity introduced by the resort to retain more guests during the COVID-19 pandemic was their '#membermoments' on Facebook and Instagram. Followers were asked to share their favourite moments at the resort, and the resort would share their favourite member stories across all SM platforms. By asking guests to participate in fun SM activities such as photo contests, caption contests, trivia questions about the resort, or even by tagging other guests, the resort subsequently builds a rapport with guests. As Yeoman et al. (2016) infer, by doing this, the resort is then retaining and learning about the guests and encouraging online engagement. Tourism Fiji, for example, also develops a market re-entry plan to coordinate a return to the market across all its international destinations in the wake of COVID-19, in which the first stage of market re-entry focuses on Fiji staying in the hearts and minds of tourists through its 'sota tale' (see you again) messaging. However, as the destination enters its second stage, Tourism Fiji's marketing message has changed to 'Our "Bula Spirit" awaits you' (Vavaitamana 2020). SM has become a medium by which people get involved in online activities, especially to create and share information.

Learning

The resort placed heavy reliance on their loyalty programme to learn about their guests prior to the COVID-19 pandemic, as this enabled the resort to learn about guest preferences, interests, and behaviour. This information was valuable to the resort, as it attracted new guests. However, during the pandemic, the resort used SM to understand guests' preferred levels of engagement. For instance, users interacted more with the resort and asked more questions regarding posts that advertised discount promotions. Another way the resort continued to learn about their guests was through gathering feedback from guest complaints on TripAdvisor. It is, thus, vital that managers take responsibility for service failure and recognize its cause. According to Bitner et al. (1994), service failure occurs because of the failure of the service system, where it is likely that customers will change the service provider in the long run.

Relate back

A major challenge for marketers is ensuring that the customers feel like they can relate to the organization. The principle of 'personalization' was derived with reference to the DMF presented by Kierzkowski et al. (1996). The 'Global Web Index' (GWI 2020) found that 87 of US consumers and 80% of UK consumers were consuming more SM content since the COVID-19 outbreak. To some extent, the resort broached the pandemic issue in their SM communication by addressing COVID-19 in 17% (16) of the TripAdvisor reviews uploaded and in one post uploaded on Facebook. However, before the pandemic, the resort personalized their responses to 96% (633) of the TripAdvisor reviews by addressing the reviewer by name. This has not fully been the case during Stage 2, in which only 53% (46) of the 86 TripAdvisor reviewers were personally addressed by their names. Other ways in which the resort tried to relate back to its users on SM was by uploading posts that celebrate special holidays such as Easter and Mother's Day as well as paying tribute to Anzac Day, which is a national day of remembrance in Australia and New Zealand. The resort also tried to personalize its packages by offering flexible options such as early check-in and late check-out times, discounted food and drink, and weekly and weekend specials on resort activities and day spas.

Discussion

The hotel industry in the South Pacific experienced more than 90% of its bookings being cancelled in 2020 and experienced a loss of potential bookings (SPTO 2020). Consequently, Fiji's hotel industry needed to look at marketing efforts to connect and engage with customers and increasingly focus on the domestic market, as also showcased in this study. Against this background, the authors deployed an adapted DMF to assess the SMM performance of a five-star resort in Fiji before and during the pandemic. Empirical results revealed that the resort used promotions, photos, and videos to attract users during both assessment periods. The results support Leung's (2012) study on the marketing effectiveness of hotel Facebook pages, which found that picture content generated more positive attitudes toward the hotel than when the message content focused on the hotel brand. However, during the pandemic, these promotional efforts reflected the resort's shift towards the domestic market. The findings for the elements of 'reach', 'volume', and 'influence' further demonstrate that the resort's Facebook pages have a larger reach, more content and influence compared to the resort's Instagram account. The Sprinklr Business Index (2021) analyzed the SMM performance of hotel brands across all platforms for 30 days and found that the best hotel brands such as Sandals Resorts and Hilton Hotels had very high follower counts of over one million each. Their competitors, such as Sheraton Hotels, Accor, and Hyatt, reached close to 400,000 followers on average across platforms

Compared on a global scale, the reach, volume, and influence of the resort under study are extremely low, even with increased follower count and engagement This may be because the resort focuses almost exclusively on what it produces. Posts concern discounted offers ranging from food and beverage, accommodation, and events. While there is a lot of engagement from users about special offers, these discount postings are done without prior examination of the preferences of the audience. Liu et al. (2020) found that SM played a crucial role in the following components of the tourist decision-making process: destination, transportation, accommodation, food and dining activities, attractions, and shopping and leisure activities. Hotels need to actively engage with their SM platforms to encourage current guests to provide detailed reviews on SM platforms to attract new customers. Reach, volume, and influence are helpful in analysing SM metrics; however, these attributes of the DMF do not highlight the differences between users who share content or gain insights and users who are not sharing content at all. Heinonen (2011) states the importance of examining the motivations behind customers' SM activities, as these can inform SMM strategies. SM metrics can be useful for hotels, but they give no information about connections, engagement and building relationships between people and content (Schrock 2016).

Interviews with the director of sales and the marketing executive manager offered insights into engagement activities and SMM strategies used by the resort. Some ways in which the resort engaged with the customer was by responding to guests' queries and reviews and handling complaints. A level of trust can be developed between the guest and the resort by providing enough information to satisfy the customer's immediate needs, handling negative reviews, and enhancing positive SM reviews. Other ways the resort generates guests' interests during the pandemic are by introducing new activities, promoting local packages, and uploading visual content. Although videos can stimulate the senses, time is required to watch the content, which may explain why photographs generate more engagement. Kwok and Yu's (2013) research also supports the popularity of posts with photographs but found that text posts and status updates receive more reactions, likes, and comments than other categories such as images or videos. The resort's SM engagement during the pandemic has also been characterized by periods of inconsistencies and inactiveness, in which reviews or posts have not been systematically responded to by management.

Retaining guests involves building relationships and maintaining contact with such customers (Wen et al. 2014), especially through communicating feelings and images of 'islandness' that are unique in the Fijian context. Such text, images or videos are shared by hotels via social media platforms but also by other tourism stakeholders. Tourism Fiji, for example, is trying to maintain contact with international tourists by launching a digital series. Accordingly, the 'Bula Series' is airing across all SM channels, along with a news segment called 'Happy Hour TV', which highlights the 'Bula Spirit' that Fiji is renowned for (Vavaitamana 2020). The campaign aimed to attract and rebuild confidence in tourists. Similarly, the resort had initiated their #membermoments to retain guests during the pandemic. The initiative worked well, as content was constantly updated with promotions and posts relating to resort events and activities. However, the interview participants felt that more time and effort could be spent on SMM. More specifically, participants stressed that the resort needed to spend a substantial amount of time and effort to train people to have them fully dedicated to SMM. A study conducted by Phelan et al. (2013) found that some hotels consistently engage in SMM and either become inactive on SMM or irregular with their postings. This is also the case for the resort, as SMM content was updated habitually prior to the pandemic, whilst during the pandemic the resort was posting SMM content irregularly. The International Finance Corporation (IFC) collaborated with the MCTTT to survey businesses in Fiji. The key findings revealed that 50% of tourism businesses surveyed were hibernating or fully closed, and 35% were active but with reduced staff. Also, 27% of staff from tourism businesses were on reduced hours/days, 25% on leave without pay, and 8% were made redundant (IFC 2020). The staff shortages at the resort could have contributed to irregular SMM posting due to reduced working hours or staff being trained to multitask rather than being dedicated to SMM.

The resort utilized SM to learn about customer preferences by monitoring the comments from the guests about the products and services. Hotels that frequently update their content on their SM pages tend to have active participation from users. The netnographic findings are consistent with research conducted by Hsu (2012) and Jung et al. (2013), which identified that one of the critical advantages of using SM is that hotels can acquire knowledge about their guests, enabling them to provide a personalized service and enhance the guest experience. The current study further shows how a crisis such as the current pandemic changes the resort's customer base, which impacts the business environment and hotel strategies. He and Harris (2020) state that hotel marketers need to reflect on such changes and re-evaluate their vision and objectives to account for such transformations. It is important for the resort to relate back to its guests, which has proven to be a major challenge for the marketing managers at the resort. Findings show that the resort tried to customize their SMM efforts by responding with personalized messages on TripAdvisor, though this was a less frequent activity during the pandemic. The resort was also more flexible to guests' needs by offering personalized package deals. Customization in the service sector is important and expected to be achieved through SM (Ahmad et al. 2019). Hotels will need to be flexible and update their packages with interesting SM content, such as advertising health and safety protocol as well as product and service quality.

Conclusion and research implications

The practical and theoretical implications aim at contributing information that will be of value to the hotel industry in the South Pacific, especially in the context of the global pandemic. The findings also prompt hoteliers to define and differentiate their hotels if they wish to compete in a complex regional and global marketplace. The study provides evidence on how Fiji's hotels use SM, especially insights into how these hotels operate and what barriers they face in adopting SM as part of their daily operations. The research provides valuable lessons in how hotels communicate with (potential) customers in times of crisis and highlights the importance of developing new tourism revenue streams during the pandemic. The findings also shed light on how Fiji's hotel industry can better strategize and tap into the opportunities of the digital world. Crucially, netnographic research can offer hotel marketers unprecedented opportunities to access real-life online data about users, information that is crucial for hoteliers and marketers to understand the ever-changing consumer landscape.

Further research can examine the impact of SM performance and its direct link to return on investment for hotel companies. Future research can also extend this study by expanding the hotel sample and applying the DMF to another type of hotel or accommodation, such as hostels or mid-range hotels. Theoretically, through the further development and application of the DMF, this study has contributed to the field of SMM in the hotel industry. Accordingly, a new DMF has been adapted and applied to a five-star island resort in Fiji. As Pacific Islands can experience limited ICT and internet connectivity resources, SMM can be challenging. Considering that there is limited research into SM in the context of island states in the Pacific, this study suggests the need for further research in analysing SMM, explicitly focusing on strategic challenges that SM implementation involves, including measuring SM performance and engagement results. SM is critical during the current time of social and physical distancing in the context of the COVID-19 pandemic, especially as it offers essential tools to stay engaged with customers.

References

Ahmad, S., A. Bakar and N. Ahmad (2019) 'Social media adoption and its impact on firm performance: The case of the UAE', *International Journal of Entrepreneurial Behavior and Research*, 25(1): 84–111.

Bitner, M.J., B.H. Booms, and L.A. Mohr (1994) 'Critical service encounters: The employee's viewpoint', *Journal of Marketing*, 58(4): 95–106.

Brey, E., D.K. Klenosky, X. Lehto, and A. Morrison (2011) 'Understanding resort marketing practices', *Journal of Tourism Insights*, 2(2): 1–22.

Buhalis, D., and S.H. Jun (2011) *E-tourism, Contemporary Tourism Reviews Series*, Oxford: Goodfellow Publishers.

Chan, N.L., and B.D. Guillet (2011) 'Investigation of social media marketing: How does the hotel industry in Hong Kong perform in marketing on social media websites?', *Journal of Travel and Tourism Marketing*, 28(4): 345–368.

Dave, S. (2019) 'Where to stay in Fiji', *Santorini Dave*. Available online at https://santorinidave.com/best-places-fiji (accessed 23 February 2022).

Davidow, M. (2000) 'The bottom line impact of organisational responses to customer complaints', *Journal of Hospitality and Tourism Research*, 24(4): 473–490.

Dwyer, L., and C. Kim (2010) 'Destination competitiveness: Determinants and indicators', *Current Issues in Tourism*, 6(5): 369–414.

Fiji Bureau of Statistics (2020) 'Fiji's tourism earnings- September Quarter 2020', Fiji Bureau of Statistics. Available online at https://www.statsfiji.gov.fj/latest-releases/tourism-and-migration/earnings-from-tourism (accessed 25 September 2021).

Fotis, J.N., D. Buhalis, and N. Rossides (2012) 'Social media use and impact during the holiday travel planning process'. In: M. Fuchs, F. Ricci and L. Cantoni (eds.) *Information and Communication Technologies in Tourism 2012*, Vienna, Austria: Springer-Verlag, pp. 13–24.

GWI (Global Web Index) (2020) *Coronavirus Research, Series 4. Media Consumption and Sport*, April. Available online at https://bluesyemre.files.wordpress.com/2020/04/gwi-coronavirus-findings-april-2020-media-consumption-release-4.pdf (accessed February 23, 2022).

He, H., and L. Harris (2020) 'The impact of Covid-19 pandemic on corporate social responsibility and marketing philosophy', *Journal of Business Research*, 116: 176–182.

Heinonen, J. (2011) 'Consumer activity in social media: Managerial approaches to consumers' social media behavior', *Journal of Consumer Behaviour*, 10(6): 356–364.

Hoisington, A. (2020) '5 insights about how the COVID-19 pandemic will affect hotels', *Hotel Management*, 18 March. Available online at https://www.hotelmanagement.net/own/roundup-5-insights-about-how-covid-19-pandemic-will-affect-hotels (accessed February 23, 2022).

Hsu, Y.L. (2012) 'Facebook as international eMarketing strategy of Taiwan hotels', *International Journal of Hospitality Management*, 31(3): 972–980.

IFC (International Finance Corporation) (2020) *Fiji COVID-19 Business Survey: Tourism Focus: Impacts, Responses, and Recommendations*, July, World Bank Group. Available online at https://www.ifc.org/wps/wcm/connect/4fc358f9-5b07-4580-a28c-8d24bfaf9c63/Fiji+COVID-19+Business+Survey+Results+-+Tourism+Focus+Final.pdf?MOD=AJPERES&CVID=ndn pJrE (access 23 April 2022).

Jiang, Y., and J. Wen (2020) 'Effects of COVID-19 on hotel marketing and management: A perspective article', *International Journal of Contemporary Hospitality Management*, 32(8): 2563–2573.

Jones, N., R. Borgan, and E. Ulusoy (2015) 'Impact of social media on small businesses', *Journal of Small Business and Enterprise Development*, 22(4): 611–632.

Jung, T.H., E.M. Ineson, and E. Green (2013) 'Online social networking: Relationship marketing in UK hotels', *Journal of Marketing Management*, 29(3–4): 393–420.

Kiani, G.R. (1998) 'Marketing opportunities in the digital world', *Internet Research*, 8(2): 185–194.

Kierzkowski, A., S. McQuade, R. Waitman, and M. Zeisser (1996) 'Marketing to the digital consumer', *McKinsey Quarterly*, 3: 4–21.

Kozinets, R.V. (2002) 'The field behind the screen: Using netnography for marketing research in online communities', *Journal of Marketing Research*, 39(1): 61–72.

Krishant, N. (2020) 'Survey reveals average hotel occupancy rate was approximately 55 percent – Koya', *Fiji Village*. 30 July. Available online at https://www.fijivillage.com/news/Survey-reveals-average-hotel-occupancy-rate-was-approximately-55-percent--Koya-fx48r5/ (accessed 23 February 2022).

Kwok, L., and B. Yu (2013) 'Spreading social media messages on Facebook: An analysis of restaurant business-to-consumer communications', *Cornell Hospitality Quarterly*, 54(1): 84–94.

Lee, C.C., and C. Hu (2004) 'Analysing hotel customers' e-complaints from an internet complaint forum', *Journal of Travel and Tourism Marketing*, 17(2/3): 167–181.

Lepkowska-White, E., and A. Parsons (2019) 'Strategies for monitoring social media for small restaurants', *Journal of Foodservice Business Research*, 22(4): 351–374.

Leung, X.Y. (2012) 'The marketing effectiveness of hotel Facebook pages: From perspectives of customers and messages', University of Nevada, Las Vegas: ProQuest Dissertations Publishing.

Li, J., W.G. Kim, and H.M. Choi (2021) 'Effectiveness of social media marketing on enhancing performance: Evidence from a casual-dining restaurant setting', *Tourism Economics*, 27(1): 3–22.

Liu, X., F. Mehraliyev, C. Lui, and M. Schuckert (2020) 'The roles of social media in tourists' choices of travel components', *Tourist Studies*, 20(1): 27–48.

Lockington, F. (2020) 'Tourism talanoa: Tourism's budget support', *Fiji Hotel and Tourism Association*, 23 July. Available online at https://fhta.com.fj/category/news/covid-19/ (accessed 28 July 2020).

Mate, M.J., A. Trupp, and S. Pratt (2019) 'Managing negative online accommodation reviews: Evidence from the Cook Islands', *Journal of Travel and Tourism Marketing*, 36(5): 627–644.

Matikiti, R., M. Kruger, and M. Saayman (2016) 'The usage of social media as a marketing tool in two southern African countries', *Development Southern Africa*, 33(5): 740–755.

Pani, C., and S. Moharana (2013) 'Banking on social media bandwagon - Analyzing social media sustenance of banks', *Kushagra International Management Review*, 3(1): 97–110.

Peters, K., Y. Chen, A.M. Kaplan, B. Ognibeni, and K. Pauwels (2013) 'Social media metrics - A framework and guidelines for managing social media', *Journal of Interactive Marketing*, 27(4): 281–298.

Phelan, K.V., H.-T. Chen, and M. Haney (2013) '"Like" and "check-in": How hotels utilize Facebook as an effective marketing tool', *Journal of Hospitality and Tourism Technology*, 4(2): 134–154.

Roult, R., M. Gaudette, D. Auger, and J.M. Adjizian (2016) 'Site management and use of social media by tourism businesses: The case of Quebec', *Czech Journal of Tourism*, 5(1): 21–34.

Sajid, S. (2016) 'Social media and its role in marketing', *Business and Economics Journal*, 7(1): 1–5.

Sami, J. (2021) 'The response of hotel room occupancy rate in Fiji to shocks: Empirical evidence from unit root tests with endogenous multiple structural breaks', *International Journal of Economics and Financial Issues*, 11(5): 11–16.

SBI (Sprinklr Business Index) (2021) 'The most successful hospitality brands on social media', 25 January. Available online at https://www.sprinklr.com/blog/hotel-social-media-marketing/ (accessed 23 April 2022).

Schrock, A.R. (2016) 'Exploring the relationship between mobile Facebook and social capital: What is the "mobile difference" for parents of young children?', *Social Media + Society*, July–September, 1–11.

SPTO (Pacific Tourism Organisation) (2020) 'SPTO COVID 19 Recovery Strategy', Suva, Fiji. Available online at https://www.pacifictradeinvest.com/media/1522/spto-covid-19-recovery-strategy.pdf (accessed 23 February 2022).

Turaga, S. (2020) 'Hotel occupancy rate have averaged approximately 55%', *Fiji village*, 2 September. Available online at https://www.fijivillage.com/news/Hotel-occupancy-rates-have-averaged-approximately-55--f4rx58/ (accessed 23 February 2022).

Vavaitamana, G. (2020) 'Our bula spirit awaits you', Fiji Hotel and Tourism Association, 28 May. Available online at https://fhta.com.fj/our-bula-spirit-awaits-you/ (accessed 23 February 2022).

Wen, D.M.-H., D.J.-W. Chang, Y.-T. Lin, C.-W. Liang, and S.-Y. Yang (2014) 'Gamification design for increasing customer purchase intention in a mobile marketing campaign app'. In: F. Nah (ed.), *HCI in Business: First International Conference*, Crete: Springer, pp. 440–444.

Xiang, Z., and U. Gretzel (2010) 'Role of social media in online travel information search', *Tourism Management*, 31(2): 179–188.

Yeoman, I., J. Oskam, and A. Postma (2016) 'The future of hotels vacation marketing, service design and management', *Journal of Vacation Marketing*, 22(3): 197–198.

Zhan, L., X. Zeng, A.M. Morrison, H. Liang, and J.A. Coca-Stefaniak (2022) 'A risk perception scale for travel to a crisis epicentre: Visiting Wuhan after COVID-19', *Current Issues in Tourism*, 25(1): 150–167.

16

OPPORTUNITIES AND CHALLENGES OF ECOTOURISM DEVELOPMENT IN FIJI

Evidence from an eco-resort and a marine park

Marica Mafi-Stephens, Marika Kuilamu, and Alexander Trupp

Introduction

In order to manage the negative impacts of tourism development, the government of Fiji, over the past five decades, in collaboration with non-governmental organizations (NGOs), formulated sustainable tourism growth plans and policies. These plans encouraged and promoted ecotourism initiatives that maximize positive environmental and socio-economic impacts and minimize negative impacts on the host community (Belt, Collins and Associates Ltd 1973; Government of Fiji 2007). In an endeavour to differentiate itself from its competitors, Fiji tourism needed to diversify its destination image by placing more emphasis on its unique cultural and environmental attributes (Bricker 2003).

While many definitions of ecotourism exist (Fennell 2008), it can broadly be conceptualized as a form of tourism which relies on the sustainable use of natural resources whilst ensuring economic and socio-cultural returns for local communities (Khanra et al. 2021). As defined by The International Ecotourism Society (TIES), ecotourism is 'responsible travel to natural areas that conserves the environment and improves the welfare of the local people' (TIES 2006). Ecotourism involves the promotion of nature-based tourist activities, ecological and cultural conservation by educating tourists and the maximization of benefits for local stakeholders (Fennel 2008). While ecotourism initiatives have increased globally, various concerns do prevail, especially the extent to which ecotourism, as a Western construct (Cater 2006), can be translated into other geographical and socio-cultural contexts. Thaman (1994) argues that the idealized Western view of ecotourism does not necessarily work in the Pacific Islands, where residents rely heavily on the sustainable utilization of natural resources to meet their needs, earn their living, and maintain cultural integrity. Accordingly, there is a need to view ecotourism from a non-Western perspective and make use of existing local notions of sustainability. For example, the concept of a 'no-take zone' in Fiji's protected areas works better than other 'imported' ideas. This localized concept reflects the Indigenous population's traditional practice concerning *tabu ni qoliqoli* which refers to 'reserving a traditional fishing ground in order to increase the fish population for a traditional ceremony' (World Wildlife Fund Nature, quoted in Cater 2006: 32).

 DOI: 10.4324/9780429019968-18

The chapter will initially provide a succinct outline of the concept of ecotourism in the Pacific context and will discuss the development of tourism and ecotourism in Fiji over the past five decades. The work then examines public and private sector perspectives concerning ecotourism principles and how these principles have been implemented. The chapter also draws attention to two case studies, an eco-resort accommodation and a national marine park, which are used to illustrate the challenges faced in implementing ecotourism in Fiji.

Ecotourism in Fiji – the prelude

International tourist arrivals reached about 900,000 visitors in 2019 (UNWTO 2021), and the total contribution of tourism toward Fiji's gross domestic product (GDP) was about 40% in 2018 (WTTC 2018). Shortly after the COVID-19 outbreak, Fiji closed its borders for most international visitors and international tourist arrivals, during which time, 279 hotels closed and 95% of flights were grounded by Fiji Airways, the country's national airline, causing 25,000 people to lose their jobs (Guardian 2020). Tourism in Fiji has generally been characterized by a high dependency of Western markets and foreign investments, resulting in uneven tourism development and high economic leakage (Soh et al. 2019). Whilst the economic impacts of the COVID-19 pandemic are undeniable, the crisis can also be seen as an opportunity for the public sector and all stakeholders to revisit previous tourism developmental goals and to foster a more local and home-grown tourism development agenda (Trupp and Dolezal 2020) as is also reflected in the notion of ecotourism.

Fiji's tourism industry has largely been based upon its natural attractions, for which there have been growing environmental, social-cultural and economic concerns regarding the adverse impacts of tourism development. Negative environmental impacts in Fiji include the degradation and destruction of ecosystems through sewage outfall in shallow waters and reef flats as well as vegetation and mangrove clearing for resort and golf course development (Hall 1996; Movono et al. 2018). Socio-economic and cultural impacts relate to foreign ownership and employment, including related leakages and the commodification of culture (Milne 1992; Pigliasco 2010). Ecotourism can be seen as a tool to address these impacts and push for sustainable development.

Historically, an important report produced by Belt, Collins and Associates Ltd. (1973) highlighted tourism's key role in Fiji's economic development and the need to balance development with conservation. Although the need for conservation was discussed in various development forums, special attention was given to sustainable development at a more global scale only when the report entitled 'Our Common Future' (WCED 1987) was published. In the 1980s, the government of Fiji developed an ecotourism and nature-based tourism strategy to promote environmental conservation in tourism development (Harrison 1998). The strategy was influenced by Fiji's Ninth National Development Plan (1986–1990) which started to promote local (indigenous) participation in tourism development and emphasize the need to protect the natural environment (Korth 2016). The Cleverdon and Brook report published in 1988 highlighted the importance of developing secondary tourism activities and products, many of which were nature-based (e.g., diving, surfing, boating and trekking) and culture-based (e.g., village tourism, farm visits and cultural entertainment) (Korth 2016). These initiatives were not labelled as ecotourism at the time, but such products and activities reflect a form of alternative tourism development as compared to the dominant form of mass resort tourism (Harrison 1998).

The continuous growth in the tourism sector saw Fiji develop its Tourism Master Plan in 1989 (Government of Fiji 1989). Key objectives in this document included sustained growth in visitor expenditure, appropriate levels of local participation, equitable distribution of benefits, cultural harmony between hosts and guests and appropriate environmental management

(Government of Fiji 1989). During the end of the 1980s, 'Fiji's tourism industry was moving closer towards what would later be regarded as ecotourism' (Korth 2016: 257), but it was by no means a complete movement in that direction and, thus, was probably more of an acknowledgement that such a possibility could become more of a reality. Harrison (1998) emphasized that there were four main reasons why such a realization was needed: first, to promote ecotourism and nature-based tourism as alternatives to prevailing large-scale tourism; second, to develop the (almost non-existent) coordination and communication amongst key players involved in these segments; third, to establish agreed criteria to determine what qualifies as an 'ecotourism product'; and fourth, to create a formal or recognized institution to represent ecotourism and nature-based tourism stakeholders.

Ecotourism recognition

In the early 1990s, Fiji became more determined to develop an ecotourism policy (Bricker 2003) and later utilized 'strategic environmental assessment' initiatives as part of the 1998–2005 Tourism Development Plan. Here, four key issues were identified to ensure sustainability: first, the sustainable use of resources should be given a priority; second, tourism should be developed within its environmental carrying capacity framework because irreversible damage can happen if large scale development continues to be encouraged; third, the pace and scale of development should be according to the resources that the country has and the constraints it faces, for which small scale development is encouraged, as it is more aligned to the resources that Fiji possesses; and finally, socio-economic benefits derived from tourism should be diversified and leakage be reduced. The plan strongly supported community-based tourism (CBT) initiatives to ensure the economic benefits have greater impact on local communities (Levett and McNally 2003: 40). In 1999, Fiji adopted its National Ecotourism Policy (NEP) which can also be seen as a response to the needs voiced by different stakeholders, such as local communities, tourism associations, and NGOs (Bricker 2003). The NEP defined ecotourism as:

> A form of nature-based tourism that involves responsible travel to relatively undeveloped areas to foster an appreciation of nature and local cultures, while conserving the physical and social environment, respecting the aspirations and traditions of those who are visited, and improving the welfare of the local people.
>
> *(Harrison 1998: 5)*

The NEP policy document promoted five principles of ecotourism development in Fiji:

1. The principle of complementarity – stating that ecotourism can add a new dimension to what Fiji can offer to its guests, and ecotourism cannot replace mass tourism.
2. The principle of environmental conservation – tourism should be avoided or restricted in places vulnerable to environmental damage or considered important to the national heritage.
3. The principle of social cooperation – communities engaged in village-based tourism need to meet regularly.
4. The principle of centralized information – updated information about ecotourism and village-based tourism products need to be centrally located and readily available.
5. The principle of strong and effective institutions – Fiji needs a formally recognized institution to represent key ecotourism and village-based stakeholder interests in government and non-government gatherings (Harrison 1998: 8–9).

While the NEP formulated important basic principles towards more sustainable tourism development in Fiji, the authors of this chapter identified the following gaps. First, more attention was put on environmental protection and sustainable use of natural and cultural resources with less focus on the economic sustainability of ecotourism development. Second, the implementation of the NEP initiative went ahead without proper planning, consultation and research from government and its key partners. This resulted in a number of failed projects. For instance, there were no clear guidelines on the selection of ecotourism grants given by the government. In addition, the lack of business knowledge, skills and experience contributed to the challenges that businesses faced after receiving assistance from the government. Third, the policy ambitiously promoted the need for a solid and effective institution to represent ecotourism and village-based stakeholders. However, ecotourism is a minor segment in Fiji's tourism sector and does not have strong representation at the government level, especially when compared to other tourism sectors (also see Verebalavu and Kuridrani 2006).

In order to implement ecotourism, the Fijian government provided a grant to encourage and assist resource owners in establishing their tourism businesses as part of the Strategic Development Plan (SDP) (2003–2005). The two key policy objectives of the SDP were to increase resource owner participation in the tourism industry, in particular ecotourism, and to promote sustainable ecotourism development and public awareness. The SDP used the Integrated Human Resource Development Programme for Employment Promotion to provide a conducive environment for investment as well as to create jobs and develop skills so that employment and businesses become more sustainable. The programme started in 2002 and created significant interest amongst resource owners and local tourism business owners (Government of Fiji 2002). Through the government's affirmative action programme, 140 projects in 12 out of 14 provinces were assisted by the Ministry of Tourism from 2002 to 2005 (Verebalavu and Kuridrani 2006). The assistance was in the form of grants which encouraged the establishment of new tourism businesses, expansion of existing operations, and support in marketing and information technology services. These projects generated 508 jobs. Relatively remote provinces outside Fiji's tourism belt such as Lau, Namosi, Serua, Tailevu, Rotuma and Naitasiri could not fully benefit from this government initiative. Not all ecotourism businesses in Fiji started from financial assistance provided by the Ministry of Tourism. Several individuals and communities started through their own savings and self-initiated projects.

Verebalavu and Kuridrani's (2006) evaluation of ecotourism projects in Fiji found benefits and challenges faced by the government in assessing, assisting and monitoring the way that grants were distributed and utilized (see Table 16.1). Burns (2005) points to further challenges and impediments to the development of ecotourism in Fiji. These include lack of coordination between stakeholders, high numbers of business failures in rural areas, lack of commitment of the Fiji Visitor Bureau to ecotourism principles, and too much 'bandwagon jumping' in the sense that any rural or nature-based tourism is simply labelled as ecotourism.

Recent developments in ecotourism

As reported in the Fiji Tourism Development Plan 2007–2016 (Government of Fiji 2007), approximately 170 ecotourism/nature-based activities and tour products were in Fiji in 2003. Therefore, with increasing demand for these activities, it was crucial to strengthen this market by encouraging the inclusion of the 'eco' component in accommodation facilities and local participation of Indigenous people. Recent research on ecotourism in Fiji identified various barriers such as political vulnerabilities, overdependence on the Australian and New Zealand market and access and infrastructure issues (Prentice et al. 2021). An interesting obser-

Table 16.1 Benefits and limitations of ecotourism projects in Fiji

Benefits	*Challenges*
Economic benefits	*Uneven distribution of financial assistance*
Creation of new jobs resulting in trickle down effects on individual households. *Mataqali* or clan owned resorts employ other family members to maximize benefits accrued from these ventures.	By region, the Yasawa group of islands received the most assistance because they had the greatest number of resorts that applied for funding. Consequently, they had greater economic benefits from the project than other regions.
A portion of the profit is donated to the church and to schools.	*Lack of monitoring system*
Social benefits	Although the Ministry of Tourism vetted the ecotourism project applications and ensured that the ecotourism grants assisted those that deserved help, it did not have the capacity to monitor and evaluate how assistance was utilized in the various regions. Ventures in isolated regions struggled to attract guests and became unproductive, failing to generate income and employment.
Mataqali owned projects allow members to meet regularly, keeping everyone informed about the development taking place in the villages.	*Lack of training*
Other village members are offered casual jobs.	Training in business management and marketing should be provided to all that receive assistance. This will ensure that locals know how to manage a small business. Basic training on how to operate a small business is important for the survival of any business.
Businesses are able to meet educational needs of clan members.	

Source: Adapted from Verebalavu and Kuridrani (2006: 9–14).

vation concerning the recent tourism development plan, 'Fijian Tourism 2021' (Ministry of Industry, Trade and Tourism 2017), is the shift from achieving sustainability through the sustainable use of the environment and its natural and cultural resources to a focus on investment, greater tourism yield, better collaboration amongst stakeholders with effective monitoring and enforcement of existing legislation. This tourism development plan only briefly refers to ecotourism as a niche segment and focuses more on facilitating and promoting investment in tourism. Moreover, emphasis is placed on strategies to ensure development is within a sustainable framework enforced by legislation and policies to reduce and prevent negative impacts on ecology, society and culture. Nonetheless, there seems to be a noticeable shift in focus away from ecotourism as a tool for sustainable development in the Fijian Tourism 2021, in which this plan does recognize that Environmental Impact Assessments are not significantly conducted on tourism developments in Fiji and thus there is a need for the enforcement of environmental compliance.

Ecotourism in practice

Two case studies will now be discussed: an eco-resort and a marine park, particularly to understand the implementation of ecotourism projects and the challenges faced by such projects at local levels.

The eco-resort

This section introduces one of Fiji's ecotourism resorts, located on Fiji's fourth largest island, Kadavu Island, which is situated 100 kilometres south of the main island of Viti Levu.

Development in terms of infrastructure to cater for tourists within Kadavu Island is scarce. The island is inaccessible by roads, so the main mode of transport around the island is by boat. There is road access only around Vunisea, which is where the airport is located. The island is so remote that there is limited internet access in most Kadavu areas, including the resort. There are three weekly flights to Kadavu via the domestic airline Fiji Link, which seats 14 passengers. The eco-resort attracts the backpacker market and can cater up to 22 people, though most residents and villagers on the islands rely on a subsistence way of life (Mafi 2018). Kadavu Island is surrounded by the Great Astrolabe Reef, the third-largest barrier reef in the world.

Three Americans privately own the resort, but local workers from neighbouring villages carry out the resort's day-to-day operations. Most of the decisions, for example, decisions concerning daily menus, diving schedules and hotel activities, are made by the owners who partly reside in the United States. Two of the foreign owners are also the resort managers, and they take turns to be at the resort, once every two to three months each. The main staff at the resort, such as the duty manager, the farm manager and two dive instructors, are Fijian, but they are not all from neighbouring villages but from all over Fiji. They are employed because of their expertise in hotel management and diving experience, possessing diving certifications such as 'Master Diver'. The resort also offers dive certification known as PADI (Professional Association of Diving Instructors), in which the resort ensures an available dive instructor. Resort managers are generally recruited from the mainland or overseas.

Because of the resort's remoteness and limited tourist access, the resort sells package deals only: six nights' stay, inclusive of three daily meals. As bookings and payments are made online by guests prior to travel, most of the money earned by the resort for accommodation and meals remains with the owners in America. Mafi (2018), who is the first author of this chapter, highlights that because the resort is privately owned, payments for accommodation, meals and activities at the resort are all done online and directly go to the bank accounts of the owners with some income trickling down to neighbouring villages. Income for villagers is mainly generated through village tours and tourists visiting the waterfalls. The resort buys fresh fish from the local fishers when needed, but this is not a regular source of income. Because the resort has its own well-maintained vegetable garden and poultry farm from which it sources the ingredients for the main meals, local suppliers have very limited economic opportunities to sell to the resort. The resort engages the services of local community members for 'traditional' performances and other purchases of local produce which cannot be obtained from the resort gardens. Sometimes the resort engages school children or dance groups from the local village to perform dances to guests, and they are paid a fee for their performance.

The resort has aligned itself as eco-accommodation and has put in place eco-oriented policies over the years. It has ascribed itself as 'Fiji's premier ecotourism resort' and is the first resort to be affiliated with The International Ecotourism Society (TIES). The resort's environment policy adopts various responsible tourism practices, such as using local vendors when possible and providing ecotourism education to local people, staff, and guests. At the same time, the resort has been instrumental in encouraging and maintaining marine protected areas on various reef environments around the island with neighbouring villages. Usually this is not an easy task because of 'customary marine tenure' (CMT). CMT is the formal and informal ownership of sea space by a Fijian group ranging from family units (*tokatoka*) to clans (*mataqali*) and districts (*tikina*). The neighbouring villages have been willing to support the conservation of their reef environments, as they do not rely on sales from fish to support their families. Accordingly, villagers appreciate conservation efforts and see ecotourism education as a viable economic alternative.

On the other hand, because the resort did not have access to electricity and water, there is no choice but to go 'green' or adopt 'eco' friendly practices such as solar power. Water for resort use is sourced from an underground spring and also rainwater. Over the years, the resort has won various awards in recognition of its efforts towards ecotourism development, for instance the 2011 Asia Pacific PADI Green Star Award, the 2015 Fiji Excellence in Tourism Award, silver award for World Responsible Tourism 2015 and the 2017 TripAdvisor Certificate of Excellence. Globally, the demand for eco-friendly leisure activities and packages has increased dramatically over the years, especially for remote ecotourism resorts (Mafi et al. 2020). Fiji is well-positioned as an attractive dive destination because of its pristine reefs and marine life and is actively promoted as a niche segment by Tourism Fiji, the country's national tourism organization (Chambers 2018). The resort is a remote ecotourism dive resort that has no white sandy beaches, so marine-based activities such as snorkelling, diving, and fishing are the main activities. However, such forms of business diversification can threaten the reef environment. Research shows that such dive resorts need to adopt management strategies that look to counter or lessen damage to reef environments, such as limiting the number of visits to a dive or snorkelling site (Fitzsimmons 2008).

Waitabu Marine Park

Waitabu is a small coastal village located on the northeastern end of the island of Taveuni, the third largest island in Fiji. According to a village study conducted by Bibi (2017), there are 21 households and approximately 120 residents. The households in this village rely on small-scale subsistence fishing, root crop farming, and the income from selling of taro and yaqona, which can be unstable at times (Lin 2012). Many residents have left their homes vacant and reside in urban centres for work and education opportunities, whilst those who remain carry out daily routines associated with agricultural activity (Bibi 2017). The Waitabu Marine Park is a conservation and ecotourism development and one of Fiji's first 'community-based marine protected areas' (CBMPA). The traditional concept of *tabu* also marks the fishing grounds off limits, which helps in preserving the reef. Bibi (2017) also notes that the marine park is popular for day trips, with activities including snorkelling, swimming and kayaking.

Waitabu is part of the Bouma National Heritage Park (BNHP), a community-based ecotourism management programme and the most visited tourist destination on the island. Waitabu is one of the first villages in Fiji to establish a CBMPA in 1998. In 2002, the BNHP won first place in the British Airways Tourism for Tomorrow Awards in the Protected Areas and National Park category. It has since been referenced as a model programme by tourism bodies, development organizations and the government for sustainable development and conservation efforts in rural communities in Fiji (Farelly 2011). The BNHP is made up of four land-owning communities within Bouma, and they each have their own ecotourism initiatives, which are the Tavoro Waterfalls, the Vidawa Rainforest, the Lavena Coastal Walk and Lodge, and the Waitabu Marine Park and Campground (Farelly 2011). All these projects involving conserving natural resources are supported by direct tourism income and are managed by communities as cooperatives.

The idea of marine conservation in Waitabu was formed when the villagers decided to take tourists for various marine-based activities such as diving and snorkelling in the reefs outside of the village (Lin 2012). In 1997, the Waitabu community invited New Zealand's Tourism Resources Consultants to conduct a study of the reef and marine ecosystem. A marine biologist conducted two surveys and found out that the reef ecosystem was badly damaged by overuse, which affected the fish population and other marine life. After many consultations and workshops with the Waitabu community and tourism development experts, a marine protected area

(MPA) or the 'no take zone' was established outside the village of Waitabu in 1998 (Sykes 2008). The MPA stretches 1,100 meters along the village coastline and goes beyond the reefs to open and deep waters. The MPA was closed off using the traditional system of closure known as *tabu*, with no fixed term on the period of closure (Sykes and Reddy 2009). Fishing activities of any kind are not allowed in the *tabu* area. Visitors and locals are also discouraged from walking or stepping on coral reefs within the *tabu* area. In order to foster more awareness of the programme, constant educational training has been conducted at the community level with villagers by NGOs, government bodies, scientists and business management specialists. Importantly, through the inclusion of the local community in annual biological monitoring, the management and monitoring of the MPA ensures that Waitabu villagers fully own and operate the project for themselves and generations to come (Lin 2012). In addition to this, the community is a founding member of Fiji Locally Managed Marine Area (FLMMA), which assists villagers with implementing policies and practices of MPA management.

Like any MPA, poaching is a significant problem. Accordingly, in 2009, six community members were trained by the Ministry of Fisheries to monitor the 'no take zone' (Sykes and Reddy 2009). Proceeds from visitors were used as wages for the guides, boat captain, the MPA manager and other people who work for the project. Care has been taken to ensure that everyone involved has received a fair wage. Moreover, scientists have been conducting biological monitoring with locals annually since the MPA was established and have shown that fish populations, invertebrate numbers and hard coral cover within the MPA have subsequently increased (Sykes 2008). Lin (2012) indicates that, as a pioneer site of marine reserves and a member of FLMMA, the village collaborates with marine scientists and foreign NGOs and continues to have annual biological monitoring and training activities. The income generated from this ecotourism venture is not enough for locals to rely on entirely. Waitabu and other villages in Taveuni rely heavily on dalo, yaqona sales and subsistence scale fishing sales to support their families. The venture does not significantly contribute to the wealth of the community but does provide some additional cash (Sykes 2008). The ecotourism enterprise is not managed or developed like any other business, as it is a joint community-based venture. Poor communication and organization skills seemingly contribute to the marine park's low economic returns. Nevertheless, even with low income returns, tourism or, rather, Waitabu Marine Park is viewed by the local community as a positive income-generating means whilst, at the same time, preserving a subsistence way of life.

Conclusion and research implications

Tourism in Fiji is primarily based upon its natural attractions, and there have been growing environmental, social-cultural and economic concerns regarding the adverse impacts of tourism development. In the late 1990s, the NEP document launched Fiji's ecotourism initiatives at a more formalized level. This framework and those which followed led to the evolvement of ecotourism/nature-based activities and products. The most recent tourism development plan, Fijian Tourism 2021, reflects a shift from pure implementation of ecotourism projects toward strengthening a legal and risk management framework that can better monitor and assess existing projects to foster and encourage sustainable tourism development.

The two introduced case studies – the privately owned eco-resort and the national marine park – reflect some of the achievements and challenges of ecotourism implementation in Fiji. The small-scale eco-resort implemented a number of nature/marine and cultural-based activities on a remote island and, thus, differs from large-scale and enclave-based resort developments as observed in Fiji's tourist hot spots of Denarau and the Coral Coast, which are located close to the country's main international airport. However, although some benefits of the eco-resort trickle down to local Indigenous communities, the majority of income remains with foreign owners. The all-inclusive package, which has to be purchased by all resort guests, leaves limited opportunities for local communities to fully offer their own services and products. Although the resort has won several sustainable development awards and has implemented ecotourism initiatives, a framework systematically assessing ecotourism's socio-economic and environmental impacts is not in place. As Luke (2013: 89) indicates, businesses and their corporate social responsibility (CSR) policies are more interested in the 'sustainability of profitable growth', as opposed to broader applications of sustainable development and the participation and support of local communities (see Scheyvens and Hughes 2019).

Waitabu Marine Park, on the other hand, managed to increase local participation in the management and monitoring of the MPA, thereby ensuring that Waitabu villagers fully own and operate the project for themselves and the generations to come. Notably, the village community declared its fishing grounds a *tabu* zone based on a local traditional management system aimed at creating no-take areas. Whilst *tabu* zones are fully recognized and observed by local communities, they do not necessarily have formal recognition in law (Mangubhai et al. 2020). In order to monitor the environmental sustainability of the marine park, the village collaborates with marine scientists and foreign NGOs. Although the returns from the MPA are not enough to sustain the community, the project ensures protection of the marine environment. Ecotourism and the research-related activities in the Waitabu Marine Park help diversify the local community's income (Shah et al. 2022).

The juxtaposition between the cultural and environmental attributes of Fiji as an ecotourism destination and the challenges in implementing sustainable tourism policies, exemplifies the realities of ecotourism development in Pacific Island states. Future research can assess how eco- and sustainable tourism initiatives designed at international and national levels translate and trickle down to the local and community level. This requires additional empirical research and case studies to examine the economic, socio-cultural and environmental benefits (and costs) of ecotourism ventures across Fiji and Pacific Island states. In order to strengthen the legal and risk management framework aimed at better monitoring existing ecotourism projects, all key stakeholders associated with tourism development ought to be fully involved. While the representation of local Indigenous communities is crucial in this process, public and private sector representatives, including the transport sector as well as accommodation and food and beverage businesses, need to collaborate closely. A key future challenge concerns identifying additional strategies in which remote Pacific Island destinations can balance economic development and socio-cultural and environmental sustainability. This is also necessary to ensure that ecotourism is more than a developmental buzzword used for tourism in remote and nature-based settings.

The COVID-19 pandemic led to a collapse of international tourism with multiple effects such as increased unemployment and closure of hotels and tourism businesses throughout island economies (Connell 2021). The negative impacts on livelihoods from ecotourism have greatly affected rural communities, service deliveries to remote locations and standards of living. Many people who worked in the tourism and hospitality industry lost their jobs and returned to their home villages (see Chapter 11). Economic losses also further endangered protected or conserved areas for biodiversity, as local communities looked for alternative resources and ways to make a living. However, the sharp decrease in visitors to ecotourism destinations also reduces the environmental and carbon footprint. This crisis can be an opportunity for tourism stakeholders to align tourism developmental goals that support eco-friendly forms of tourism development, especially to address adverse environmental impacts. Simultaneously, and because of the absence of international tourists, the crisis placed significant socio-economic value on domestic travellers, leading to the need to re-evaluate and re-address local tourism needs (Trupp et al. 2022). Moreover, this is also an opportunity to examine ways in which it is possible to transform tourism's relationship with nature and the economy, especially to address ways to strengthen structural and economic linkages within and across the tourism industry in Fiji.

References

Belt, Collins and Associates Ltd (1973) *Tourism Development Program for Fiji*, Washington: United Nations Development Program/International Bank for Reconstruction and Development.

Bibi, P. M. (2017) 'Tourism, indigenous women and empowerment: A case study of Taveuni, Fiji', *Master of Business Thesis*, Southern Cross University.

Bricker, K. (2003) 'Ecotourism development in Fiji: Policy, practice, and political instability'. In: D. A. Fennell and D. K. Ross (eds.) *Ecotourism Policy and Planning*, Wallingford: CABI, pp. 187–203.

Burns, P. (2005) 'Ecotourism planning and policy "Vaka Pasifika"?', *Tourism and Hospitality: Planning and Development*, 2(3): 155–169.

Cater, E. (2006) 'Ecotourism as a western construct', *Journal of Ecotourism*, 5(1–2): 23–39.

Chambers, C. (2018) 'Dive expo a niche market', *Fiji Sun*, 9 March. Available online at https://fijisun.com.fj/2018/03/09/dive-expo-a-niche-market/ (accessed 14 May 2021).

Connell, J. (2021) 'COVID-19 and tourism in Pacific SIDS: Lessons from Fiji, Vanuatu and Samoa?', *The Round Table*, 110(1): 149–158.

Farelly, T. A. (2011) 'Indigenous and democratic decision making: Issues from community-based ecotourism in Bouma National Heritage Park, Fiji', *Journal of Sustainable Tourism*, 19(7): 817–835.

Fennell, D. A. (2008) *Ecotourism*, London: Routledge.

Fitzsimmons, C. (2008) 'Why dive? And why here? A study of recreational diver enjoyment at a Fijian eco-tourist resort', *Tourism in Marine Environments*, 5(2–3): 159–173.

Government of Fiji (1989) *Fiji Islands Tourism Master Plan: Volume 1 – Strategies*, Suva: Government of Fiji.

Government of Fiji (2002) *Strategic Development Plan; Rebuilding Confidence for Stability and Growth for a Peaceful, Prosperous Fiji*, Parliament Paper No. 72 of 2002.

Government of Fiji (2007) *Fiji Tourism Development Plan 2007–2016: Tourism – Fiji's Opportunity*, Fiji: Government of Fiji.

Guardian (2020) '"It's catastrophic": Fiji's colossal tourism sector devastated by coronavirus', *The Guardian*, 15 April. Available online at https://www.theguardian.com/world/2020/apr/16/its-catastrophic-fijis-colossal-tourism-sector-devastated-by-coronavirus (accessed 1 May 2022).

Hall, C. M. (1996) 'Environmental impact of tourism in the Pacific'. In: C. M. Hall and S. J. Page (eds.) *Tourism in the Pacific: Issues and Cases*, London: International Thomson Business Press, pp. 65–80.

Harrison, D. (ed.) (1998) *Ecotourism and Village-Based Tourism: A Policy and Strategy for Fiji*, Suva, Fiji: Ministry of Tourism and Transport.

Khanra, S., A. Dhir, P. Kaur and M. Mäntymäki (2021) 'Bibliometric analysis and literature review of ecotourism: Toward sustainable development', *Tourism Management Perspectives*, 37. https://doi.org/10.1016/j.tmp.2020.100777.

Korth, B. (2016) 'Ecotourism and the politics of representation in Fiji'. In: A.H. Akram-lodhi (ed.) *Confronting Fiji Futures*, Canberra: ANU Press, pp. 249–268.

Levett, R. and R. McNally (2003) *A Strategic Environmental Assessment of Fiji's Tourism Development Plan*, May, Suva: Word Wildlife Fund for Nature and Asian Development Bank.

Lin, H.-L. (2012) 'Colonial uneven development, Fijian Vanua, and modern ecotourism in Taveuni, Fiji', *Pacific Asia Inquiry*, 3(1): 41–57.

Luke, T. W. (2013) 'Corporate social responsibility: An uneasy merger of sustainability and development', *Sustainable Development*, 21(2): 83–91.

Mafi, M. S. (2018) 'Determining tourist satisfaction attributes of ecotourism: A case study of Matava Eco Resort', *MA Thesis, Master of Commerce in Tourism and Hospitality Management*, Suva, Fiji: The University of the South Pacific.

Mafi, M., S. Pratt and A. Trupp (2020) 'Determining ecotourism satisfaction attributes–a case study of an ecolodge in Fiji', *Journal of Ecotourism*, 19(4): 304–326.

Mangubhai, S., H. Sykes, M. Manley, K. Vukikomoala and M. Beattie (2020) 'Contributions of tourism-based Marine conservation agreements to natural resource management in Fiji', *Ecological Economics*, 171: 1–8.

Milne, S. (1992) 'Tourism and development in South Pacific microstates', *Annals of Tourism Research*, 19(2): 191–212.

Ministry of Industry, Trade and Tourism (2017) *Fijian Tourism 2021*, Suva, Fiji: Ministry for Industry, Trade and Tourism, Government of Fiji. Available online at mitt.gov.fj/wp-content/uploads/2019/04/FT2021.pdf (accessed 24 April 2022).

Movono, A., H. Dahles and S. Becken (2018) 'Fijian culture and the environment: A focus on the ecological and social interconnectedness of tourism development', *Journal of Sustainable Tourism*, 26(3): 451–469.

Pigliasco, G. C. (2010) 'We branded ourselves long ago: Intangible cultural property and commodification of Fijian firewalking', *Oceania*, 80(2): 161–181.

Prentice, C., S. Kundra, M. Alam, M. A. Alam and M. Nguyen (2021) 'Utopia or dystopia–deterrents to eco-tourism development in Fiji', *Tourism Geographies*. https://doi.org/10.1080/14616688.2021.2016931.

Scheyvens, R. and E. Hughes (2019) 'Can tourism help to "end poverty in all its forms everywhere"? The challenge of tourism addressing SDG1', *Journal of Sustainable Tourism*, 27(7): 1061–1079.

Shah, C., A. Trupp and M. L. Stephenson (2022) 'Deciphering tourism and the acquisition of knowledge: Advancing a new typology of "Research-related Tourism (RrT)"', *Journal of Hospitality and Tourism Management*, 50: 21–30.

Soh, A. N., C. H. Puah and M. A. Arip (2019) 'Forecasting tourism demand with composite indicator approach for Fiji', *Business and Economic Research*, 12(2): 477–490.

Sykes, H. (2008) *Waitabu Marine Park: Biological and Economic Report 2006/2007/2008*, Fiji: Marine Ecology Fiji. Available online at https://www.marineecologyfiji.com/publications-papers-journals/ (accessed 12 May 2021).

Sykes, H. and C. Reddy (2009) '"Sacred water"; 10 years of community managed marine protection supported by ecotourism-based income generation at Waitabu Marine Park, Fiji Islands', Fiji: Marine Ecology Consulting. Available online at https://marineecologyfiji.com/wp-content/uploads/2010/04/PSI2009_584_Waitabu_MPA_-Sykes_H.pdf (accessed 2 February 2022).

Thaman, K. H. (1994) 'Environment-friendly or the new sell? One woman's view of ecotourism in Pacific island countries'. In: A. Emberson-Bain (ed.) *Sustainable development or Malignant Growth?: Perspectives of Pacific Island Women*, (pp. 183–198), Suva, Fiji: Marama Publications.

TIES (The International Ecotourism Society) (2006) *Global Ecotourism*. Available online at http://www.ecotourism.org (accessed 14 May 2021).

Trupp, A. and C. Dolezal (2020) 'Tourism and the sustainable development goals in Southeast Asia', *Austrian Journal of South-East Asian Studies*, 13(1): 1–16.

Trupp, A., S. Pratt, M. L. Stephenson, I. Matatolu and D. Gibson (2022) 'Representing and evaluating the travel motivations of Pacific islanders', *International Journal of Tourism Research*, 24: 653–666.

UNWTO (United Nations World Tourism Organization) (2021) *UNWTO Tourism Highlights*, 2020 edition. Available online at e-unwto.org/doi/epdf/10.18111/9789284422456 (accessed 12 May 2021).

Verebalavu, J. and L. Kuridrani (2006) *Evaluation Reports of Ecotourism Projects in Fiji for Ministry of Tourism*, Unpublished Report, Suva, Fiji.

WCED (World Commission on Environment and Development) (1987) *Our Common Future*. Available online at https://sustainabledevelopment.un.org/content/documents/5987our-common-future.pdf (accessed 8 March 2019).

WTTC (World Travel and Tourism Council) (2018) *Travel and Tourism Economic Impact 2018 Fiji*. Available online at https://wttc.org/Research/Economic-Impact (accessed 8 May 2019).

PART III

Tourism and island states in Micronesia

17

TOURISM DEVELOPMENT IN THE MARSHALL ISLANDS

Examining the challenges and opportunities

Ai Ling Tan and Vijaya Malar Arumugam

Introduction

Crompton (1979: 18) defines destination image as 'the sum beliefs, ideas and impressions that a person has of a destination', whereas Mayo and Jarvis (1981: 22) conceptualize it as 'the perceived ability of the destination to deliver individual benefits'. Kim and Perdue (2011) define destination attractiveness by internal psychological factors such as tourist motivations, attitudes and experiences, together with external destination factors such as destination image and situational constraints. Their research indicates that affective images of a destination, such as the destination being 'fun', 'comfortable' and 'upscale', feature heavily in determining destination attractiveness. The image of the Marshall Islands, formally known as the Republic of the Marshall Islands (RMI), has been reshaped throughout history, from Western mythologies about 'deserted' islands to perceptions concerning the drastic alteration of the landscape caused by nuclear testing in the 1940s and 1950s (Davis 2005). Despite the country's widespread association with nuclear testing and nuclear waste, there has been an attempt to counteract popular perceptions by reinforcing the Westernized destination image of such islands as places where slow-paced island life can be experienced along with hospitable Indigenous islanders (Cheer et al. 2018). Such conceptualization is being consistently reinforced in tourism brochures and through media activities. The ever-popular tourism website, Lonely Planet, for instance, is indicative of this in that the Marshall Islands is described as possessing 'perfect white sands' and a 'dreamy seascape' (Lonely Planet n.d.).

Harrison (2004) notes that, although the stereotypes of the tropical Pacific Island paradise contain some truth, there is also much inaccuracy. This is especially true in the RMI where the United States conducted 67 nuclear tests at Bikini and Enewetak Atolls from 1946 to 1958, which brought tremendous change for the Marshallese (Harrison 2004). Even though the United States (US) formally terminated its 40 years of trusteeship of the RMI in July 1987 (Thomas 2019), the fatal effects of colonization continued to impact the nation beyond the period of the Cold War and into the present (Ramirez 2018). Since the 1950s, researchers, from such disciplines as anthropology, journalism, law and science, conducted studies to investigate the impacts of US militarism in the Marshall Islands (Smith-Norris 2016). The focus of the studies has been diverse, looking at the environment, the economy, society, cul-

DOI: 10.4324/9780429019968-20

ture, health issues and the impacts of nuclear colonialism in the Marshall Islands (Carter 2020; Daborn 2013; Smith-Norris 2016; Sutow et al. 1965; Thomas 2019). Building on the work of past studies, this chapter initially seeks to examine how the Cold War impacts tourism development in the RMI, providing an overview of the country by focusing on its physical and demographic characteristics and then identifying the significance and impacts of nuclear colonization. The latter part of the chapter contextualizes the needs of tourists in relation to tourism development and concentrates more on the opportunities of tourism development through capacity building as well as the advancement of proactive marketing activities and strategies.

The Republic of Marshall Islands – an overview

The Marshall Islands is located in the north Pacific Ocean and comprises 29 atolls and more than 1,000 islands positioned halfway between Hawaii and Japan in the Micronesia region (Daborn 2013). The atolls and islands are in two parallel island chains, the Ratak (Sunrise) group and the Ralik (Sunset) group with a total land area of 181 square kilometres (CIA 2020) (Figure 17.1). In 2020, it had a population of 59,190 (World Bank 2022a), and the two official spoken languages are Marshallese and English. The country has attracted European explorers from the 1600s, notably Spanish explorers, British whalers and German merchants (Gittelsohn et al. 2003). Protestant missionaries from the US and Hawaii started to convert the islanders to Christianity from the 1850s, and in 1886, Germany formulated a treaty with island chiefs to establish a protectorate over the Marshall Islands (Berta 2021). Japan appropriated the islands in 1914 and formally administered the country from 1919, following a decree from the League

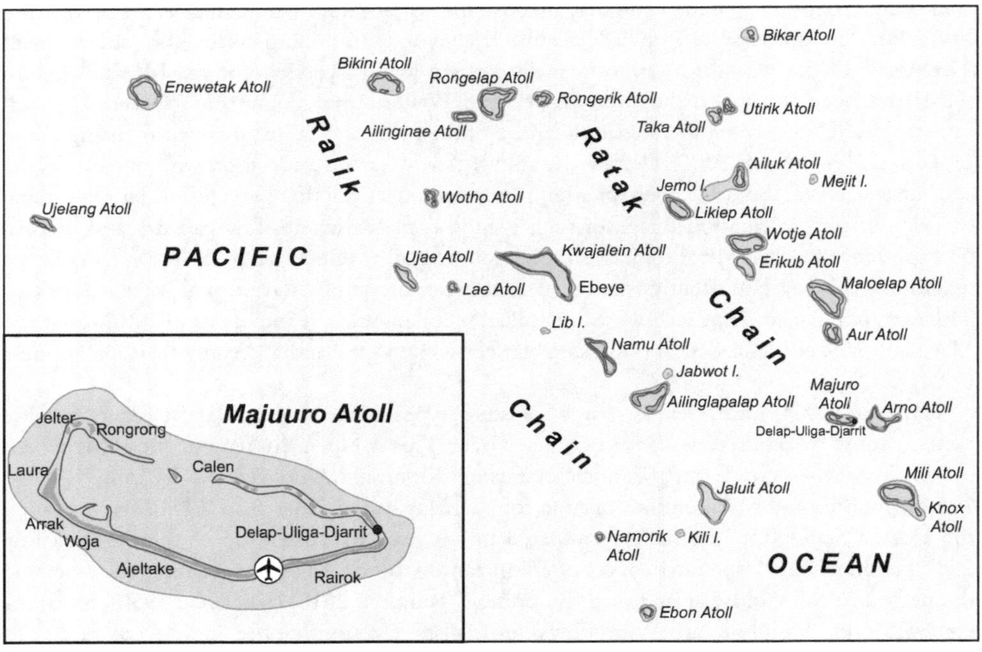

Figure 17.1 Map of the Marshall Islands, including the capital city of Majuro – an atoll. Source: Rainer Lesniewski (www.shutterstock.com/image-vector/marshall-islands-map-164903012).

of Nations, which lasted until World War II when the country was captured by the US, thereby establishing the Trust Territory of the Pacific Islands (TTPI) in 1947 (Kiste 2021). Under jurisdiction of the US, the Trust included Northern Marianas, Palau, Ponape, Truk and Yap (Thomas 2019).

As one of the low-lying island nation-states in the world, the RMI is exposed to rising sea-levels, flooding, limited freshwater supplies, low agricultural production and poor education and healthcare systems. These problems have caused substantial migration flows between islands, specifically from outer islands to Majuro, the capital city situated on the Majuro Atoll, the largest atoll in the nation, which houses more than 50% of the country's population (van der Geest et al. 2020). Besides internal migration, many Marshallese are migrating internationally, particularly to the US because of the bilateral agreement between the RMI and the US, the number of which increased dramatically from 6,700 in 2000 to 22,434 in 2010 (McElfish 2016).

The country's economic development is directly challenged by climate change and global warming that significantly effect vulnerable small Pacific islands (IOM 2019). The global sea-level has risen 8–9 inches since 1880, with a record increase of 3.6 inches from 1993 to 2020 (Lindsey 2022). If there are no pre-emptive actions taken, the Marshall Islands could be uninhabitable within the next 30 years (Letman 2018) or even as early as 2030 (Germanos 2019). The country's trade deficit in 2019 was estimated to be around US$10 million (Statista 2020), affected by the continuing high imbalance of commercial trade in the country. The COVID-19 pandemic has further strained the RMI's economy, especially the tourism sector, for which the country's gross domestic product (GDP) recorded a deficit of 5.5% (ADB 2021). Nonetheless, the RMI's government adopted strong containment measures to prevent the spread of COVID-19. As of 25 April 2022, there have been only 14 total cases of virus transmission and no recorded deaths (Pacific Community 2022).

Significance and impacts of nuclear colonization

The challenges faced by the RMI are inevitably associated with many years of US military interests (Thomas 2019). Although the articles of the 1947 trusteeship called for the obligation from the US to promote self-government and social as well as educational advancement of the inhabitants of Marshall Islands, the US was permitted to use TTPI for military purposes (Kiste 1986), including nuclear testing. The Marshallese were forced to leave their original atolls where they had lived for centuries and relocate to other atolls. The United States helped to clean up the leftover radioactive and non-radioactive debris, support soil rehabilitation and restructure the damaged infrastructure on the Enewetak Atoll (US General Accounting Office 1980). The US government believed that, with living restrictions and strict control of the dietary consumption of local foods to below 30%, supplemented through provisions of imported foods, various islands would be habitable (Johnson and Takala 2016). However, contamination levels, especially in the Bikini Atoll, had prevailed which led to re-evacuation in 1978, as the radioactivity from the nuclear contamination was deemed too high for human living (Niedenthal 1997).

The burdensome legacy of nuclear colonization of the RMI is evident in its agriculture, living conditions, culture, tourism and other economic activities. Despite the separation of the RMI from the Trust Territory in 1978, the legacy of the Cold War continues to threaten the life of the Marshallese. Bikinians remain scattered around the Marshall Islands, as the radiological clean-up of their atoll is not yet fully complete. Ironically, the Bikini Islands was enlisted on the UNESCO World Heritage List in 2010 as an outstanding 'example of a nuclear test site and as a source of globally significant cultural symbols and icons of the twentieth century', though this recognition is not representative of Bikinian heritage and history (Brown 2012). Brown's enquiry expressed concern that the monuments and icons of nuclear testing were formally perceived as valid heritage rather than the lived experiences of Bikinians and the atoll's ecology.

Long-term exposure to radioactive fallout has taken its toll on the Marshallese, where people have suffered from numerous health issues, including miscarriages and higher rates of thyroid cancer and leukaemia (Land et al. 2010). One national report indicated that nuclear testing contributes extensively to the reduction of soil fertility in the islands (OEPPC 2017). Therefore, the country has a high dependence on imported food, and a significant amount of the land is not suitable for agriculture activities due to the impact of radioactivity (Barnett 2011). Native plants, such as breadfruit, pandanus and coconut, which are central to Marshallese culture and contain valuable nutritional sources, carry high levels of radiation since studies were conducted from the mid-1970s (Thaman 1988). The shifted consumption of traditional foods to processed and imported food, remote access to healthcare, reduced physical activities, and cultural acceptance of 'bigger bodies as beautiful' contribute to the country's high prevalence of obesity and diabetes (Senthilingam 2015). This is a significant concern as the obesity epidemic has a direct association with high medical costs, low productivity because of high absenteeism and presenteeism (low productivity while present at work), high insurance costs, and premature mortality (Hammond and Levine 2010). Therefore, due to ongoing nuclear waste concerns and the related environmental impact on Marshallese society, it is of no surprise that the realities of life in the Marshall Islands do not equate to a well-developed tourism destination.

Tourism development and the needs of tourists

While environmental factors continue to dominate the discussion on why the RMI is not perceived as an attractive destination compared to other island nations, it is pertinent that the nation comprehends tourists' needs. The country is known for its marine life and diving sites such as at Rongelap. There are at least 160 species of coral surrounding the islands. Another attraction point, as mentioned previously, is the World Heritage Site of Bikini Atoll that was the nuclear detonation site in the 1940s and 1950s (Atomic Heritage Foundation 2019). The main tourism activities offered in the RMI are fishing, diving, sailing and surfing. Tourist arrivals in the Marshall Islands saw a 11.48% decrease from 2018 to 6,100 visitors in 2019 (World Bank 2022b). The biggest market share was from East Asia and the Pacific, especially Japan, which appears to be one of the largest growing markets, increasing from 2.1% in 1998 to 23% in 2007, because of the interests of Japanese tourists in diving and the expansion of charter flights to service this market (Brown et al. 2010).

Within the RMI, the most visited attractions are the city of Majuro and the island of Ebeye, which is the most populous island of the Kwajalein Atoll (Brown et al. 2010). The unique environment and the perception of island paradise makes such a destination more exotic and, thus, attractive to tourists. This is evident in the Marshall Islands, where the natural ecology of the islands and atolls provide ambient conditions for sea-related tourism activities. Nonetheless, the destination balance between such inherent characteristics as 'climate, ecology, natural resources, culture' and such tourist-inclined attributes as 'hotels, catering, transportation and entertainment' (Lee et al. 2010: 812) is integral to the successful placement of the RMI as a 'to-visit destination'. While the country has ample possession of innate destination attributes, it is still lacking in tourist-inclined attributes.

Tourism development is constrained by limited natural resources, poor accessibility and inadequate air connectivity. The RMI is not an attractive choice for tourists because of the limited access to the island. The only two international airlines which fly to the country are United Airlines, which flies from Japan via Honolulu (Hawaii), and Nauru Airlines, which connects Nauru and the RMI, with weekly flights between the atolls operated by Air Marshall Islands – the national carrier (United Nations n.d.). Due to limited flights, costs are comparatively high. For example, a flight from Tokyo (Japan) to the Marshall Islands in 2018 totalled US$3,227 (Maplandia.com n.d.). The destination is not competitive compared to other destinations because of the high costs involved and low awareness of the country to the outside world (Holden and Holden 2003). Despite these challenges, the main concern faced by the tourism industry is how to capture the market once international movement and mobility become more fluid, as the world starts to deal with the pandemic through effective mass vaccination and eventual suppression of the COVID-19 virus. Subsequently, there is a need to map out actions to effectively market the destination once international travel and mobility becomes fully normalized. Around 14 accommodation establishments are advertised to visitors in the Marshall Islands, many of which are in Majuro city (Tripadvisor 2021). Tourism development in the outer atolls is difficult to advance as Air Marshall Islands only provides domestic flights every other week to most outer atolls, and especially as tourists depend on aircraft as the primary mode of transport in the RMI (Collison and Spears 2015).

In recognition of the potential development of tourism, the government established a statutory authority, the Marshall Islands Visitor Authority (MIVA), in 1997 to develop and promote tourism. MIVA identified the potential of five niche markets: (1) Scuba diving (both wreck and general sport diving); (2) sport fishing in various forms; (3) World War II historic tourism; (4) cultural tourism, especially annual cultural events; and (5) yachting/cruising (Treloar and Hall 2005: 220). However, with a shift in the government's decision to invest in other more important sectors, such as education and healthcare, the budget allocated for MIVA's activities was downsized (ADB 2011). The government's National Strategic Plan 2015–2017 indicates that tourism development should be driven by the private sector, though with the lack of hospitality products, such as natural resources, recreational facilities, skilled personnel and limited access to the islands, the country is less appealing to foreign investors. However, the Marshall Islands has its unique culture, marine resources, history, arts, traditional handicrafts, navigation techniques and outrigger sailing canoes, as well as World War II and colonial artefacts – all of which can be developed and repackaged as authentic cultural heritage tourism products (Collison and Spears 2010: 136). Nevertheless, regardless of the country's heritage tourism potential, the deficiencies in tourism planning and support from the government impede tourism development prospects.

Tourism development strategies

The threat posed by climate change colours any effort undertaken to enhance tourism activities in the country, which may consider several strategies that are not just ecologically sustainable but also economically viable. There are both broader macro and micro strategies that may be pursued to better place the RMI as a tourism destination. These strategies comprise actions in two critical areas: capacity building and marketing.

Capacity building

A key strategy that must be developed in an operative manner relates to capacity building. The lack of tourism and hospitality infrastructure in the RMI needs to be addressed, particularly in terms of accessibility and accommodation. A more concerted effort at regional cooperation among the Pacific Island nations must also be reviewed to ensure a mutually beneficial outcome for all stakeholders in island nation tourism.

While air travel may be limited and cost intensive, the RMI can consider the use of cruises to access its tourism offerings. Cooperation with regional cruise and charter operators has the potential to provide the island nation with the opportunity to access important new markets, such as Chinese tourists. Charter operations should prove to be successful again in the region, as with Fiji and the Cook Islands in the pre-COVID 19 era when there was an increase in charter flights from China (Everett et al. 2018). However, because of the outbreak of the pandemic in early 2020, consumer confidence in cruise holidays dwindled significantly, particularly given reported cases of virus outbreaks onboard cruise ships where passengers had to be quarantined. In fact, from 1 March to 10 July 2020, the Centers for Disease Control and Prevention in the US discovered 3,000 cases of COVID-19 or suspected COVID-19, together with 34 deaths across 123 ships (Glusac 2021). Images abound of several Pacific Islands 'turning back cruise ships' from entering their shores (Walsh 2020). Nonetheless, strategies to attract this market in the post-COVID-19 era should be proactively addressed, in which the RMI should consider collaborating with private sector organizations to explore the opportunities provided by cruise ports. The development of these ports would allow larger cruise ships to access the country and its offerings. Everett et al.'s (2018) research highlighted the development of waterfront sites as a means of cultivating the popularity of a destination. While such a development primarily caters to the cruise crowd, it also acts as the loci for tourists to experience the RMI's colourful artisanal products, culture and food. Creation of waterfront attractions may have a positive effect on the local economy through increased visitor spending, creation of business opportunities for traders and an enhanced quality of life (Kostopoulou 2013).

Another area for capacity building that the country may consider is regional cooperation. The United Nations World Tourism Organization's (UNWTO) (2014) report on Tourism in Small Island Developing States highlights the need for regional cooperation as a means of gaining economies of scale in tourism, sharing of information and knowledge and achieving a higher profile in the marketplace as integrated regional destinations. This is echoed by other researchers who indicate the benefits of collaboration among tourism destinations to include the reasons above as well as other such benefits as increased sustainability, intensified destination competitiveness and the creation of a larger tourist region (van der Zee and Vanneste 2015). Capacity building should arguably foster public and private sector collaborations and partnerships in tourism development, especially to offset its financial limitations in building capacity. Private sector organizations may be co-opted in various tourism-related developments, including in the building of transportation infrastructure, management of tourism facilities (such as ports) and in creating employment opportunities (Everett et al. 2018). To enable these outcomes, the Marshall Islands needs to encourage a positive climate for investment in its tourism activities, especially in such areas as land leasing and foreign ownership of tourism businesses. Nonetheless, as Bordner and Ferguson (2020: n.p.) note:

> But their nation's colonial history has made it hard for them to act by leaving them dependent on foreign aid. And, to date, outside funders have been unwilling or unable to invest in projects that could save the nation.

Bordner and Ferguson (2020) place significant importance on dealing primarily with climate change, a point which implies that investment confidence and a sustainable future could be a reality if environmental solutions are found. Therefore, these writers observe that one of the crucial ways forward is 'radical adaptation' through controlling flooding because of rising seas, though there is a dilemma in that this process is very expensive.

Importantly, the RMI should review its tourist-inclined infrastructure. Accommodation is an issue, as there are very few locations that cater to the tourist crowd. However, the solution to this is not necessarily the building of four- or five-star hotels, which can prove daunting given space and financial constraints, but the expansion of bed-and-breakfast and homestay accommodation establishments. The onslaught of the pandemic further reiterated the need for the country to emphasize regenerative forms of tourism integral to capacity building. The greater emphasis on ecotourism may encompass the advancement of environmentally sustainable accommodation. Unlike tourism resort enclaves, the revenue received from the informal sector of hospitality and tourism could trickle down more to Marshallese communities, as this localized sector is more economically and spatially connected to the community and host society (Opperman 1993).

Marketing activities and strategies in the post-COVID era

The COVID-19 pandemic placed severe constraints on an already struggling tourism market in the RMI. However, the pandemic has provided the country with an opportunity to revisit its approaches to destination marketing, emphasizing high value over high volume. One such strategy that may be pursued concerns promoting the 'visiting friends and relatives' (VFR) market. There should be strong encouragement or incentives for members of the Marshallese diaspora to not only travel periodically to the islands but also to contribute significantly to the tourism economy, particularly through indirect investments such as additional spending and kinship networking. Additionally, VFR travel is culturally and economically sustainable as individuals are emotionally invested in the communal wellbeing of their homeland and can act as its word-of-mouth ambassadors to further encourage others to visit and experience their home. As Gershon (2007) highlights, families are important in sustaining the diaspora's identity and strengthening feelings of reconnection influencing non-resident Marshallese to return to their island. VFR also has an added element of seasonal resilience in which visits by family members and friends are fairly evenly distributed throughout the year (Asiedu 2008) rather than aligned to typical tourism seasons.

The RMI may consider an adaptation of Bhutan's 'high value, low impact' approach to tourism policy, in which principles of 'sustainability, ecological stability and cultural acceptability' (Dorji 2001; Tourism Council of Bhutan 2019) are the basis for tourism activities. This landlocked nation faces a similar scenario to the Marshall Islands, namely pristine ecological conditions, and a lack of tourist-inclined attributes. However, Bhutan has been able to leverage its natural setting as its central attribute of destination attractiveness. A comparable approach may be adopted by the RMI, one which imposes high government tariffs on tourists which creates an image of 'high value tourism' that appeals to a select segment of the tourist market (Gurung and Seeland 2008). It is also likely to appeal to tourists who are environmentally responsible, thus, firmly placing the country as a sacrosanct location to visit in one's lifetime rather than a 'tourist dive'. In addition, this strategy counters the stress that could be placed upon the population and land area as a result of enhanced tourism activities. Small island nations have constrained space to accommodate high levels of tourism, and, if left unchecked, eco-systems and existing infrastructure could be adversely impacted. However, by adopting a 'high value, low impact' strategy, the RMI would be able to not only maintain its pristine environment but also capitalize on the economic opportunities offered by tourism activities. This requires a concerted effort from the government and community leaders to establish a destination identity that synergizes with the natural and cultural environments of the island destination (McElroy 2003).

The country could utilize the concept of 'fear of missing out' (FOMO) to project itself as a destination to visit. This concept can be applied to connote the need to visit a destination prior to any potential degradation of its pristine locations, which may lose out to rapid development. If the externally initiated FOMO appeals are strategically instituted, they can arouse emotional dissonance among potential tourists. As Hodkinson (2016) pointed out, missing out on the social experience and a competitive feeling (what others are getting to do while one is not) are compelling emotions that may trigger a visit. The feeling of not being able to participate in meaningful tourism must be activated in the minds of the tourists. Crisis situations such as COVID-19 further underline FOMO, which may be capitalized on by the RMI as its marketing approach to increase tourist visits. The RMI can leverage on its various atolls to create a tourist who can 'delight in the fact that "their" island is not shared with any other tourism operators' (Scheyvens and Momsen 2008: 498). The RMI's 'not to be missed, exclusive destination' pitch should target tourists who are adventurous and are attracted to 'rustic' locations that offer authentic tourism experiences. This unique selling proposition must be clearly perceptible in tourists' mind set.

The MIVA, which shoulders the responsibility for tourism development, marketing, and promotions, should develop tourism literature and marketing efforts that stimulate FOMO and proactively market the niche tourism it offers. Outlying atolls such as the Ebon Atoll are likely to benefit from this approach, given that Majuro and Ebeye (Kwajalein Atoll) are densely populated (Collinson and Spears 2010). The central approach that should be taken by the RMI is to build its resilience and rebrand its tourism activities. Such an approach would also help to counteract the country's image being crudely associated with 'nuclear testing', 'nuclear waste', 'poverty' and 'sinking islands' and, thus, not an interesting place to visit. In an endeavour to foster alternative and progressive destination perceptions of the Marshall Islands, the country should build a cultural theme into their tourism offerings, in which unique elements of Marshallese culture, such as the native navigational techniques, traditional art of self-defence (*maanpa*), 'way-finding' sticks, culturally significant sites, Manit ('culture') Day and rides on outrigger canoes offer a unique experience to the traveller (Collinson and Spears 2010). Such efforts not only showcase how the country and the Marshallese are represented but also provide a platform for the island state to stand out among other destinations in the region.

The strategies suggested above require funding and greater coordination between the public and private sector. Moreover, inter- and intra-atoll collaboration and coordination are necessary for the RMI to successfully pursue economic and ecological sustainability. The nation has an abundance of natural marine resources and a strong cultural identity. Therefore, if these assets are strategically managed with marketing foresight, then the country's tourism endeavour has a bright future, particularly in contemplating the structure and nature of tourism in the post-COVID-19 era.

Conclusion and research implications

The Marshall Islands face a multitude of challenges in attracting tourists to its shores, notably a lack of tourism infrastructure, ineffective marketing and destination branding activities and a relatively piecemeal approach to tourism development. Moreover, there is a considerable lack of tourism planning in the RMI, as seen in its passive approach to pursuing the VFR segment and limited endeavours to progressively promote the country as a FOMO destination. Nonetheless, there is scope for the country to pursue the benefits of the tourism sector. What is critical is a focused tourism policy that would govern activities and efforts in positioning itself as a destination of choice, particularly in a manner that is both ecologically sustainable and economically viable. Legislators in the RMI need to perform a balanced scorecard assessment of current tourism-related practices and use these to determine the next step forward as the country addresses current challenges as well as the fallout from the pandemic. The role of capacity building is also undeniably important, as it forms the basis of rejuvenating the country's tourism endeavours. A concerted effort by all the stakeholders in ensuring the existence of the relevant capacities to support tourism development and growth is of absolute necessity. Such endeavours are clear in the recent National Strategic Plan 2020–2030 in which sustainability and resilience are emphasized as the driving forces behind the revitalization of the future of the Marshall Islands.

Further research is necessary to identify the strategic agenda surrounding tourism policy in the RMI. It is necessary for the development of a concerted tourism agenda that considers stakeholders' concerns, tourism resilience and climate change strategies. In addition, the development of these strategic directions must also factor in the 'new norms' arising out of the COVID-19 phenomenon, especially well-regulated standard operating procedures in the country to ensure that tourists perceive the destination as a safe and secure place. The RMI has to adopt a critical evaluation of its tourism policies to ensure that it is able to progress for the betterment of the country, economy and society. Involvement of government stakeholders, local businesspeople, and regional tourism bodies or associations can all ensure that this nation state develops in a feasible strategic direction vis-à-vis its tourism future. In an endeavour to sustain and maximize the benefits of tourism activities, the Marshall Islands has to step up its efforts in securing resources to support its growth, improving its infrastructure, marketing itself as a destination for new experiences and providing a supportive regulatory environment for foreign investments. Its tourism survival is highly dependent on such engagement.

References

ADB (Asian Development Bank) (2011) *Development Effectiveness Brief: Marshall Islands, Together, Forging a Better Future*, Metro Manila: Asian Development Bank. Available online at https://www.adb.org/sites/default/files/publication/28749/decb-rmi.pdf (accessed 20 February 2021).

ADB (Asian Development Bank) (2021) *Marshall Islands and ADB*, Metro Manila: Asian Development Bank. Available online at https://www.adb.org/countries/marshall-islands/economy (accessed 9 June 2021).

Asiedu, A.B. (2008) 'Participants' characteristics and economic benefits of visiting friends and relatives (VFR) tourism - an international survey of the literature with implications for Ghana', *International Journal of Tourism Research*, 10(6): 609–621.

Atomic Heritage Foundation (2019) 'Marshall Islands'. Available online at https://www.atomicheritage.org/location/marshall-islands (accessed 3 March 2020).

Barnett, J. (2011) 'Dangerous climate change in the Pacific islands: Food production and food security', *Regional Environmental Change*, 11(1): 229–237.

Berta, O.G. (2021) 'From arrival stories to origin mythmaking: Missionaries in the Marshall Islands', *Ethnohistory*, 68(1): 53–75.

Bordner, A. and C.E. Ferguson (2020) 'The Marshall Islands could be wiped out by climate change – and their colonial history limits their ability to save themselves', *The Conversation*, 11 December. Available online at https://theconversation.com/the-marshall-islands-could-be-wiped-out-by-climate-change-and-their-colonial-history-limits-their-ability-to-save-themselves-145994 (accessed 17 February 2021).

Brown, K.G., J. Cave, F.M. Collison and D.L. Spears (2010) 'Marketing cultural and heritage tourism: The Marshall Islands', *International Journal of Culture, Tourism and Hospitality Research*, 4(2): 130–142.

Brown, S. (2012) 'Poetics and politics: Bikini atoll and world heritage listing'. In: S. Brockwell, S. O'Connor and D. Byrne (eds.) *Transcending the Culture-Nature Divide in Cultural Heritage: Views from the Asia-Pacific Region*, Canberra, Australia: ANU E Press, pp. 35–52.

Carter, G. (2020) 'Pacific island states and 30 years of global climate change negotiations'. In: G. Carter (ed.) *Coalitions in the Climate Negotiations*, Abingdon: Routledge, pp. 73–90.

Cheer, J.M., S. Pratt, D. Tolkach, A. Bailey, S. Taumoepeau and A. Movono (2018) 'Tourism in Pacific island countries: A status quo round-up', *Asia and the Pacific Policy Studies*, 5(3): 442–461.

CIA (Central Intelligence Agency) (2020) *Australia - Oceania: Marshall Islands — The World Factbook*. Available online at https://www.cia.gov/the-world-factbook/countries/marshall-islands/ (accessed 4 May 2022).

Collison, F.M. and D.L. Spears (2010) 'Marketing cultural and heritage tourism: The Marshall Islands', *International Journal of Culture, Tourism and Hospitality Research*, 4(2): 130–142.

Collison, F.M. and D.L. Spears (2015) 'Marshall Islands, tourism', In: J. Jafari and H. Xiao (eds.) *Encyclopedia of Tourism*, Cham: Springer, pp. 660–661.

Crompton, J.L. (1979) 'An assessment of the image of Mexico as a vacation destination and the influence of geographical location upon that image', *Journal of Travel Research*, 17(4): 18–24.

Daborn, M. (2013) 'Blown to hell: The health legacies of US nuclear testing in the Marshall Islands', *Constellations*, 5(1): 26–35.

Davis, J.S. (2005) 'Representing place: "Deserted isles" and the reproduction of Bikini Atoll', *Annals of the Association of American Geographers*, 95(3): 607–625.

Dorji, T. (2001) 'Sustainability of tourism in Bhutan', *Journal of Bhutan Studies*, 3(1): 84–104.

Everett, H., D. Simpson and S. Wayne (2018) 'Tourism as a driver of growth: A pathway to prosperity for Pacific island countries', *Issues in Pacific Development*, No. 2, June. Available online at https://www.adb.org/sites/default/files/publication/430171/tourism-growth-pacific.pdf (accessed 20 February 2021).

EPPSO (Economic Policy, Planning and Statistics Office) (2020) *National Strategic Plan 2020–2030, Government of the Republic of the Marshall Islands*, Majuro, Republic of the Marshall Islands. Available online at https://rmieppso.org/rmi-national-strategic-plan/ (accessed 23 January 2022).

Germanos, A. (2019) 'Pacific island nations declare climate crisis, fear being uninhabitable by 2030', *Eco Watch*, 1 August. Available online at https://www.ecowatch.com/pacific-islands-climate-crisis-2639602416.html (accessed 20 February 2021).

Gershon, I. (2007) 'Viewing diasporas from the Pacific: What Pacific ethnographies offer Pacific diaspora studies', *The Contemporary Pacific*, 19(2): 474–502.

Gittelsohn, J., H. Haberle, A.E. Vastine, W. Dyckman and N.A. Palafox (2003) 'Macro-and microlevel processes affect food choice and nutritional status in the Republic of the Marshall Islands', *The Journal of Nutrition*, 133(1): 310S–313S.

Glusac, E. (2021) 'When can we cruise again – and will it be safe?' *AARP*, 19 January. Available online at https://www.aarp.org/travel/vacation-ideas/cruises/info-2020/cruise-status-during-pandemic.html (accessed 20 February 2021).

Gurung, D.B. and K. Seeland (2008) 'Ecotourism in Bhutan: Extending its benefits to rural communities', *Annals of Tourism Research*, 35(2): 489–508.

Hammond, R.A. and R. Levine (2010) 'The economic impact of obesity in the United States', *Diabetes, Metabolic Syndrome and Obesity: Targets and Therapy*, 3: 285–295.

Harrison, D. (2004) 'Tourism in Pacific islands', *The Journal of Pacific Studies*, 26(1): 1–28.

Hodkinson, C. (2016) '"Fear of missing out" (FOMO) marketing appeals: A conceptual model', *Journal of Marketing Communications*, 25(1): 1–24.

Holden, P. and S. Holden (2003) *Republic of the Marshall Islands – A Private Sector Assessment: Promoting Growth through Reform*, The Enterprise Research Institute, Report Prepared by ERI for the Asan Development Bank. Available online at https://www.adb.org/sites/default/files/institutional-document/32214/rmi-psa.pdf (accessed 20 February 2021).

IOM (International Organization for Migration) (2019) *Climate Change and Migration in Vulnerable Countries: A Snapshot of Least Developed Countries, Landlocked Developing Countries and Small Island Developing States*, Geneva: International Organization for Migration.

Johnson, B.R. and B. Takala (2016) 'Environmental disaster and resilience: The Marshall Islands experience continues to unfold', *Cultural Survival Quarterly*, September. Available online at https://www.culturalsurvival.org/publications/cultural-survival-quarterly/environmental-disaster-and-resilience-marshall-islands-0 (accessed 4 May 2022).

Kim, D. and R.R. Perdue (2011) 'The influence of image on destination attractiveness', *Journal of Travel and Tourism Marketing*, 28(3): 225–239.

Kiste, R.C. (1986) 'Termination of the US trusteeship in Micronesia', *The Journal of Pacific History*, 21(3): 127–138.

Kiste, R.C. (2021) 'Marshall Islands', *Encyclopedia Britannica*. Available online from https://www.britannica.com/place/Marshall-Islands (accessed 15 February 2021).

Kostopoulou, S. (2013) 'On the revitalized waterfront: Creative milieu for creative tourism', *Sustainability*, 5(11): 4578–4593.

Land, C.E., A. Bouville, I. Apostoaei and S.L. Simon (2010) 'Projected lifetime cancer risks from exposure to regional radioactive fallout in the Marshall Islands', *Health Physics*, 99(2): 201–215.

Lee, C.F., H.I. Huang and H.R. Yeh (2010) 'Developing an evaluation model for destination attractiveness: Sustainable forest recreation tourism in Taiwan', *Journal of Sustainable Tourism*, 18(6): 811–828.

Letman, J. (2018) 'Rising seas give island nation a stark choice: Relocate or elevate', *National Geographic*, 19 November. Available online at https://www.national geographic.com/environment/2018/11/rising-seas-force-marshall-islands-relocate-elevate-artificial-islands/ (accessed 20 June 2020).

Lindsey, R. (2022) 'Climate change: Global sea level', *Climate.gov*, 19 April. Available online at https://www.climate.gov/news-features/understanding-climate/climate-change-global-sea-level (accessed 4 May 2022).

Lonely Planet (n.d.) 'Marshall Islands travel'. Available online at https://www.lonelyplanet.com/marshall-islands (accessed 4 April 2020).

Maplandia.com (n.d.) 'Flights from Japan to Marshall Islands – compare prices'. Available online at http://www.maplandia.com/japan/flights/marshall-islands/ (accessed 9 June 2021).

Mayo, E.J. and L.P. Jarvis (1981) *The Psychology of Leisure Travel: Effective Marketing and Selling of Travel Services*, Boston: CBI Publishing.

McElfish, P.A. (2016) 'Marshallese COFA migrants in Arkansas', *The Journal of the Arkansas Medical Society*, 112(13): 259–262.

McElroy, J.L. (2003) 'Tourism development in small islands across the world', *Geografiska Annaler: Series B, Human Geography*, 85(4): 231–242.

Niedenthal, J. (1997) 'A history of the people of Bikini following nuclear weapons testing in the Marshall Islands: With recollections and views of elders of Bikini Atoll', *Health Physics*, 73(1): 28–36.

OEPPC (Office of Environmental Planning Policy Coordination) (2017) *Republic of the Marshall Islands Fifth National Report: Convention on Biological Diversity*. Available online at https://www.cbd.int/doc/world/mh/mh-nr-05-en.pdf (accessed 20 June 2020).

Opperman, M. (1993) 'Tourism space in developing countries', *Annals of Tourism Research*, *20*(4): 535–556.

Pacific Community (2022) 'COVID-19 pacific community updates', 25 April. Available online at https://phd.spc.int/covid-19 (accessed 2 May 2022).

Ramirez, N. (2018) 'Toxic colonialism: Nuclear materials in the Pacific islands', *The Gallatin Research Journal*, pp. 1–10. Available online at https://wp.nyu.edu/compass/2018/04/24/toxic-colonialism-nuclear-materials-in-the-pacific-islands/ (accessed 20 June 2020).

Scheyvens, R. and J. Momsen (2008) 'Tourism in small island states: From vulnerability to strengths', *Journal of Sustainable Tourism*, 16(5): 491–510.

Senthilingam, M. (2015) 'How paradise became the fattest place in the world', *CNN.com*, 1 May. Available online at https://edition.cnn.com/2015/05/01/health/pacific-islands-obesity/index.html (accessed 20 April 2020).

Smith-Norris, M. (2016) *Domination and Resistance: The United States and the Marshall Islands During the Cold War*, Honolulu, HI: University of Hawai'i Press.

Statista (2020) 'Trade balance of Marshall Islands 2019', 12 August, New York: Statista. Available online at https://www.statista.com/statistics/731693/trade-balance-of-marshall-islands/#statisticContainer (accessed 15 February 2021).

Sutow, W.W., R.A. Conard and K.M. Griffith (1965) 'Growth status of children exposed to fallout radiation on Marshall Islands', *Pediatrics*, 36(5): 721–731.

Thaman, R.R. (1988) 'Health and nutrition in the Pacific islands: Development or underdevelopment?', *GeoJournal*, 16(2): 211–227.

Thomas, P. (2019) 'The economics of dependency in the Marshall Islands', *Pacific Economic Bulletin*, 2(2): 25–30.

Tourism Council of Bhutan (2019) 'Tourism policy'. Available online at https://www.tourism.gov.bt/about-us/tourism-policy (accessed 2 June 2020).

Treloar, P. and C.M. Hall (2005) 'Tourism in the Pacific islands'. In: C. Cooper and C.M. Hall (eds.) *Oceania: A Tourism Handbook*, Clevedon: Channel View Publications, pp. 173–294.

Tripadvisor (2021) 'Majuro hotels and places to stay'. Available online at https://en.tripadvisor.com.hk/Hotels-g301393-Majuro-Hotels.html (accessed 9 June 2021).

United Nations (n.d.) 'Travel to Marshall Islands', *Permanent Mission of the Republic of the Marshall Islands to the United Nations*. Available online at https://www.un.int/marshallislands/marshallislands/travel-marshall-islands (accessed 9 June 2021).

UNWTO (United Nations World Tourism Organization) (2014) *Tourism in Small Island Developing States (SIDS) – Building a More Sustainable Future for the People of Islands*, Madrid: UNWTO. Available online at https://www.e-unwto.org/doi/epdf/10.18111/9789284416257 (accessed 3 June 2020).

U.S. General Accounting Office (1980) 'Report to Congress: Enewetak Atoll- cleaning up nuclear contamination'. Available online at https://www.osti.gov/opennet/servlets/purl/16364899.pdf (accessed 9 June 2021).

van der Geest, K., M. Burkett, J. Fitzpatrick, M. Stege and B. Wheeler (2020) 'Climate change, ecosystem services and migration in the Marshall Islands: Are they related?', *Climatic Change*, 161(1): 109–127.

van der Zee, E. and D. Vanneste (2015) 'Tourism networks unravelled: A review of the literature on networks in tourism management studies', *Tourist Management Perspectives*, 15: 46–56.

Walsh, M. (2020) 'Coronavirus fears are seeing Pacific islands turn back cruise ships: Here's what you need to know', *ABC News*, 6 March. Available online at https://www.abc.net.au/news/2020-03-06/coronavirus-fears-are-seeing-pacific-islands-turn-back-cruises/12019916 (accessed 17 February 2021).

World Bank (2022a) 'Population, total'. Available online at https://data.worldbank.org/indicator/SP.POP.TOTL (accessed 11 January 2022).

World Bank (2022b) 'International tourism, number of arrivals - Marshall Islands'. Available online at https://data.worldbank.org/indicator/ST.INT.ARVL?locations=MH (accessed 11 January 2022).

18

DECIPHERING THE ENVIRONMENTAL CHALLENGES AND ADVANCEMENTS OF TOURISM DEVELOPMENT IN PALAU

Anuradha Vyas and Marcus L. Stephenson

Introduction

The Republic of Palau (Figure 18.1) is situated in the western Pacific Ocean and in the southwest part of Micronesia, with New Guinea located 650 kilometres to the south, the Philippines 890 kilometres to the west and Guam 1,330 kilometres to the northeast (Shuster and Foster 2021). The country's population is estimated at 18,238 (Worldometer 2022), and the country's capital is Ngerulmud, which is located on the largest island of Babeldaob (also Bebelthuap), though the most populous island and commercial hub is Koror (Abbot 2016). Palau pursues a democratic political system and is in a free association agreement with the United States (US), which provides defence and access to federal services (e.g., postal and meteorological services) as well as granting permission to Palau citizens to study, work and reside in the US and its territories (DFAT n.d.). Palau's political history is diverse. Spain claimed sovereignty over the territory from 1686 until 1899 when the island was sold to Germany following the Spanish-American war, who then governed the island until Japan occupied the territory from 1914 until after World War II (Treloar and Hall 2005). The country came under US administration following the war, gaining independence in 1994 (Carlile 2000). Palau's economy has been historically dependent on agriculture and fishing activities, though over the past two decades tourism has become the most dominant sector in the country's economy, particularly until the outbreak of the COVID-19 pandemic in early 2020. Although travel was suspended in March 2020 as borders were closed, there was a short-lived attempt from 1 April 2021 to open a travel bubble with Taiwan, until this was suspended on 19 May 2021 because of a spike in COVID-19 infections in Taiwan (IMF 2021).

The work initially presents an overview of tourism development in Palau, drawing attention to the country's environmental resources and the historical development of tourism and tourism market segments pertaining to nationality. The discussion indicates the reasons for the change of direction concerning tourism development, particularly from a mass to a high-end, low-volume approach. The chapter then moves towards a succinct outline of the environmental

DOI: 10.4324/9780429019968-21

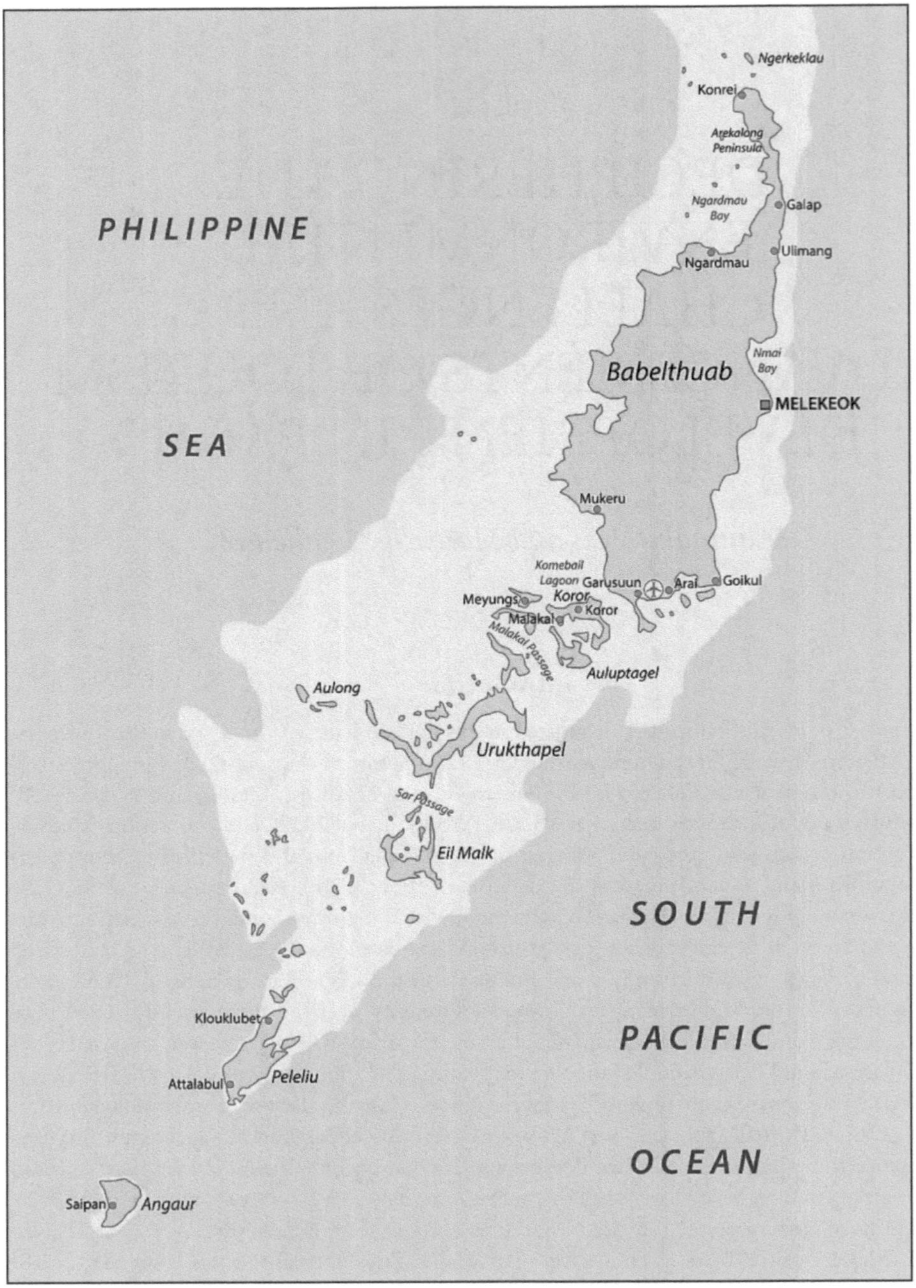

Figure 18.1 Map of Palau. Source: Rainer Lesniewski (www.shutterstock.com/image-vector/palau-map -166657325).

impacts of tourism, especially in terms of water supply. This leads to a discussion which positively highlights how the government is recently addressing environmental concerns, especially through establishing a range of progressive strategies and sustainable directives to protect its natural and marine environments. The latter section re-visits the high-end tourism market in terms of contemplating how Palau should move forward in the post-COVID-19 context.

Tourism development in Palau

It is claimed that Palau enjoys certain natural advantages compared to other Micronesian countries, especially in terms of being less prone to typhoons, having more species of marine life in its waters and more land for tourism-related developments (Palau Visitors Authority 2019). The coral reef's importance for the tourism industry is crucial. Around 86% of Palau's tourists visit the country for scuba diving and snorkelling, especially as the marine environment is exceptionally diverse and home to 425 coral species, 1,700 species of fish, 235 species of crustaceans and 302 species of molluscs (Golbuu et al. 2018).

The early 1970s witnessed the development of tourism. The Palau Tourist Commission (now the Palau Visitors Authority) was established in 1971 by the US Trust Territory Government, realizing the value of tourism development in line with economic development policy (Treloar and Hall 2005). However, Palau became popular in the 1990s with Japanese tourists interested in participating in non-mass forms of tourism, notably diving (Yamashita 2000). Yamashita suggests that, although an interest in the physical environment through diving activities was popular for Japanese tourists, some were already travelling to Palau because of 'memorial tourism'. Hence, from the 1960s, older Japanese veterans and their families travelled to Palau for memorial services, including visiting sites commemorating more than 12,000 Japanese soldiers who lost their lives fighting the US army. Although there are some symbolic attractions in Palau associated with World War II nostalgia and commemoration, the country does not have many significant cultural sites developed for tourism. Nonetheless, the Rock Islands Southern Lagoon was inscribed as a United Nations Educational, Scientific and Cultural Organization (UNESCO) World Heritage Site in 2012, classified as a mixed cultural and natural property with such archaeological remains as stone villages, rock art and cave deposits. The area demonstrates small communities existing on marine resources and how climate change and population growth impact the area (Smith 2015).

In addition to the increased attention in the late twentieth century from Japanese tourists who could afford high-end customized packages, Japan also started to invest in tourism development in Palau through the purchase of several hotels (Carlile 2000). This author notes that Palau's political independence encouraged more control over air traffic in and out of the country, especially as the country did not have to abide by the bilateral aviation engagements that were founded on the basis of the Japan–US aviation treaty, thus, encouraging more autonomy on flight negotiations with Japan and other countries. Nonetheless, Palau's push towards mass tourism began in the 1990s with the development of large hotels, such as the Palasia Hotel with 165 rooms (Ueki 2000), thereby adding to such existing hotel stock as the Palau Pacific Resort which opened in 1984 with Japanese investment (Treloar and Hall 2005). Ueki (2000) observed that one of the most significant developments in the 1990s was the construction of a US-financed loop road around the island of Babeldaob, which threatened the environment by opening up areas that naturally existed on subsistence economies. Nonetheless, the road project was welcomed in some spheres, as it was seen as a catalyst for the social modernization of local societies, particularly as the new road would not only interconnect states and communities but also encourage families who temporarily reside in the urban centre of Koror to re-settle in their home villages in Badeldaob (Graham and Idechong 1998).

The government of Palau established the '2009–2014 Medium Term Development Strategy' which pursued economic, social, cultural and environmental objectives (Government of Palau 2009). One key objective was to encourage environmentally conscious visitors to travel to Palau, especially for the country to become a high-end destination through establishing eco-tourism initiatives and formulating partnerships in the pursuit of environmental management and responsibility. The management of natural resources was slow to evolve because of limited resources allocated to reform of the tourism sector, little uniformity of policy across government and disequilibrium between national government strategy and its implementation at the state level (IMF 2016). More recently, however, Palau has seriously embraced environmental stewardship practices, as will be highlighted in subsequent discussions.

The composition of Palau's tourist-base has changed remarkably in terms of nationality over the past two decades. In 2000, more than 50% of tourists were from Japan and the Republic of Korea, around 20,000 and 13,000 respectively, with only 1,100 tourists coming from the People's Republic of China (ADB 2017). The highest growth market segment concerned Chinese tourist arrivals, which increased from 608 in 2008 to 88,476 in 2015, with the total arrivals peaking at 163,905 in the same year (Government of Palau 2020). However, in 2017, Palau's gross domestic product (GDP) contracted by 3.7%, mainly due to a 17% decrease in tourist arrivals from the preceding year (IMF 2019). There was a 66% decline in overall tourist arrivals from 2015 to 2019, reducing to 30,147 tourists (Government of Palau 2020a). In November 2017, it was alleged that tour operators were informed by the Chinese government not to sell package tours to Palau, as the country was not willing to compromise its relationship with Taiwan, as it was receiving US$10 million in aid annually from Taiwan (Lyons 2018).

Nonetheless, there has been a growing realization over the past two decades that Palau needs to proactively advance a sustainable tourism approach at both public and private sector levels. Therefore, in 2007, the private sector, represented by the Belau Tourism Association, the Palau Chamber of Commerce and the public sector, represented by the Palau Visitors Authority, came together to critically assess the tourism industry and formulate the Tourism Action Plan (Republic of Palau 2010). The challenges that were identified concerned a range of issues: the downgraded value of tourism due to budget-based package tourism, lack of transportation flexibility due to over-reliance on one airline, limited range of tourism products and services, limited infrastructural development for cruise ships and private vessels, socio-cultural concerns relating to the promotion of mass tourism, lack of management attention to tourism's reliance on natural resources and challenges relating to the under-representation of Palau citizens working in the tourism industry (Republic of Palau 2010: 99-100). The Tourism Action Plan indicated that a form of high-end, low-volume tourism would be a better way forward than pursuing mass tourism, thereby appeasing both public sector tourism planners and private sector operators. This approach partly represents a reaction to mass tourism and the impacts associated with overtourism. It also reflects the high-value, low-impact tourism dimension that is inherent in the sustainable tourism branding plan, 'Pristine Paradise Palau', launched in 2016 (Bureau of Tourism 2016). High-value tourists are seen to be 'experiential' in terms of their tourism experiences (wishing to break away from the mass tourist experience, for instance) and are not necessarily fixated on high spending activities (ADB 2017: 18).

Environmental challenges and the problematic role of high-end tourism

Palau needed to reorientate its tourism products away from mass tourism, lessening the impact of overtourism on Palau's fragile environments. More than two decades ago, Yamashita (2000:

457) warned of the ecological ramifications of pursuing an unsustainable form of tourism, noting:

> The most critical problem of Palauan tourism is the question of where the breaking point lies in the ecological balance of the people, the land, and the sea ... what is clear is that if the balance is broken, Palau will be unable to sustain not only tourism but even the nation itself. The sustainability of Palauan tourism depends on this delicate ecological balance.

Wabnitz et al. (2018: np) is also concerned with environmental impacts, particularly with regards to dive tourism, stating:

> However, such large numbers of, often inexperienced, divers have raised concerns about overcrowding at dive sites and poor diver behaviour (e.g., coral holding or kicking, full-body layouts on corals, etc.) contributing to coral reef decline ... Visitation rates have also led to increased pressure on resources through greater freshwater usage, mostly reef-sourced seafood consumption in restaurants, rapid coastal development to accommodate and provide for a greater influx of people, and increased waste generation.

The country changed track to promote a high-end and high-value form of tourism rather than promote a form of mass tourism popularly perceived to be at odds with the environment. However, this approach is arguably myopic given that high-end tourism and hospitality in Palau can actually be overly consumptive and negatively impactful to the environment. Palau's focus on the luxury tourism market is acknowledged in its tourism investment policy, instructing foreign investors to no longer invest in low-status accommodation but only in high-end (five-star) hotel properties, in which they are encouraged to invest through the provision of a tax credit of up to 40% (Carreon 2018). Water demand may increase significantly in proportion to luxury hotel developments (see Ioannides and Holocombe 2003), as annual water consumption per guest in luxury hotels in tropical countries is around 820 litres per guest (Greenhotelier.com 2019). Apart from water consumption for drinking, bathing, cooking and laundry purposes, significant amounts of water evaporate in hotel swimming pools. A typical unoccupied pool of a general size (9.3 x 4.5 metres) can lose approximately 37,260 litres of water annually during an average air temperature of around 25 degrees Celsius (Shah 2011). The average temperature in Palau is recorded between 25 and 30 degrees Celsius.

Although Palau's water production capacity is fairly adequate, operational improvements in the long-term are necessary, which should be taken seriously in light of the constant risk of droughts. In 2016, a severe drought led to water supply disruption for a significant period, and the island of Koror had to ration the daily supply of water for three hours per day. The 2017 Palau State of the Environment report indicated that droughts in the region will intensify in future years because of climate change (Carreon 2017). Palau's water supply is vulnerable, as it can take only five days to empty existing reservoirs (Nath et al. 2005). Lack of water supply can increase the cost burden on service providers to arrange alternative sources of water during the initial infrastructural development, such as the provision of inbuilt water purification plants or alternative supply sources. Over the past few decades, urbanization in Palau has played a crucial role in contributing to water shortage, and these areas host a greater proportion of tourists through accommodation establishments and restaurants (Mason et al. 2020).

Progressive tourism environment directives

Palau is the first nation to ban sunscreens which produce such 'reef-toxic' chemicals as oxybenzone, octinoxate and triclosan, legislating a fine of US$1,000 for any attempt to retail this product, which will also be seized from tourists on arrival (Peters 2018). UNESCO and the government of the state of Koror supported a study which focused on sunscreen accumulation in Jellyfish Lake (known locally as Ongeim'l Tketau), which is the World Heritage property of Rock Islands Southern Lagoon, and concluded that, as sunscreen compounds have a high presence in the lake and in the golden jellyfish medusa, the use of UV filters needs to be addressed in order to stabilize the jellyfish population (Bell et al. 2017). At the request of traditional chiefs and leaders, the lake closed in 2016 due to declining jellyfish stock and did not reopen until 2019 (Salinas 2019). Researchers also found that young coral on the ocean floor can also be impeded by low concentrated amounts of sunscreen (Xu 2018). The sunscreen ban was effective in January 2020 and is part of Palau's Responsible Tourism Education Act of 2018, emphasizing the duty of residents as 'custodians' of the natural environment 'to encourage responsible treatment of these landmarks' (Office of the President 2018: 1).

The Palau government developed progressive directives to protect the marine environment from the tourism industry and other industries. Since January 2020, a commercial fishing ban stretching over 190,000 square miles of the surrounding ocean has come into existence in which the government has closed 80% of the exclusive economic zone to fishing and marine activities (Braine 2020), thereby establishing a marine sanctuary. There is an alignment with a traditional cultural practice of protecting the marine ecosystem known as 'bul', in which certain parts of the reef are banned from fishing, thus encouraging various fish species to thrive naturally (World Economic Forum 2019). Palau had already established the world's first shark sanctuary to protect hammerheads, leopard sharks and oceanic whitetip sharks, as well as more than 130 other species fighting extinction in the Pacific Ocean, where shark fishing is an illegal activity (Heilprin 2009). The protection of shark communities brings economic dividends through the expansion of a particular type of ecotourism associated with the shark-diving industry. Vianna et al. (2010: iii) conducted research concerning reef sharks as an ecotourism asset in Palau, estimating that the lifetime value to the tourism industry of a reef shark represents US$1.9 million and shark diving represents US$1.5 million as annual tax income to the government. Although this research was conducted prior to significant increases in tourism numbers, the economic data also indicates that yearly salary incomes received by the local community from the shark-diving industry amounted to US$1.2 million, not forgetting the additional positive impact of shark tourism on tourism-related employment in such sectors as hotels, restaurants and souvenir shops.

Palau's efforts to protect the marine environment are constructive. Palau's National Environmental Protection Council report, entitled 'The 2019 State of the Environment', indicates that the country's shallow coral reefs are in 'good condition' (International Institute for Sustainable Development 2020). Though this report acknowledges that the reefs are vulnerable to such pressures as global climate change, sedimentation and over-fishing, Palau's accomplished approach to coral reef management should also be applied in the management of nearshore fisheries (International Institute for Sustainable Development 2020). However, some states within Palau are more positive than others in controlling and protecting marine resources. The state of Koror, for instance, possesses a portfolio of licenses and fees for diving and fishing (Graham and Idechong 1998).

Since April 2018, Palau started to charge a 'green fee' for all non-Palauan citizens travelling to Palau (SPREP 2018), involving a US$100 visitor fee incorporated into the airline ticket which helps to finance the marine sanctuary (as discussed above) and Palau's Protected Area Network (PAN) (von Saltza 2019). This approach signifies the 'polluter pays principle' that ethically stipulates that those who are responsible for impacting the environment, whether individuals or companies, should be liable for potential costs; even if the impact is intentional or not (see Fennel and Ebert 2004). Visitors entering Palau must guarantee to respect the environment and culture through the signing of a pledge, which states:

> Children of Palau … I take this Pledge (sic) … To preserve and protect your beautiful and unique island home … I vow to tread lightly, act kindly, and explore mindfully … I shall not take what is not given … I shall not harm what does not harm me … The only footprints I shall leave are those that will wash away.
>
> *(von Saltza 2019: 7)*

This pledge is stamped in visitors' passports prior to entering the country, and any violation of this pledge could result in individuals being fined. The pledge, which may be instrumental in encouraging the desired behavioural change in tourists, demonstrates how tourists can be conceived as environmental citizens with direct responsibilities towards protecting the environment (Raouf 2011). It is an affirmation by tourists of their responsibilities as environmental citizens, whilst enjoying the right to consume or appreciate tourism products and services. The idea of environmental citizenship is manifest in the virtuous aspirations of human beings, irrespective of national boundaries (Dobson 2007), and the passport pledge invokes an urge in tourists to be virtuous by being conscious of their ecological footprints. Therefore, the idea of environmental citizenship in the context of the tourism industry centres on tourists' roles and responsibilities in protecting the environment, involving practices and accountabilities that transcend national boundaries. However, as tourism is an activity encouraging individuals to escape daily routine, people may not demonstrate the same level of commitment towards environmentally friendly practices as they do in their home environment (Miao and Wei 2013), which may well mean that the philosophy of environmental citizenship should require consistent reinforcement so that tourists rigorously pursue such practices. Cheng and Wu (2015) observe that tourists become proactively engaged in environmentally friendly practices when they experience high levels of attachment with places of significance. Accordingly, the pledge and proactive engagement in such practices and activities may encourage tourists to develop a strong sense of 'place attachment'.

Apart from the initiatives discussed, the government takes a proactive role in the protection of marine resources through setting up the PAN fund as a result of the 2003 Protection Areas Network Act. The fund operates as a non-profit entity and ensures that the government retains significant decision-making power (von Saltza 2019). The intention of PAN is to ensure that a regulatory framework of non-governmental organizations (NGOs) and local governments is established to coordinate the conservation of marine resources and preserve Palau's marine biodiversity (Marino et al. 2008). Nonetheless, there are further directives that can still be employed to strengthen the sustainability of the marine environment, notably monitoring and decreasing diver numbers for each site, making sure that dive operators work in compliance with responsible dive programmes (for example, Blue Star and Green Fins), ensuring that divers and snorkellers enter the water at high tide or in deep water and also at sandy areas to avoid fragile reef and coral areas, and encouraging diver operators to be proactively involved in marine conversation, especially as it is within their livelihood interests to ensure marine resources are protected and that they provide marine education to tourists (Wabnitz et al. 2018).

High-end tourism and the post-COVID-19 era

In an endeavour to strengthen the tourism economy and, at the same time, taking measures to protect the environment, Palau has been shifting its focus away from a mass tourism strategy towards luxury tourism. The country is optimistically looking at the future of high-end tourism as a sustainable path to the development of its economy and for establishing more employment opportunities for locals. In terms of shifting the focus from 'volume growth' to 'value growth', the policy entitled 'Palau Responsible Tourism Policy Framework' promotes Palau as a niche destination offering nature-based tourism, cultural heritage tourism, culinary tourism, agritourism, sports tourism, weddings and honeymoons, and adventure tourism. This policy emphasizes the importance of tourists seeking out 'authentic experiences', with particular focus on the tourism industry catering for the 'engaged traveler' (Bureau of Tourism 2016: 7). Appropriately, the 2018 Responsible Tourism Education Act legislated the need for aircraft and vessels to ensure that passengers travelling to Palau are aware of the country's environmental policies, communicated through video and literature sources (Island Times 2018).

Nonetheless, there is lack of clarity among government and tourism agencies on the conceptual and practical application of 'high-end' tourism, including how this may be accomplished and whether 'high-end' implies the targeting of tourists from specific places or on the basis of higher spending per day (IMF 2016). Although the shift from mass tourism to luxury tourism reduces tourist arrivals, implying a reduction of environmental pressure on resources, the actual impact of high-end tourism on Palau's environment ought to be rigorously assessed. However, Lee et al.'s (2020) empirical study concerning the attitudes of the host community in Palau toward the trade-off between tourism development and environmental conservation is insightful. The work emphasizes that, although there are significant differences of opinions between the community and the government, there is a general consensus that tourism strategy should run harmoniously with environmental conservation through ensuring a shift to high-end tourism.

Shifting perspectives and practices towards tourism development in Palau have further been augmented by the COVID-19 pandemic. The first quarter of 2020 began optimistically, as tourism arrivals grew by nearly 30% (32,255 visitors) compared to the same period in 2019 (Graduate School USA 2020: 1). Tourism-related sectors contributed more than one third of the GDP in 2019 (IMF 2021), though GDP decreased by 8.7% in 2020 (Countryeconomy.com 2020). Although tourism is anticipated to slowly recover in 2022, there is an expectation that tourism-related activities will still be lower than pre-pandemic activities, and this is likely to continue through to 2023 (IMF 2021). The International Monetary Fund (IMF 2021) report acknowledges that the pandemic provided an opportunity for Palau to look more at reforming the tourism sector, especially in terms of diversifying source markets and orientating towards high-value tourism as well as targeting ecotourism, as this would help to reduce environmental impact, increase revenue per tourists and improve the tourism experience as a whole. For this form of tourism to avoid being negatively impactful, however, it is important to critically consider initiatives and strategies which are environmentally friendly and where there continues to be strong governance concerning environmental controls, enforcement and legislation. Crucially, Romagosa (2020: n.p.) suggests that the implementation of sustainability in the tourism sector should be facilitated by 'appropriate forms of governance, integrating the public and private sectors in a co-ordinated manner'.

Conclusion and research implications

This chapter discussed the development of Palau's tourism industry and the environmental risks it has posed. The challenges brought about by Palau's subsequent shift from mass tourism towards high-end tourism and the steps taken by the government to address particular environmental impacts have also been critically acknowledged. The work strongly suggests that this form of tourism should be approached cautiously, as it will continue to have a significant environmental impact relative to those types of tourists who are attracted to such niche and luxury forms of tourism. Nonetheless, as the chapter has illustrated, the government of Palau took significant strides to protect and conserve its natural environment, and other independent Pacific Island states should, thus, take heed of the various directives instituted: stipulations against tourists utilizing environmentally toxic chemicals, policy protection of marine ecology, employment of environmental management practices, 'green' visitor fees and enforcement of a responsible tourist ethos. However, in light of the current COVID-19 pandemic, there is a prevailing need to revisit the country's high-end tourism strategy. Expansion of luxurious forms of tourism and hospitality should be challenged, and forms of tourism that are experiential, sustainable, nature-driven and conservation-based should be prioritized far more than overly consumptive forms of tourism and hospitality.

It is crucial to critically assess the threats of high-end tourism on the environment, especially if it requires more luxury facilities and services. The research objective would be to engage with key stakeholders to ascertain a collective consensus on how new forms of tourism can be developed in a manner which strengthens natural resources and at the same time ensures that the physical environment is both protected and resilient to tourism development. This chapter thus implicates an operational framework that acknowledges the importance of pursuing environmental citizenship directives for tourism development in which tourists can be encouraged to be environmentally responsible and accountable for their potentially impactful behaviour.

Critical enquiry is necessary to comprehend how the private sector can play a major role in producing and promoting environmental citizenship, including ways to seriously mitigate environmental impact through ensuring that business operations are adhering to sustainable practices at every level and that corporate social responsibility strategies are environmentally focused. As the tourism and hospitality industry can provide services and activities that can affect the fragile environment of small island states, there is a need for companies to make sacrifices in the provision of particular activities and services, especially if sustainable targets and practices are to be achieved (de-Miguel-Molina et al. 2014). For an environmental citizenship approach to be applied to tourism development and for it to take full effect, there needs to be more applicability beyond pursuing the role of tourists as environmental citizens in which tourism and hospitality employees can be empowered to foster and support acts of environmental responsibility (see Luu 2018). This has been demonstrated in this chapter in terms of addressing the implementation of sustainable business practices in Palau, through the induced role of transport employees and residents in brokering information and advice to tourists concerning the adoption of sustainable practices and protocols whilst on vacation. Accordingly, it would be useful to empirically track and monitor such initiatives, especially in terms of the effectiveness of the desired outcomes.

References

Abbott, L. (2016) *Palau: History and Culture*, Abidjan: Sonit Education Academy.

ADB (Asian Development Bank) (2017) *Private Sector Assessment for Palau: Policies for Sustainable Growth Revisited*, Mandaluyong City: ADB. Available online at https://www.adb.org/sites/default/files/institutional-document/230131/palau-psa-2017.pdf (accessed 7 June 2020).

Bell, L.J., G. Ucharm, S. Patris, M.S. Diaz-Cruz, M.P.S. Roig and M.N. Dawson (2017) *Final Report: Sunscreen Pollution Analysis in Jellyfish Lake*, Koror, Palau: Coral Reef Research Foundation. Available online at https://coralreefpalau.org/wp-content/uploads/2017/10/CRRF-UNESCO-Sunscreen-in-Jellyfish-Lake-no.2732.pdf (accessed 23 June 2020).

Braine, T. (2020) 'First ban on coral-bleaching sunscreen chemicals goes into effect in Palau, a world first', *New York Daily News*, 1 January. Available online at https://www.nydailynews.com/news/world/ny-palau-sunscreen-ban-coral-bleaching-unesco-20200102-x466faoebbgcndbbga2rxj2ltm-story.html (accessed 8 June 2020).

Bureau of Tourism (2016) *Palau Responsible Tourism Policy Framework: Ensuring a Pristine Paradise: Palau for Everyone 2017–2021*, Republic of Palau: Ministry of Natural Resources, Environment and Tourism, December. Available online at https://www.palaugov.pw/wp-content/uploads/2017/04/Final_Palau-Responsible-Tourism-Framework1.pdf (accessed 6 March 2022).

Carlile, L. (2000) 'Niche or mass market: The regional context of tourism in Palau', *The Contemporary Pacific*, 12(2): 415–436.

Carreon, B. (2017) 'Emerging from the drought', *The Pacific Island Times*, 1 June. Available online at https://www.pacificislandtimes.com/single-post/2017/06/01/Emerging-from-drought (accessed 22 March 2022).

Carreon, B.H. (2018) 'New Palau pushes for "high end" tourism with tax breaks', *Pacific Note*, 30 March. Available online at https://www.pacificnote.com/single-post/2018/04/03/New-Palau-pushes-for-high-end-tour ism-with-tax-breaks (accessed 22 March 2022).

Cheng, T. and H. Wu (2015) 'How do environmental knowledge, environmental sensitivity, and place attachment affect environmentally responsible behavior? An integrated approach for sustainable island tourism', *Journal of Sustainable Tourism*, 23(4): 557–576.

Countryeconomy.com (2020) 'Palau GDP – gross domestic product'. Available online at https://countryeconomy.com/gdp/palau (accessed 5 March 2022).

de-Miguel-Molina, B., M. de-Miguel-Molina and M.E. Rumiche-Sosa (2014) 'Luxury sustainable tourism in small island developing states surrounded by coral reefs', *Ocean and Coastal Management*, 98: 86–94.

DFAT (Department of Foreign Affairs and Trade) (n.d.) 'Republic of Palau country brief', Available online at https://www.dfat.gov.au/geo/palau/republic-of-palau-country-brief (accessed 5 March 2022).

Dobson, A. (2007) 'Environmental citizenship: Towards sustainable development', *Sustainable Development*, 15(5): 276–285.

Fennel, D.A. and K. Ebert (2004) 'Tourism and the precautionary principle', *Journal of Sustainable Tourism*, 12(6): 461–479.

Golbuu, Y., M. Gouezo, G. Mereb, V. Nestor and D. Olsudong (2018) 'Republic of Palau'. In: C. Moritz, J. Vii, W. Lee Long, J. Tamelander, A. Thomassin and S. Planes (eds.) *Status and Trends of Coral Reefs of the Pacific*, Global Coral Reef Monitoring Network, pp. 189–193.

Government of Palau (2009) *Actions for Palau's Future the Medium-Term Development Strategy 2009 to 2014*, Government of Palau with the Assistance of the Facility for Economic and Infrastructure Management and the Support of the Asian Development Bank. Available online at http://extwprlegs1.fao.org/docs/pdf/pau179609.pdf (accessed 21 March 2022).

Government of Palau (2020) 'Visitor arrivals'. Available online at https://www.palaugov.pw/visitor-arrivals/ (accessed 1 May 2021).

Graduate School USA (2020) '*Assessing the Impact of COVID-19 on the Palauan economy*', *EconMAP Technical Note*, 31 March. Available online at https://pitiviti.org/storage/dm/2021/05/palau-econimpact-covid-19-mar2020-web-20210525190 012351.Pdf (accessed 23 March 2022).

Graham, T. and N. Idechong (1998) 'Reconciling customary and constitutional law', *Ocean & Coastal Management*, 40(2–3): 143–164.

Greenhotelier.com (2019) 'Annual water consumption in hotels', 5 March. Available online at https://www.greenhotelier.org/know-how-guides/water-management-and-responsibility-in-hotels/attachment/figure-1-annual-water-consumption-in-hotels/ (accessed 21 July 2022).

Heilprin, J. (2009) 'Palau creates world's first shark sanctuary', *Phys.org*, 25 September. Available online at https://phys.org/news/2009-09-palau-world-shark-sanctuary.html (accessed 14 June 2020).

International Institute for Sustainable Development (2020) '2019 state of the environment report reveals positive trends in Palau's shallow coral reefs', 30 January. Available online at https://sdg.iisd.org/commentary/policy-briefs/2019-state-of-the-environment-report-reveals-positive-trends-in-palaus-shallow-coral-reefs/ (accessed 22 March 2022).

IMF (International Monetary Report) (2016) 'Republic of Palau, 2016 Article IV Consultation – Press release; staff report; and statement by the executive director for the Republic of Palau', *IMF Country Report no. 16/328*, October, Washington, D.C. Available online at https://www.imf.org/en/Publications

/CR/Issues/2016/12/31/Republic-of-Palau-2016-Article-IV-Consultation-Press-Release-Staff-Report-and-Statement-by-44344 (accessed 20 July 2020).

IMF (International Monetary Fund) (2019) 'Republic of Palau, 2018 Article IV Consultation – Press release; staff report; and statement by the executive director for the Republic of Palau', *IMF Country Report No. 19/4*, February, Washington, D.C. Available online at file:///C:/Users/courtsq/Downloads/cr1943(6).pdf (accessed 23 March 2022).

IMF (International Monetary Fund) (2021) 'Republic of Palau, 2021 Article IV Consultation – Press release; staff report; and statement by the executive director for the Republic of Palau', *IMF Country Report* No. 21/263, December, Washington, D.C. Available online at https://www.imf.org/en/Publications/CR/Issues/2021/12/09/Republic-of-Palau-2021-Article-IV-Consultation-Press-Release-Staff-Report-and-Statement-by-510871 (accessed 20 March 2022).

Ioannides, D. and B. Holcomb (2003) 'Misguided policy initiatives in small-island destinations: Why do up-market tourism policies fail?' *Tourism Geographies*, 5(1): 39–48.

Island Times (2018) 'Remengesau inks law on responsible tourism education', 31 October. Accessed online https://islandtimes.org/remengesau-inks-law-on-responsible-tourism-education/ (accessed 27 May 2022).

Lee, P.Y., L. Qi and P. Li (2020) 'Host community attitude toward trade-off between tourism development and environmental conservation: A case study of Palau', *Business and Management Research*, 9(1): 21–34.

Luu, T.T. (2018) 'Activating tourists' citizenship behavior for the environment: The roles of CSR and front-line employees' citizenship behavior for the environment', *Journal of Sustainable Tourism*, 26(7): 1178–1203.

Lyons, K. (2018) '"Palau against China!": The tiny island standing up to a giant', *Guardian*, 8 September. Available online at https://www.theguardian.com/global-development/2018/sep/08/palau-against-china-the-tiny-island-defying-the-worlds-biggest-country (accessed 20 July 2020).

Marino, S., A. Bauman, J. Miles, A. Kitalong, A. Bukurou, C. Mersai, E. Verheij, L. Olkerill, K. Basilius, P. Colin, S. Patris, S. Victor, W. Andrew, J. Miles and Y. Golbuu (2008) 'The state of coral reef ecosystems of Palau'. In: J. E. Waddell and A.M. Clarke (eds.) *The State of Coral Reef Ecosystems of the United States and Pacific Freely Associated States*, Silver Spring: NOAA Technical Memorandum, NOS NCCOS, pp. 511–539.

Mason, D., A. Iida, S. Watanabe, L.P. Jackson and M. Yokohari (2020) 'How urbanization enhanced exposure to climate risks in the Pacific: A case study in the Republic of Palau', *Environmental Research Letters*, 15(11). https://doi.org/10.1088/1748-9326/abb9dc.

Miao, L. and W. Wei (2013) '"Consumers" pro-environmental behavior and the underlying motivations: A comparison between household and hotel settings', *International Journal of Hospitality Management*, 32: 102–112.

Nath, D., M. Mudaliar and D. Dengokl (2005) *Republic of Palau: Water Supply System Description Koror/Airai*, Republic of Palau: Water Safety Plan Programme - A project funded by AusAID. Available online at http://www.pacificwater.org/userfiles/file/Palau%20 Water%20System%20Description.pdf (accessed 15 June 2020).

Office of the President (2018) 'Signing Statement SB No. 10-135, SDl, HDl (The Responsible Tourism Education Act of 2018)', 25 October, Republic of Palau. Available online at https://www.divenews-wire.com/wp-content/uploads/2018/11/Trans.-Ltrs.-to-OEK-with-RPPL-No.-10-30-re.-The-Responsible-Tourism-Education-Act-of-2018-Final.pdf (accessed 13 June 2022).

Palau Visitors Authority (2019) 'Pristine Paradise Palau'. Available online at https://www.pristineparadisepalau.com (accessed 23 March 2022).

Peters, A. (2018) 'The Pacific island nation of Palau just became the first country to ban reef-killing sunscreen', *Fastcompany*, 11 February. Available online at https://www.fastcompany.com/90261652/the-pacific-island-nation-of-palau-just-became-the-first-country-to-ban-reef-killing-sunscreen (accessed 23 March 2022).

Raouf, M.A. (2011) 'Environment, citizenship and sustainability', *Gulf News*, 8 April. Available online at https://gulfnews.com/opinion/op-eds/environment-citizenship-and-sustainability-1.788846 (accessed 20 August 2020).

Republic of Palau (2010) *Mauritius +5 Status Report: Republic of Palau*, Produced by the Office of the Vice President in cooperation with ESCAP, January, Palau: Republic of Palau. Available online at https://sustainabledevelopment.un.org/content/documents/1280Palau-MSI-NAR 2010.pdf (accessed 1 July 2020).

Romagosa, M. (2020) 'The COVID-19 crisis: Opportunities for sustainable and proximity tourism', *Tourism Geographies*, 22(3): 690–694.

Salinas, J. (2019) 'Jellyfish lake reborn', *Pacific Island Times*, 9 August. Available online at https://www.pacificislandtimes.com/single-post/2019/08/09/Jellyfish-Lake (accessed 5 July 2020).

Shah, M.M. (2011) 'Simplified method of calculating evaporation for swimming pools', *HPAC Engineering*, 2 October. Available online at https://www.hpac.com/humidity-control/article/20927124/simplified-method-of-calculating-evaporation-from-swimming-pools (accessed 26 July 2019).

Shuster, D.R. and S. Foster (2021) 'Palau', *Encyclopedia Britannica*, 7 May. Available online at https://www.britannica.com/place/Palau (accessed 5 March 2022).

Smith, A. (2015) 'World heritage and outstanding universal value in the Pacific islands', *International Journal of Heritage Studies*, 21(2): 177–190.

SPREP (Secretariat of the Pacific Regional Environment Programme) (2018) *Practical Guide to Solid Waste Management in Pacific Island Countries and Territories*, Apia, Samoa: SPREP, March. Available online at https://www.sprep.org/attachments/j-prism-2/SWM_GUIDEBOOK_.pdf (accessed 25 July 2020).

Treloar, P. and C.M. Hall (2005) 'Palau'. In: C. Cooper and C. M. Hall (eds.) *Oceana: A Tourism Handbook*, Clevedon: Channel View Publication, pp. 239–243.

Ueki, M.F. (2000) 'Eco-consciousness and development in Palau', *The Contemporary Pacific*, 12(2): 481–487.

Vianna, G.M.S., M.G. Meekan, D. Pannell, S. Marsh and J.J. Meeuwig (2010) *Wanted Dead or Alive? The Relative Value of Reef Sharks as a Fishery and an Ecotourism Asset in Palau*, Perth: Australian Institute of Marine Science and University of Western Australia. Available online at https://www.pewtrusts.org/~/media/assets/2011/05/02/palau_shark_tourism.pdf (accessed 15 August 2021).

von Saltza, E. (2019) *Green Passport: Innovative Financing Solutions for Conservation in Hawai'i': Improving the Visitor Experience and Protecting Hawai'i's Natural Heritage, A Report Prepared for Conservation International*, October. Available online at https://www.conservation.org/docs/default-source/publication-pdfs/final-full-report_green-passport_10-2-2019.pdf?sfvrsn=77fe4ae4_2 (accessed 4 April 2022).

Wabnitz, C.C., A.M. Cisneros-Montemayor, Q. Hanich and Y. Ota (2018) 'Ecotourism, climate change and reef fish consumption in Palau: Benefits, trade-offs and adaptation strategies', *Marine Policy*, 88: 323–332.

World Economic Forum (2019) 'This Pacific island has banned fishing to allow the marine ecosystem to recover'. Available online at https://www.weforum.org/agenda/2019/12/palau-pacific-marine-protected-area-fishing-environment/ (accessed 15 August 2020).

Worldometer (2022) 'Palau population'. Available online at https://www.worldometers.info/world-population/palau-population/ (accessed 5 March 2022).

Xu, V.X. (2018) 'Palau bans many kinds of sunscreen, citing threat to coral', *The New York Times*, 2 November. Available online at https://www.nytimes.com/2018/11/02/world/asia/palau-sunscreen-ban-coral.html (accessed 8 June 2020).

Yamashita, S. (2000) 'The Japanese encounter with the south: Japanese tourists in Palau', *Contemporary Pacific*, 12(2): 437–463.

19

UTILIZING SUSTAINABLE DEVELOPMENT GOALS TO GUIDE THEMATIC TOURISM DEVELOPMENT IN KIRIBATI

Evanthie Michalena, Jeremy M. Hills, and Jale Samuwai

Introduction

Kiribati, or the Republic of Kiribati, is a nation comprising 33 islands, including 20 inhabited islands, with a land area of 811 square kilometres. The country extends 2,900 kilometres from the Gilbert Islands in the west (Figure 19.1), where the majority of the population is concentrated, to the Line Islands in the east (Macdonald and Foster 2020). Kiribati was formed in 1979, when independence from Britain was achieved for the Gilbert Islands. The country's population for 2022 is estimated at around 122,849, and 57% of the population in 2020 lived in an urban environment (Worldometer 2022). More than half of the population live on the atoll of Tarawa in the Gilbert Islands, representing a total land area of 26 square kilometres that is separated into two districts: South Tarawa Urban District and North Tarawa District (Duvat 2013). South Tarawa is the capital of Kiribati, with government centres at Ambo, Bairiki and Betio (Macdonald and Foster 2020).

As a less developed country, Kiribati faces many challenges that are unique to small island nations, including remoteness from markets, limited natural resources and a widely dispersed population as well as being vulnerable to climate change and economic shocks. With many islands situated at 2 or fewer metres above sea level, Kiribati has witnessed first-hand the impacts of global climate change. According to South Pacific Regional Environment Programme, two small, uninhabited Kiribati islets, Tebua Tarawa and Abanuea, disappeared underwater in 1999 (Kirby 1999). An economic evaluation of the costs of climate change impacts was estimated to represent 35% of Kiribati's gross domestic product (GDP) (Republic of Kiribati 2017). Extreme environmental issues have affected communities in Kiribati and generated social change (Lundsgaarde 1966; Watters and Banibati 1984). The government authorities have been considering a twofold approach to deal with this predicament: first, the fortification of at least one Kiribati island so that the country's physical presence does not entirely disappear (Worland 2015); and second, the relocation of local people to other Pacific Island states such as Fiji (Laurence 2014). The Kiribati Adaptation Program has been established to strengthen local adaptation capacity. This is a US$5.5 million initiative to introduce measures to reduce Kiribati's vulnerability to the effects of climate change and sea level rise, particularly by raising awareness

DOI: 10.4324/9780429019968-22

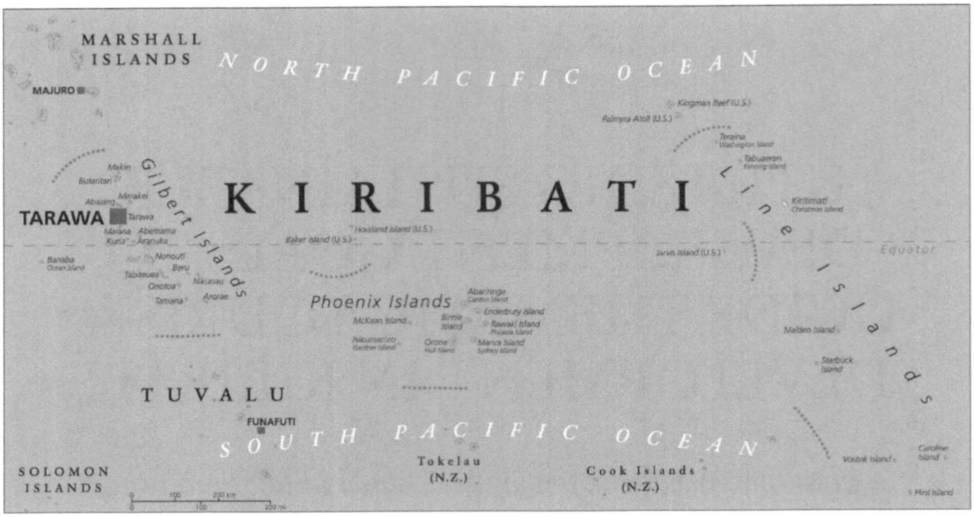

Figure 19.1 Map of Kiribati, indicating the capital city of Tarawa – an atoll. Source: Peter Hermes Furian (www.shutterstock.com/image-vector/kiribati-political-map-capital-tarawa-republic-457806091).

of climate change, assessing and protecting available water resources and managing inundation (Astriviany 2020). Kiribati is a recipient of large amounts of overseas development assistance targeting climate change, for which Australia and the United Arab Emirates (UAE) are among the main donors. The country has received at least US$31 million of climate finance since 2014 (Atteridge and Canales 2017). However, adaptation measures cannot suppress all climate-related impacts (Warner and Van der Geest 2013), especially in the case of small islands which are disproportionately vulnerable to climate-induced disasters.

The aim of this chapter is to assess how the United Nations (UN) 2030 Agenda for Sustainable Development can provide an opportunity to reframe tourism development in Kiribati and bring about social, economic and environmental benefits. Inherent in the UN 2030 Agenda is the interlinkage or 'indivisibility' between all the areas of development, and this chapter explores the connotation for tourism development from this perspective. Sector-specific tourism initiatives in Kiribati have had minimal effect to date. Therefore, this work provides a broader framework to reposition tourism development by looking at how the country can develop thematic tourism. The concern is that Kiribati has not experienced significant levels of tourism development. Of the data that is openly available, the number of arrivals in Kiribati for 2017 was only around 5,800, which can be compared to 2004 where the country recorded only 3,400 arrivals (Index Mundi n.d.). Although tourism development in Kiribati is not a major industry, the country recently has been impacted by 'undertourism' (see Smith 2020) during the COVID-19 pandemic. Therefore, in the context of such low visitation, the chapter indicates the importance of developing thematic tourism. Thematic tourism development concerns a destination which is looking at developing new, specialized and novel products to elevate the visitor experience (see Đurašević 2014) and can also be likened to a type of special interest tourism (Douglas et al. 2001). Accordingly, as the paper implicates, energy tourism and forms of marine-based tourism are useful illustrations of thematic tourism and closely interconnected to the sustainable tourism theme, which could be seriously pursued in a post-pandemic context.

The following section highlights the challenges and concerns relating to tourism planning and tourism and development strategy. The discussion introduces Kiribati's tourist status with an emphasis on the policy nexus around the Kiribati National Tourism Action Plan and related Kiribati development plans, leading to the contextualization and application of the Kiribati 20-Year Vision (KV20) 2016–2030. Other challenges are duly highlighted, notably the COVID-19 pandemic and its impacts, the lack of diversity of tourism products and an undeveloped approach to destination marketing. The work then details the UN 2030 Agenda for Sustainable Development and the 17 Sustainable Development Goals (SDGs) (see Chapter 2), with a specific emphasis on the interlinkages between SDGs and tourism. The discussion then moves to an evaluation of the development and advancement of thematic tourism in Kiribati, with special attention to the role of energy tourism and marine-based tourism, including their connections to the SDG framework. The work then discusses thematic tourism as a catalyst for strengthening destination reputation, which is considered necessary in the context of Kiribati having a low tourism profile. The final section identifies prerequisites for achieving SDGs through tourism, especially in the context of constructing a post-pandemic approach to tourism development. The conclusion suggest that the SDGs can provide a useful lens through which to view the broader development role of tourism and identifies how further research on this topic may be productive in comprehending the unique tourism and development needs of places such as Kiribati.

Tourism planning and strategic vision: challenges and concerns

From a governance perspective, the Kiribati government has produced various policies that intersect with tourism development. The earlier Kiribati Development Plan (KDP) 2008–11 was a planning tool incorporating tourism as a 'key economic priority'. Relevant goals concerned the development of the Line and Phoenix group of islands as tourism growth centres, stimulation and enhancement of the private sector and public enterprises, improvement of infrastructure, enhancement of the quality of local products and their promotion to tourists, development and promotion of ecotourism opportunities and the strengthening of wildlife conservation (Republic of Kiribati 2009). As a more specialized document, the Kiribati National Tourism Action Plan 2009–2014 was a five-year plan designed to provide the necessary strategies to stimulate a sustainable tourism industry. Its vision was '[f]or tourism to become the largest and most sustainable economic sector driving employment, growth and the Kiribati economy' (Republic of Kiribati 2009: 10), in which eight key thematic areas of interventions were identified: (1) infrastructure investments; (2) enterprise investments; (3) communications; (4) aviation access and route development; (5) establishment of a tourism culture; (6) outer island development; (7) regulation, taxation and legislation; and (8) education and training (2009: 19–20). Although some positive developments materialized, notably the recovery of the tourism industry in Kiritimati (also known as Christmas Island) in the northern Line Islands and a small but successful tourist fishing venture in Nonouti in the southern Gilbert Islands, there was no significant progress overall because of the inadequate development and limited marketing of the Kiribati tourism industry to regional and international markets (Government of Kiribati 2012). This government report also identified key economic challenges, including limited support for private sector development, inadequate development of local industries for value-added products, limited capacity in maximizing the benefits of marine resources for sustainable development and the limited flow of funds into the country (2012: 31).

The KDP 2016–2019 represents the tenth development plan for Kiribati, focusing on six priority areas: (1) human resource development, (2) economic growth and poverty reduction, (3) health, (4) environment, (5) governance and (6) infrastructure (Government of Kiribati 2016:

i). However, tourism is not significantly evoked within this policy document and is incorporated only in the priority areas of 'environment' and 'infrastructure', rather than interconnected to such crucial areas as 'economic growth' and 'poverty reduction'. Nonetheless, this plan directs attention to gender through integrating issues and concerns relating to gender equity and the empowerment of women, whereas previous national development plans were not consistently attentive to gender issues.

Tourism is considered a key component in achieving the KV20, founded on four critical pillars: (1) wealth, (2) peace and security, (3) infrastructure and (4) governance, together with an overall emphasis on socio-economic development (Government of Kiribati 2018: 6). Accordingly, the development of the country's tourism industry is a significant focus of KV20, emphasizing that the advancement of sustainable forms of tourism could be beneficial for employment, productive for investment in the improvement of services and useful to reduce increased urbanization, with a realization that there is a need to invest in tertiary skills training and transport infrastructure (Webb 2020). KV20 envisaged the tourism sector to focus on the development of natural, human and cultural capital as well as to improve economic growth and reduce poverty, thereby aiming to boost revenue in the tourism sector and increase its GDP contribution from 3.6% in 2016 to 50% by 2036 (Government of Kiribati 2018). This strategy further envisages the importance of achieving such high-value tourism products as cultural tourism and ecotourism, improving human capacity in delivering high-end tourist products, strengthening institutional and regulatory frameworks, improving marketing strategies and upgrading infrastructure connectivity and facilities. Webb (2020) confirms that the challenge Kiribati faces concerning low levels of tourism visitation is due to such problems as inconvenient transport routes to Kiribati, environmental degradation (e.g., the unhealthy state of Tarawa Lagoon) and sub-standard tourist experiences. To make matters worse, the movement control orders and international tourism immobility during the pandemic intensified the prevailing problem.

Kiribati has been impacted by the COVID-19 pandemic, especially in terms of construction, remittances and export channels. Due to restrictions on the movement of labour and capital equipment, various infrastructure projects have been delayed and, thus, have affected the construction and hospitality sectors (Homasi and Wainiqolo 2020). Those government funded projects which were delayed included work on outer island roads and airstrips (DFAT 2020). Homasi and Wainiqolo (2020) further note that Kiribati experienced 1,040 estimated job losses, including 140 tourism workers, of which 69% of the total losses came from the domestic market and 31% were offshore. On 14 January 2022, Kiribati opened its borders to encourage tourism and travel. This trial run was to enable citizens who had been stranded oversees to return home and for technical assistance workers to enter the country, though more than two-thirds of the first passenger plane that visited Kiribati tested positive for COVID-19, leading to the spread of the virus to the mainland population and provoking the government to declare a 'state of disaster' on the 22 January 2022 (Gunia 2022). This is a challenge for Kiribati, placing significant pressure on an already strained health system, exacerbated by inadequate sanitation in such densely populated areas as South Tarawa (World Bank 2022).

Despite the Kiribati government aiming to refocus the country's development programmes in the pre-COVID-19 era, the tourism sector has been restricted to narrow tourism products and markets that are mainly centred in South Tarawa and Kiritimati. The dominant category of visitor travel has tended to be business-related, while pleasure travel was largely composed of visitors who were focused on fishing (Government of Kiribati 2018). Although the impacts and potentiality of tourism activity in Kiribati were studied three decades ago (Milne 1992), updated scientific data remains extremely limited and not representative. This creates issues of transparency and limited capacity for validation and statistic elaboration. Also, air transportation still represents a major constraint for the tourism industry, as Kiribati is not located on major air routes, and international flights take place only on certain days of each week. This problem is further exacerbated as Kiribati has an underdeveloped infrastructure, which is critical to supporting the industry (Save Kiribati 2019). In addition, Kiribati's approach to destination marketing is underdeveloped (Government of Kiribati 2018), as regional and international markets have been significantly lagging behind other more popular Pacific Island states, notably Cook Islands, Fiji, Palau, Samoa and Vanuatu, thereby hindering the visibility of the country as an ideal tourist destination. Nonetheless, as tourism encourages extensive interactions between human and natural systems (Jones et al. 2017; Liu et al. 2015), the tourism sector contains many opportunities to enhance sustainability regarding job creation, economic growth and environmental protection (Scheyvens 2018). For 'lesser developed countries' (LDCs), it is already appreciated that tourism can provide other such benefits as image enhancement, poverty reduction and the consolidation of value chains (UNWTO 2017: 15–16).

Conceptualizing tourism and SDGs

The UN 2030 Agenda for Sustainable Development and the accompanying SDGs established the trajectory of world development that is endorsed by almost all countries in the world, including Kiribati. SDGs aim to promote peaceful and inclusive societies and justice through building accountable institutions in diverse societies. Tourism can potentially contribute directly and indirectly to advancing the 17 SDGs (see Scheyvens 2018). Under the umbrella of the SDGs, tourism can aim to generate quality jobs for sustainable growth without undue pressure on biodiversity. In the private sector, for example, Jones et al. (2017) argue that the tourism and hospitality industry is in a powerful position to contribute to the successful achievement of the SDGs. This framing of the importance of tourism for progressing the SDGs is mirrored in the work of the United Nations World Tourism Organization (UNWTO), which has tracked possible tourism-based contributions across all 17 SDGs (see Table 19.1). Many benefits across the spectrum of the SDGs are attainable through more thematic forms tourism as opposed to mass tourism.

The desire for economic growth with simultaneous environmental, cultural and social protection has been an integral part of Kiribati's political agenda (Throsby 2011). However, this goal has not been achieved due to such reasons as inefficiency of the public sector, limited export base and the lack of a dynamic private sector. By viewing the SDGs through the lens of the vision of the KDP 2016–2019, which emphasizes the importance of moving 'towards a better educated, healthier, more prosperous nation with a higher quality of life' (Government of Kiribati 2016: 4), it is possible to identify links to societal and economic benefits as well as the subsequent SDG components. More specifically, the KDP 2016–2019 vision directly tracks SDGs 3 (health and well-being), 4 (education), 8 (economic growth) and 10 (economic diversification and prosperity), in which such economic objectives can help to 'reduce inequalities' (SGD 10) and strengthen public and private 'partnerships' (SDG 17). The prior tracking of the benefits of tourism to the SDGs (see Table 19.1) can be used to identify the specific dimensions that could be included in tourism development and planning, especially to achieve SDG co-benefits. In this way, tourism in Kiribati can be used as a tool for sustainable development and achieving beneficial SDG outcomes on a national scale.

The development and advancement of thematic tourism in Kiribati

Energy tourism and its intersection with the SDGs

There are many economic, social, cultural and product-related factors that are likely to influence holiday-taking (Stephenson and Hughes 1995). One example of thematic tourism is energy

Table 19.1 Possible contributions of tourism to the implementation of SDGs

SDG	Possible contribution of tourism
SDG 1 – End poverty in all its forms everywhere	Tourism provides income through job creation at local and community levels, and potential involvement in national poverty reduction strategies and entrepreneurship.
SDG 2 – End hunger, achieve food security and nutrition, promote sustainable agriculture	Tourism can spur sustainable agricultural practices by promoting sustainable production of goods and their supply to hotels and tourists.
SDG 3 – Ensure healthy lives and promote well-being for all age groups	Tax income generated from tourism can be reinvested in health care and services, improving maternal health, reducing child mortality and preventing diseases.
SDG 4 – Ensuring inclusive and equitable quality education and promoting lifelong learning for all	Tourism has the potential to promote inclusiveness, producing a skilful workforce and opportunities for direct and indirect jobs for youth, women, and those with special needs, who should benefit through educational means.
SDG 5 – Achieve gender equality and empower all women and girls	Tourism can empower women, particularly through the provision of direct jobs and income-generation from micro and medium enterprises in the tourism- and hospitality-related sectors.
SDG 6 - Ensure availability and sustainable management of water and sanitation for all	Tourism investment requirement for providing utilities can play a critical role in achieving water access and security as well as hygiene and sanitation for all.
SDG 7 – Ensure access to affordable, reliable, sustainable and modern energy for all	As an energy intensive sector, tourism can accelerate the shift towards increased renewable energy shares in the global energy mix and by promoting investments in clean energy sources – in which tourism can help to mitigate climate change and contribute to energy access for all.
SDG 8 – Promote sustained, inclusive and sustainable economic growth, employment and decent work for all	Decent work opportunities in tourism, particularly for women and the youth, and policies that favour better diversification of the economy through tourism value chains, which can encourage the tourism industry to produce positive socio-economic impacts.
SDG 9 – Build resilient infrastructure, promote inclusive and sustainable industrialization and foster innovation	Tourism development can influence public policy for upgrades and retrofitting infrastructure, making it more sustainable, innovative and resource-efficient and moving towards low-carbon growth.
SDG 10 – Reduce inequality within and among countries	Tourism can be a powerful tool for reducing inequalities if it engages local populations and all key stakeholders in its development as well as acting as an effective means for economic integration and diversification.
SDG 11 – Make cities and human settlements inclusive, safe, resilient and sustainable	Tourism can advance urban infrastructure and accessibility, promote regeneration and preserve cultural and natural heritage, as investment in green infrastructure (e.g., through more efficient transport and reduced air pollution) should result in smarter and greener cities for tourists and residents.
SDG 12 – Ensure sustainable consumption and production patterns	The tourism sector needs to adopt sustainable consumption and production modes, accelerating the shift towards sustainability, which includes monitoring sustainable development impacts of tourism through such means as energy efficiency, water conservation, waste management, biodiversity protection and job creation, resulting in enhanced economic, social and environmental outcomes.
SDG 13 – Take urgent action to combat climate change and its impacts	Tourism contributes to and is affected by climate change, and tourism stakeholders should play leading roles in the global response to climate change through reducing its carbon footprint in the tourism-related sectors.

(*Continued*)

Table 19.1 (Continued)

SDG 14 – Conserve and sustainably use the oceans, seas and marine resources for sustainable development
Coastal and maritime tourism rely on healthy marine ecosystems, for which tourism development ought to help conserve and preserve fragile marine ecosystems and serve as a vehicle to promote a blue economy by contributing to the sustainable use of marine resources.
SDG 15 – Protect, restore and promote sustainable use of terrestrial ecosystems and halt biodiversity loss
Tourism can play a major role if sustainably managed in fragile zones not only in conserving and preserving biodiversity and natural heritage but also in generating revenue for local communities.
SDG 16 – Promote peaceful and inclusive societies, provide access to justice for all and build inclusive institutions
Given the cultural diversity of tourists, the sector can foster multicultural and interfaith tolerance and understanding, laying the foundation for more peaceful societies and strengthening harmony in post-conflict societies.
SDG 17 – Strengthen the means of implementation and revitalize the global partnership for sustainable development
Tourism's cross-sectoral nature can strengthen private and public partnerships and engage multiple stakeholders at local, national, regional and international levels to mutually achieve the SDGs, as public policy and innovative financing are crucial for achieving the 2030 Agenda.

Source: Adapted from UNWTO and UNDP (2017: 14–15).

tourism, which is a recent concept concerning visitors who are attracted by site innovations and energy plants as well as concepts that include monitoring of the environmental footprint (Frantál and Urbánková 2017). These authors also infer that the energy–tourism nexus can be conceptualized in terms of energy to produce sustainable consumption and supply clean electricity as well as energy supplied for promoting economic activities. The potential benefits from this relationship can be spread across those SDGs indicated above. In this case, visitor interest can be attracted by two significant destination 'pull factors':

(i) Industrial energy heritage sites (which are generated energy sites of the past) (responding to SDGs 4, 8, 9, 11 and 16);
(ii) New energy landscapes that can attract visitors who are interested in the reduction of the environmental footprint when innovative energy practices are applied (responding to SDGs 3, 4, 8, 9, 11, 12 and 14 or 15, depending on where the energy plant is located – water or land).

This growing energy–tourism nexus is partly predicated on the 'clean' renewable nature of the supply, which is contrary to the past when energy generating landscapes were often perceived as polluting sites that visitors would have little interest in visiting. However, energy tourism can be much more than the branding of a site but can also change human behaviour and improve energy literacy. Energy tourism can, thus, simultaneously satisfy the SDGs for 'environmental education' (SDG 4), 'clean energy' (SDG 7), 'economic growth' (SDG 8) and 'sustainable cities' (SDG 11). A visit to an energy plant can, thus, represent a form of 'energy citizenship', which is conceptualized in terms of individuals interconnecting with low-carbon energy practices and sustainable energy transitions (Ryghaug et al. 2018), and this can be mobilized through education and the increase of local energy literacy (SDG 4 and 7). Towards this goal, sustainable energy development programmes have been combined with ecological centres, thus, interconnecting education with the potential for scientific and innovative forms of tourism (Frantál and Urbánková 2017). In the case of an Ocean Thermal Energy Conversion (OTEC) power plant, for instance, international institutes of botany can visit aquaculture and hydroponics that are supplied with clean and nutrient-rich wastewater from OTEC generation. The planned development of the land-based OTEC plant in South Tarawa, involving a partnership between the government of Kiribati and the Korean Research Institute of Ships and Ocean Engineering, would signify significant achievement, as the aim of this project is to be a 'role model' for conceptualizing 'interconnected geoscience' (Petterson and Kim 2020: 3). Therefore, this initiative can be framed under the tourism typology of thematic tourism, with the potential to develop 'energy tourism' as a sub-form that also interconnects with 'research-related tourism' (see Shah et al. 2022).

In 2015, the government of Kiribati designed a comprehensive 10-year energy road map, with intentions to reduce imported energy supply and electricity expenditures and meet future electricity demands that should simultaneously satisfy 'nationally determined contributions' (NDC) which demand the reduction of fossil fuels and the increase of energy efficiency (Michalena et al. 2018). In the road map, other energy uses also offer opportunities for improvement, such as continued solar photovoltaic (PV) power deployment, renewable cooling and transport solutions and solar home systems and solar desalination, as well as a combination of wind power with PV and battery storage. Crucially, biofuels and electric vehicles could make land-based transport more sustainable. The clean energy nexus can become a demonstration project, which is diverse and includes world-leading technology, such as OTEC. The portfolio of renewable energy approaches can attract specific groups of visitors to the island while satisfying many SDGs, including 7, 8, 9 and 12, among others (see Table 19.1).

Marine-based tourism and its intersection with the SDGs

One clear example of the potential contribution to thematic tourism to the SDGs is the Phoenix Islands Protected Area (PIPA), a 408,250 square kilometre ocean expanse of marine and terrestrial habitats (Carreon 2021). PIPA, established in 2006 and made a United Nations Educational, Scientific and Cultural Organization (UNESCO) World Heritage Site in 2010, is the largest and deepest marine protection area in the world. PIPA has been closed to commercial fishing since January 2015, which can attract various unique high-end tourism products, such as snorkelling and diving and catch and game fishing (Teroroko 2016). Therefore, Kiribati could also aspire to advance a niche diving market and transform its tourism industry to cater exclusively to this market based on the vast marine eco-system, thereby satisfying SDGs 3 (good health), 8 (employment), 13 (climate change) and 14 (life under water). This niche market would exemplify the potentiality of thematic tourism in Kiribati to be multifaceted. Although marine activities can generate income for Kiribati, they would have to be sensitive to the environment, as the region has been recognized as an 'underwater Eden' (Stone and Obura 2012). The PIPA Conservation Trust provides a steady flow of financial support needed to manage and sustain the protected area's operations. Financial support for the trust includes grants from the Kiribati government, New England Aquarium and Conservation International, Waitt Foundation and Ocean 5 (Teroroko 2016). Increasing financial support for the trust should help to encourage low-volume, high-end and ecologically sound tourism development. The whole experience potentially satisfies SDGs for 'quality education' (SDG 4), 'sustainable production' (SDG 12) and 'sustainable life under water' (SDG 14). The lifting of the closure of the Phoenix Islands Protected Area as a no-take zone in late 2021 and the introduction of marine spatial planning (MSP), which includes licensing of commercial fishing (UNESCO 2021), creates an opportunity for coherent planning for tourism, fishing, and other viable economic activities but only if sustainable directives are rigorously pursued. The challenge is to ensure that governance of the protected area fully reflects the interests of Kiribati and its citizens, especially as the PIPA 'conservation trust fund' has US representation and also as the vast ocean area is protected with clear surveillance systems in place (Mallin et al. 2019).

Thematic tourism as a catalyst for destination reputation

Kiribati could eventually gain a strong reputation as a destination for tourists to pursue sustainable forms of thematic tourism. Despite their small size, the islands that comprise the Republic of Kiribati can be highly instructive and informative about wider lessons relating to coping with environmental vulnerability and climate impact, especially those struggling with rising sea levels. Kiribati as a destination could thus become a living laboratory illustrating observable multiple climate impacts, attracting a certain category of visitors who are interested in seeing and experiencing development challenges but at the same time are captivated by the way in which tourism could be utilized to produce positive development outcomes for the destination. The same principle applies in the energy, technology and innovation sectors in which Kiribati can transfer lessons learnt on green growth and entrepreneurship because of the plans to develop the OTEC plant (Michalena 2017). Despite the political and geographical complexities of renewable energy governance, lessons learnt from the installation of renewable energy technologies will be valuable (Michalena and Hills 2013). Therefore, these thematic forms of tourism can promote SDG 4 ('quality education'). Moreover, cultural tourism is another thematic option for Kiribati, but accessing and learning about traditional ways of life could prove to be a difficult process because of remoteness and possible resistance to the acceptance of innovation and/or new ways of thinking.

Identifying the prerequisites for achieving SDGs through tourism: a post-pandemic approach

Given that this chapter illustrates the potential of tourism advancement in Kiribati, indicating the prospects of developing such emerging markets as thematic tourism, the COVID-19 pandemic has represented a challenge to Kiribati's strategic vision for tourism and development. Nonetheless, it would be crucial for future strategies to reorientate Kiribati's tourism development objectives to consider those forms of tourism that have been explicated in this chapter, helping to reframe tourism development in the context of post-pandemic directives – though paying greater attention to the needs of the country and the islanders themselves. Moreover, the need to develop strategies and forms of thematic tourism in the wider context of the post-COVID-19 era is crucial, particularly in terms of advancing types of tourism that are regenerative in nature and form, possessing ecological, social and/or educational value as well as helping to boost the local and national economy. As implicated in this chapter, for this advancement to take place, it is necessary for Kiribati to work within the SDG guidelines. Thus, for multiple SDGs to be achieved, it is proposed that nine key guidelines should be developed:

1. As a low-income country Kiribati should maintain its political stability and facilitate the arrival of international tourists from a range of countries through enhancing visa-free policies which encourage unrestricted movement (Chung et al. 2020).
2. Just prior to the pandemic outbreak the country established diplomatic ties with China, which then designated Kiribati as an 'approved destination' for its tour groups (Pacific Island Times 2020). Therefore, it is anticipated that this national market would be targeted in the post-pandemic era – depending of course on Kiribati's ability to pursue thematic forms of tourism in a consistent and strategic manner that would be of interest to specific national segments.

3. Tourism development requires a suitable enabling-context for investment as private sector investments have been a 'mixed bag' across small island developing states, challenged by climate change and economic vulnerability (Briguglio 1995; Dornan and Shah 2016). Once private sector investment gathers momentum through increased confidence in Kiribati's ability to develop such key sectors as the tourism industry, the future role of this industry in stimulating Kiribati's economy could be bright.
4. Investments can be made to ensure tourism management skills are developed in communities, and this can be implemented through existing training infrastructure (such as the University of the South Pacific campus in South Tarawa). Thematic tourism will bring income to households, thus, providing opportunities for strengthening sustainability and resilience of livelihoods and communities (Collins 2015; Elrick-Barr et al. 2016).
5. Tourism development should be integrated into natural resource protection policies, as conservation efforts (e.g., protected areas) may contribute to balancing the benefits and risks of tourism development (Connell 2018).
6. Kiribati's tourism resources, notably the ocean, marine diversity and technology innovations, need to be harnessed through government, non-government and international organizations and strengthened through public and private sector partnerships.
7. Academic research through fieldwork should be intensified, and research outcomes in the form of tourism statistics (e.g., the number of arrivals per type of visitor, age groups of visitors and levels of satisfaction from tourism services) must be rigorously sought, thoroughly assessed and openly available. This is necessary for reasons of transparency and when it comes to dealing with health standards, safety and other issues that concern visitors as well as helping to inform sustainable tourism planning and policy directives. Monitoring is requested to track sustainable development impacts and practices which promote local jobs and ensure sustainable consumption and production patterns – though some of this is already underway through SDG reporting to the UN.
8. Kiribati's attraction potential of its tourism assets can be enhanced with interactive material and dialogue. OTEC, for instance, can be accompanied by awareness-raising concerning climate change and its impacts on the land and the cultural aspects of traditional marine usage.
9. A marketing focus is needed for specific types of tourism that might be interested in specific local attractions. In the cases of OTEC or PIPA, for instance, target groups could be environmentalists, scientists, research tourists, volunteer tourists and energy experts.

Conclusions and research implications

In Kiribati, social adaptation to tourism attitudes is not only shaped or influenced by awareness and motivations of local and governmental stakeholders but also by regional and international donors and their implementing agencies (Throsby 2011). Kiribati, which is supported by considerable development partner funding, has established a national tourism policy framework and submitted ambitious international intentions, such as the 'nationally determined contributions'. The past and present dependence on external donor finance is not sustainable, nor is it reliable in the long-term, and is seemingly unable to create self-generating community prosperity. Kiribati is a country with special resources that can attract the interest of international visitors. On the other hand, it is a country that needs to meet challenges such as low income, high dependence on external donor finance, issues of health and safety, environmental protection, climate change and remoteness. Despite these limitations, this chapter argues that tourism, if promoted through the framework set by the 17 SDGs, can act as a pathway for economic development and environmental protection while boosting private business, community welfare and export opportunities. There are strong opportunities for Kiribati to undertake high-value tourism for small numbers of visitors in thematic areas specifically related to its geography, bio-physical attributes and national actions, such as the marine protected area designation.

Further research is needed in the use of the SDG framework for development in the tourism sector. It has been shown that tourism links with many of the SDGs, but the nature of those linkages is poorly understood. For example, linkages between tourism and the SDGs can vary in forms of association (positive, negative or neutral), magnitude (small or massive effect) and situational context (always linked or linked only in certain situations). For more detailed planning to take place, further research must interrogate these complexities and in light of development challenges and opportunities in the post-pandemic era. For Pacific Islands states, such research is imperative, as their small size and relatively weak administrative structures mean that sectoral fragmentation across development areas can lead to inefficient resource mobilization and diluted development benefits. Research approaches which help to integrate tourism as part of the development agenda, rather than separate it out as an isolated and dislocated agenda, should be prioritized.

The SDG approach also lends itself to the management and control of the private-sector tourism industry, whereby private sector investment is channelled into tourism developments which have tangible and demonstrable benefits across the SDG spectrum. Such channelling can be done though the licensing process and financial levers (taxes and subsidies) as well as other planning tools such as Environment Impact Assessments. Strengthening tourism planning around wider SDG benefits would have the potential to closely align tourism development with national development aspirations. In addition, with limited government finance in many island countries in the Pacific, effectively shaping private-sector investments in tourism can be used as an alternative credit line and as an indication of a movement away from long-term reliance and dependence on external donor aid.

References

Astriviany, M. (2020) 'Kiribati's strategy in facing the problem of sea level rise through the Kiribati adaptation program (KAP)', *Nation State Journal of International Studies*, 3(1): 48–59.

Atteridge, A. and N. Canales (2017) 'Climate finance in the Pacific: An overview of flows to the region's small island developing states', *Working Paper* 2017–04, Stockholm: Stockholm Environment Institute.

Briguglio, L. (1995) 'Small island developing states and their economic vulnerabilities', *World Development*, 23(9): 1615–1632.

Carreon, B. (2021) 'Kiribati to open one of the world's largest marine protected areas to commercial fishing', *The Guardian*, 15 November. Available online at https://www.theguardian.com/world/2021/nov/16/kiribati-to-open-one-of-worlds-largest-marine-protected-areas-to-commercial-fishing (accessed 20 February 2022).

Chung, M.G., A. Herzberger, K.A. Frank and J. Liu (2020) 'International tourism dynamics in a globalized world: A social network analysis approach', *Journal of Travel Research*, 59(3): 387–403.

Collins, R. (2015) 'Keeping it in the family? Re-focussing household sustainability', *Geoforum*, 60: 22–32.

Connell, J. (2018) 'Islands: Balancing development and sustainability?' *Environmental Conservation*, 45(2): 111–124.

DFAT (Department of Foreign Affairs and Trade) (2020) 'Kiribati', *Marketing Insights: Connecting Australian Business to the World*, December, Australian Government. Available online at https://www.dfat.gov.au/sites/default/files/kiribati-market-insights-2021.pdf (accessed 15 January 2021).

Dornan, M. and K.U. Shah (2016) 'Energy policy, aid, and the development of renewable energy resources in small island developing states', *Energy Policy*, 98: 759–767.

Douglas, N., N. Douglas and R. Derrett (eds.) (2001) *Special Interest Tourism*. Milton: John Wiley and Sons Australia, Ltd.

Đurašević, S. (2014) 'Thematic tourism as an important segment in the business of modern tour operators', *Turističko Poslovanje*, 13: 109–117.

Duvat, V. (2013) 'Coastal protection structures in Tarawa Atoll, Republic of Kiribati', *Sustainability Science*, 8(3): 363–379.

Elrick-Barr, C.E., T.F. Smith, B.L. Preston, D.C. Thomsen and S. Baum (2016) 'How are coastal households responding to climate change?' *Environmental Science and Policy*, 63: 177–186.

Frantál, B. and R. Urbánková (2017) 'Energy tourism: An emerging field of study', *Current Issues in Tourism*, 20(13): 1395–1412.

Government of Kiribati (2012) *Kiribati Development Plan 2012–2015: Enhancing Economic Growth for Sustainable Development - A Vibrant Economy for the People of Kiribati*. Available online at http://prdrse4all.spc.int/system/files/kiribati-development-plan-2012-2015_0.pdf (accessed 12 January 2021).

Government of Kiribati (2016) *Kiribati Development Plan 2016–19: Towards a Better Educated, Healthier, More Prosperous Nation with a Higher Quality of Life*, Tarawa, Kiribati: National Economic and Planning Office. Available online at https://policy.asiapacificenergy.org/node/2789 (accessed 12 January 2020).

Government of Kiribati (2018) *Kiribati 20-Year Vision 2016–2036*, Tarawa, Kiribati: National Economic and Planning Office. Available online at http://www.mfed.gov.ki/sites/default/files/KIRIBATI%2020-YEAR%20VISION%202016-2036%20.pdf (accessed 12 February 2020).

Gunia, A. (2022) 'COVID-free Pacific nation opened it border a crack. The virus came rushing in', *Time*, 31 January. Available online at https://time.com/6143260/covid-19-pacific-islands-kiribati/ (accessed 8 February 2021).

Homasi, L.A. and I. Wainiqolo (2020) 'Impact of COVID-19 on small economies–Kiribati and Tuvalu: Recasting essential reforms', *Pacific Monitor*, July, pp. 12–14. Available online at https://www.adb.org/sites/default/files/publication/622976/pem-july-2020.pdf (accessed 20 April 2022).

Index Mundi (n.d.) 'Kiribati – International tourism, number of arrivals'. Available online at https://www.indexmundi.com/facts/kiribati/indicator/ST.INT.ARVL (accessed 16 January 2021).

Jones, P., D. Hillier and D. Comfort (2017) 'The sustainable development goals and the tourism and hospitality industry', *Athens Journal of Tourism*, 4(1): 7–18.

Kirby, A. (1999) 'Islands disappear under rising seas', *BBC News*, 14 June. Available online at http://news.bbc.co.uk/2/hi/science/nature/368892.stm (accessed 12 February 2021).

Laurence, C. (2014) 'Besieged by the rising tides of climate change, Kiribati buys land in Fiji', *The Guardian*, 1 July. Available online at https://www.theguardian.com/environment/2014/jul/01/kiribati-climate-change-fiji-vanua-levu (accessed 23 March 2021)

Liu, J., V. Hull, J. Luo, W. Yang, W. Liu, A. Viña, C. Vogt, Z. Xu, H. Yang, J. Zhang, L. An, X. Chen, S. Li, Z. Ouyang, W. Xu and H. Zhang (2015) 'Multiple telecouplings and their complex interrelationships', *Ecology and Society*, 20(3): 1–17.

Lundsgaarde, H.P. (1966) *Cultural Adaptation in the Southern Gilbert Islands*, Eugene: Department of Anthropology, University of Oregon.

Macdonald, B.K. and S. Foster (2020) 'Kiribati', *Encyclopedia Britannica*, 9 September. Available online at https://www.britannica.com/place/Kiribati (accessed 8 February 2021).

Mallin, M.A.F., D.C. Stolz, B.S. Thompson and M. Barbesgaard (2019) 'In oceans we trust: Conservation, philanthropy, and the political economy of the Phoenix islands protected area', *Marine Policy*, 107: 1–12.

Michalena, E. (2017) 'Building green growth and entrepreneurship in the Pacific through knowledge and innovation'. In: S. Sindakis and P. Theodorou (eds.) *Global Opportunities for Entrepreneurial Growth: Competition and Knowledge Dynamics Within and Across Firms*, London: Emerald Publishing Limited, pp. 475–506.

Michalena, E. and J. Hills (2013) 'Renewable energy governance: Is it blocking the technically feasible?' In: E. Michalena and J.M. Hills (eds.) *Renewable Energy Governance: Complexities and Challenges*, London: Springer Publisher Ltd, pp. 3–8.

Michalena, E., V. Kouloumpis and J.M. Hills (2018) 'Challenges for Pacific small island developing states in achieving their nationally determined contributions (NDC)', *Energy Policy*, 114: 508–518.

Milne, S. (1992) 'Tourism and development in South Pacific microstates', *Annals of Tourism Research*, 19(2): 191–212.

Pacific Island Times (2020) 'Following diplomatic reunion with Beijing, Kiribati eyes windfall from China's travel market', 13 January. Available online at https://www.pacificislandtimes.com/post/2020/01/13/amid-restored-ties-with-beijing-kiribati-hopes-to-catch-windfall-from-china-s-travel-mark (accessed 16 February 2021).

Petterson, M.G. and H.J. Kim (2020) 'Can ocean thermal energy conversion and seawater utilisation assist small island developing states? A case study of Kiribati, Pacific island region'. In: A.S. Kim and H.J. Kim (eds.) *Ocean Thermal Energy Conversion (OTEC): Past, Present, and Progress*, London: IntechOpen Limited, pp. 3–30.

Republic of Kiribati (2009) *Kiribati National Tourism Action Plan, 2009–2014*, Tarawa, Republic of Kiribati: Ministry of Communications, Transport and Tourism Development. Available online at https://www.mfed.gov.ki/sites/default/files/Kiribati%20National%20Tourism%20Action%20Plan.pdf (accessed 14 January 2020).

Republic of Kiribati (2017) *Intended Nationally Determined Contribution*. Available online at https://policy.asiapacificenergy.org/sites/default/files/INDC_KIRIBATI.pdf (accessed 14 February 2021).

Ryghaug, M., T.M. Skjølsvold and S. Heidenreich (2018) 'Creating energy citizenship through material participation', *Social Studies of Science*, 48(2): 283–303.

Save Kiribati (2019) *Tourism*. Available online at http://savekiribati.com/tourism.php (accessed 14 February 2021).

Scheyvens, R. (2018) 'Linking tourism to the sustainable development goals: A geographical perspective', *Tourism Geographies*, 20(2): 341–342.

Shah, C., A. Trupp and M.L. Stephenson (2022) 'Deciphering tourism and the acquisition of knowledge: Advancing a new typology of "research-related tourism (RrT)"', *Journal of Hospitality and Tourism Management*, 50: 21–30.

Smith, J.R. (2020) 'Travel's new threat is undertourism', *Lonely Planet*, 9 August. Available online at https://www.lonelyplanet.com/articles/travel-destinations-suffer-during-covid (accessed 16 February 2021).

Stephenson, M.L. and H.L. Hughes (1995) 'Holidays and the UK Afro-Caribbean community', *Tourism Management*, 16(6): 429–435.

Stone, G.S. and D. Obura (2012) *Underwater Eden: Saving the Last Coral Wilderness on Earth*, Chicago and London: University of Chicago Press.

Teroroko, T. (2016) 'Rising Phoenix: The GEF's contribution to sustainable development in Kiribati', Global Environment Facility, 11 August. Available online at https://www.thegef.org/news/rising-phoenix-gefs-contribution-sustainable-development-kiribati (accessed 17 April 2021).

Throsby, D. (2011) 'The Kiribati economy: Performance and prospects', *Pacific Economic Bulletin*, 16(1): 1–18.

UNESCO (United Nations Educational, Scientific and Cultural Organization) (2021) 'UNESCO expresses concern over the lifting of fishing no-take zones in Kiribati's phoenix islands protected area', Available online at https://whc.unesco.org/en/news/2370 (accessed 27 December 2021).

UNWTO (United National World Tourism Organisation) (2017) *Tourism for Sustainable Development in the Least Developed Countries – Leveraging Resources for Sustainable Tourism with the Enhanced Integrated Framework*, Madrid: World Tourism Organization, Geneva: International Trade Centre and Enhanced Integrated Framework. Available online at https://www.e-unwto.org/doi/book/10.18111/9789284418848 (accessed 17 January 2021).

UNWTO (United Nations World Tourism Organization) and UNDP (United Nations Development Programme) (2017) *Tourism and the Sustainable Development Goals – Journey to 2030, Highlights*, Madrid: UNWTO. Available online at https://www.e-unwto.org/doi/book/10.18111/9789284419340 (accessed 17 April 2022).

Warner, K. and K. Van der Geest (2013) 'Loss and damage from climate change: Local-level evidence from nine vulnerable countries', *International Journal of Global Warming*, 5(4): 367–386.

Watters, R. and K. Banibati (1984) *Abemama. Atoll economy: Social change in Kiribati and Tuvalu*, No. 5, Canberra: Development Studies Centre, Australian National University.

Webb, J. (2020) 'Kiribati economic survey: Oceans of opportunity', *Asia and the Pacific Policy Studies*, 7(1): 5–26.

Worland, J. (2015) 'Meet the president trying to save his island nation from climate change', *Time*, 9 October. Available online at http://time.com/4058851/kiribati-cliamte-change/ (accessed 23 March 2021).

World Bank (2022) 'US$14 million world bank project puts universal health care within reach for all I-Kiribati', 8 April, Washington, D.C.: World Bank Group. Available online at https://www.worldbank.org/en/news/press-release/2022/04/08/us-14-million-world-bank-project-puts-universal-health-care-within-reach-for-all-i-kiribati (accessed 25 April 2022).

Worldometer (2022) 'Kiribati population'. Available online at https://www.worldometers.info/world-population/kiribati-population/ (accessed 20 April 2022).

20

DECIPHERING NAURU AS A NON-TOURISM DESTINATION

Current challenges and potentialities

Nazia Ali and Marcus L. Stephenson

Introduction

Nauru is located in the western Pacific Ocean, lying northeast of Australia and located 59 kilometres south of the equator. It was formerly known as 'Pleasant Island', which was named by an English ship captain, John Fearn, who sighted the island in 1798 (Gale 2016). Nauru is a small island state with an estimated population of 10,834 in 2020, representing a high population density of 542 people per square kilometre (Countryeconomy.com 2022). The country is considered to be the second smallest United Nations (UN) state behind Monaco (Amos 2020), with a total landmass of 21.3 square kilometres (Commonwealth of Nations 2015). The chapter proceeds by introducing the development context of Nauru and identifying the past and current development challenges that impact the economic, environmental and social landscape of the island, particularly the effects of both the phosphate and asylum industries. The discussion will draw attention to how the former industry negatively impacts the physical and ecological environment. The work then implicates Nauru's limited potential to attract tourists and advance its tourism industry, partly influenced by its role as an offshore processing centre for asylum seekers and refugees. The construction of an asylum industry has economically compensated, in part, for the decline in the phosphate industry. As the chapter illustrates, Nauru is not recognized as a tourism destination as such but, indeed, a destination where refugees and asylum seekers are relocated as a consequence of not gaining access to Australia. Crucially, the work aims to conceptualize 'non-tourism' within a socio-political context which limits the freedom of movement of migrants and strategically immobilizes them to the status of 'non-tourists'. Unlike tourists, who are more seasonal and who stay in a destination for a shorter period, asylum seekers and refugees stay for longer and indefinite periods. Curtis and Pajaczkowska (1994: 213) state:

> The predicament of the migrant worker and refugee is largely an inversion of the experience of the tourist – their journeys are not circular, they are neither an escape from work nor a pursuit of the identification of sensory experience.

Consequently, this chapter deciphers non-tourism as a consequence of Australia's asylum and refugee policy, known as the 'Pacific Solution', whereby migrants travelling to Australia by boats

DOI: 10.4324/9780429019968-23

or vessels are intercepted and transferred to offshore processing centres in the Pacific region, notably Nauru and Manus Island (Papua New Guinea) (Phillips 2012). The discussion contends that the perception of Nauru as a place where individuals have been incarcerated reinforces an undesirable image of the country as a tourist destination. In the case of Nauru, the discussion critically inspects how Australia's offshore immigration practices have led to the emergence of new island diasporas, which may well influence future tourism and travel mobilities. Nonetheless, the latter part of the chapter endeavours to highlight three important ways forward for Nauru to develop a tourism industry: (1) the development of a new visiting friends or relatives (VFR) market along with recognition of the island's ethnic and cultural diversities, which could imbue the destination with a wider cosmopolitan appeal; (2) the opportunity to develop industrial heritage tourism in Nauru, especially given its historic links with the phosphate industry; and (3) the advancement of high-quality fishing tourism experiences.

Nauru: background context and development challenges

Nauru (Figure 20.1), or more formally the Republic of Nauru, became independent from Australia in 1968, though bilateral relations were established with Australia for trade, investment and developmental assistance (DFAT 2019a). Nauru has been economically dependent on (and rich in) the mineral resource of phosphate, which has been significantly exhausted due to excessive mining. Phosphate was discovered on the island in 1899 by a British prospecting company known as the Pacific Islands Company (Williams and Macdonald 1985). By 1939, a consortium representing Australia, Britain and New Zealand was exporting 1.25 million tonnes of phosphate from Nauru annually (Treloar and Hall 2004: 224). Without access to soil nutrients from Nauru, Banaba Island (Kiribati Republic) and Christmas Island (Australia), both Australia and New Zealand could not have established thriving industrial agricultural systems fundamental to their demographic, economic and social expansion (Williams and Macdonald 1985).

Nauru started to gain more control of the phosphate industry when Nauru became independent. Pre-independence had witnessed significant levels of 'anti-colonial agitation', and Australia eventually succumbed to such concerns raised by partner governments and the UN (Williams and Macdonald 1985: 556). These writers note that such levels of agitation gained momentum since World War II when Nauruans formally voiced concerns over the failure of Australian officials and British Phosphate Commissioners (BPC) to acknowledge 'their status as owners of the land and their right to benefit from its resources' (1985: 397). The Court of Justice in The Hague concluded that BPC was unsuccessful in advancing the development of Nauru society, an objective that was embedded in the League of Nations Mandate, and Nauru was subsequently awarded compensation money amounting to AU$210 million (currently US$145.7 million) under the Nauru/Australia Compact of Settlement Treaty (Pollock 2014). Following independence in 1968, Nauru reaped significant economic benefits, as the price of phosphate rose from AU$10 (US$6.9) per ton to more than AU$65 (US$45) per ton in the 1970s (Takahashi 2015). Writing in the *New York Times* in the early 1980s, Trumbull (1982: n.p.) comments: 'Nauru comes close to being the ultimate welfare state'. He notes that this was due to the range of public services freely available and paid for by the government, such as education and medical and dental care, as well as subsidized housing, electricity, transport and telephone services. In comparison to other Pacific Island countries, Nauru was a relatively rich nation which consistently recorded significant trade surpluses, valued at US$80 million in 1989 (Fairbairn 1994). However, by the end of the twentieth century, phosphate rock production reduced from one million tonnes per annum in 1968 to 266,000 tonnes in 2001 and 11,000 tonnes in 2005 (Arrowsmith and Parker 2015: 11). Mining impacted the economic viability

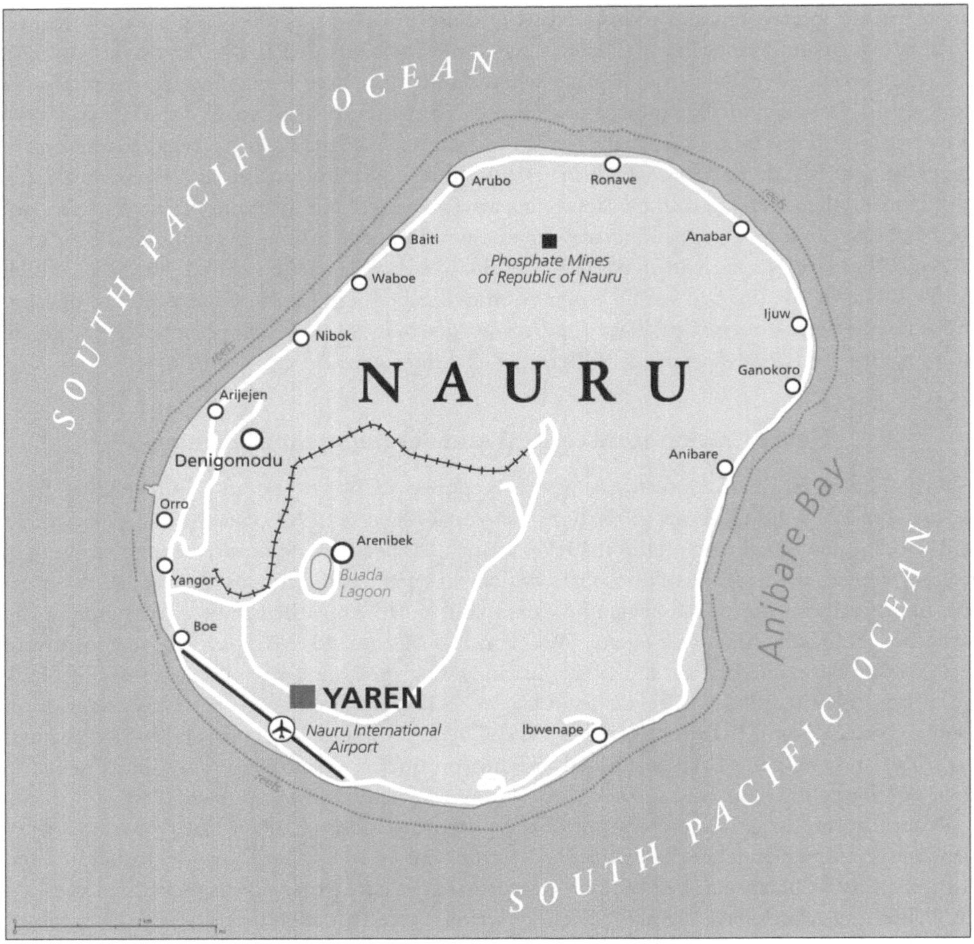

Figure 20.1 Map of Nauru, indicating the district of Yaren, the country's capital. Source: Peter Hermes Furian (www.shutterstock.com/image-vector/nauru-political-map-de-facto-capital-453588187).

of the small island, especially as there were no other development alternatives (Fagence 1997). Connell (2006: 49) suggests that Nauruans are 'victims of a "resource curse"', where:

> rapid and wasteful spending of windfall revenues from resource booms (usually in minerals) is difficult to avoid, and that the over-rapid absorption of these revenues produces various problems including wage inflation and exchange rate appreciation. This leads to decline in production in other sectors of the economy, such as agriculture or industry, where costs rise relative to the cost of local production.
>
> *(2006: 48)*

The nation faces problems with poor education standards, truancy and public health concerns, as non-communicable diseases associated with high levels of obesity and diabetes are prevalent (Maclellan 2020). In terms of human development indicators associated with economic and social conditions, the UN System in the Pacific (2017: 103) emphasized that Nauru has the 'worst human development indicators in the sub-region and the highest incidence of food poverty'. Life expectancy at birth is 57.5 years for males and 63.2 years for females (ADB 2017). While Nauru's gross domestic product (GDP) for 2020 stands at US$115 million (Countryeconomy.com 2022), decreasing modestly from the pre-COVID-19 pandemic amount of US$118.3 million in 2019 (World Bank 2021), its national debt in 2020 was US$70 million (Countryeconomy.com 2022). Nauru, thus, needs to develop new sources of economic development (IMF 2020). Despite the global pandemic, halting travel and tourism mobility across the world did not appear to have a major negative impact on Nauru due to low levels of tourism development, including a low-profile VFR market by the Nauruan community.

Tourism challenges and environmental concerns

The likely contributing factors as to why Nauru has not experienced tourism development concern its geographical remoteness, distance and peripheral location (Cassidy et al. 2006). Pacific Islands are often symbolically associated with exotic places and destination paradises, though in the case of Nauru its interior is damaged by intense mining, leaving 'a moonscape of jagged limestone pinnacles unfit for agriculture or even building' (Davies and Doherty 2018a). Davies and Doherty (2018a) note of the Nauru Rehabilitation Corporation, which was established following a financial resolution in 1993 for Australia to pay US$135 million in compensation for this environmental catastrophe. Although 400 hectares of land was originally targeted for rehabilitation, only a small section known as 'Pit 6' has been developed and currently used as the jail. Around 80% of the island is uninhabitable, with the population largely restricted to the coast (Watanabe 2018).

Nauru increased visa tariffs to dissuade unwelcome visitors, with business visas rising from AU$400 (currently US$290) to AU$6,000 (US$4,349) and visas for foreign journalists rising from AU$200 (US$144) to AU$8,000 (US$5,798) (Takahashi 2015). In 2018, there were 77 employees in the tourism industry, representing 2.53% of the country's total employment, along with eight hotels, motels and guest houses (SPTO 2019). In terms of transport, there is only one airport – Nauru International Airport (INU) and the national airline is Nauru Airlines, which operates flights to Australia, Fiji, Kiribati and the Republic of the Marshall Islands (Miles and De Marchi 2020). The main visitors to Nauru have been the 'fly-in-fly-out' advisors from Australia, including Australian workers in government positions and those providing immigration and refugee management services (Morris 2020).

The environmental concerns and threats will logically have an impact on tourist perceptions and expectations. It is ironic, therefore, that an environment deemed unsuitable for tourism development and tourist activities would be considered appropriate to house refugees and asylum seekers. Accordingly, destinations such as Nauru are not able to prioritize tourism development, as the island has largely engaged with the 'mobility poor' rather than the 'mobility rich' (Bianchi and Stephenson 2014: 87) but, nonetheless, identified a niche group that can actually be commodified through the processing fees received from the Australian government. Admittedly, the attraction of a vibrant tourism industry is often seen as a socio-economic panacea for vulnerable destinations and small islands, but there are other (more contentious) options available if challenges to attract the tourist dollar are simply not viable.

The development of an asylum industry and the emergence of 'non-tourists'

The decline of the phosphate industry coincided with the development of the asylum industry in Nauru. In August 2001, a ship from Norway, MV Tampa, rescued 438 asylum seekers from a distressed boat floating off the coast of northwest Australia. The ship was then refused entry into the country, and the refugees were then transported to a detention centre in Nauru (Muller 2016: 6). The Nauru government agreed with the Australian government on the basis of receiving AUS$30 million (currently US$21.36 million) of aid, that the island would receive and accommodate asylum seekers who arrive by sea while their applications for asylum were being assessed in Australia (Commonwealth of Nations 2015). Two-thirds of Nauru's total revenue for 2017–2018 was provided by the Australian government for direct aid and fees for the Regional Processing Centre as well as visa and resettlement fees for refugees (Davies and Doherty 2018b). While the centre is now empty, the contract with the Australian government still remains profitable, with the government of Nauru receiving around US$150 million in fees and taxes in 2020 (Howes and Surandiran 2021).

Asylum seekers either remained in Nauru as detainees of the Australian government or resettled in other nations as part of an offshore processing regime. The Australian asylum and refugee policy became known as the 'Pacific Solution' and is a legacy of the Howard government, initiated in 2001 and practised by successive Australian government administrations (Lohana et al. 2016; Phillips 2012). The metaphoric reference to Australia's 'Pacific Solution' as a 'ring of steel' (Shen 2018) functions to intercept migrants arriving by sea on vessels and diverting them to Nauru. The freedom of movement of those who have been incarcerated on Nauru has been restricted and policed, especially as people are not able to travel freely beyond the island's borders. This assertion is reflective of a socio-political argument that some people's mobilities can be feared as threats to national (and international) security, especially on the basis of such determinations as ethnicity, race, religion and nationality (Bianchi and Stephenson 2014; Stephenson and Ali 2010). The migrant as an asylum seeker becomes a 'non-tourist' for the duration of their immobility in Nauru, without the right to travel from the island. Nauru has not been a popular destination of choice but more of a destination of necessity and one that has been imposed on individuals beyond their direct control. This lack of choice is a key determining characteristic of being a 'non-tourist', a status that is directly juxtaposed to the status of a tourist – one who has high levels of consumer choice and sovereignty.

As a consequence of the 'Pacific Solution', Nauru has become a place of residence for asylum seekers and refugees, whether temporary or permanent, pending the outcome of claims for asylum administered by Australia. There are three possible formal outcomes of asylum application: (1) to settle in Nauru if in need of international protection, (2) to settle in a third safe country if in need of international protection, and (3) to return to the country of origin if not in need of international protection, or to a third country if the right to enter and reside is permitted (DFAT 2019b: 4). In 2018, the United States (US), as a third country, agreed to resettle some asylum seekers, including the reunification of children with parents (BBC News 2019). Nonetheless, by remaining in the custody of Nauru, not a destination of choice, individuals become simultaneously non-tourists or non-citizens. Experiences of island incarceration and feelings of not being able to move to a welcoming place are indicative of the 'non-tourism' phenomenon. Non-tourism thus has a coercive, immobilizing and destabilizing effect, as one Pakistani refugee living in Nauru evoked:

> I think Australians deliberately forced us to live under such conditions, so that we would spread the word: don't try to enter Australia illegally, you will go to hell instead.
> *(Sokhin 2015)*

The 'Pacific Solution' strategy has been accused of being inhumane and dehumanizing and in violation of laws pertaining to international human rights (Banks and McGregor 2011). The destination developed a negative 'organic image' (see Gunn 1988) due to deleterious narratives and impressions pursued by the media industry over the 'asylum issue'. Media perceptions converge on the socio-economic and political hardships and conditions experienced by the asylum-seeking community. Such concern clearly surfaced in 2016 when 2,000 leaked incident reports from Nauru's Australian detention camp reached the international press, which subsequently exposed the atrocities experienced by the asylum seekers. The Australian government's 'Nauru files' comprises 8,000 pages of incidents concerning self-harm, assault, sexual abuse, child abuse and poor living conditions (Farrell et al. 2016). Subsequent media-fuelled responses resulted in such headline stories as: 'Australia's dumping ground for refugees' (Doherty 2016); 'Pitiful plight of refugees locked up in Australia island internment camps' (Jones 2017); and 'Nauru refugees: The island where children have given up on life' (Harrison 2018). However, the conservative press had a different approach. The United Kingdom's (UK) *Daily Mail*, for instance, stated: 'These solutions involve hardship for the migrants. But at least people are not drowning in their thousands, and the immigration policies are consistent' (Burleigh 2015).

In addition to the US, departure from Nauru has been routed to Afghanistan, Albania, Bangladesh, Cambodia, Egypt, India, Iran, Iraq, Jordan, Lebanon, Pakistan, Somalia, Sri Lanka, Sudan and Syria (Parliament of Australia 2016). Men, women and children with asylum seeker or refugee status who remain in Nauru have now moved to dwellings in the local community rather than reside in the Regional Processing Centre, which ceased to house refugees in 2019 (Refugee Council of Australia 2022). Offshore processing data shows that there are 112 people remaining in Nauru, of which 83 are refugees, 17 non-refugees and 12 still in Refugee Status Determination (RSD) (Refugee Council of Australia 2022). Although the Australian government intends to exit its programme with Papua New Guinea, Nauru will still continue to host refugees in the future (Refugee Council of Australia 2022).

Morris (2020) emphasizes that, through the process of becoming asylum seekers, individuals need to prove their suffering, which is a traumatizing experience. She believed that this process was problematized further when the vetting process produced a Nauruan (not Australian) refugee visa, which was not the reason why individuals risked their lives to travel to the Pacific region in the first place. Hence, the process of becoming a refugee, i.e., a non-tourist, stands in stark contrast to the process of becoming a tourist. The tourist has the decision-making power associated with travel in terms of choice of destination for leisure purposes and can be greeted with pleasurable gestures of hospitality, albeit commodified. Tourists, as expected guests, are welcomed, whereas non-tourists (asylum seekers and refugees), as unexpected guests, are unwelcomed or treated with suspicion. Although the contract to process asylum seekers and refugees in Nauru prevails, the state arguably needs to think of constructive economic development beyond this industry, especially if the Australia government changes direction with regard to its immigration policy. Therefore, it is crucial to consider Nauru's tourism potential, especially to politically and economically counteract the asylum industry. Possible ethical and sustainable tourism development markets can focus on future growth of the VFR market, landscape and heritage tourism and high-quality fishing experiences. This draws upon the unique social, cultural and environmental landscapes of Nauru and presents a 'new' narrative for the island beyond the 'Pacific Solution' policy.

Nauru's tourism potential

The future growth of the VFR market and a cosmopolitan environment

Individuals are displaced on Nauru in a country that they had no intention of visiting, as their original objective was to travel to the developed world. Nonetheless, asylum seekers and refugees living in Nauru could work towards advancing a 'sense of co-ethnicity with others of a similar background' (Cohen 1997: ix), as they share similar histories and experiences of migration and displacement (Hall 2006). For these communities, there can be a desire to maintain links and connections with family members and friends in their homeland or in other countries. This is reflected in such travel behaviour as ethnic reunion tourism (Stephenson 2002), return migration (Ali and Holden 2006) or VFR (Seaton 1994). Iranian, Sri Lankan, Pakistani and Afghani nationalities in Nauru could eventually form diasporas as current generations settle and future generations are born, which would inevitably materialize into diaspora tourism markets. There is potential here for a VFR tourism market to develop, with friends and relatives travelling from other nations to reunite with kin in Nauru to uphold friendship and family ties.

Tourist arrivals to Nauru are low compared to other Pacific Islands, with annual arrivals at 3,038 in 2016 and 3,002 in 2017, and with an overall tourism performance labelled as 'negative' (SPTO 2018: 3). It has been noted, different to other Pacific Island states, that Nauru has surprisingly not experienced significant waves of outward migration and, therefore, the extent to which a VFR market can be developed is negligible (Arrowsmith and Parker 2015). Nevertheless, there may well be an eventual opportunity to establish a tourist market for Nauru concerning VFR travel, especially if asylum seekers arrange temporary or permanent residency to remain in Nauru, and also taking into consideration migrants who resettled elsewhere who may want to return to visit friends and relatives who did not leave Nauru. Such a development could help contribute to the economic and social fabric of Nauru, especially as the island's tourism industry is not significantly well-developed. It may also help to reconcile the difficult and past experiences of living in Nauru. However, irrespective of Nauru's unique natural environment, the island has little tourist activity and limited tourist facilities (e.g., transport, accommodation, entertainment and restaurants) to support a vibrant infrastructure. Therefore, the tourism and transport infrastructure would need significant improvement, and this would have to be integrated carefully into a feasible development plan.

Winsor (2017) draws attention to ways in which newcomers influence the cultural diversity and landscape of Nauru, where Bengali, Iranian, Iraqi, Lebanese, Pakistani, Rohingya and Somali cuisines are evident – indicating how the island is witnessing a drive towards international forms of culinary entrepreneurship and consumption. The island experienced several new restaurants and cafes opening in the past decade, such as Iranian, Lebanese and Pakistani restaurants, adding to more ethnic diversity to the restaurant scene which was largely based on Nauruan, Chinese and Western influences. Nauru Airlines has been promoting the trend towards more ethnically diverse choices of restaurants and cafes on the island, declaring:

> In recent years, Nauru's dining scene has drastically expanded thanks to refugees opening businesses around the island. Some other longer-term restaurants have also employed refugees as chefs, kitchen hands and wait staff which has seen standards change.
>
> *(Nauru Airlines 2017)*

The cosmopolitan element of community or society can be a compelling feature of tourism destinations, in which migrants, as cultural cosmopolitans, can be seen 'using their skills and resources they carried with them from their origin countries in order to support their economic, social and political transnational practices' (Horst and Olsen 2021: 13) in their host country. Therefore, Nauru's cultural cosmopolitan environment can be adaptive to a broad range of travellers and settlers but at the same time retaining the Nauruan sense of cultural identity through the backdrop of its natural environments and industrial heritage.

Unique destination landscape and industrial heritage tourism

Gale (2016: 342) contends that the perception of Nauru's landscape as 'environmental horror' is a popularized view that does not account for how the destination can re-establish themselves through growth of indigenous vegetation and through the natural restoration of soils in mined-out lands. He progressively argues that Nauru's karrenfields, with their pinnacles reaching over 14 metres in height, symbolize 'the landscape of the island as it must have appeared a quarter of a million years ago' (2016: 343). Gale (2016) argues that the landscape should be preserved and recognized globally as a travel destination, likening the karrenfields to the UN Educational, Scientific and Cultural Organization (UNESCO) World Heritage Sites of Gunung Mulu in Sarawak, the stone forests of Lunan in China and the stingy karrenfields in Madagascar, for instance. The preservation and protection of Nauru's unique heritage landscape can retain the historical narrative but can also present an opportunity for placemaking in a niche tourism marketing context.

Nauru's National Sustainable Development Strategy 2005–2025 identifies heritage sites as having significant niche market potential (along with the development of small-scale but high-quality fishing – see the following section) (Republic of Nauru 2009). Nauru's role in phosphate production is internationally well-known, and there is significant opportunity to develop industrial heritage interpretation programmes and presentations concerning the mining of this valuable mineral, both within the context of its traditional role as the major industrial focus of the island's historic development and in terms of ongoing phosphate extraction. The heritage tourism product would be educationally orientated and, thus, have significant social and economic value. There are a range of sites that can be developed as part of the industrial heritage park; for instance, there is a 3.9-kilometre stretch of railway that was built in 1907 to transport the mined phosphate (Smith 2018) (Figure 20.1). The story behind Nauru's role in the international arena of phosphate extraction has been well acclaimed, especially as the country has been known to produce phosphate which is of 'extremely high quality' (Viviani 1970: 32).

Industrial heritage tourism in Nauru is significant in positioning phosphate extraction within narratives concerning the representation and interpretation of the colonial period and independence. Heritage tourism experiences associated with the mining of phosphate can thus focus on the narrow-gauge railway reported as the 'lynchpin of the island's economy for decades' (Dickenson and Smith 2022). Consequently, the phosphate mining industry is the main industrial heritage attraction and could be enhanced by experiences such as railway tours, views of natural industrial landscapes, education expositions, exhibitions, museums and a visitor centre. The industrial heritage tourism market could also add value to VFR-orientated return visits by the Nauruan diaspora as they attempt to re-connect with their past through landscapes of cultural, emotional, national and social importance. Moreover, the development of industrial heritage tourism is vital to the economic regeneration of Nauru following several failed economies. As Edwards and Llurdés i Coit (1996: 345) assert: 'heritage has been the root of many regeneration projects and, in some cases, has contributed to economic development through the expansion of tourism'.

High-quality fishing tourism experiences

There is potential for niche small-scale high-quality fishing experiences in Nauru, as the island is renowned for game fish, which can be consumed and experienced for recreational use of marine resources. Game fishing is a year-round charter attraction on the island, with packaged experiences for tourists wanting a big catch of species such as yellowfin tuna, marlin, wahoo and sailfish in the 2,000-metre deep seas (Fishing Getaways 2020). It has been reported that there are two main deep-sea fishing companies on the island, Nauru Game Fishing and Equatorial Fishing Charters, and locals also rent their boats to visitors to fish (iExplore, n.d.). Although fishing tourism experiences are currently being offered to tourists, the intention here is not to replicate this but to offer an alternative high-quality experience that targets professional/elite fishers from across the globe. These professional/elite fishers belong to fishing associations, clubs, federations or trusts and travel to Nauru to train in the deep waters ahead of local, regional, national and international competitions. These sport tourists can also be accompanied by local fisherfolk to learn about different traditional styles of fishing and techniques and exchange knowledge to develop their practice. Other tourists visiting the island can spectate the professional/elite fishers in training with local expert fisherfolk, further enhancing their tourist experiences. High-quality fishing tourism will seek to be a niche experience, underpinned with sustainable practices in terms of being a training facility for professional/elite fishers, with proceeds benefiting local people and communities.

A possible proposal for an ecotourism-led fishing activity is appropriate and feasible, as this aims to benefit local people, curb economic leakage and complement the existing sustainable practices associated with the fisheries industry (i.e., the Nauru Agreement, see Pacific Islands Forum Fisheries Agency (2022)). It is not uncommon for small islands in the Pacific to plan, develop and manage such niche markets as ecotourism for economic growth based on principles and practices of sustainable development, as sustainable tourism is reflective of ecotourism (Panakera et al. 2011). It has been noted by de Hass (2002) that many ecotourism destinations can be found in developing countries because of the sensitive nature of untouched and unspoilt physical environments. Therefore, a fundamental aspect of sustainability is 'responsible development' on small islands where natural resources are extremely sensitive (Jędrusik 2014). Given that outdoor forms of sport tourism (in this case fishing) in peripheral locations depend on natural environmental resources such as mountains, lakes and rivers to deliver the sport tourism experience (Higham and Hinch 2018), then tourism development needs to be sustainable and responsible.

Conclusion and research implications

This paper has aimed to decipher the conceptual significance of Nauru as a non-tourism destination not only because very few tourists visit the island but because it caters to non-tourists, notably asylum seekers and refugees who have been stranded on the island along with other such non-tourists as industrial workers and immigration personnel. By deciphering non-tourism mobilities, one can further understand articulations of tourism, identifying tourism markets such as VFR-orientated mobilities that complement the cultural, social and political landscapes of host nations. As a destination with little tourist activity and tourism infrastructure, Nauru can learn from neighbouring islands in terms of best practices to implement markets such as VFR, heritage tourism and fishing tourism, especially to avoid the pitfalls associated with mass/overtourism and accelerated growth which compromise the socio-cultural, economic, environmental and social landscape of the island. In light of the potentialities of these new markets identified in this chapter, the future tourism development of Nauru should be seen in the context of a wider post-COVID tourism strategy, one which capitalizes on the renewed mobilities of tourists and appropriate visitor categories. Moreover, as a nation with only three confirmed COVID-19 cases and no deaths from 3 January 2020 to 3 May 2022 (WHO 2022), there is an ongoing risk that virus transmissions could increase as tourism-related migration rejuvenates.

Accordingly, there is a need to balance the importance of economic development with the priority to protect the health of the nation.

Industrial heritage tourism offers the potential to stimulate tourism-related mobilities from the global Nauruan diaspora and those in search of historically significant times related to the mining industry. Ecotourism led development, underpinned by principles and practices of sustainability, presents a viable high-quality experience for professional/elite fishers participating in training activities that are planned, developed and managed by local fisherfolk. The potential tourist markets can arguably merge together 'people', 'place' and 'planet' by working with the destination's existing socio-cultural and environmental resources to stimulate the tourism industry. In doing so, Nauru can be reimaged and reimagined in its 'old' context as a 'Pleasant Island' rather than an island associated with the suffering that has taken place as a consequence of the asylum industry.

This paper has deciphered Nauru as a non-tourist destination and, in doing so, has identified potential markets that can transition the island into a tourism-generating nation to serve a cultural, economic, environmental and social purpose for both hosts and guests. Although this paper has deciphered non-tourism and tourism mobilities in Nauru, this has been from a conceptual standpoint rather than through empirical investigation accounting for the perceptions of island stakeholders (e.g., local people, local businesses and government). This presents directions for future research, involving the employment of both qualitative and quantitative enquiries. Quantitative research could focus on the conduct of market feasibility surveys to assess Nauru's long-term potential of the VFR market, heritage tourism and high-quality fishing tourism before pursuing all or one of these markets. This could further benefit from a benchmarking activity that is qualitative in nature to learn from other destinations with similar demography, history and geography to then inform a sustainable tourism development strategy for Nauru. It is suggested that future qualitative and quantitative research studies need to include community engagement to establish the aspirations, expectations and perceptions of tourism development from the perspectives of host communities.

References

Ali, N. and A. Holden (2006) 'Post-colonial Pakistani mobilities: The embodiment of the "myth of return" in tourism', *Mobilities*, 1(2): 217–242.

Amos, O. (2020) 'Coronavirus: Where will be the last place to catch Covid-19?' *BBC News*, 3 April. Available online at https://www.bbc.com/news/world-52120439 (accessed 25 May 2020).

Arrowsmith, J. and J. Parker (2015) *Situational Analysis of Employment in Nauru*, International Labour Organisation Publications, Office for Pacific Island Countries. Available online at https://www.unescap.org/sites/default/files/Situational%20Analysis%20of%20Employment%20in%20Nauru.pdf (accessed 4 May 2022).

Asian Development Bank (ADB) (2017) 'Poverty, social and gender assessment', *Nauru Port Development Project*, 17 April. Available online at https://www.adb.org/sites/default/files/project-documents/48480/48480-001-sprss-en_0.pdf (accessed 1 February 2021).

Banks, G. and A. McGregor (2011) 'Pacific "solutions" and imaginaries: Reshaping Pacific relations or re-colonising the "sea of islands"?' *Asia Pacific Viewpoint*, 52(3): 233–235.

BBC News (2019) 'Nauru migrants: Last four children to leave island for US', 3 February. Available online at https://www.bbc.co.uk/news/world-australia-47107179 (accessed 18 October 2019).

Bianchi, R.V. and M.L. Stephenson (2014) *Tourism and Citizenship: Rights, Freedoms and Responsibilities in the Global Order*, London: Routledge.

Burleigh, M. (2015) 'We must reclaim Europe's borders to stop such strategies repeating themselves', *Daily Mail*, 21 April. Available online at https://www.dailymail.co.uk/news/article-3048032/we-reclaim-europe-s-borders-stop-tragedies-repeating-michael-burleigh.html (accessed 1 June 2020).

Cassidy, F., L. Brown and B. Prideaux (2006) 'Why are the outer islands in the South Pacific popular with some tourists and not with others? The Vanuatu example', *Academy of World Business, Marketing, and Management Development Conference Proceedings*, 2(14): 170–177.

Cohen, R. (1997) *Global Diasporas*, London: Routledge.

Commonwealth of Nations (2015) 'Nauru'. Available online at https://www.commonwealthofnations.org/yb-pdfs/nauru_country_profile.pdf (accessed 15 August 2022).

Connell, J. (2006) 'Nauru: The first failed Pacific state?', *The Round Table*, 95(383): 47–63.

Countryeconomy.com (2022) 'Nauru'. Available online at https://countryeconomy.com/countries/nauru (accessed 15 January 2022).

Curtis, B. and C. Pajaczkowska (1994) 'Getting there: Travel, time and narrative'. In: G. Robertson, M. Mash, L. Tickner, J. Bird, B. Curtis and T. Putnam (eds.) *Travellers' Tales: Narrative of Home and Displacement*, London: Routledge, pp. 199–215.

Davies, A. and B. Doherty (2018a) 'Corruption, incompetence and a musical: Nauru's cursed history', *The Guardian*, 3 September. Available online at https://www.theguardian.com/world/2018/sep/04/corruption-incompetence-and-a-musical-naurus-riches-to-rags-tale (accessed 3 January 2019).

Davies, A. and B. Doherty (2018b) 'Nauru: A nation in democratic freefall propped up by Australia', *The Guardian*, 2 September. Available online at https://www.theguardian.com/world/2018/sep/03/nauru-a-nation-on-the-cusp-of-democratic-calamity (accessed 28 January 2020).

de Hass, H.C. (2002) 'Sustainability of small-scale ecotourism: The case of Niue, South Pacific', *Current Issues in Tourism*, 5(3–4): 319–337.

DFAT (Department of Foreign Affairs and Trade) (2019a) 'Nauru country brief', Australian Government. Available online at https://dfat.gov.au/geo/nauru/Pages/nauru-country-brief.aspx (accessed 13 October 2019).

DFAT (Department of Foreign Affairs and Trade) (2019b) 'Memorandum of Understanding between the Republic of Nauru and the Commonwealth of Australia, relating to the transfer to and assessment of persons in Nauru, and related issues', Australian Government. Available online at https://www.dfat.gov.au/sites/default/files/nauru-mou-20130803.pdf (accessed 14 October 2019).

Dickenson, G. and O. Smith (2022) '12 Facts about Nauru, the tiny island without a single Covid case', *The Telegraph*, 7 March. Available online at https://www.telegraph.co.uk/travel/destinations/oceania/12-facts-nauru-tiny-island-without-single-covid-case/ (accessed 1 April 2022).

Doherty, B. (2016) 'A short history of Nauru: Australia's dumping ground for refugees', *The Guardian*, 9 August. Available online at https://www.theguardian.com/world/2016/aug/10/a-short-history-of-nauru-australiasdumping-ground-for-refugees (accessed 28 January 2020).

Edwards, J.A. and J.C. Llurdés i Coit (1996) 'Mines and quarries: Industrial heritage tourism', *Annals of Tourism Research*, 23(2): 341–363.

Fairbairn, T. (1994) '"Pacific islands economies: Trade patterns and some observations on trade policy issues", trade and environment', 25 September. Available online at https://nau tilus.org/trade-and-environment/pacific-islands-economies-trade-patterns-and-some-observ ations-on-trade-policy-issues-4/ (accessed 28 January 2020).

Fagence, M. (1997) 'An uncertain future for tourism in microstates: The case of Nauru', *Tourism Management*, 18(6): 385–392.

Farrell, P., N. Evershed and H. Davidson (2016) 'The Nauru files: Cache of 2,000 leaked reports reveal scale of abuse of children in Australian offshore detention', *The Guardian*, 10 August. Available online at https://www.theguardian.com/australia-news/2016/aug/10/the-nauru-files-2000-leaked-reports-reveal-scale-of-abuse-of-children-in-australian-offshore-detention (accessed 1 March 2021).

Fishing Getaways (2020) 'Fishing charters - Nauru: Isles of Micronesia'. Available online at https://www.fishinggetaways.com.au/charters-nauru.html#:~:text=The%20unique%20depths%20off%20up,you%20could%20have%20in%20Nauru (accessed 1 April 2022).

Gale, S.J. (2016) 'The mined-out phosphate lands of Nauru equatorial western Pacific', *Australian Journal of Earth Sciences*, 63(3): 333–347.

Gunn, C. (1988) *Vacationscape: Designing Tourist Regions*, 2nd edition, New York: Van Nostrand Reinhold Company.

Hall, S. (2006) 'Once more around – Cultural identity', *Public Lecture*, 15 November, London: Queen Mary University of London.

Harrison, V. (2018) 'Nauru refugees: The island where children have given up on life', *BBC News*, 1 September. Available online at https://www.bbc.com/news/world-asia-45327058 (accessed 25 May 2020).

Higham, J. and T. Hinch (2018) *Sport Tourism Development*, Bristol: Channel View Publications.

Horst, C. and T.V. Olsen (2021) 'Transnational citizens, cosmopolitan outlooks? Migration as a route to cosmopolitanism', *Nordic Journal of Migration Research*, 11(1): 4–19.

Howes, S. and S. Surandiran (2021) 'Nauru: Riches to rage to riches', *Devpolicybog*, 12 April. Available online at https://devpolicy.org/nauru-riches-to-rags-to-riches-20210412/ (accessed 16 January 2022).

iExplore (n.d.) 'Nauru – things to do'. Available online at https://www.iexplore.com/articles/travel-guides/australia-and-south-pacific/nauru/things-to-do (accessed 3 April 2022).

IMF (International Monetary Fund) (2020) 'Republic of Nauru: 2019 Article IV Consultation – press release; staff report; and statement by the executive director for the Republic of Nauru', 29 January, Washington, D.C. IMF. Available online at https://www.imf.org/en/Publications/CR/Issues/2020/01/29/Republic-of-Nauru-2019-Article-IV-Consultation-Press-Release-Staff-Report-and-Statement-by-49001 (accessed 1 February 2021).

Jędrusik, M. (2014) 'The elusive sustainable development of small tropical islands', *Miscellanea Geographica - Regional Studies on Development*, 3(18): 26–30.

Jones, S. (2017) 'Pitiful plight of refugees locked up in Australian island "internment camps" who have provoked Trump's wrath', *Mirror*, 2 February. Available online at https://www.mirror.co.uk/news/world-news/pitiful-plight-refugees-locked-up-9743949 (accessed 28 October 2020).

Lohana, K., A.Z. Khaskhely, A.R. Khan and S. Razzaq (2016) 'Policy analysis of the "Pacific solution/offshore processing" component of immigration and asylum seeker/refugee policy of Australia', *The Government - Annual Research Journal of Political Science*, 5(5): 173–184.

Maclellan, N. (2020) 'Nauru', *The Contemporary Pacific*, 32(1): 213–225. *Project MUSE*.

Miles, D. and M. De Marchi (2020) '*Post COVID-19 Pacific Short-Term Aviation Strategy – A Scoping Study*', October, L&B Worldwide Australia Pty Ltd. Available online at https://theprif.org/sites/default/files/documents/ADB%20PRIF%20Post%20COVID-19%20Pacific%20Short-term%20Aviation%20Strategy%20-%20Final%20Report%20-%20Public%20Version_1.pdf (accessed 28 February 2021).

Morris, J. (2020) 'Refugee extractivism: Law and the mining of a human commodity in the Republic of Nauru', *Saint Louis University Law Journal*, 64(5): 59–90.

Muller, D. (2016) *'Islamisation' and Other Anxieties: Voter Attitudes to Asylum Seekers*, Melbourne: The University of Melbourne.

Nauru Airlines (2017) 'Fish, dive and explore', *Pacific Island Living*, Winter, Issue 20, Nauru Airline.

Pacific Islands Forum Fisheries Agency (2022) 'Nauru agreement'. Available online at https://www.ffa.int/nauru_agreement (accessed 3 April 2022).

Panakera, C., G. Wilson, C. Ryan and G. Liu (2011) 'Considerations for sustainable tourism development in developing countries: Perspectives from the South Pacific', *Tourismos: An International Multidisciplinary Journal of Tourism*, 6(2): 241–262.

Parliament of Australia (2016) 'Australia's offshore processing of asylum seekers in Nauru and PNG: A quick guide to statistics and resources'. Available online at https://www.aph.gov.au/About_Parliament/Parliamentary_Departments/Parliamentary_Library/pubs/rp/rp1617/Quick_Guides/Offshore#_Total_number_of (accessed 16 October 2019).

Phillips, J. (2012) 'The "Pacific solution" revisited: A statistical guide to the asylum seeker caseloads on Nauru and Manus island', Department of Parliamentary Services, Parliament of Australia. Available online at https://parlinfo.aph.gov.au/parlInfo/download/library/prspub/1893669/upload_binary/1893669.pdf;fileType=application/pdf (accessed 5 May 2019).

Pollock, N.J. (2014) 'Nauru phosphate history and the resource curse narrative', *Journal de la société des Océanistes*, 138–139: 107–120.

Refugee Council of Australia (2022) 'Offshore processing statistics'. Available online at https://www.refugeecouncil.org.au/operation-sovereign-borders-offshore-detention-statistics/ (accessed 6 May 2022).

Republic of Nauru (2009) *National Sustainable Development Strategy 2005–2025*, Republic of Nauru: Aid Management Unit/Development Planning and Policy Division Ministry of Finance and Economic Planning. Available online at https://www.adb.org/sites/default/files/linked-documents/cobp-nau-2016-2018-nsds.pdf (accessed 13 January 2022).

Seaton, A.V. (1994) 'Are relatives friends? Reassessing the VFR category in segmenting tourism markets'. In: A.V. Seaton, C.L. Jenkins, R.C. Wood, P.U.C. Dieke, M.M. Bennett, L.R. MacLellan and R. Smith (eds.) *Tourism: The State of the Art*, England: John Wiley and Sons, pp. 316–321.

Shen, T. (2018) 'The Nauru question: Australia and the South Pacific's "ring of steel"', *Journal of International Relations*, 2 December. Available online at http://www.sirjournal.org/op-ed/2018/12/2/copy-of-the-nauru-question-australia-nauru-relations-and-the-south-pacifics-ring-of-steel (accessed 22 October 2019).

Smith, O. (2018) '11 amazing facts about Nauru, the least visited, most obese nation on earth', *The Telegraph*, 31 January. Available online at https://www.telegraph.co.uk/travel/destinations/ oceania/ articles/nauru-facts/ (accessed 16 January 2022).

Sokhin, V. (2015) 'No way but Nauru', *Roads & Kingdoms*, 31 March. Available online at https://roadsandkingdoms.com/2015/no-way-nauru/ (accessed 25 May 2020).

SPTO (South Pacific Tourism Organisation) (2018) *Annual Review of Visitor Arrivals in Pacific Island Countries 2017*, Suva, Fiji: South Pacific Tourism Organisation. Available online at https://www.corporate.southpacificislands.travel/wp-content/uploads/2017/02/2017-AnnualTourist-Arrivals-Review-F .pdf (accessed 14 October 2019).

SPTO (South Pacific Tourism Organisation) (2019) *2018 Annual Visitor Arrivals Report*, Suva, Fiji: South Pacific Tourism Organisation. Available online at https://pic.or.jp/ja/wp-content/uploads/2019/07 /2018-Annual-Visitor-Arrivals-ReportF.pdf (accessed 28 February 2021).

Stephenson, M.L. (2002) 'Travelling to the ancestral homelands: The aspirations and experiences of a UK Caribbean community', *Current Issues in Tourism*, 5(5): 378–425.

Stephenson, M.L. and N. Ali (2010) 'Tourism, travel and Islamophobia: Post 9/11 journeys of Muslims in non-Muslim states'. In: J. Jafari and N. Scott (eds.) *Tourism and the Muslim World*, Bingley: Emerald, pp. 235–251.

Takahashi, K. (2015) 'An economic tale from a country that had no Plan B', *Nikkei Asian Review*, 28 May. Available online at https://asia.nikkei.com/Economy/An-economic-tale-from-a-country-that-had-no-Plan-B (accessed 3 January 2019).

Treloar, P. and C.M. Hall (2004) 'Tourism in the Pacific islands'. In: C. Cooper and C.M. Hall (eds.) *Oceania: A Tourism Handbook*, Clevedon: Channel View Publications, pp. 173–294.

Trumbull, R. (1982) 'World's richest little isle', *The New York Times Magazine*, 7 March. Available online at https://www.nytimes.com/1982/03/07/magazine/world-s-richest-little-isle.html (accessed 3 January 2020).

United Nations System in the Pacific (2017) *United Nations Pacific Strategy 2018–2022: A Multi-Country Sustainable Development Framework in the Pacific Region*. Available online at https://unsdg.un.org /sites/ default/files/2019-12/UNDP_WS_FINAL_UNPS_2018-2022.pdf (accessed 1 February 2021).

Viviani, N. (1970) *Nauru: Phosphate and Political Progress*, Canberra: Australian National University Press.

Watanabe, A. (2018) 'Nauru: From economic goldmine to refugee "hell"', *Kyodo News*, 16 September. Available online at https://english.kyodonews.net/news/2018/09/0fdc626a0cdb-feature-nauru-from -economic-goldmine-to-refugee-hell.html (accessed at 26 May 2020).

Winsor, B. (2017) 'Entrepreneurial refugees are starting restaurants on Nauru, but few are keen to remain on the island prison', *SBS News*, 10 April. Available online at https://www.sbs.com.au/news/nauru-s -culinary-boom-locals-reap-the-rewards-of-stranded-refugees (accessed 15 November 2019).

Williams, M. and B. Macdonald (1985) *The Phosphateers: A History of the British Phosphate Commissioners and the Christmas Island Phosphate Commission*, Australia: Melbourne University Press.

WHO (World Health Organization) (2022) 'Nauru'. Available online at https://covid19.who.int/region/ wpro/country/nr (accessed 5 May 2022).

World Bank (2021) 'Nauru'. Available online at https://data.worldbank.org/country/NR (accessed 1 February 2021).

21

UNDERSTANDING TOURISM DEVELOPMENT IN THE FEDERATED STATES OF MICRONESIA

Reshaping sustainable ways forward for a post-pandemic future

Marcus L. Stephenson, Alexander Trupp, and Chetan Shah

Introduction

The Federated States of Micronesia (FSM) is geographically located between the Philippines and Hawaii. The country can be commonly mistaken for the region of Micronesia. The FSM consists of four distinct states: Chuuk, Kosrae, Pohnpei and Yap, ensuring that the country is geographically and socio-culturally diverse. This chapter provides a critical overview of the relationship between tourism and development in the country, exposing the main concerns and identifying future potentialities. Because the challenges and opportunities for tourism need to be understood in a broader context, the first part of this chapter provides an overview and background of the country, including the geographical, historical and socio-economic context. This is followed by a succinct outline of tourism development prior to detailing the country's socio-cultural assets and attributes aligned to each of the country's four states. The latter part of the chapter focuses more specifically on the critical challenges to tourism, particularly concerning land management, connectivity and transportation, heritage representation and interpretation, climate change, and the COVID-19 pandemic and its impact on the tourism industry. Here, the discussion identifies sustainable opportunities and ways forward for the FSM as a tourism destination, particularly within the context of the post-pandemic era.

Country background

Geographical context

The FSM is located about 1,500 kilometres north of Papua New Guinea, with a landmass of 700.8 square kilometres (Treloar and Hall 2005: 184). The country has a range of natural resources, notably deep-seabed minerals, marine products, phosphate and timber. From estimated

DOI: 10.4324/9780429019968-24

data in 2018, the FSM's agricultural land comprises 25.5% of the landmass, which is mainly for permanent crops (19.7%) and lesser for permanent pasture (3.5%) and arable land (2.3%) (CIA 2022). From the latest figures from United Nation's (UN) World Population Prospects, the FSM has an estimated population of 117,282 (World Population Review 2022). The country is compartmentalized into four geographically separate island states. From east to west, these states are Kosrae (formerly Kusraie), Pohnpei (formerly Ponape), Chuuk (formerly Truk) and Yap, with the capital being Palikir which is located on the island of Pohnpei. The total landmass of each state is relatively small: Kosrae has a landmass of 109 square kilometres, Pohnpei 344, Chuuk 127 and Yap 119 (Dunford and Ridgell 1996). Pohnpei is also the highest state in the FSM, with significant annual rainfall supporting 40 rivers (Haga et al. 2012). Data available for 2020 estimate that Kosrae has a population of just over 6,000, making it the least populated state, and Chuuk has more than 45,000, making it the most populated state (HRSA 2021).

The FSM consists of nine terrestrial ecological zones: (1) coastal/beach, (2) mangrove forest, (3) swamp forest, (4) freshwater marsh, (5) grassland, (6) secondary forest, (7) primary forest, (8) rain forest and (9) crest forest. Forests occupy 63% of the total land area in Kosrae, 56% in Pohnpei, 33% in Yap, and 10% in Chuuk (Haga et al. 2012). These writers note that, although large forest areas only exist in Kosrae and Pohnpei, the country has generally witnessed the conversion of natural upland forests to agroforestry and secondary vegetation.

Given its geographical vastness, the FSM is ethnically diverse, both linguistically and culturally. Although the English language is the official language, there are eight main Indigenous languages: Chuukese, Kapingamarangi, Kosraean, Nukuoro, Pohnpeian, Ulithian, Woleaian and Yapese (HRSA 2021). Kosrae upholds ethnic and linguistic unity, the people of Yap speak Yapese, whose language has some similarities to Chuuk, and whilst Chuuk and Pohnpei have several dialects, two of Pohnpei's atolls host Polynesian inhabitants who communicate differently (Hezel and Foster 2020). From the latest data available, Chuukese represent 48.8% of the population, Pohnpeian 24.2%, Kosraean 6.2%, Yapese 5.2%, Outer Yapese 4.5%, Asian 1.8% and Polynesian 1.5%. Ethnic 'others' represent 6.4% (mainly expatriates from Australia, Europe and the United States (US)), and those designated as 'unknown' represent 1.4% (World Population Review 2022).

Colonial history and 'free association' with the US

The FSM was originally colonized by the Spanish, who conceded colonial rule to Germany after the Spanish-American War in 1899. Chuuk was directly administered by Germany, along with the islands of Palau and Northern Mariana Islands. Japan took over from the outset of World War I and remained the colonial power until the end of World War II, entrusted with administering the Islands of Micronesia by the League of Nations. In fact, the League of Nations decreed that Japan should control all Pacific Islands north of the equator (except Hawaii) under the South Seas Mandate (Poole 2018). Following World War II, the US endorsed a UN trusteeship agreement that enabled close involvement in the Trust Territory, comprising six districts: Northern Marianas, Marshall Islands, Palau, Ponape, Truk and Yap. The US was authorized to be involved in the region's military concerns (Michal 1993), accelerating its involvement in Micronesia under the Kennedy government in the 1960s (Hezel 2017). The country formed its own constitutional government in 1979, although the group of islands of Chuuk, Kosrae, Pohnpei and Yap only became legally recognized in 1986, as the Compact of Free Association (COFA) was confirmed by the US, which ceased its occupation in 1982. Following legal recognition of independence, the FSM aligned itself to the global stage. The country was admitted to the South Pacific Forum in 1987, the International Civil Aviation

Organization in 1988, the Asian Development Bank in 1990 and the UN in 1991 (Michal 1993).

Although the FSM is an independent nation, it continues to be connected to the US through COFA, managing the state's national defence, providing economic assistance and allowing FSM citizens visa-free travel, including the right to reside in the US. There is a substantial reliance on US aid, which expire in 2023. The FSM also significantly depends on the allocation of fishing permits for economic development (Thompson 2013). Nonetheless, as Thompson (2013) further observes, despite limited resources for the export trade, the country does have substantial cultural and historical resources to expand heritage tourism.

Socio-economic context

Prior to the COVID-19 pandemic, the agricultural sector was estimated to represent 26.3% of the gross domestic product (GDP), the industrial sector to represent 18.9% and the service sector to represent 54.8% (Lloyds Bank 2019). The FSM's exclusive economic zone, spreading over 2.7 million kilometres, provides great potential by offering leasing rights and fishing licenses to countries such as China (DFAT 2021). China has been assisting the FSM in the development of various infrastructure projects, including schools, public buildings and farming initiatives (Puas and D'arcy 2021). The government relies on assured US financial aid until the expiry of COFA, providing US$1.3 billion in grants and aid from 1986 to 2001 and around US$2.1 billion for a second (20-year) contract period until 2023 (CIA 2022). Nonetheless, as the Centra Intelligence Agency (CIA) report acknowledges, the country's medium-term economic prospects are considered bleak given economic dependence on the US and lack of proactivity by the private sector, as around two-thirds of the labour force are government employees (CFE-DM 2019). As this pessimistic economic outlook was expressed prior to the COVID-19 pandemic, the additional socio-economic challenges are of fundamental concern for the FSM.

Prior to the pandemic, the World Bank (2018) noted that the FSM's poverty headcount rate stood at 41.2%, and poverty levels were highest in Chuuk (45.5%), slightly lower in Yap (39.4%) and Pohnpei (39.2%) and the lowest in Kosrae (21.0%). The country's free association with the US ensures that its citizens have rights to access the US in the same way Cook Islanders can live and work in New Zealand (Bedford and Hugo 2012). By 2013, about 50,000 Micronesians emigrated to the US mainland, including US jurisdictions (e.g., Guam and the Commonwealth of the Northern Mariana Islands) (Hezel 2013). Consequently, 'brain drain' is a significant concern for the country's socio-economic development, as there is a shortage of educated professionals (IOM 2016).

Tourism development – overview

In the post-war era of the 1950s, tourism mobility patterns to the islands were controlled and special permission to visit had to be granted. However, one of the main drivers to tourism and travel was the establishment of Continental Air Micronesia in 1968, with a new airstrip built in Pohnpei in February 1970 (Hezel 2017). Currently, each of the states has its own international airport: Kosrae International Airport, Pohnpei International Airport, Chuuk International Airport and Yap International Airport (HRSA 2021).

The FSM's 'National Tourism Development Strategy' was identified in 1997 by the central government, especially to develop a robust commercial sector in tourism and ensure that social and economic benefits intersect with culture and the environment (Treloar and Hall 2005: 188–189). Nonetheless, the tourism industry has not grown significantly over the past three decades, with tourism numbers decreasing from 23,171 in 1990 (2005: 190) to 18,019 in 2019, of which 60% visited Pohnpei, 28% visited Chuuk, 6% visited Yap, and 6% visited Kosrae (Embassy of the Federated States of Micronesia 2022). However, there has been continuous government recognition to develop a clear strategy for tourism development, as reflected in the 'FSM 2023 Action Plan', indicating that tourism represents a 'key driver of the growth strategy' and, to develop the tourism industry, it is necessary to encourage investment in a large-scale hotel development (Government of the Federated States of Micronesia 2015a: 8). Nonetheless, it is initially necessary to present a description of the country's destination attributes, especially in recognition of its unique socio-cultural and historical resources.

Socio-cultural assets and destination attributes

The FSM has affinities to prehistoric, pre-European and European-Asian histories, in which each of the four FSM states features distinct tourism attractions and products that are the country's central attributes.

Kosrae

Kosrae's significant cultural attraction is its stone city, known as Lelū, located in the northeast bay and was the home of the high chiefs and the king. Lelū was still occupied in the 1850s and was perceived as the capital of Kosrae, and, although later claimed by the jungle, it is characterized by walled compounds (Cordy 1982). Cordy notes the historical attributes of the feudal nature of Kosrae, in which Kosraen warriors overthrew the Saudeleur dynasty at Nan Madol (the stone city in Pohnpei Island), establishing themselves as the new ruling dynasty between the years 1510–1575 AD. There are ongoing renovations to the Lelū ruins and the Kosrae Museum, encouraged by the downturn in tourism numbers due to the COVID-19 pandemic, which could encourage a surge towards cultural heritage tourism once more people travel freely and safely (Embassy of the Federated States of Micronesia 2022), especially if the Kosrae Visitors Bureau (KVB) works closely with the federal government entities to proactively promote and advance this form of tourism. KVB estimates that, prior to the pandemic, the island received approximately 2,100 visitors annually (FSM Department of Resources and Development 2020). Although Lelū is a major attraction, there is also the Menka Ruins, a sacred site in the centre of the island dedicated to Sinlaku, the Breadfruit Goddess and Prophet Spirit (Beardsley 2014). This traditional cultural heritage site is set within a broader ecological experience encompassing the reef, mangroves, waterfalls, lagoons and the Yela Ka 'Terminalia' Forest, along with the trails attached to the mountains of Olum, Poro, Finkol, Oma and Mutunte (FSM Department of Resources and Development 2020).

Pohnpei

Pohnpei state has a larger and more renowned stone city, known as the Nan Madol Ceremonial Centre of Eastern Micronesia, representing one of the few UN Educational, Scientific and Cultural Organization (UNESCO) World Heritage Sites in the Pacific Islands. Nan Madol (Figure 21.1) was the centre of the Saudeleur empire, built on 92 artificial islets spread over 200 acres and connected by walled canals and subterranean tunnels. It was built from the thirteenth to the seventeenth centuries by the Saudeleurs, involving islets formed by walls of columnar basalt infilled with coral rubble (Dahl 1993). One intricate building is the Nandowas, the royal mortuary, surrounded by walls 25 feet high that were constructed with cornerstones weighing 50 tons (Pala

Figure 21.1 A channel and town walls in Nan Madol – a stone city built of stone slabs. Source: Maloff (www.shutterstock.com/image-photo/channel-town-walls-nan-madol-prehistoric-1216711867).

2009). The mysteries and enchantment surrounding Nan Madol inspired the American writer, Howard Philips Lovecraft (1880–1937), to describe the fictitious lost city of R'lyeh, appearing in the short story: 'The Call of Cthulhu'. The aquatic ruins and largely built formations characterized his mythical city (Smith 2017). Advantageously, Nan Madol is protected by Pohnpei's 'Historic and Cultural Preservation Act of 2002', acknowledging that Pohnpei 'contains a wealth of historic, archaeological, and cultural properties' which are 'important to the maintenance and development of the identity, pride, and integrity of the people of Pohnpei, and to the world's understanding of Micronesian history and culture' (Angyal et al. 2021: 78). However, upon its UNESCO designation in 2016, Nan Madol was added to the UNESCO List of World Heritage in Danger for which problems concerning vegetation overgrowth and siltation of the waterways have contributed to the uncontrolled growth of mangroves and destabilization of building structures (UNESCO n.d.).

Chuuk

Chuuk has a strong association with World War II, with sunken remnants of the fleet belonging to the Imperial Japanese Navy, also known as the 'ghost fleet', initially popularized by Phillippe Cousteau who visited the island in 1969 and produced a film on shipwrecks (Hezel 2017). More than 60 of its ships were sunk in the lagoon by American planes (Dunford and Ridgell 1996). Chuuk has Japanese coastal fortifications seen through pillboxes, antiaircraft gun embankments and aviation infrastructure; such infrastructure is well represented in the Etten airfield on the island of Etten, indicating the ambitions of the Japanese empire to dominate the Pacific region. In terms of Chuuk's tourism landscape, Pruett (2020: n.p.) summarizes the state's historic and international significance, emphasizing:

> During World War II, Chuuk was considered Imperial Japan's 'Gibraltar of the Pacific' because of its fortifications' reputed impregnability and its 'Pearl Harbor', because its wide and deep lagoon was the Imperial Japanese Navy's largest forward operating base. Following the Anglo-American 'Operating Hailstone' attacks of February 1945, the lagoon became the largest ship graveyard in the world.

Yap

This state is generally perceived as the most traditional state, and local society is structured by a complex organization of castes, councils and chiefs (Stumpf and Cheshire 2019). Yap is renowned for its giant stone money (Figure 21.2), which has socio-economic meaning. These flat rounds of stones with large holes in the middle were mined in Palau, several hundred miles away to the west of the state, and the more arduous the journey to bring the stones back to Yap, the more they were worth (Dunford and Ridgell 1996). In addition, material Yapanese culture, such as community houses, shell money, sailing canoes and fish traps, form part of the state's ethnic tourism attractions (Mansperger 1992). Other attractions in Yap are linked to its marine resources, manta ray diving and deep-sea fishing (Stumpf and Cheshire 2019). One unique destination is the Ulithi Atoll, which is administered by Yap and consists of 41 islets totalling 4.5 square kilometres and located 191 kilometres east of Yap (Ongaro 2016). As Ongaro notes in her study, although the place is at risk of continuous socio-cultural change, there are opportunities to develop its cultural and heritage assets, notably 'indigenous knowledge and practices, from fishing techniques to nutrition habits to dances to the art of navigation' (2016: 29).

Figure 21.2 Stone money in Yap. Source: Iurii Kazakov (www.shutterstock.com/image-photo/stone-money-yap-island-1101482138).

Tourism development challenges

The FSM faces a range of challenges, though the critical concerns are land management, connectivity, heritage interpretation, climate change and the COVID-19 pandemic.

Land management

The complexities of the land ownership system are a challenge to tourism development, making access to land difficult and risk prone, which has 'led to underutilised and unproductive lands, hampering the growth of the tourism industry' (ADB 2015: n.p.). This report observes land ownership concerns in Pohnpei and emphasizes that, although non-citizens cannot buy land it can be leased, but the process can be time-consuming, compounded by the fact that there may be indeterminate land boundaries and titles. The government of the FSM needs to resolve legal claims to land. The case of an American couple who had to close their hotel on the island of Pohnpei after 40 years of operation is an example of problems concerning the land lease and lease extensions between tourism operators and local landowners (Dorney 2013). Increasing the lease period from 25 years to 99 years, similar to other Pacific Island states, will encourage trust and safety of investment by foreign individuals (Francis and Hezel 2006). However, there is also a need to move from top-down approaches focusing on liberalized land holding laws and foreign investment to bottom-up approaches to understand what land and existing land tenure systems mean for different individuals and communities in the FSM (Stumpf and Chesire 2019).

Connectivity

Stumpf and Cheshire (2019) further implicate that one of the FSM's main challenges in tourism development is the limited policy coordination between tourism and air transportation. These writers make note of the well-intentioned attempt from the 1960s by Continental Airlines to not only develop air passenger transport but also to develop a chain of hotels across Micronesia, starting with the Continental Hotel in Chuuk, which was built in 1970. However, the company's proposal to build a hotel in Pohnpei was thwarted at the local government level. Consequently, the endeavour for an 'integrated supply chain' approach did not reach fruition, and United Airlines subsequently bought Continental Airlines in 2011, which is now the current and main provider of international air transportation to and from the FSM. Connectivity is a challenge, as the country is geographically far from principal markets in Asia, which is compounded by higher airfares (Hezel 2017). Currently, the FSM relies on a single airline for connectivity to other nations. The major route to Chuuk International Airport is through Guam, which connects to the international destinations of Hong Kong, New York, Seoul, Singapore and Tokyo. United Airlines originates in Guam and passes through Chuuk, Pohnpei and Kosrae before continuing to the Marshall Islands and then ending in Hawaii, which involves the Island Hopper service that has connected the islands – with a separate connecting flight from Guam to Yap. However, due to the COVID-19 pandemic, the service was downgraded from a regular service to a monthly service in 2020 (Rust 2020).

For the FSM to connect to the lucrative Chinese market, Chinese tourists would have to travel from Beijing to Palau and then onto the FSM via Guam, involving three flights. Therefore, the country faces a double challenge of longer flying hours (based on routes) and higher fares, as tourists have to return to Guam for their onward return journey. Over-reliance on one airline has also led to monopolistic market positioning, which can be a challenge for developing countries (Taumoepeau 2007). Therefore, the country needs to evaluate other airlines and routes for improved and more economical forms of connectivity if it intends to surge forward with increased tourist arrivals in the post-pandemic era. Improved connectivity through new routes and airlines can also offer additional markets which may have been restrained due to the requirement of a US visa to fly to Guam.

Internal mobility and connectivity are also a challenge, together with undeveloped infrastructure, including poor roads and power and water supply (Stumpf and Cheshire 2019). Although the FSM has the potential to tap into the cruise market, there are limited ports of call for cruise ships. Most inter-island travel within the FSM is by boat, as the prices compared to airfares are more affordable. There are two vessels that service inter-island and inter-state travel, although this form of travel is not popular for tourists. Furthermore, due to unregulated private operators and limited government services, reliable and safe transportation services are a significant challenge (World Bank 2019). In addition, the FSM has no public transport on its islands for visitors. However, accommodation providers can arrange shuttle buses, and the main islands have numerous car rental companies and taxi services (PSDI 2021).

Heritage representation and interpretation

Current day cultural practices, customs and legends are associated with canoe building and sailing, especially as the initial inhabitants of the FSM were coastal dwellers and well known for their navigational and boat building skills. The FSM contains a variety of tangible forms of heritage relating to dwellings and material remains of past foreign trade and successive colonizers, including World War II remnants. Tangible heritage sites include tidal stone-walled fish weirs (*aech*), seventeenth-century Spanish ships, nineteenth-century trading vessels and whaling vessels and Japanese and American World War II remains (aircraft, naval and merchant ships) (Jeffery 2014). According to Jeffery's (2006) study of Chuuk Lagoon's submerged heritage sites, the tangible heritage sites of World War II have more personal significance to the Japanese and Americans than to the Chuukese, and this is despite dive tourism having economic value to destinations.

Therefore, the key to the preservation of local culture and heritage is to ensure that the country's traditions, ethnic diversities, and localized histories are carefully and accurately represented by the tourism industry. Intangible heritage plays a crucial role in society and in daily life, through evoking oral histories, participating in singing and being involved in rhythmic recitations about communities, places and traditional practices (e.g., navigation) (Diettrich 2015). These activities express tradition, history and social continuity. Furthermore, the country's culturally relevant places can be better preserved and narrated. For instance, at cultural sites in Yap, the traditional houses are overgrown, and the ancient paths and the dock which delivered the stone money lack clear heritage interpretation, with no effective management plan to preserve cultural assets (Morris 2017). There is also a colonial narrative that arguably needs to be fully represented, in which some of the stone money left the country during the colonization era. The money was collected or gifted through colonial occupations as well as acquired by other countries and displayed in such museums as the British Museum (UK), the Übersee-Museum in Bremen (Germany), the Peabody Museum of Archaeology (Harvard University) (US) and the Tokyo National Museum (Japan) (Gillilland 1975: 37–55). Therefore, one related concern is how heritage tourism in the FSM can critically narrate its political history of colonial diversities from Europe, the Far East and America – though, at the same time, seriously recognize its own cultural diversities as well as more localized internal and regional political histories.

Climate change

The FSM experiences a yearly sea-level increase of around 10 millimetres compared to the worldwide sea-level rise of 3.1 millimetres per year (Byrnes and Harrington 2019). As much of the FSM's population lives near the coastal areas, the economy is tied to coastal resources and infrastructure. Fletcher and Richmond (2010) noted more than a decade ago that no specific coastal management policy was in place in the FSM, leading to problems in managing risks associated with climate change. These writers also observe that coastal erosion continues to threaten groundwater supplies, agroforestry production, habitable dwellings, roadways and beaches, all interfering with plans for tourism development. The impact of climate change and natural disasters on the country's road network is a major challenge. This has motivated the World Bank (2021) to approve US$40 million for the FSM to upgrade its road network and infrastructure, encouraging more travel and tourism mobility within and across the islands. This project intends to ensure that travel is safe and reliable.

As the FSM is highly susceptible to climate change impacts, it is likely to experience environmental, social and economic losses due to climate change-induced threats, affecting fisheries-related revenue, agricultural production, livelihoods and food security. Extreme weather (such as hurricanes and typhoons) increases the vulnerability of infrastructure, which will impact access to basic services, reduce labour productivity and capital accumulation and affect human health, thereby critically impacting the FSM's GDP (IMF 2019). Successfully achieving climate adaptation within the FSM may be facilitated by two steps: (1) strengthening international partnerships to aid adaptation efforts and (2) continuing the development of internal policies focused on building resilient and sustainable communities (Fletcher and Richmond 2010; IMF 2019) (see Chapter 27).

The COVID-19 pandemic, its impact and sustainable ways forward

Although the FSM has not attracted mass tourists or resort-based tourism, as seen in such Pacific Island countries as Fiji, the COVID-19 pandemic has impacted the country's economy since the closure of its international borders in March 2020. Therefore, without tourism revenues, the FSM's economy contracted by 5.4% in 2020, although this is not as significant as the neighbouring country of Palau, where the economy was affected by 13.8% (Clarke et al. 2022). Considering the relatively small size of the FSM, employment retraction in the hospitality industry was significant. There were 669 projected jobs losses in the hotel and restaurant sector, and 609 jobs in the transport sector (EconMap 2020). The real GDP is estimated to have declined by 1.8% in 2020, as COVID-19 restrictions on the movement of personnel and materials delayed major investment projects, resulting in a decline in domestic service activity (IMF 2021).

Despite the economy being impacted, the situation could have been far worse if support systems were not in place. Therefore, the government initiated an economic stimulus package supporting businesses impacted by the lack of tourism activities, enabling individuals to receive wage subsidies and gross tax revenue rebates, which commenced through the 'Tourism Sector Mitigation Fund' (Government of the Federated States of Micronesia 2020: 74). Accordingly, the FSM received US$118 million in US aid in less than one year of the lockdown (Westerman 2021). The Asian Development Bank (ADB) and the government of the FSM authorized US$5.5 million grants to fund transport projects in the country, with ADB contributing US$5 million of this amount to help strengthen road maintenance and development as well as improve connectivity by ensuring roads and bridges are able to sustain climate change and natural hazards (ADB 2021). Other support systems concern the role of local communities, where residents (especially those in Kosrae) focused on rural life, with some evidence of increased agricultural production, gardening and taro patching (LMMA Network/KCSO/TRCT 2020).

The FSM government must take the opportunity to re-address the challenges faced by tourism development, working with key stakeholders (including the private sector) to establish sustainable ways forward for the betterment of tourism development, particularly within the context of a post-pandemic society and economy. These sustainable dimensions should be framed with regard to socio-cultural, environmental/ecological and economic forms of sustainability. Importantly, sustainable tourism is recognized as one of the FSM's main economic pillars (Government of the Federated States of Micronesia 2020), though its social and cultural dimensions need to be applied carefully if this form of tourism is going to be a significant component for future development. The government has been acutely aware of the need to intersect sustainable tourism with 'other niche types of tourism like ecotourism, responsible tourism, ethical tourism, and pro-poor tourism' (Government of the Federated States of Micronesia 2015b: 17). However, as implicated in this chapter, it is necessary to pursue heritage tourism as a culturally sustainable form of tourism, as long as the cultural histories of the country and the four states are carefully and genuinely represented. Moreover, tourism planning ought to ensure that traditional cultural heritage sites synchronize with the wider ecological experience.

In Micronesian societies, the local population is strongly tied to their land and cultural heritage. Early studies on the impacts of tourism in the Yap show an overall positive picture, with little perceived tourism-induced change on the environment and local culture (Mansperger 1992). However, plans for a mega-project backed by Chinese investment and involving the construction of a 1,500 room hotel caused deep concerns and divisions within the local communities (Puas and D'arcy 2021), with some locals fearing that they would be 'reduced to just doing cultural dances' for tourists; whilst others actually welcomed outside investment in tourism development (Bohane 2016: n.p.). Nonetheless, local landowners subsequently opposed the project, as they were not properly consulted of the plans (Stumpf and Cheshire 2019), which suggests that land has much higher value and meaning than being perceived as a purely economic asset.

Future approaches to tourism development in the FSM, including socially sustainable forms of tourism, ought to embrace community engagement and consultation and be consistent with local perspectives and needs, as long as community residents are willing to carry any possible social and communal costs that such forms of tourism may actually entail (Ringer 2004). Hofmann's (2018) study of the local reception of climate change in Chuuk found that, although this concept was not well known to residents, residents observed and experienced such environmental changes as sea-level rise and flooding of houses and pathways in coastal areas. However, they were more concerned about socio-cultural change and viewed climate change as a consequence of the transformation of cultural values (e.g., sharing and respect). Accordingly, climate change is conceived of as one of many changes. Hence, there is a need to contextualize local interpretations and concepts of sustainability, ensuring that the socio-cultural component is central to tourism planning and development. Stumpf and Cheshire (2019) indicate that the importance of localizing the UNs sustainable development goals (SDGs). For instance, in relation to SDG 15 ('life on land'), these authors illustrate the importance of a sense of place as embedded in the ownership of land, stating:

> Life on Land (SDG 15) is predicated not on liberalised regulations that may facilitate sustainable development via tourism in Pacific Islands, but on retention of Indigenous land holding systems as the lifeblood of the Indigenous livelihoods tied directly to it.
>
> *(2019: 14)*

Sustainable tourism needs to be sensitive to local livelihoods and be respectful of Indigenous rights. In addition to socio-cultural and environmental forms of sustainability, tourism needs to be economically sustainable in the long-term. The FSM's economy mainly consists of subsistence agriculture and fishing, with relatively limited private sector activity. In order to diversify its economy and add more sustainable tourism options, the country ought to focus on advancing available and viable markets aligned to sustainable principles and practices, especially as the tourism industry has not significantly grown in the past two decades. One such option concerns cultured pearl farming. Oyster shells have been sold in the FSM since the eighteenth century, and the Japanese occupation of Micronesia further fuelled interest in pearl oyster resources, though in the past three decades the country has witnessed substantial efforts to develop commercial and community-based pearl farming. Cultivation of black cultured pearls started to evolve in the FSM, where pearl farming and related economic activity are known for their development value, as can be seen in French Polynesia and the Cook Islands (Cartier et al. 2012). The demand for cultured pearls from the FSM is growing in international markets, especially in Japan. Also, through purchasing pearls, tourists contribute financially to the country's GDP and foreign exchange, as the pearl industries create both formal and informal employment. The sale of pearls as souvenirs in the form of jewellery and as 'inlays' in handicrafts and furniture is gaining traction, including pearl farm tours such as the ones in Fiji (Chand et al. 2015). Crucially, as Cartier and Ali (2012: 30) assert:

> cultured pearl farming is one of the few economic activities in which sound environmental management and conservation are a prerequisite to economic success, the sector offers some interesting insights with regard to sustainability in the marine realm.

Consequently, the FSM needs to pursue niche forms of tourism, especially to showcase its unique attributes pertaining to socio-cultural and ethnic diversities as well as environmental and geographical uniqueness.

Conclusion and research implications

While the FSM is an independent nation, its association with the United States has been crucial for providing economic assistance and enabling greater freedom of mobility of FSM citizens. Nevertheless, limited connectivity remains one of the main challenges in tourism development, both domestically and internationally. Although the FSM is characterized by a rich cultural heritage, both in terms of tangible and intangible cultural resources, there are long-term concerns relating to heritage conservation. Nan Madol was placed on the World Heritage List in 2016. However, it was simultaneously inscribed onto the List of World Heritage in Danger because it is threatened by vegetation overgrowth, erosion, siltation and human activities. Sustainable conservation strategies are needed which can be informed by remote sensing methods such as Airborne LiDAR, which involves laser scanning to assess the conditions and structures of the heritage site (Comer et al. 2019).

Climate change and the lack of specific coastal management policies threaten groundwater, agricultural production, residential housing, roadways and beaches, all interfering with plans for the development of tourism. Whilst the FSM government strongly recognizes tourism as one of the main economic drivers for the country, there is no agreement on tourism development among different stakeholders. Therefore, it is imperative that future research employs both a multistakeholder perspective and involves methodologies that pursue multistakeholder engagement activities, empirically engaging with the public and private sector and tourist and host communities. This is necessary in order to consider the range of perspectives and areas of consensus concerning the development and advancement of specific niche types of sustainable tourism, as long as the interests of the local community are centralized, as this is particularly crucial for the more traditional societies such as Yap. For sustainable tourism initiatives to be successful, the complexities and political and cultural dynamics of land ownership must be acknowledged and understood. As this is a complex field, tourism research should focus on the different ethnic, cultural and island nuances relating to land ownership, land management and Indigenous land holding systems, especially when the focus is on the built environment or when tourists utilize or access land resources. The ultimate intention is to seriously contextualize how sustainable forms of tourism can be developed and advanced but in socio-cultural harmony with the land and physical environment.

Given that there is arguably a need to stimulate the FSM economy because of the recognized challenges associated with the COVID-19 pandemic, it would be useful to also look at market research which considers the economic viability of particular kinds of sustainable tourism. However, such an approach should incorporate a social marketing perspective which encourages tourism segments that are beneficial to local society and culture (and the environment). Another form of tourism that would benefit from employing rigorous research would be to assess the FSM's VFR market, through addressing how this market segment could be significantly enhanced within a post-pandemic context, especially as this market has much potential due to strong emigration patterns of FSM citizens to US territories. VFR tourism is arguably economically sustainable in tourism destinations, as these tourists have a tendency to have a greater dispersal of spending patterns and can reflect stable demand (Griffin 2013). It is also socio-culturally sustainable in that it can strengthen ethnic and cultural ties (Stephenson 2002).

References

ADB (Asian Development Bank) (2015) 'Understanding land issues and their impact on tourism development: A political economy analysis of Pohnpei, Federated States of Micronesia', September. Available online at https://www.adb.org/sites/default/files/publication/176085/land-issues-tourism-development-fcas.pdf (accessed 26 April 2022).

ADB (Asian Development Bank) (2021) 'ADB, FSM sign 45.5 million grant to develop road projects', 15 November. Available online at https://www.adb.org/news/adb-fsm-sign-5-5-million-grant-develop-road-projects (accessed 1 March 2022).

Angyal, D.C., L.M. Cupps, and B. Myers (2021) *Federated States of Micronesia Legal & Policy Framework Assessment Report*, 27 August, Blue Prosperity Coalition. Available online at https://gov.fm/files/FSM_Legal_Framework_Assessment_V4_Final_27AUG21.pdf (accessed 6 February 2022).

Beardsley, F. (2014) 'Temple architecture in the sacred site of Menka, Kosrae, Federated States of Micronesia', *Monuments and People in the Pacific*, 20: 191–217.

Bedford, R. and G. Hugo (2012) *Population Movement in the Pacific: A Perspective on Future Prospects*, New Zealand: Labour and Immigration Research Centre, Department of Labour.

Bohane, B. (2016) 'Chinese company seeks to build "mega resort" on remote Micronesian island of Yap', *ABC News*, 2 May. Available online at https://www.abc.net.au/news/2016-05-02/chinese-company-seeks-to-build-mega-resort-on-island-of-yap/7300588 (accessed 26 December 2021).

Byrnes, H. and J. Harrington (2019) 'Global warming threatens islands', *TTR Weekly*, 20 November. Available online at https://www.ttrweekly.com/site/2019/11/global-warming-threatens-islands/ (accessed 7 September 2020).

Cartier, L. and S. Ali (2012) 'Pearl farming as a sustainable development path', *Solutions*, 3(4): 30–34.

Cartier, L.E., M.S. Krzemnicki, and M. Ito (2012) 'Cultured pearl farming and production in the Federated States of Micronesia', *Gems and Gemology*, 48(2): 108–122.

CFE-DM (Center for Excellence in Disaster Management and Humanitarian Assistance) (2019) *Federated States of Micronesia: Disaster Management Reference Handbook*, November, Hawaii. Available online at https://

reliefweb.int/sites/reliefweb.int/files/resources/disaster-mgmt-ref-hdbk-FSM.pdf (accessed 19 September 2020).

Chand, A., S. Naidu, P.C. Southgate, and T. Simos (2015) 'The relationship between tourism, the pearl and mother of pearl shell jewellery industries in Fiji'. In: S. Pratt and D. Harrison (eds.) *Tourism in Pacific Islands: Current Issues and Future Challenges*, Abingdon: Routledge, pp. 148–164.

CIA (Central Intelligence Agency) (2022) 'Micronesia, Federated States of', *The World Factbook*. Available online at https://www.cia.gov/the-world-factbook/countries/micronesia-federated-states-of/ (accessed 26 April 2019).

Clarke, C., A. Wheatley, M.R. Fraser, T. Richey, and K. Ensign (2022) 'US-affiliated Pacific islands response to COVID-19: Keys to success and important lessons', *Journal of Public Health Management and Practice*, 28(1): 10–15.

Comer, D.C., J.A. Comer, I.A. Dumitru, W.S. Ayres, M.J. Levin, K.A. Seikel, and M.J. Harrower (2019) 'Airborne LiDAR reveals a vast archaeological landscape at the Nan Madol World Heritage Site', *Remote Sensing*, 11(18): 2152.

Cordy, R. (1982) 'Lelū, the stone city of Kosrae: 1978–1981 research', *The Journal of the Polynesian Society*, 91(1): 103–119.

Dahl, C. (1993) 'Tourism development on the island of Pohnpei (Federated States of Micronesia): Sacredness, control and autonomy', *Ocean and Coastal Management*, 20(3): 241–265.

DFAT (Department of the Foreign Affairs and Trade) (2021) 'Federated States of Micronesia country brief'. Available online at https://www.dfat.gov.au/geo/federated-states-of-micronesia/federated-states-of-micronesia-country-brief (accessed 23 December 2021).

Diettrich, B. (2015) 'Performing arts as cultural heritage in the Federated States of Micronesia', *International Journal of Heritage Studies*, 21(7): 660–673.

Dorney, S. (2013) 'Pacific tourism icon closes after 40 years', *Australia Network News*, 18 April. Available online at https://www.abc.net.au/news/2013-04-18/an-fsm-tourism-icon-the-village-closes/4636220 (accessed 27 December 2021).

Dunford, B. and R. Ridgell (1996) *Pacific Neighbours: The Islands of Micronesia, Melanesia and Polynesia*, Honolulu: Pacific Resources for Education and Learning.

EconMAP Program (2020) 'Assessing the impact of COVID-19 on the Federated States of Micronesia economy', *Economic Monitoring and Analysis Program (EconMAP) Technical Notes*, 3 June, Graduate School USA's Pacific and Virgin Islands Training Initiatives. Available online at https://pitiviti.org/storage/files/econmap/FSM_EconImpact_COVID-19_June2020_Web_Remediated.pdf (accessed 26 April 2022).

Embassy of the Federated States of Micronesia (2022) 'The FSM's tourism destinations are preparing for open borders, and the FSM's agriculture sector hurts from lack of funding', 31 January. Available online at https://fsmembassy.fm/the-fsms-tourism-destinations-are-preparing-for-open-borders-the-fsms-agriculturesector-hurts-from-lack-of-funding-day-4-of-the-2nd-resources-development-conference/ (accessed 3 February 2022).

Fletcher, C.H. and B.M. Richmond (2010) *Climate Change in the Federated States of Micronesia: Food and Water Security, Climate Risk Management, and Adaptive Strategies*, Center for Island Climate Adaptation and Policy, Honolulu: University of Hawai'i Sea Grant College Program.

Francis, X. and S.J. Hezel (2006) *Is That the Best You Can Do? A Tale of Two Micronesian Economies*, Honolulu: East West Centre.

FSM Department of Resources and Development (2020) *Federated States of Micronesia: Forest Action Plan 2020–2030*, 31 December. Available online at https://www.stateforesters.org/wp-content/uploads/2021/02/FSM-Forest-Action-Plan-2020-2030-Jan24-FINAL.pdf (accessed 4 February 2022).

Gillilland, Cora.L.C. (1975) *The Stone Money of Yap: A Numismatic Survey*, Washington, D.C.: Smithsonian Studies in History and Technology, Smithsonian Institution Press.

Government of the Federated States of Micronesia (2015a) *FSM 2023 Action Plan*, 2023 Planning Committee, Palikir, Pohnpei, February. Available online at https://dofa.gov.fm/wp-content/uploads/2018/12/FSM-2023-Action-Plan.pdf (accessed 1 March 2022).

Government of the Federated States of Micronesia (2015b) *Federated States of Micronesia, National Tourism Policy Volume I. Final Report*, June 2015. Available online at https://www.fsmstatistics.fm/wp-content/uploads/2019/10/4-FSM-National-Tourism-Policy-Vol-1a_W-Foreword_8July.pdf (accessed 27 December 2021).

Government of the Federated States of Micronesia (2020) *Federated States of Micronesia, First Voluntary National Review on the 2030 Agenda for Sustainable Development*, June 2020, SDG Working Group, Department of Resource and Development, Palikir, Pohnpei. Available online at https://pacificdata.org/data/dataset/oai-www-spc-int-5bc2b7d1-ee93-48c3-b0d4-157c6e8b7522 (accessed 26 April 2022).

Griffin, T. (2013) 'Visiting friends and relatives tourism and implications for community capital', *Journal of Policy Research in Tourism, Leisure and Events*, 5(3): 233–251.

Haga, M., O. Hanada, N. Takahashi, N. Kusakabe, H. Gotoh, Y. Maeno, and M. Takezawa (2012) 'Development and infrastructure of the Federated States of Micronesia', *WIT Transactions on Ecology and the Environment*, 166: 169–179.

Hezel, F.X. (2013) *Micronesians on the Move: Eastward and Upward Bound*, Honolulu, HI: East-West Center.

Hezel, F.X. (2017) *On Your Mark, Get Set… (Tourism's Take-Off in Micronesia)*, Honolulu, HI: East West Centre.

Hezel, F.X. and S. Foster (2020) 'Micronesia', *Encyclopedia Britannica*, 6 February. Available online at https://www.britannica.com/place/Micronesia-republic-Pacific-Ocean (accessed 12 July 2020).

Hofmann, R. (2018) 'Localizing global climate change in the Pacific knowledge and response in Chuuk, Federated States of Micronesia (FSM)', *Sociologus*, 68(1): 43–62.

HRSA (U.S. Department of Health and Human Services) (2021) *Overview of the state – Federated States of Micronesia – 2021*. Available online at https://mchb.tvisdata. hrsa.gov/Narratives/Overview/5287367a-ff2c-4459-b22a-5344998a0548 (accessed 25 April 2022).

IMF (International Monetary Fund) (2019) *Federated States of Micronesia: Climate Change Policy Assessment*, 6 September. Available online at https://www.imf.org/~/media/Files/Publications/CR/2019/1FSMEA2019002.ashx (accessed 14 March 2022).

IMF (International Monetary Fund) (2021) *Federated States of Micronesia: 2021 Article IV Consultation – press release; staff report; and statement by the executive director for the Federated States of Micronesia*, Asia Pacific Department. Available online at https://www.elibrary.imf.org/view/journals/002/2021/237/article-A001-en.xml (accessed 3 February 2022).

IOM (International Organization for Migration) (2016) '*Migration in the Federated States of Micronesia: A country profile 2015*', 5 May, Geneva. Available online at https://publications.iom.int/books/migration-federated-states-micronesia-country-profile-2015 (accessed 26 April 2022).

Jeffery, B. (2006) 'A CRM approach in investigating the submerged World War II sites in Chuuk Lagoon', *Journal of the Humanities and Social Sciences*, November, 137–155.

Jeffery, B. (2014) 'The underwater cultural heritage of the Federated States of Micronesia', *Asia and the Pacific Conference on Underwater Cultural Heritage*, May 12–16, Honolulu, Hawaii.

LMMA Network (Locally Managed Marine Area Network International), KCSO (Kosrae Conservation and Safety Organisation) and TRCT (Tamil Resources Conservation Trust) (2020) 'COVID-19 impacts on fishing and coastal communities: Update #3: Federated States of Micronesia', 15 June. LMMA Network. Available online at https://www.peump.dev/sites/default/files/2020-08/LMMA%20Network%2C%20KCSO%:20and%20TRCT.%20Covid%20Update%20%233%20FSM%2015.06.2020.pdf (accessed 26 April 2022).

Lloyds Bank (2019) 'Federated States of Micronesia: Economic outline'. Available online at https://www.lloydsbanktrade.com/en/market-potential/federated-states-of-micronesia/economy?vider_sticky=oui (accessed 6 November 2019).

Mansperger, M. (1992) 'Yap: A case of benevolent tourism', *Practicing Anthropology*, 14(2): 10–13.

Michal, E.J. (1993) 'Protected states: The political status of the Federated States of Micronesia and the Republic of the Marshall islands', *The Contemporary Pacific*, 5(2): 303–332.

Morris, D. (2017) 'A remote Pacific island faces up to China: Can Yap Island both protect its unique heritage and capitalise on Chinese investment?', *The Diplomat*, 26 June. Available online at https://thediplomat.com/2017/06/a-remote-pacific-island-faces-up-to-china/ (accessed 21 September 2020).

Ongaro, A. (2016) 'Challenges and opportunities for sustainable tourism development in Ulithi Atoll', *Ara: Journal of Tourism Research*, 6(1): 25–37.

Pala, C. (2009) 'Nan Madol: The city built on coral reefs', *Smithsonian Magazine*, 3 November. Available online at https://www.smithsonianmag.com/history/nan-madol-the-city-built-on-coral-reefs-147288758/ (accessed 5 December 2009).

Poole, R.M. (2018) 'Yap, the Pacific island Japan has almost forgotten', *Japan Times*, 25 May. Available online at https://www.japantimes.co.jp/life/2018/05/25/travel/yap-pacific-island-japan-almost-forgotten/ (accessed 25 April 2022).

Pruett, R.K. (2020) 'Where is Chuuk heading?' 3 June. Washington: Centre for Australian, New Zealand and Pacific Studies, Georgetown University. Available online at https://canzps.georgetown.edu/2020/06/03/where-is-chuuk-heading/ (accessed 3 February 2022).

PSDI (Pacific Private Sector Development Initiative) (2021) *'Federated States of Micronesia. Pacific tourism sector snapshot'*, November, Sydney: Asian Development Bank. Available online at https://www.pacificpsdi.org/assets/Uploads/PSDI-TourismSnapshot-FSM. pdf (accessed 26 April 2022).

Puas, G. and P. D'arcy (2021) 'Micronesia and the rise of China: Realpolitik meets the reef', *The Journal of Pacific History*, 56(3): 274–295.

Ringer, G. (2004) 'Geographies of tourism and place in Micronesia: The "sleeping lady" awakes', *Journal of Pacific States*, 26(1&2): 131–150.

Rust, S. (2020) 'As COVID-19 cases climb in the U.S., there are still none in the Marshall Islands', *Los Angeles Times*, 16 July. Available online at https://www.latimes.com/world-nation/story/2020-07-16/as-covid-19-cases-climb-in-the-u-s-there-are-still-none-in-the-marshall-islands (accessed 25 April 2022).

Smith, J.R. (2017) 'Son of providence: The weird legacy of H.P. Lovecraft', *CNN Travel*, 25 October. Available online at https://edition.cnn.com/travel/article/providence-h-p-lovecraft/index. html (accessed 21 September 2020).

Stephenson, M.L. (2002) 'Travelling to the ancestral homelands: The aspirations and experiences of a UK Caribbean community', *Current Issues in Tourism*, 5(5): 378–425.

Stumpf, T.S. and C.L. Cheshire (2019) 'The land has voice: Understanding the land tenure-sustainable tourism development nexus in Micronesia', *Journal of Sustainable Tourism*, 27(7): 957–973.

Taumoepeau, S.P. (2007) 'A blueprint for the economic sustainability of the small national airlines of the South Pacific', *Unpublished Doctoral Thesis*, University of the Sunshine Coast, Queensland.

Thompson, A. (2013) 'Tourism in Yap and Micronesia: Will China Run the Show?' *Asia Pacific Bulletin*, February, 7. No. 199.

Treloar, P. and C.M. Hall (2005) 'Tourism in the Pacific islands'. In: C. Cooper, and C.M. Hall (eds.) *Oceania: A Tourism Handbook*, Toronto: Channel View Publications, pp. 173–294.

UNESCO (United Nations Educational, Scientific and Cultural Organization) (n.d.) 'Nan Madol: Ceremonial Centre of Eastern Micronesia', *UNESCO World Heritage Centre 1992–2022*, United Nations. Available online at https://whc.unesco.org/en/list/1503/ (accessed 27 December 2021).

Westerman, A. (2021) 'The pandemic wiped out tourism on Pacific island nations. Can they stay afloat?' *The World*, 28 January. Available online at https://theworld.org/stories/2021-01-28/pandemic-wiped-out-tourism-pacific-island-nations-can-they-stay-afloat (accessed 1 March 2022).

World Bank (2018) 'Federates States of Micronesia poverty and equity brief', Spring. Available online at http://documents.worldbank.org/curated/en/398101528203022792/Federated-States-of-Micronesia-poverty-and-equity-brief-spring-2018 (accessed 19 November 2019).

World Bank (2019) 'Combined project information documents/integrated safeguards datasheet (PID/ISDS)', *Federated States of Micronesia Maritime Investment Project (P163922)*. Available online at https://documents1.worldbank.org/curated/en/596691553250751468/pdf/Project-Information-Document-Integrated-Safeguards-Data-Sheet-Federated-States-of-Micronesia-Maritime-Investment-Project-P163922.pdf (accessed 14 March 2022).

World Bank (2021) 'Federated States of Micronesia receives $40 million boost to improve reliance of roads network', 17 May. Available online at https://www.worldbank.org/en/news/press-release/2021/05/17/federated-states-of-micronesia-receives-40-million-boost-to-improve-resilience-of-roads-network (accessed 3 February).

World Population Review (2022) 'Micronesia population 2022'. Available online at https://worldpopulationreview.com/countries/micronesia-population (accessed 21 September 2020).

PART IV

Tourism and island states in Polynesia

22

SUSTAINABLE TOURISM PLANNING IN SAMOA

Identifying the challenges and opportunities

Brent Lovelock, Lenara Tuipoloa-Utuva and Anna Carr

Introduction

Samoa (formerly known as Western Samoa until 1997) is located in Polynesia about halfway between New Zealand and Hawaii, its nearest South Pacific Ocean neighbours being Niue, Tokelau and Tonga. Samoa was the first Pacific Island state to gain independence (in 1962) from its colonial administrators, New Zealand (see Table 1.1, Chapter 1). Geographically, Samoa has a land area of 2,842 square kilometres (roughly one-sixth the size of Fiji) consisting of two main islands, Upolu and Savai'i, and a number of smaller islands (CIA 2019) (Figure 22.1). The census in 2016 estimated the population of Samoa at 196,000 (Samoa Bureau of Statistics 2019), the majority of whom live on the island of Upolu and in coastal areas, including in the capital of Apia. The country's economy has traditionally been based on semi-subsistence agriculture, which along with fishing, provides for 90% of exports. There is a high dependency on overseas remittances from family members and, increasingly, tourism, which has played an important role in recent economic activity. While in 2014 Samoa graduated from 'least developed country status', as recently as 2013 an estimated 18.8% of Samoa's population lived below the national poverty line (ADB 2019). External debt is approximately 56% of gross national income (The Global Economy.com 2021), and the Samoan government has called for deregulation of the country's financial sector and encouragement of investment (CIA 2019).

This chapter provides an overview of the tourism products and markets and goes on to illustrate current challenges and opportunities for sustainable tourism development. The chapter draws upon relevant literature, field observations and interviews in Samoa undertaken by two of the authors, one of whom is Samoan. Policy and planning for Samoa tourism is addressed, including key planning and policy instruments and processes. Some recent developments are discussed, including the case of the proposed waterfront development in Apia in order to illustrate sustainable planning processes and challenges in practice. A brief case study discusses cultural tourism, which is an important component of Samoa's tourism product, especially in relation to opportunities to integrate the local transportation network into the cultural tourism experience. The chapter concludes with some remarks about the future challenges for sustainable tourism development in Samoa.

DOI: 10.4324/9780429019968-26

Figure 22.1 Map of Samoa, indicating the capital city of Apia, Upolu. Source: Peter Hermes Furian (www.shutterstock.com/image-vector/samoa-political-map-capital-apia-important-438669727).

Visitor arrivals

More than two decades ago, Pearce (2000) described tourism in Samoa as being at a very 'youthful' stage of development, predominantly small scale, and ranking as a 'middle-order' South Pacific destination, well behind the regional leaders Fiji and French Polynesia. Reflective of growth in tourism across the South Pacific, visitor arrivals have, however (at least up to the COVID-19 pandemic (see below)), grown substantially from 68,000 arrivals recorded in 1997 to 182,858 total arrivals for the year 2019 (Central Bank of Samoa 2021; STA 2021). International tourism is a major employer, providing jobs for just over 10% of national employment (STA 2014) and contributing approximately SAT$493 million (currently US$187 million) to the Samoan economy and worth around 25% of the gross domestic product (GDP) (STA 2018). New Zealand has consistently been the main source market (47% in 2018) for Samoa, followed by Australia (20% in 2018) (STA 2019a). Opportunities for growth in both of these markets have recently been enhanced by the entry of Samoa Airways in 2018, with a route to Sydney, and increased flights and capacity from Air New Zealand.

One particular feature of Samoa's international arrivals is that a large proportion are visiting friends and relatives (VFR), between 33% and 39% over the past five years. Holiday arrivals account for marginally more, comprising 40% in 2018 (STA 2018). While there may be a tendency to underplay the importance of the VFR market, the return of overseas nationals can have significant economic, socio-cultural and political benefits for small island destinations such as Samoa (Scheyvens 2007). By the end of 2018, the value of the VFR market stood at approximately SAT$246 million (US$93 million), while holiday arrivals contributed an estimated SAT$168 million (US$64 million) into Samoa's economy (STA 2018).

Impact of KOVITI-19

COVID-19, or KOVITI-19 as it is known in Samoa, has had a profound impact upon visitation and the Samoan economy. Samoa declared a 'state of emergency' on 22 March 2020, and imposed restrictions within Samoa, closing all international travel to and from Samoa (including flights to/from American Samoa). Visitor numbers fell to a total of just 21,673 in 2020 (Central Bank of Samoa 2021). This ban is still in place at the time of writing, and a decline in visitor arrivals of 56.9% (of 2019 arrivals) had been forecast for the 2021/2022 season (STA 2021). Samoa's experience with a severe measles epidemic in 2019 has, to some extent, helped prepare the country for the pandemic and hardened its resolve to fight KOVITI-19. The local restrictions on activities and gatherings, alongside a 21-day quarantine for repatriating Samoans, have been effective in that Samoa remained one of the few nations in the Pacific Islands (and indeed countries in the world) to record no cases of community transmission of KOVITI-19, all cases being associated with inbound repatriation flights and managed through quarantine. However, community transmission emerged in February 2022, with approximately 250 community cases recorded (March 2022).

Samoa's tourism and wider economy continue to feel the consequences of the pandemic: tourism earnings dropped to zero, compared to earnings of over $US200 million to mid-2019, with Samoa's economy experiencing an 8.6% decline (Godfrey 2021). Some smaller tourism businesses (e.g., beach *fales* and motels) closed and only opened upon demand. However, in the absence of international visitors, Samoans have been supporting their own tourism industry through the KOVITI-19 pandemic, encouraged by the Samoa Tourism Authority's (STA) 'Try Something Different' campaign, targeted at the domestic market to stimulate tourism demand while borders are closed (SPTO n.d.). Further initiatives on the part of the tourism industry have helped, with accommodation providers pricing their products and services to attract local people. Prior to the pandemic it was usually considered too expensive for most Samoans to spend a night at a resort, and domestic tourism has now become more accessible, especially as Samoans cannot travel overseas. Consequently, even the island of Savai'i which is less visited by tourists, is now considered a getaway for Upolu residents. Alongside this trend, restaurants and other facilities in the hotels/resorts are also being better utilized by local people. Along with this domestic focus, a further response has been the STA's *Destination Marketing Plan FY2020–2025* (STA 2020), which sets out the marketing goals and strategies needed to achieve growth in international visitor arrivals and expenditure back to levels that are comparable with the period prior to the pandemic.

In addition to the profound impact of the absence of inbound tourism, the impact of KOVITI-19 on other countries, particularly New Zealand and Australia, has had flow-on effects for Samoa. For example, border restrictions imposed by New Zealand have meant that the Recognized Seasonal Employer (RSE) programme was substantially reduced. Typically, many Samoans travel to New Zealand and Australia to engage in relatively lucrative seasonal horticultural employment. This outbound travel from Samoa has been an important source of employment and repatriation of funds there. However, due to the shortage of seasonal workers in New Zealand, from August 2021, the New Zealand Government allowed RSE workers from Samoa, Tonga and Vanuatu to enter New Zealand without going through managed isolation and quarantine (MIQ). Flights to collect RSE workers from Samoa, along with those repatriating citizens, have been among the few flights permitted to land in Samoa during the KOVITI-19 pandemic.

Transport networks

Samoa's road network is in generally good repair, with main routes sealed; however, seasonal cyclonic storms typically cause some disruption to transport. Land ownership issues can sometimes delay infrastructural development, for example the process of tar-sealing roads through villages. This requires village representatives, together with the government, to foster a relationship that allows such infrastructure improvements. Samoa's two main islands (Upolu and Savai'i) are serviced by ferries taking both passengers and cargo, and once a week, a ferry service is available to American Samoa from Apia, while an air service to Pago Pago operated from Fagali'i Airport until its closure in late 2019 due to safety concerns. American Samoa is Samoa's closest neighbour, and there has been a strong visitor flow between the islands for the purpose of VFR, religious purposes or in transit to other destinations (STA 2014). Aviation capacity overall has increased since 2016–17, complementing the upgrade of Samoa's Faleolo International Airport in May 2018, a project designed and constructed by the Chinese government at an estimated cost of approximately SAT$147 million (US$55,492 million) (Ah-Hi 2018a). The Faleolo International Airport upgrade started in 2014 as a priority development and its purpose was threefold: (1) to reverse the declining trend of passenger and aircraft movement; (2) to trigger economic activities to positively impact Samoa's economy, and (3) to contribute to boosting development in Samoa's tourism industry (Ministry of Finance 2015).

The cruise industry has been an inconsistent contributor to the Samoan tourism sector, with the number of ships varying from a high of 14 in 2014–15 to only five in 2017–18 (STA 2018), with the number of cruise visitors varying over this period from a high of 18,000 to a low of 3,200, with 13,212 cruise visitor arrivals being recorded in 2019 (Central Bank of Samoa 2021). The STA (2014) acknowledges that the cruise market to Samoa is relatively undeveloped compared to some Pacific destinations. Samoa is also disadvantaged because of its relative geographic isolation from Australia and New Zealand, thus, affecting the types of cruises attracted, with the majority of cruises arriving in Samoa being trans-Pacific or round-the-world cruises (STA 2018). The cruise sector was one of the earliest and most severely impacted by the outbreak of the KOVITI-19 pandemic, with no cruise ships visiting Samoa from early 2020.

Attractions, activities and accommodation

Samoa's visitor attractions are mainly based on the natural environment and/or cultural products. The most popular attractions in Samoa are predominantly beaches and marine-based activities, alongside land-based natural sites such as the lava fields and waterfalls. Historic attractions such as Robert Louis Stevenson's Grave and Museum are also popular. Overall visitor satisfaction levels with Samoa's tourism services are high with the most appealing elements reported to be the weather, environmental cleanliness, local people, activities, attractions, entertainment and events and culture (Milne et al. 2018). Unlike resort-dominated destinations, such as Fiji and New Caledonia, Samoa's tourism is characterized by small-scale operations owned and run by local people and communities (Scheyvens 2005). While deluxe accommodations are present and have the highest occupancy rate of approximately 67% (STA 2019b), beach *fales* are an example of a common accommodation type which are typically small-scale, family-run operations, with emphasis often placed on tourists having an authentic '*fa'a* Samoa' (Samoan way of life) experience (Scheyvens 2005) (see Figure 22.2). And importantly, almost one-third of the

Figure 22.2 Beach *fales* at Lalomanu Beach, Upolu Island, Samoa. Source: corners74 (https://www.shutterstock.com/image-photo/lalomanu-beach-upolu-island-samoa-october-749143831).

substantial VFR segment uses commercial accommodation, generally at the lower-priced end of the accommodation range. Only a relatively small proportion of visitors stay in lower cost beach *fales* (STA 2018), although researchers have noted that the potential value of *fales* to the local economy appears to remain an unrecognized opportunity (c.f. large-scale hotel development), meeting the demand for visitors who wish to experience locally controlled, moderately priced and authentic beachfront accommodation that also offers the opportunity for a unique cultural experience (Schanzel 2018; Scheyvens 2005).

In total, there are 2,305 hotel rooms in commercial accommodation of deluxe, superior, standard and budget categories and 2,724 rooms in the specialized accommodation category which includes Samoa's beach *fales* (STA 2019b). As is the case in many developing destinations, reliance on imported goods is an issue in Samoa with this being apparent in the more upmarket accommodation grouping. A study of linkages of tourism and agriculture in Samoa found that 70% of fruit and vegetables for three-, four- and five-star hotels and restaurants were imported (Sofield and Tamasese 2011).

Threats: climate change, cyclones and tsunamis

In recent years, Samoa has suffered the impact of a tsunami (2009) and Cyclone Evan (2012), both of which had substantial impact in terms of loss of lives and damaged infrastructure as well as negative effects on destination image for Samoa (STA 2014). The tsunami led to the loss of 143 lives, including a number of overseas visitors, with damages estimated at SAT$310.11 million (US$124.04 million), of which SAT$79.1 million (US$29.59 million) was associated with the tourism sector (World Bank 2009). Three years later, Cyclone Evan caused nationwide damages and losses of approximately SAT$465 million (US$175.54 million), of which the tourism sector endured around SAT$27.7 million (US$10.46 million) in damages to physical assets and SAT$21.71 million (US$8.20 million) in revenue loss (World Bank 2013).

While these events are unpredictable and infrequent, the impact of climate change is likely to be more persistent, and Samoan villagers have been observed as resilient and adaptive in response to such challenges (Latai-Niusulu et al. 2020). Nevertheless, Samoa has been identified as a 'climate-tourism hotspot' where climate change is predicted to have a major adverse effect on tourism (Jiang et al. 2015; Parsons et al. 2018; Wong et al. 2013). Importantly, tourism operations in Samoa rely on 'pristine marine resources and coastal infrastructure' (Wong et al. 2013: 136), both of which are threatened by climate change (Parsons et al. 2018). Interestingly, the impact of climate change on visitor originating destinations also affects visitor demand for tourism; for instance, earlier declines in arrivals to Samoa (in 2012) have been attributed to a warmer than usual autumn period in New Zealand, which suppressed demand by New Zealanders for overseas tropical holidays that year. This raises the issue of the indirect effects of climate change on the long-term tourism prospects for Samoa, which markets to and relies upon cold climate tourism-generating regions.

Samoa's tourism planning framework and tourism vision

Samoa recognized in the early 1990s that tourism development must be both environmentally responsible and culturally sensitive (Scheyvens 2008). The initial approach to tourism was low-volume, high-yield because of Samoa's determination to protect the *fa'a* Samoa and land ownership, in which small-scale tourism operations were deemed suitable (Samoa Tourism Authority 2014; Scheyvens 2008). Today, the Global Sustainable Tourism Council considers Samoa to have systems and policies in place to practice, manage and monitor sustainable tourism, with Samoa being the first in the Pacific to do so (Samoa Tourism Authority 2014).

The STA is the leading agency for tourism sector coordination, progressing the country's tourism vision: 'By 2019, Samoa will have a growing tourism sector, which engages our visitors and people and is recognized as the leading Pacific destination for sustainable tourism' (STA 2014: 18). This vision supports the overall sustainable development framework for Samoa as well as the sustainable tourism plan to ensure a quality visitor experience and quality of life for the local people. The STA, as a state-owned enterprise, provides policy advice to the government, including progress reporting on the various tourism projects and activities. While the increased accessibility of Samoa may be attributable to its re-developed international airport and the acquisition of its own national airline in 2017 (STA 2017a), some local hoteliers have yet to realize improved occupancy rates (Nataro 2018). Furthermore, some tourism industry commentators have observed that, despite the rhetoric about tourism being the backbone of Samoa's economy, there are a lack of real incentives and opportunities for tourism businesses to grow, with tourism businesses on the main island of Upolu struggling, and the situation being worse for Savai'i (Nataro 2018). Some argue that significant improvements to this situation can only be brought about by placing greater attention on attracting greater numbers of holiday visitors, pointing out that, despite VFR visitors comprising the largest group of travellers to Samoa and providing the greatest economic impact overall, the holiday traveller benefits the local tourist economy more per capita (Samoa Observer 2019).

Government support for tourism growth and enhancing the benefits of tourism to Samoans is not simply limited to securing resources from international partners (e.g., New Zealand and the European Union) but also involves developing policies, such as those which, for example, may empower locals in the tourism business space. Enabling this, tourism planning aims to be transparent and consultative, including with the local community which involves tourism officials visiting the villages and reaching out through the *matai* (chiefs).

Guiding sustainable tourism development

In the *Samoa Tourism Sector Plan 2014–2019* (STSP), STA outlined the country's tourism vision which relates to the United Nations World Tourism Organization's principles of sustainable tourism development (STA 2014). These principles are used by the STA to develop indicators for tourism expenditure and employment, market size and composition, ensuring visitor satisfaction and community engagement while, at the same time, endeavouring to maintain environmental sustainability. The STSP identified the priorities for tourism sector development within five main categories, focusing on 'marketing and research', 'facilitation of an investment and business enabling environment', 'value product development', 'human resource development and training' and 'stabilisation of tourism infrastructure and accessibility'. These thematic priorities align with the priorities of the current *Strategy for the Development of Samoa* (SDS) 2016/2017–2019/2020 (Ministry of Finance 2016), involving the vision to ensure all its citizens achieve 'an improved quality of life (for all)'. The intended strategic outcomes for tourism in Samoa are reflected in the STSP as well as supporting the United Nations Global Sustainable Goal 8: 'Promote sustained, inclusive and sustainable economic growth, full and productive employment and decent work for all'. This last goal will assume particular relevance in the light of the impact of KOVITI-19 on the economy and employment in Samoa.

In order to support the implementation of the STSP priority areas, for example, 'enhancing the resilience of tourism reliant communities to climate change risks', management plans are developed which allow progress to be governed. Different areas usually have their own plans because of the dynamics of the resources available in the area, the vulnerability of the area, appropriate adaptation measures and possible developments for the area (STA 2015). As noted above, throughout these processes, stakeholder involvement in creating and implementing these plans is valued (STA pers. comm. 2019), and collectively, tourism stakeholders are considered to be the 'arms and legs' in the sector plan implementation (STA 2017b). A wide cross-sectorial representation of tourism stakeholders from both public and private sector organizations, includes: the government; the industry, which is comprised of tourism-related businesses; education and training organizations; communities (the villages) and the donors with direct or indirect impact on sector activities. The stakeholders collaborate on strategic issues such as analyzing the tourism environment in terms of accommodation, the threat of climate change, land lease, skilled workforce, market awareness and regional and global environment to name a few of the ongoing discussions (STA 2017b: 18–30). While such an inclusive process is laudable, there is potential for greater external representation; for example, the inclusion of tourists as stakeholders through consultation and enhanced demand-side information gained through market research.

Apia waterfront development

Among many other projects, a current example that reflects sector collaboration and a 'whole of Samoa' approach, calling on the involvement of not only tourism but all sectors, is the Apia Waterfront Development Project. This is an initiative of the STA and the Ministry of Natural Resources and Environment. Since its endorsement by the Samoan Cabinet in 2014, the New Zealand (NZ) government, through its Ministry of Foreign Affairs and Trade, and the NZ Aid Programme have provided financial and technical support (Ministry of Natural Resources and Environment and Samoa Tourism Authority 2016). This project invests in enhancing Samoa as a tourist destination, revitalizing the capital by showcasing significant historical and cultural spaces, with activities for locals and visitors, innovative climate resilient infrastructure and business opportunities for locals.

The intention is to also grow various areas of the tourism industry, including cruise tourism, which is one of the high priority focus areas for product development (STA 2014: 42). The Apia Waterfront plan specifies a cruise ship function at the port in Apia (Ministry of Natural Resources and Environment and Samoa Tourism Authority 2016), thus, allowing convenient access to a number of land and marine activities suitable for the cruise visitor itinerary and, importantly, positioning Samoa as an accessible and attractive destination for cruise ship operators to visit (STA 2014). As indicated earlier, cruise tourism has been an inconsistent sector in Samoa and, therefore, the Apia Waterfront Development is a logical step forward to increase visitor numbers. The challenge, however, concerns negative popular perceptions towards cruise ship tourism as a form of travel that can easily spread the KOVITI-19 virus (Pan et al. 2021).

Despite the whole-of-government approach to this project (Ministry of Natural Resources and Environment and Samoa Tourism Authority 2016: 2), there have been a range of other concerns raised by the local community. These include doubts over the climate change resilience capacity of the proposed infrastructure (Samoa Observer 2016). There have also been concerns that greater efforts were needed to achieve consensus among all the villages impacted rather than agreement from just a few highly ranked *matai* (Keresoma 2015a) as well as concerns including impact upon coastal resources, water quality, public safety, traffic and parking (Tauafiafi 2016). In addition to this, in mid-2021, Samoa's Prime Minister-elect Fiame Naomi Mataafa pledged to cancel the US100 million Chinese-backed port development, calling it excessive for the small Pacific Island that is already heavily indebted to China (Barrett 2021).

Sustainable tourism and the local community

The *fa'a* Samoa plays a practical role in community cooperation in developments, mainly because approximately 80% of Samoa is customary land (Meleisea and Schoeffel 2015). As a social structure, *fa'a* Samoa encompasses the 'complex set of traditional values and behavior patterns influencing the way Samoans think, act and go about their daily lives' (Twining-Ward and Tuailemafua 2004: 78). In a recent qualitative study of the tourism potential of Apolima Island (a small island located between Savai'i and Upolu, currently not developed for tourism), the importance of community consensus and its potential to delay or even prevent any development was clearly demonstrated (Tuipoloa-Utuva and Lovelock 2017). One of the study's findings observed that, although there is support for (and an overall positive attitude towards) tourism development, the community's perception of tourism 'may be limited to the experiences they have had with visitors, a few foreigners' (2017: 68).

This situation, which is also likely to be the case for many communities, hints at a need for a precautionary approach for Samoa as an emerging destination in the Pacific and which, as a nation, is engaging in tourism growth strategies such as 'China Ready' (Ah-Hi 2018b). In this sense, Samoa is similarly placed with other South Pacific destinations, with the STA and the South Pacific Tourism Organisation (SPTO) having launched the *China Pacific Tourism Year 2019* in Samoa. Similar to the perspective expressed by the community residents on Apolima Island, tourism development for many communities in Samoa is not a priority on their list of developmental needs. Nonetheless, if tourism development is pursued, it ought to directly benefit community members and preferably be in a form and scale that they themselves can manage (Tuipoloa-Utuva and Lovelock 2017). Tourism is recognized not simply as another potential source of income for Apolima but also more for the social and cultural benefits that allow communities to display pride in the Samoan way of life. Ford et al. (2019) highlight this prioritization of community benefits as a necessity for successful sustainable development.

Case study: local bus travel as sustainable cultural tourism in Samoa

Could public transport be developed as a type of sustainable cultural tourism experience – as a means for visitors to the islands to experience *fa'a* Samoa? For international visitors arriving at Samoa's Faleolo International Airport, the cultural differences are first apparent with the new terminal building, which has been designed in a traditional '*fale*' style. The cultural experience extends to interaction with local airport employees and instant exposure to the local language, dress and other cultural elements that abound (Keresoma 2015b). However, it is the prominent, local '*aiga*' buses that are most noticed by visitors to the islands as a unique form of transport echoing island life (Figure 22.3). The friendliness of local Samoan people has been noted in the 2017–18 International Visitor Survey (IVS) as a top influential reason for those travelling to Samoa. Accordingly, 'culture and history' was the sixth most influential reason, while 'activities involving cultural interaction' ranked the highest for visitor satisfaction (Milne et al. 2018: 21). Nevertheless, survey respondents reported a low engagement with local buses, with less than 1% of visitor expenditure per day devoted to this form of transport. While public transport experiences appear to have low user engagement from visitors, the buses have potential to optimize cultural engagement. Attending church services and markets were the types of cultural interactions reported as popular by IVS participants, and there is a practical necessity for visitors to use local ferries and buses when other transport options are not available – in Samoa many of the buses are owned by village collectives or village churches. Therefore, while the IVS report did not mention visitor experiences associated with the island buses, they are ideally situated to serve the dual purpose of enabling interactions between locals and visitors as well as transportation.

Figure 22.3 Local bus in Upolu, Samoa. Source: peacefoo (www.shutterstock.com/image-photo/upolusamoa_19-sep-2019-local-samoa-iconic-1530125756).

Hall (2007: 94) notes the existence of a 'transport enthusiast leisure sub-culture', though cautioned that such enthusiasts tend to be in pursuit of a 'male-dominated specialist, esoteric, often far from fashionable, interest'. Nonetheless, there is arguably an increasing awareness to reduce carbon footprints and a potential to broaden the appeal of heritage or cultural experiences through the use of public bus transport. Heritage transport experiences as cultural tourism attractions continue to be popular, for instance San Francisco's trams, horse and carriage rides in Europe and North America, sleigh rides in Scandinavia and heritage train trips in many parts of the world (Page and Ge 2009) are the equivalent of this. Therefore, for travellers seeking an authentic and memorable experience of Samoan culture, the brightly painted *aiga* village buses of Samoa have scope for developing such heritage transport experiences. Travellers can immerse themselves into local transport networks in which interaction with locals on a daily basis is possible. Such visitors may also represent the 'incidental cultural tourist' (McKercher and Du Cros 2003: 46), whose intentions may be motivated by other aspects of Samoa's visitor offerings but who utilize the buses for accessing the surroundings and the people.

As an integral part of local Samoan transport, there is no need to directly market the buses to tourists, particularly as their colourful presence located around markets, ferry terminals and town centres is very apparent. There are few 'official' bus stops, with journeys between villages being located around the main market areas or churches. As the drivers choose the music and volume levels, the buses are often heard before they are seen. Social media commentators are spreading the word about the cultural interactions on Samoan buses. The *Lonely Planet Guide* (2019) promotes the buses to its readers:

> The vibrantly painted, wooden-seated vehicles (more often than not blasting Samoan pop music at deafening volumes) each have their own character. Drivers are often as eccentric as the vehicles, and services operate completely at their whim: if a driver feels like knocking off at 1pm, he does, and passengers counting on the service are left stranded.

However, experiences vary with one blogger commenting on such practicalities as the affordability and convenience of the buses (Nolf 2018), while others comment on their personal experiences:

> We saw a Bon Jovi and Guns 'n' Roses themed one, a lottery themed one and several that appear to be named after the driver's mother. The drivers appear to make their own rules and leave whenever they're ready or when the bus is full, and if they want to knock off early and go and watch the rugby, they will.
>
> *(Watson 2014)*

There is a two-way cultural exchange occurring: the tourists being seated towards the front of the buses can be a convenience for the driver keeping an eye on non-local passengers, while at the same time, tourists themselves also become an object of entertainment. In being placed towards the front, the '*palagi*' (Western) visitors are on full view to the locals riding the buses. Alternatively, being seated at the front is a gesture of respect to the visitors from the local users. Crowding and jostling for space on the buses is part of the localized authentic experience, though such behaviour could be seen by policymakers as potential safety and public image concerns. Consequently, from a planning perspective, there have been suggestions about the need to modernize Samoa's buses. This was particularly pertinent in early 2018 following a serious bus accident involving three fatalities and many injuries (RNZ 2018). Integrating the island buses within a more modern, centralized transport infrastructure scheme, rather than the current local-owned offering, could be tempting. However, 'sanitizing' the village buses through modernization could destroy the cultural appeal not only for visitors but also in terms of a local sense of ownership. Making improvements, such as enhancing bus timetabling and providing shelters to increase ease of use, including strengthening feelings of safety and reliability (Thompson and Schofield 2007), could homogenize the service and take away the essential Samoan character. While such Western notions of an efficient bus service could potentially improve visitors' satisfaction with their bus trip, such interventions through 'over-planning' may threaten the fun-filled character of bus travel in Samoa. By encouraging visitors to utilize public transport, there is the additional contribution to local sustainability actions, reducing the number of vehicles driven by non-locals on roads in terms of fewer carbon emissions and traffic congestion.

The Samoa Tourism Sector Plan 2014–2019 advocates for a type of tourism that will 'respect the socio-cultural authenticity of host communities, conserve their built and living cultural heritage and traditional values, and contribute to inter-cultural understanding and tolerance' (STA 2014: 18). Considering bus transportation as part of the cultural heritage mix of Samoa's tourism offerings may well be an acknowledgement of the more authentic offerings of contemporary Samoan lifestyle that can be experienced by visitors.

Conclusion and research implications

This chapter has indicated that while Samoa is one of the smaller players in South Pacific tourism, it has, until the KOVITI-19 pandemic, managed to maintain its competitiveness and cultural integrity – *fa'a* Samoa remains strong despite setbacks in recent years from the tsunami, cyclones and measles epidemic. These events have had significant impact on human life, tourism infrastructure and destination image. There is no doubt, however, that the tourism industry will be expected to lead the recovery of Samoa's economy. The Economic and Social Commission for Asia and the Pacific (ESCAP 2020: 5) identifies 'strong prospects for robust recovery in the tourism industry' in Samoa but stresses the need for adequate government/donor funding for effective destination marketing to help achieve this. ESCAP also stress the need for care during the re-opening of international tourism, particularly in terms of managing the quality of the visitor experience. When international tourism does return to Samoa, there is no reason to suspect that it will be substantially different to the pre-KOVITI-19 period, however, the pressure for a tourism-led economic recovery may challenge aspects of Samoa's sustainable tourism aspirations. The recently (post-KOVITI-19) released national economic strategy, *Transforming Samoa to a Higher Growth Path* (Government of Samoa 2021), identifies KOVITI-19 as an opportunity 'to "reset" and recalibrate priorities' (2021: 9). While noting the role of Samoa's pristine natural environment and unique way of life as key aspects of Samoa's tourism product, the 2040 vision identifies the need to address some of the structural issues that have been constraining growth, namely limited strategic investment in the sector (both public and private), inadequate hospitality and service standards, poor transport connectivity and high prices for consumables. Up to this point, Samoan tourism has been generally of a scale and style that has allowed for a degree of circumstantial alternative tourism (Weaver 2000). Nonetheless, the nation has set its sights on tourism growth.

This chapter has outlined the challenges to achieving that growth while maintaining and building upon the essential cultural elements that characterize Samoa's tourism, i.e., linked to *fa'a* Samoa, and which is local, small scale and friendly as well as predominantly managed by local communities. Encouragingly, Samoa has sound tourism planning arrangements and appears to be committed to transparent planning processes, involving communities in discussions over the future of tourism (Ford et al. 2019). To some extent the likelihood of destructive growth may be curbed by the high percentage of ownership of customary lands, leading to more considered tourism development processes.

Scheyvens and Momsen (2008a) called for us to re-conceptualize such small island nations, from occupying positions of economic and physical vulnerability to, instead, possessing considerable strengths, including, for example, high levels of cultural, social and natural capital; respect for traditional, holistic approaches to development and strong international linkages. Recent developments, such as the Faleolo International Airport and the Apia Waterfront Plan, highlight that Samoa has strengths and capacity to embark with international partners on substantial projects that will enhance its tourism offerings sustainably with communities. However, addressing Samoa's vulnerability to climate change must also remain a key focus of tourism planning, including the actual implementation of plans that may mitigate risk and potentially enhance the islands in various ways.

To advance sustainable tourism in Samoa, tourism research should focus on how to capitalize on such strengths and how they can contribute to community resilience in relation to the economic and physical shocks that will inevitably arise (see Tuipoloa-Utuva and Lovelock 2017). The measles epidemic in 2019 and the current KOVITI-19 pandemic have both significantly affected visitation numbers to Samoa. They represent two more examples of how tourism can be implicated in and affected by such shocks which have profound socio-economic impacts. In relation to KOVITI-19, further research may investigate the extent to which re-invigorated forms of domestic tourism and associated consumption patterns are sustained in the post-pandemic period. Likewise, a better understanding is needed concerning how KOVITI-19 restrictions in other South Pacific countries, in particular New Zealand and Australia, have impacted the *outbound* travel prospects of Samoans for seasonal work with flow-on economic impacts to all.

In support of Scheyvens and Momsen's (2008b) call for the social sustainability of tourism in small island states to achieve more attention, research that strengthens our understanding of how tourism (in its many forms, from community driven, pro-poor tourism to resort-level developments) can contribute to measurable wellbeing and livelihood outcomes must be at the forefront of the tourism research agenda for Samoa. Within a Pasifika context, and specifically a Samoan context, such research needs to be culturally appropriate (see Palaamo 2018), approaching research questions with methods and values that embrace *fa'a* Samoa and thus address community and individual benefits from tourism.

Acknowledgements

The authors wish to thank the staff of the Samoa Tourism Authority for their time and helpful advice. Lenara wishes to acknowledge the National University of Samoa for their support and encouragement for this research.

References

ADB (Asian Development Bank) (2019) *Poverty in Samoa*, Manila, Philippines: ADB. Available online at https://www.adb.org/offices/south-pacific/poverty/samoa (accessed 5 April 2019).

Ah-Hi, E. (2018a) 'Samoa's $147 million international gateway', *Samoa Observer*, 12 May. Available online https://www.samoaobserver.ws/category/samoa/2355 (accessed 5 April 2019).

Ah-Hi, E. (2018b) 'China ready is global ready', *Samoa Observer*, 9 August. Available online at https://www.samoaobserver.ws/category/samoa/24740 (accessed 5 April 2019).

Barrett, J. (2021) 'Samoa to scrap China-backed port project under new leader', *Reuters*, 20 May. Available online at https://www.reuters.com/world/asia-pacific/samoa-shelve-china-backed-port-project-under-new-leader-2021-05-20/ (accessed 17 March 2022).

Central Bank of Samoa (2021) 'Visitor earnings and remittances'. Available online at https://www.cbs.gov.ws/index.php/statistics/tourism-earnings-and-remittance/ (accessed 29 April 2021).

CIA (Central Intelligence Agency) (2019) *Australia - Oceania: Samoa, The World Factbook*. Available online at https://www.cia.gov/library/publications/the-world-factbook/geos/ws.html (accessed 5 April 2019).

ESCAP (Economic and Social Commission for Asia and the Pacific) (2020) *Micro, Small and Medium-sized Enterprises' Access to Finance in Samoa: COVID-19 Supplementary Report and Recommendations*, Bangkok: ESCAP United Nations.

Ford, A., A. Carr, N. Mildwaters, D. Fonoti and G. Jackmond (2019) *Promoting Cultural Heritage for Sustainable Tourism Development: Samoa*, University of Otago, Report Commissioned by New Zealand Institute of Pacific Research/Ministry Foreign Affairs and Trade.

Godfrey, D. (2021) 'From pandemic to economic crisis, Samoa's covid journey one year on', *RNZ (Radio New Zealand)*, 16 March. Available online at https://www.rnz.co.nz/international/pacific-news/438422/from-pandemic-to-economic-crisis-samoa-s-covid-journey-one-year-on (accessed 29 April 2021).

Government of Samoa (2021) *Samoa: 2040 - Transforming Samoa to a Higher Growth Path*, Apia: Ministry of Finance, Economic Policy and Planning Division. Available online at https://www.mof.gov.ws/wp-content/uploads/2021/03/Samoa-2040-Final.pdf (accessed 29 May 2021).

Hall, D. (2007) 'Transport tourism travelling through heritage and contemporary recreation'. In: M. Novelli (ed.) *Niche Tourism: Contemporary Issues, Trends, and Cases*, Oxford: Elsevier, pp. 89–98.

Jiang, M., E. Calgaro, L.M. Klint, D. Dominey-Howes, T. Delacy and S. Noakes (2015) 'Understanding climate change vulnerability and resilience of tourism destinations: An example of community-based tourism in Samoa'. In: S. Pratt and D. Harrison (eds.) *Tourism in the Pacific Islands*, Abingdon: Routledge, pp. 263–280.

Keresoma, L. (2015a) 'History and culture need to be key factors to the city of Apia Waterfront Development', Samoa: Talamua Media and Publications, 23 October. Available online at http://www.talamua.com/history-culture-need-to-be-key-factors-to-the-city-of-apia-waterfront-development/ (accessed 4 April 2019).

Keresoma, L. (2015b) '$590 Million facelift for Faleolo International Airport terminal building', Samoa: Talamua Media and Publications. Available online at http://www.talamua.com/50-million-facelift-for-faleolo-international-airport-terminal-building/ (accessed 29 March 2019).

Latai-Niusulu, A., T. Binns and E. Nel (2020) 'Climate change and community resilience in Samoa', *Singapore Journal of Tropical Geography*, 41(1): 40–60.

Lonely Planet (2019) 'Getting around Samoa by bus'. Available online at https://www.lonelyplanet.com/samoa/transport/getting-around/bus (accessed 30 March 2019).

McKercher, B. and H. Du Cros (2003) 'Testing a cultural tourism typology', *International Journal of Tourism Research*, 5(1): 45–58.

Meleisea, M. and P. Schoeffel (2015) 'Land, custom and history in Samoa', *The Journal of Samoan Studies*, 5: 22–34.

Milne, S., M. Sun, J. Yi, H. Qi and B. Bakker (2018) *Samoa International Visitor Survey Report, January 2018 to June 2018*, Auckland: New Zealand Tourism Research Institute (NZTRI), Auckland University of Technology.

Ministry of Finance (2015) *Economic Analysis for the Faleolo International Airport Upgrade*, Samoa: Government of Samoa - MOF.

Ministry of Finance (2016) *Strategy for the Development of Samoa 2016/2017–2019/2020*, Samoa: Government of Samoa - MOF.

Ministry of Natural Resources and Environment and Samoa Tourism Authority (2016) *Apia Waterfront Development Project: Waterfront Plan 2017–2016*, Samoa: Government of Samoa.

Nataro, I. (2018) 'Tourism industry is struggling', *Samoa Observer*, 23 November. Available online at https://www.samoaobserver.ws/category/samoa/25854 (accessed 20 January 2020).

Nolf, M. (2018) 'Upolu island bus schedule and fares – Western Samoa independent travel', *thinkoholic.com a blog by Markus Nolf*. Available online at http://www.thinkoholic.com/2018/06/08/samoa-independent-travel-bus-schedule-fares-upolu-island-january-2018/ (accessed 29 March 2019).

Page, S. and Y. Ge (2009) 'Transportation and tourism: A symbiotic relationship?' In: T. Jamal and M. Robinson (eds.) *The SAGE Handbook of Tourism Studies*, London: Sage Publications pp. 371–395.

Palaamo, A. (2018) 'Tafatolu (three-sides): A Samoan research methodological framework', *Aotearoa New Zealand Social Work*, 30(4): 19–27.

Pan, T., F. Shu, M. Kitterlin-Lynch and E. Beckman (2021) 'Perceptions of cruise travel during the COVID-19 pandemic: Market recovery strategies for cruise businesses in North America', *Tourism Management*, 85: 104275.

Parsons, M., C. Brown, J. Nalau and K. Fisher (2018) 'Assessing adaptive capacity and adaptation: Insights from Samoan tourism operators', *Climate and Development*, 10(7): 644–663.

Pearce, D.G. (2000) 'Tourism plan reviews: Methodological considerations and issues from Samoa', *Tourism Management*, 21(2): 191–203.

RNZ (Radio New Zealand) (2018) 'Samoa's colourful buses look set to go', *RNZ*, 1 February. Available online at https://www.radionz.co.nz/international/pacific-news/349426/samoa-s-iconic-colourful-buses-look-set-to-go (accessed 29 March 2019).

Samoa Bureau of Statistics (2019) 'Population and demography indicator summary'. Available online at http://www.sbs.gov.ws/index.php/population-demography-and-vital-statistics (accessed 20 January 2020).

Samoa Observer (2016) 'Waterfront development explained at chamber', 3 October. Available online at https://www.samoaobserver.ws/category/samoa/11358 (accessed 4 April 2019).

Samoa Observer (2019) 'The tourism situation: Letters to the editor by Kevin'. Available online at https://www.samoaobserver.ws/category/article/11188 (accessed 20 January 2020).

Schanzel, H.A. (2018) 'Slow hospitality experiences: A case study of family holidays at beach fale in Samoa'. In: *Critical Hospitality Symposium II: Hospitality IS Society*. Available online at https://ojs.aut.ac.nz/critical-hospitality-symposium/index.php/CHS/CHSII/paper/view/15 (accessed 2 April 2020).

Scheyvens, R. (2005) 'Growth of beach fale tourism in Samoa: The high value of low-cost tourism'. In: C.M. Hall and S.W. Boyd (eds.) *Nature-based Tourism in Peripheral Areas: Development or Disaster?* Clevedon: Wiley, pp. 188–202.

Scheyvens, R. (2007) 'Poor cousins no more: Valuing the development potential of domestic and diaspora tourism', *Progress in Development Studies*, 7(4): 307–325.

Scheyvens, R. (2008) 'On the beach: Small scale tourism in Samoa'. In: J. Connell and B. Rugendyke (eds.), *Tourism at the Grassroots: Villagers and Visitors in the Asia-Pacific*, London: Routledge, pp. 131–147.

Scheyvens, R. and J. Momsen (2008a) 'Tourism in small island states: From vulnerability to strengths', *Journal of Sustainable Tourism*, 16(5): 491–510.

Scheyvens, R. and J.H. Momsen (2008b) 'Tourism and poverty reduction: Issues for small island states', *Tourism Geographies*, 10(1): 22–41.

Sofield, T.H.B. and E. Tamasese (2011) *Tourism-Led Poverty Reduction Programme for Samoa: Strengthening Linkages Between Tourism and Agricultural Sectors*, Geneva: United Nations International Trade Centre.

SPTO (Pacific Tourism Organisation) (n.d.) 'Samoa Tourism Authority launches "try something different" campaign'. Available online at https://southpacificislands.travel/samoa-tourism-authority-launches-try-something-different-campaign/ (accessed 17 March 2022).

STA (Samoa Tourism Authority) (2014) *Samoa Tourism Sector Plan 2014–2019*, Samoa: STA.

STA (Samoa Tourism Authority) (2015) 'Activity design document (ADD): Samoa tourism support program (STSSP) phase 2', Samoa: STA.

STA (Samoa Tourism Authority) (2017a) 'Samoa airways launches into the global market', 23 November. Available online at http://samoatourism.org/Articles/219/Samoa-Airways-Launches-into-the-Global-Market (accessed 20 January 2020).

STA (Samoa Tourism Authority) (2017b) *Corporate Plan 2017–2020*, Samoa: STA.

STA (Samoa Tourism Authority) (2018) '2018 Visitor Statistics Report', *Unpublished Report*, Samoa: STA.

STA (Samoa Tourism Authority) (2019a) 'Unpublished accommodation database from the STA', September, Samoa: STA.

STA (Samoa Tourism Authority) (2019b) 'Occupancy information mid-2019', Samoa: STA. Available online at http://samoatourism.org/articles/158/occupancy-information-mid-2019 (accessed 20 January 2020).

STA (Samoa Tourism Authority) (2020) *Destination Marketing Plan FY2020–2025*, Samoa: STA.

STA (Samoa Tourism Authority) (2021) Visitor arrivals forecast (2021/22–2024/25), Unpublished forecasts from the STA, Samoa: STA.

Tauafiafi, L.A. (2016) 'Plan turning Apia's Waterfront to a unique public space will benefit Pacific region', *Pacific Guardians: Pacific Perspectives*, 31 August. Available online at https://pacificguardians.org/blog/2016/08/31/plan-turning-apias-waterfront-to-a-unique-public-space-will-benefit-pacific-region/ (accessed 28 March 2019).

TheGlobalEconomy.com (2021) '*Samoa: external debt*'. Available online at https://www.theglobaleconomy.com/Samoa/External_debt/ (accessed 10 January 2022).

Thompson, K. and P. Schofield (2007) 'An investigation of the relationship between public transport performance and destination satisfaction', *Journal of Transport Geography*, 15(2): 136–144.

Tuipoloa-Utuva, L.L. and B.A. Lovelock (2017) 'Potential for community managed sustainable tourism development on Apolima Island', *Journal of Sāmoan Studies*, 7(1): 20–37.

Twining-Ward, L. and T.S. Tuailemafua (2004) 'Small island tourism: Monitoring sustainable tourism development in Samoa', *Journal of Pacific Studies*, 26: 77–103.

Watson, P. (2014) 'Pimp my bus ride: Samoan buses are a unique experience', *Atlas and Boots Outdoor Travel Blog*, 1 October. Available online at https://www.atlasandboots.com/samoan-buses-are-a-unique-experience/ (accessed 29 March 2019).

Weaver, D.B. (2000) 'A broad context model of destination development scenarios', *Tourism Management*, 21(3): 217–224.

Wong, E., M. Jiang, L. Klint, T. DeLacy, D. Harrison and D. Dominey-Howes (2013) 'Policy environment for the tourism sector's adaptation to climate change in the South Pacific - the case of Samoa', *Asia Pacific Journal of Tourism Research*, 18(1–2): 52–71.

World Bank (2009) Samoa - *Post Disaster Needs Assessment: Following the Earthquake and Tsunami of 29 September 2009*. Available online at http://documents.worldbank.org/curated/en/716291468025505741/Samoa-Post-disaster-needs-assessment-Following-the-earthquake-and-tsunami-of-29-September-2009 (accessed 20 January 2020).

World Bank (2013) *Samoa Post-Disaster Needs Assessment: Cyclone Evan 2012*, March, Washington, D.C. Available online at https://openknowledge.worldbank.org/handle/10986/15977 (accessed 20 January 2020).

23

TOURISM IN THE KINGDOM OF TONGA

Challenges, solutions and research

Jenny Cave and Rosemarie Fili Grover

Introduction

The Kingdom of Tonga is located within the Polynesian Triangle, lying southwest of Samoa and east of Fiji. The archipelago spans 700 square kilometres, represented by four main island groups, Ha'apai, Niuas, Tongatapu and Vava'u, where only 36 of its 169 islands are inhabited (Stanley 2004). The capital of Tonga, Nuku'alofa, is located in the country's largest island, Tongatapu (Figure 23.1), where the Fua'amotu International Airport is also located. Tonga is one of the least developed of the Pacific tourism destinations, retaining cultural strength and distinctiveness. The country has been sceptical of tourism development progressing at the expense of environmental sustainability or local culture and values. Accordingly, each island group has distinctive geographies and unique natural and cultural attributes, although, to some extent, reflecting traditions of long-distance voyaging and centuries of interaction with neighbouring Fiji and Samoa (Clark 2010). This chapter aims to identify the key challenges to the development and advancement of tourism in Tonga, especially in relation to tourism markets. The work identifies concerns relating to tourism markets, traditional practices and the ecological environment – simultaneously looking at potential solutions to the concerns raised and towards a future research agenda. There are, indeed, lessons to be learnt from challenges relating to natural disasters and the COVID-19 pandemic, which have affected the Tongan economy and prospects to move forward to expand the tourism industry. Initially, however, the chapter will provide a background context and a general overview to the evolution of Tonga as a tourism destination, making way for a more critical assessment of the concerns that this island state faces.

Background, context and tourism overview

Tonga has experienced a turbulent civil history. Portuguese, Dutch and English explorers and Wesleyan missionaries influenced internal conflicts, ending with unification in 1845. It became a monarchy in 1875 led by King George Tupou I (1845–1893), who converted to Christianity. Tonga became a British protectorate in 1900, achieving full independence in 1970 (Foster and Latukefu 2022). Major constitutional reforms occurred in 2010 following civil unrest which produced a democratically elected assembly, though including nine hereditary nobles (World Bank 2020a). Recent data indicates that 27% of the Tongan population are living in low-

 DOI: 10.4324/9780429019968-27

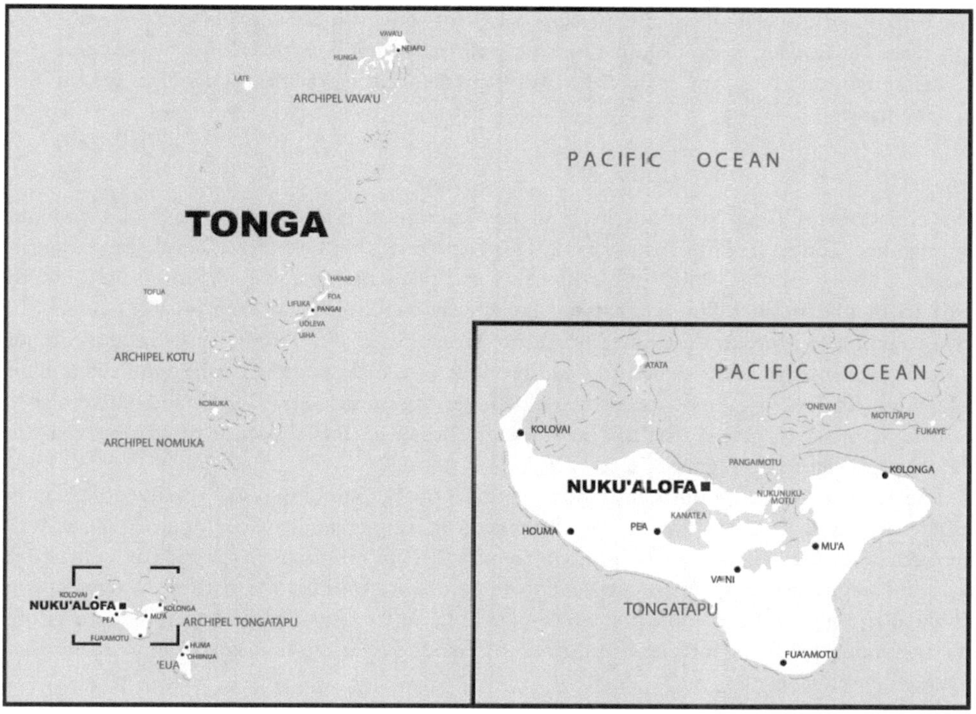

Figure 23.1 Map of Tonga, including the capital city of Nuku'alofa, Tongatapu. Source: GEOATLAS-GRAPHI-OGRE (www.shutterstock.com/image-illustration/ administrative-map-tonga-149323931).

income households, in which 33% of children and 23% of adults live in poverty (Tonga Statistics Department 2020). Women outperform men in educational attainment and are well represented in senior management positions in the workforce (World Bank 2020a). The estimated population of Tonga in 2022 is 107,890 (Worldometer 2022).

Much of Tonga's human capital (70%) lives abroad (Howes and Surandiran 2020), and its economic growth is constrained by its small economy, dispersed population and remoteness (World Bank 2020a). The economy has a narrow export base, two-thirds of which are agricultural goods such as roots and tubers, kava, vanilla beans, melon and squash as well as fish (FAO 2012). Despite out-migration, there is a strong economic subsistence base because traditional livelihood strategies in the islands have not been completely replaced (Besnier 2011). However, Tonga increasingly imports its food and is highly reliant on imported tinned goods (FAO 2012). The inward balance of trade is largely dependent on foreign aid and remittances, in which remittances from Tongan families living overseas were 43% of the gross domestic product (GDP) in 2019 (Howes and Surandiran 2020).

Tongans are happier than others in the Pacific when measured on quality of life, the environment, sustainability, economic performance, equality and cultural capital, according to Tonga's 'gross happiness index' (Pratt 2016). Although Tongan life embodies cultural heritage and traditions, some regions of the country are more customarily steadfast than others. Young-Leslie and Moor's (2012) research, for instance, indicated that Ha'apai is perceived as a place where authentic cultural practices and happiness strongly intersect, noting:

> Importantly, as a result of this perceived lack of foreign influence, Ha'apai is stereotyped by Tongans as an example of 'true' cultural practices within Tonga … an ideal place for examining the cultural basis of happiness and satisfaction judgments within the nation.
>
> *(2012: 184)*

The Kingdom of Tonga positions itself as the 'True South Pacific' amongst Pacific tourism destinations (Tonga Tourism Authority 2018). Named as the 'Friendly Islands', Tonga began to attract tourists in the 1960s, with cruise ship and plane passengers increasing from 2,866 in 1961 to 18,111 in 1969. The government proactively established the Tonga Visitor's Bureau in 1971 to promote Tonga as a tourism destination (Urbanowicz 1989). Tonga's natural and cultural resources are integral to its varied tourism product. The main island of Tongatapu offers high coral cliffs (e.g., Hufangalupe), spectacular blowholes, traditional and second-hand marketplaces, tourism resorts and globally significant heritage sites (e.g., Royal Tombs of Lapaha and the Ha'amonga Trilithon – Figure 23.2).

The Ha'apai island group is distinctive because of its remoteness and rurality, long sandy beaches and well-preserved *siaheu-lupe*, the pigeon snaring mounds (Cave et al. 2012), as well as marine-based adventure (Kessler and Harcourt 2010). 'Eua Island's spectacular cliffs, deep caves and rainforest make it an eco-adventure destination. Whilst the sheltered coves of the Vava'u island group are popular for whale-watching and a trans-Pacific halt for ocean-going yachts (Orams 2013), its northernmost group of islands, Niuas, are remote, volcanic islands that and rarely visited.

Figure 23.2 Ha'amonga 'a Maui ('The Burden of Maui') - a stone trilithon, Niutōua, Tongatapu. Source: Henryk Sadura (www.shutterstock.com/image-photo/haamonga-maui-arch-tonga-174337907).

Tonga's annual visitor numbers reached a record high of 94,000 in 2019, which contributed 11.14% to country's gross national product (GNP), which increased significantly from approximately 33,000 visitors in 1995, when tourism only contributed 4.9% to the GNP (World Data. Info 2020). The hospitality sector is a significant contributor to tourism employment (Harrison and Prasad 2013), of which 60% of the accommodation establishments are located in the main island of Tongatapu and 20% are located in Vava'u, whilst the rest (largely guest houses) are located in such islands as Ha'apai and 'Eua (Milne 2009). Tourism is the second largest source of income after remittances, representing such niche areas as 'diving and snorkelling, honeymooning, cruise shipping, fishing, sailing, kayaking, snorkelling, whale-watching, cultural tourism and nature tourism' (Scheyvens and Russell 2009: 10).

Tourism development concerns

Despite the range of opportunities for tourism expansion in Tonga, its tourism industry has faced development, socio-political and economic hinderances over the past two decades, such as the collapse of the Royal Tongan Airline in 2004 and pro-democracy riots in Nuku'alofa in 2006 (Ratuva 2019). Tonga is vulnerable to natural disasters and cyclones which significantly impact the country's economy. In 2018, for instance, Cyclone Gita caused significant damages to properties with a loss of revenues equating to 38% of gross domestic product (GDP) (Matangi Tonga Online 2018). Tonga suffered another shock when Cyclone Harold hit the island in April 2020, with damages of US$111 million that heavily affected beach resorts (Ober and Bakumenko 2020). More recently, on 15 January 2022, the undersea volcano Hunga Tonga-Hunga Ha'apai, located 65 kilometres from Tongatapu, erupted, creating an ash plume that was 30 kilometres high and 260 kilometres wide and a regional tsunami, prompting the World Bank to donate US$8 million in immediate emergency response funding to the country (World Bank 2022). This incident is anticipated to have a significant impact on the Tongan population, as contamination from the fall-out could well affect households in Tongatapu and the Ha'apai island, especially as the groundwater, crops, livestock and fisheries have been significantly impacted (United Nations 2022). Indeed, more than 30 tourism business were fully or partially destroyed by the tsunami and eruption, especially on the main island of Tongatapu (Morton 2022). It will take years to fully recover international tourism activities following the devastation caused by the tsunami and ash fall.

Ongoing financial investment in tourism development is often a challenge given the unforeseen circumstances that Tonga experiences. Fortunately, remittances (both collective and individual) are crucial to the survivability of the nation (Cave and Koloto 2015) and especially during the COVID-19 pandemic, which had serious implications to the tourism development pathway of Tonga. These recent challenges add to a range of pre-existing concerns impacting tourism development in Tonga, notably poor infrastructure, limited marketing funds and rundown tourism accommodation (Traveller 2012). In the first half of 2020, pandemic-affected arrivals in Tonga declined by 62.7% compared to the previous year and a total decline when borders closed to international visitors in March 2020, notably affecting tourism, retail and construction (Wolf et al. 2021). It is also anticipated that the rapid decline in the tourism industry will, in the long-term, affect infrastructure provision and repair (MFAT 2020). However, the border closure, state-enforced quarantine and restricted public gatherings resulted in few infections and no deaths (RNZ 2021). To offset the economic consequences of the COVID-19 pandemic, the government provided loan schemes for 'micro, small and medium enterprises' (MSMEs) and financial assistance to formal and informal businesses, including tourism. These offer bridging finance and continue to provide basic food, shelter and transport for families.

International development aid agencies also provided support during the pandemic via social welfare payments, financial support for secondary students to stay at school and wage subsidies for businesses to retain workers to reduce social disruption (World Bank 2020b). Informal, collective and traditional practices for food cultivation were revived in the absence of 'foreign exchange' from international tourism, from which Pacific people should consider whether their interests are best served by engagement in tourism (Scheyvens and Russell 2009). However, restricted mobilities reduced numbers of returnee entrepreneurs and role models whose transference of knowledge, skills and funds have, in the past, made a material difference to Tonga villages and families and to the economic 'footprints' of Tongans worldwide (Cave and Koloto 2015).

Nonetheless, in identifying and reviewing the main challenges of tourism development in Tonga, three main components emerge: the current nature of the existing tourism markets, traditional values and practices which are at odds with international tourism development and the effect that the ecological environment is having on the tourism industry and vice versa. Whilst these challenges are highlighted below, the discussion also draws reference to potential solutions to these problems.

Challenges and solutions to tourism markets

Sea-based forms of tourism have been key markets for Tonga but have not been without challenges, especially with regard to both whale-watching and cruise ship tourism. Whale-watching became a stalwart of Tonga's nature-based tourism and was already a US$5 million per year industry by 1999 (Orams 2013). The kingdom is one of a few countries that permits swim-with-whales tourism (Fiori et al. 2019), encouraged by Tongan government policy derived from a 1978 Royal Decree which banned whale harvesting (Kessler and Harcourt 2010). The 1989 Fisheries Act proscribed the killing of cetacean animals, including whales and dolphins, so breeding populations in the region have increased from near extinction to more than 2,000 whales (SPREP 2017). However, it is argued that many boat operators do not meet the 2013 Whale Watch Regulations, especially where the health and safety of tourists are concerned and that whale conservation is not being adequately enforced, as whales are being displaced from their calving grounds during whale-watching tours (Bowe 2020). Potential solutions to strengthen confidence in this market concern the following recommendations: the strengthening of regulatory compliance for boat operators, ensuring fewer licensed vessels operate, instituting compulsory breaks (Fiori et al. 2019) and zoning to create safe sanctuaries for the whales (Kessler and Harcourt 2012).

In 2017, there were 19 cruise ships transporting 22,000 passengers to Tongan ports, an amount that had doubled since 2013, of which 25% of passengers disembark from the ships and spend an average of T$200 (currently US$87.86) (Government of Tonga 2018). However, the proven correlation between ports of call and COVID-19 outbreaks (Ito et al. 2020) significantly impacted Tonga's cruise ship market as well as affecting people's confidence that this market is safe and secure. Tonga targeted this market as an industry with great potential by building the Vuna Wharf in Nuku'alofa in 2013 to accommodate cruise ships (Hardwick et al. 2013) with the hope of encouraging more localized spending. Cruise numbers have been significant to the local economy for the short visit period if passengers are encouraged to spend and buy local products. The immediate concern is to seriously consider how the cruise industry can be more sustainable to the local economy in the future. Progressive changes in environmental sustainability will need to be addressed by cruise companies themselves, assisted by intensified lobbying from government and international organizations to pursue a sustainable development agenda.

The geographic remoteness of Tonga creates logistical challenges for travel because of limited routes and infrastructure, which crucially obstructs the development of new markets. Several airlines have tried to meet the demand for residents, freight and tourism, but they typically struggle to survive the low passenger volumes, significant distance between islands, uncertain weather conditions and high operating costs (Taumoepeau 2016). Daily flights from Tongatapu to Vava'u and the weekly ferry are expensive. Real Tonga provided domestic air service for eight years but ceased in 2020 because of financial shortfalls, compounded by the pandemic (RNZ 2020). The government plans to establish two inter-island bridges to counteract air and sea transportation problems by connecting Mala Island to Talihau and Patangata to Tongatapu's Eastern Districts (Matangi Tonga Online 2021). An integrated transportation system is fundamental for the continued development of Tonga, as noted almost 15 years ago by the Asia Development Bank (2008: 21):

> If Tonga had better land, water, and air transportation services, the movement of people and goods would be more frequent and cheaper. Improving services in areas where service is limited or nonexistent would raise the incomes of remote communities by providing them with more opportunities to engage in commerce. In turn, this could reduce the incentives of the rural population to move into urban areas.

Post-pandemic Tonga should encourage markets that are financially, culturally and socially sustainable, or re-align particular products with this focus in mind. The most logical and natural market for Tonga is the 'visiting friends and relatives' (VFR) market associated with the wider Tongan diaspora. Return visits are well understood in terms of their significance to the wellbeing of the Tongan nation, its residents and retention of culture, language and traditions. Even short-term visits can ensure the continuance of social, cultural and economic activities that are crucial for the propagation of Tongan identity (Cave and Koloto 2015).

Challenges and solutions to traditional practices

Family members are often employed in the daily operation of the business so that earnings remain in the family (Gibson-Graham 2008). Informal resource exchange such as gifts of unpaid labour and international remittances also create dependencies of responsibility, obligation and reciprocity within families and social networks (Baldachinno 2011). The moral worth and fiscal value of informal exchange are evaluated and regulated within cultural and consumer marketplaces (Busse and Sharp 2019). An internationalized approach to human resource management for a globalized tourism market might not synchronize well with Tonga's localized approach to cultural values in tourism development. Reciprocity, obligation and business donations to known networks are emphasized as crucial business operational components (Fili 2017). From the Tongan perspective, there is no sense of real loss if obligations are met at the cost of the business (James 2002). One key social value of Tongan society concerns maintaining relationships and networking. Accordingly, Tongan MSMEs, in the sphere of tourism, uniquely embed kindness, reciprocal relationships and social obligations into business life (Fili 2017). Indigenous ownership of tourism enterprises is relatively high in Tonga, in part due to restrictions on foreign land ownership and a government policy accentuating Indigenous strategies. Tourism enterprise can be an avenue of opportunity for women, although neither women nor foreigners can own land because succession passes through the male line (Besnier 2011). Nonetheless, women play significant roles in business decisions and operations (Nishitani 2014) and can use land for enterprise activities (Fili 2017).

The Sunday Law, a unique piece of legislation in Tonga, restricts retail and tourist operations. Many sport and recreational activities are also prohibited on the Sabbath, aiming to encourage locals to return to traditional healthy diets and spend the day with their family (Kingdom of Tonga 1998). God and Christian beliefs are the primary cornerstones of Tongan way of life (Besnier 2011). However, if visitors are unaware of the restrictions, inter-cultural conflicts can occur. Port stays and scheduled visits for cruise ships also encounter restrictions. Tongans do feel ambivalent towards tourism, derived from pride in a constitution that prevents Tonga's land from being purchased by foreigners and the existence of a largely monocultural community, with the exception of recent Chinese migrant communities (Lee 2016). Nonetheless, the Sunday Law is an opportunity to find common ground for locals and tourists to interact, perhaps educating tourists about respect for Tongan culture. Churches could, thus, openly invite tourists to their Sunday services and family communities could invite tourists for a shared meal. Specific beaches could also be allocated to tourist picnicking on Sundays (Fili 2017).

A significant challenge is the potentially dehumanizing effect of tourism, which is at odds with Tongan national pride and the country's legacy of being autonomous regarding its internal affairs. In her observational study, Lee (2016: 398–399) states:

> This issue of sovereignty and Tongans' pride in their independence may make it impossible for Tonga ever to build the kind of tourist industry seen in Pacific countries like Fiji. Most Tongans simply don't want to be subservient to tourists; in fact, they often seem uncomfortable working for pālangi bosses in any context. When I've discussed tourism with Tongans, they also frequently mention the risk of 'losing our culture'. Since the rule of the revered Queen Sālote, Tongans have paid heed to her constant encouragement to maintain their traditions and their 'authenticity' is another source of pride. Of course, Tongan culture has changed in many ways already; the country is firmly embedded in the world economy and Tongans are exposed to global popular culture.

Research concerning village walking tours has, in the past, raised similar cultural concerns, as 'Tongans cannot tolerate being regarded as members of a "cultural zoo"' (Urbanowicz 1989: 113). Cruise operators take visitors on buses to tour the island without time to visit the town or markets, which extends cultural distancing and reinforces a form of 'no contact tourism'. Indeed, Lee (2016: 39) talks of Tonga's relationship with tourists as 'vaguely annoying intrusions' rather than being 'enthusiastically welcomed', because of tourists' immodest clothing and behaviour patterns that contravene moral codes. Tongan textile makers, who have high standing in Tongan society as upholders of tradition (Besnier 2011), can feel demeaned by the bargaining behaviour of tourists and lack of respect or cultural awareness of the significance and meaning of their work or the skill and worth of the maker (Kirch 1984). Although cultural pride and attitudes towards tourists may be seen as impediments to international tourism development, such behaviour reinforces the need for sensitivity to local culture and island identities. Accordingly, it is crucial that the tourism experience embodies cultural awareness so that tourists are clearly informed of the cultural sensitivities of Tongan society via tourist information channels and related forms of communication.

Cross-sectoral linkages would also ensure that farming, tourism and hospitality systems align to essential climate change adaptation, rural development and the protection of traditional local culture. One of the crucial challenges is to ensure local traditional foods are systematically and consistently utilized in the tourism and hotel industry. Accommodation and restaurant menus in Tonga can feature the country's high-quality fish, but its supply is inconsistent because of over-fishing due to pressures of commercial fisheries and limited ways to restore fish stocks (Hestad 2020). Food sales are often conducted informally through family/friendship network, based on socio-cultural obligations and responsibilities in Tongan food supply systems, which are often misunderstood by development agencies whose focus is largely on growth (Bryceson and Ross 2020). Meats are imported except for locally grown pigs, often used in tourist cultural events. However, menus tend to offer international foods and rarely focus on local foods and fruit drinks (FAO 2012). When they do offer locally produced foods, there is a need for training and capacity building for chefs, farmers and procurement staff to work effectively and collectively to promote local food value chains (Bibi et al. 2020) and, indeed, traditional farming systems.

Yacht tourism is another way to leverage agritourism. Tonga is known as a safe haven for visiting yachts and charter fleets to re-supply with local fresh produce as they traverse the Pacific (FAO 2012). However, the islands of 'Eua and Ha'apai are constrained by the limited range of produce available for tourism use (Milne 2009). Nonetheless, from 2007 to 2011, the number of visiting yachts recorded in Vava'u increased from 400 to 557, with some yachts staying for up to two to three months, and it was estimated that one yacht alone catered each season for around 35,000 on-board meals, aiming to utilize fresh produce from local market suppliers as much as possible (FAO 2012: 19). Tonga attracts the private yachting market, which, though smaller than the cruise ship market, is seen as a higher spending market, with 2,000 arrivals annually prior to the global pandemic (Government of Tonga 2018). Community-based tourism linkages with agriculture that are cognisant of traditional industries have great potential, such as cultural tourism, adventure trekking, visits to bush-blocks to demonstrate farming, adding a traditional meal to resort menus to participate in a cultural night and visits to the 'Ene'io Beach and Botanical Garden on Vava'u, where demonstrations of local agricultural techniques, food processing and product sales already occur (FAO 2012).

New forms of cultural expression have emerged in response to globalization (Tolkach and Pratt 2019). Handicraft sellers, for example, have taken advantage of new technologies to gain competitive advantage in the marketplace, providing unique 'added value' to traditional designs, as seen with mother-of-pearl (Naidu et al. 2014). Tongan handicraft production focuses largely on tapa, weaving and carving that are gifted for lineage and exchanged for ritual purposes (Besnier 2011). Whilst raw materials are accessible, they now cost appreciably more, especially pandanus, which limits production (Milne 2009). Handicraft production and selling slowed significantly during the pandemic. This is a cause for concern, as the handicraft industry contributed to livelihoods and the economy and given Tonga's durable relationship to tradition. However, a distinction needs to be drawn between culturally significant crafts and items created for sale as souvenirs to tourists (Cave 2009). Crafts are sold at a women's handicraft collective established by Queen Sālote III in 1953 in central Tongatapu (Besnier 2011) and distributed through temporary stalls on cruise ship day, cultural sites, second-hand markets and tourist retail stores, often targeting the VFR market (Cave and Koloto 2015). The annual 'Katoanga' (mat exchange) takes place in the summer (Evans 2001). Local 'tour/travel' agents organize groups of expatriate Tongan ladies to travel from Australia, New Zealand or the United States (US) to purchase traditional Tongan mats made by local weavers (FAO 2012). Local cultural production is a draw for tourists seeking authentic experiences (MacCannell 1999).

Challenges and solutions to the ecological environment

Climate change presents a broad challenge to livelihoods and tourism development. Tonga ranks second in the world for most-at-risk countries, incurring around US$15.5 million in annual losses due to tropical cyclones and earthquakes (Cook and Chen 2019). Climate affected re-construction of infrastructure is frequently required, and populations relocate away from low-lying ground. Increasing frequency and intensity of storms and flooding are expected to exacerbate tourism development concerns (Scheyvens and Russell 2009). Ha'apai, for instance, is a low-lying sandy coral atoll, particularly susceptible to coastal erosion, storm surges, rogue waves, earthquakes and volcanic action. Disappearing corals and coastal erosion in both Ha'apai and Vav'au have negatively impacted tourism development and growth (Wolf et al. 2021). Warming seas affect marine reef food and mangrove areas, desiccating inland crops. Tonga's traditional fisheries are changing rapidly, with a 23% decline in total production between 1990 and 2018 (World Meteorological Organization 2021). Moreover, sea-level rise is salinating crucial freshwater lenses (Government of Tonga 2016).

However, the loss of biodiversity can be offset by remediation of littoral mangroves, which preserve fish stocks and shellfish, providing a buffer zone for storm damage (Ellison 2018). Restoration of the mangroves in Tonga is crucial, as they have generally declined due to dredging, clearing and reclamation, over-use and land issues (Aholahi et al. 2017). There have been various initiatives to rehabilitate the mangroves in Tonga, with a community-oriented approach to strengthening local seawalls and buffers as well as managing the replanting and nurturing of mangrove seedlings (MEIDECC 2019). Solutions to ecological challenges are difficult to achieve given the rapidity of global warming and natural events. Localized, short-term solutions can be found, such as establishing ecotourism projects based on traditional practices, moving to new locations and raising settlement foundations and sea-wall barriers. Thus, adaptation is probably more likely than resolution. Consequently, ecotourism development might well be a focus for Tonga's post-pandemic future. However, many attractive ecotourism locations are difficult to reach year-round. Moreover, wildlife protections are few. An ecotourism resort in the Vava'u island group opened in 2013 but closed within two years because of insufficient visitors to cover operational costs (Van Der Veeken et al. 2016). Also, organizational support to boost ecotourism and establish standards rests on the 'Eua Ecotourism Association, a small but energetic group (SPREP 2017). In fact, there are concerns relating to limitations in the protection of wildlife, which is an important ecotourism resource. For instance, the population of the Polynesian Megapode (*Megapodius pritchardii*) bird, once found in a number of Tongan islands, only now exists in Vava'u and is subsequently classified as being critically endangered, thereby affecting bird watching as a niche tourism market (Van Der Veeken et al. 2016).

Because marine-based tourism is a cornerstone for Tonga's tourism development pathway, the challenge is to utilize marine resources for economic development whilst still protecting the marine ecosystem. In whale tourism, misunderstandings between operators and tourists can arise, especially when visitors exert pressure on Tongan operators to locate and experience close encounters with the whales, thus, endangering the whales, boats and tourists (Kessler and Harcourt 2010). However, a nationally accredited guide training programme for Tongans has aimed to mitigate these issues (Walker and Weiler 2017), though the key objective should be to ensure that regulations are enforced and monitored (as implied earlier).

In the pre-COVID era, there was long-established concern about seaborne plastic waste, illegal dumping of on-ship waste from cruise ships (Chesher 1984; Gregory 1991; SPREP 2018) and on-shore plastic waste, which, along with local illegal dumping of rubbish, increases the island's environmental vulnerability (Gee 2018). As some feel that the Tongan government's waste management responsibilities are fragmented and scattered across several ministries and government departments, leading to ineffective coordination, a coherent policy framework is required to ensure attainment of collective goals (Farrelley et al. 2020). Another solution is consistent and proactive regulation (and voluntary action) within tourist operations.

Conclusion and research implications

This chapter drew attention to the historical, human and natural context of tourism in Tonga, followed by an overview of tourism's development in Tonga. It identified a range of key challenges pertaining to the diasporan tourism market, development constraints and cultural and ecological issues and, at the same time, endeavoured to highlight potential solutions to deal directly with the diversity of concerns raised. Cross-sectoral policies, legislative links and sustainable planning directives are essential for the future re-development of tourism.

Tongan tourism is reliant on marine and cultural resources, and markets are easily disrupted by exogenous shocks. The COVID-19 pandemic has influenced a demographic and economic shift from externally focused tourism to local travel. The pandemic's correlation with changing socio-cultural practices, particularly in light of the fact that the country has witnessed more forms of immobility than tourism mobility, could be a productive area of study for future investigation. Gendered roles in traditional livelihoods could be re-examined, especially in terms of establishing tourism-related experiences in newly planted mangrove forests.

Nevertheless, in terms of research implications there are three core components to discuss, especially in terms of identifying future research areas that connect to the potentialities identified in this paper: market-focused research, traditional-practices research and ecological-based research.

Market-focused research

More research is necessary to fully comprehend the role and mechanisms for being responsible tourists/returnee tourists in Tonga, especially from the contrasting and competing perspectives and differing behaviour patterns of first, second and now third generation Tongans visiting their ancestral homeland. For Tonga, the international family circuits of return are integral to everyday life and livelihoods (Connell 2010), facilitating significant inward and outward flows of capital. The offshore diaspora influences the nature of tourism in the archipelago and has profound impacts on the economy, social wellbeing and emotional heath on families who remain behind, often with children in the care of older relatives. Nonetheless, the dynamic flow of visiting friends and relatives could be more deeply investigated not only in terms of non-/post-pandemic experiences but also the effect of mobile Tongans residing elsewhere on the wellbeing, poverty and cultural change of communities of youth and elderly who continue to remain on the islands. Local and international priorities concerning post-natural disasters are focused on safe water, shelter, food and health, as well as diasporan care provision rather than tourism. Another important area of needed research would concern performance of obligation when in-person visits are not permissible, for instance, because of border closures during the pandemic, which would include attention to the role of cultural hierarchies amongst globally dispersed communities as they affect obligations. Also, the registered seasonal worker schemes which bring Tongans and others from the Pacific Islands to New Zealand and Australia for work in the horticultural industries impact diasporan and homeland communities as an avenue for remittances, education and cultural experiences but may have unintended effects. A key interest area in terms of the market-focus issue relates to ways in which alternative and diverse economies emerge (Cave and Dredge 2020), especially under circumstances of border closure and COVID-19-influenced behaviours.

Traditional-practices research

The cycles of return of VFR tourism infrastructure is worthy of investigation. The Tongan population has a high proportion of youth, some of whom do not want to migrate. A recent project in Nukunuku Village highlighted the value of inter-generational dreaming about a future in which youth remain in the village, engaged in livelihoods that may support tourism, but be part of its wider ecosystem (Cave et al. 2019). Such inter-generational strategies to combat the culture of migration and retain youth and local talents would be crucial to ensuring that Tonga focuses on capacity building initiatives pertaining to strengthening the local workforce. A study of tourism development would be useful to explore empowerment mechanisms for women and, at the same time, identify the structural and inter-cultural inequities that impede advancement. Research concerning how a cultural tourism industry might develop that can deal with the challenges identified in this chapter would be beneficial, notably the importance of cultural and national pride and how to integrate cultural identity into the tourism industry in a sensitive but empowering way. Furthermore, as faith is extremely important in Tongan communities, mindful and spiritual experiences could be developed that highlight places of pilgrimage, such as where the Tu'i Tonga converted to Christianity, linked with the landing places of missionaries, the church communities and other sacred sites. These are all areas for which qualitative research should be employed, alongside empirical methodologies, to deepen understanding of the perspectives of both hosts and visitors and how these perspectives can mutually intersect for the betterment of Tongan society and culture.

Ecological research

Climate change, water and land crop salination do affect the low-lying atolls and are crucial foci for future tourism-related research, as they affect food production, water scarcity, transportation and where or whether settlements can be safely established. Research could be undertaken traversing ecotourism feasibility and the establishment of traditional horticultural trails and visits to popular habitats where there are bat colonies, pigs, crabs, doves and turtles. Research should focus on the challenges of the outer islands and ways that new forms of mindful and regenerative tourism in remote areas would benefit the livelihoods of people who live there, especially in terms of integrating the ecological environment into the tourism experience. This could be enabled through more focus on specific islands such as Ha'apai and 'Eua and the more remote areas of the Nuias to comprehend ways that natural, historical and cultural resources collectively impact tourism provision. Whatever the focus, it is essential to ground research in Tongan values and research ethos, such as the Talanoa and Kakala methods, to ensure uptake of research findings and produce lasting impacts. Western frameworks can work alongside Tongan ones in co-creative partnerships. Further, it is essential that researchers enact regenerative processes of stewardship, reciprocity, respect and deep connection.

References

Aholahi, H., P. Aleamotu'a, D.J. Butler, H. Etika, T. Faka'osi, S. Hamani, T.M. Helu, T.F. Hokafonu, U. Kaly, R.A. Kautoke, V.T. Manu, A.L. Matoto and P. Ma'u (2017) *Status of Fanga'uta Lagoon in 2016*, Report for United Nations Development Programme (UNDP), Nuku'alofa, Tonga: Department of Environment, Tonga Ministry of Meteorology, Environment, Information, Disaster Management, Energy, Climate Change and Communications. Available online at https://tonga-data.sprep.org/system/files/Fanga%27uta%20Lagoon%20Status%20Report%202016%20280917_signed_0.pdf (accessed 30 January 2022).

ADB (Asian Development Bank) (2008) *Transforming Tonga: A Private Sector Assessment – Summary* (in conjunction with P. Holden, C. Russell and W. Wicklein), Manila, Philippines: ADB. Available online at https://www.adb.org/sites/default/files/ institutional-document/32222/psa-ton.pdf (accessed 5 May 2022).

Baldachinno, G. (2011) 'Surfers of the ocean waves: Change management, intersectoral migration and the economic development of small island states', *Asia Pacific Viewpoint*, 52(3): 236–246.

Besnier, N. (2011) *On the Edge of the Global: Modern Anxieties in a Pacific Island Nation*, Stanford: Stanford University Press.

Bowe, K. (2020) 'Rise and fall of Vava'u whales watching industry', *Mantangi Tonga Online*, 27 April. Available online at https://matangitonga.to/2020/04/27/rise-and-fall-vava-u-whale-watch-industry (accessed 29 December 2021).

Bryceson, K.P. and A. Ross (2020) 'Agrifood chains as complex systems and the role of informality in their sustainability in small scale societies', *Sustainability*, 12(16): 6535.

Busse, M. and T.L.M. Sharp (2019) 'Marketplaces and morality in Papua New Guinea: Place, personhood and exchange', *Oceania*, 89(2): 126–153.

Cave, J. (2009) 'Between worldviews: Nascent Pacific tourism enterprise in New Zealand', *Unpublished PhD Thesis*, Hamilton: University of Waikato. Available online at https://researchcommons.waikato.ac.nz/bitstream/handle /10289/3281/thesis.pdf?sequence=1 (accessed 20 April 2022).

Cave, J. and D. Dredge (2020) 'Regenerative tourism needs diverse economic practices', *Tourism Geographies*, 22(3): 503–513.

Cave, J., S. Johnasson-Fua, K. Jones, A. Koloto and M. Paunga (2012) *Tongan Heritage Scoping Study* (Vol. I and II). Available online at https://scholar.googleusercontent.com/scholar?q=cache:Yw3skOH5kmkJ:scholar.google.com/+)+Tongan+Heritage+Scoping+Study&hl=en&as_sdt=0,5&as_vis=1 (accessed 20 February).

Cave, J. and A. Koloto (2015) 'Short-term visits and Tongan livelihoods: Enterprise and transnational exchange', *Population, Space and Place*, 21(7): 669–688.

Cave, J., A. Koloto, C. Cater, P. Parton and S. Dabamona (2019) *Tonga West Polynesian Education for Sustainability*, Report on GCRF RIG1029-128, Swansea University (Project Report).

Chesher, R.H. (1984) *Pollution Sources Survey of the Kingdom of Tonga*, SPREP/Topic Review, 19 November, Nouméa, New Caledonia: South Pacific Regional Environment Programme.

Clark, G. (2010) 'The sea in the land: Maritime connections in the chiefly landscape of Tonga'. In: K.V. Boyle and A. Anderson (eds.) *The Global Origins and Development of Seafaring*, Cambridge: McDonald Institute for Archaeological Research, pp. 229–237.

Cook, A.D.B. and C. Chen (2019) *Disaster Governance in the Southwest Pacific: Perspectives, Challenges, and Future Pathways for ASEAN*, Singapore: Nanyang Technological University.

Connell, J. (2010) 'Pacific island states in the global economy: The paradox of migration and culture Singapore', *Journal of Tropical Geography*, 31(1): 115–129.

Ellison, J.C. (2018) 'Effects of climate change on mangroves relevant to the Pacific islands', Pacific marine climate change report card, *Science Review*: 99–111.

Evans, M. (2001) *Persistence of the Gift: Tongan Tradition in Transnational Context*, Waterloo: Wilfrid Laurier University Press.

FAO (Food and Agriculture Organization) (2012) *Report on a Scoping Mission in Samoa and Tonga, Agriculture and Tourism Linkages in Pacific Island Countries*, Food and Agriculture Organization of the United Nations. Available online at http://www.fao.org/3/an476e/an476e.pdf (accessed 1 February 2022).

Farrelly, T., S. Borrelle and S. Fuller (2020) *Plastic Pollution Prevention in Pacific Island Countries: Gap Analysis of Current Legislation, Policies and Plans*, Environmental Investigation Agency, August. Available Online at https://reports.eia-international.org/wp-content/uploads/sites/6/2020/09/Plastic-Prevention-Gap-Analysis-2020.pdf (accessed 21 April 2021).

Fili, R.L. (2017) 'Tongan business obligation: Across a sea of enterprise', *Master of Management Studies Dissertation*, Hamilton: University of Waikato.

Fiori, L., E. Martinez, M. Orams and B. Bollard (2019) 'Effects of whale-based tourism in Vava'u, Kingdom of Tonga: Behavioural responses of humpback whales to vessel and swimming tourism activities', *PLOS ONE*, 14(4). https://doi.org/10.1371/journal.pone.0219364.

Foster, S. and S. Latukefu (2022) 'Tonga', *Encyclopedia Britannica*, 18 January. Available online at https://www.britannica.com/place/Tonga (accessed 31 January 2022).

Gee, E. (2018) 'Illegal dumping of rubbish overwhelms Tongan authorities after Gita', *Matangi Tonga Online*, 8 June. Available online at https://matangitonga.to/2018/06/08/illegal442dumping-rubbish-overwhelms-tongan-after-gita (accessed 8 June 2019).

Gregory, M.R. (1991) 'The hazards of persistent marine pollution: Drift plastics and conservation islands', *Journal of the Royal Society of New Zealand*, 21(2): 83–100.

Gibson-Graham, J.K. (2008) 'Diverse economies: Performative practices for "other worlds"', *Progress in Human Geography*, 32(5): 613–632.

Government of Tonga (2016) *Tonga Climate Change Policy: A Resilient Tonga by 2035*, Department of Climate Change, Government of the Kingdom of Tonga. Available online at http://www.lse.ac.uk/GranthamInstitute/wp-content/uploads/laws/4353.pdf (accessed 15 May 2021).

Government of Tonga (2018) *Post Disaster Rapid Assessment: Tropical Cyclone Gita*, 12 February. Available online at https://reliefweb.int/sites/reliefweb.int/files/resources/tonga-pdna-tc-gita-2018.pdf (accessed 29 December 2021).

Hardwick, L., J. Youdale and R. Frankland (2013) 'Cruise ship driven development of port and coastal infrastructure'. In: *Coasts and Ports, Combining the 21st Australasian Coastal and Ocean Engineering Conference & the 14th Australasian Port and Harbour Conference*, Sydney, Australia, 11–13 September, pp. 352–357.

Harrison, D. and B. Prasad (2013) 'The contribution of tourism to the development of Fiji and other Pacific island countries'. In: C.A. Tisdell (ed.), *Handbook of Tourism Economics: Analysis, New Applications and Case Studies*, Singapore: World Scientific, pp. 741–761.

Hestad, D. (2020) *Report Reveals Positive Outcomes of Tonga's Special Management Areas on Marine Environment*, International Institute for Sustainable Development, 6 October. Available online at https://sdg.iisd.org/commentary/policy-briefs/report-reveals-positive-outcomes-of-tongas-special-management-areas-on-marine-environment/ (accessed 18 April 2021).

Howes, S. and S. Surandiran (2020) 'Pacific remittances: Holding up despite COVID-19', *The Devpolicy Blog*, 16 November, Canberra: Development Policy Centre, Australian National University. Available online at https://devpolicy.org/pacific-remittances-covid-19-20201116/ (accessed 1 February 2022).

Ito, H., S. Hanaoka and T. Kawasaki (2020) 'The cruise industry and the COVID-19 outbreak', *Transportation Research Interdisciplinary Perspectives*, 5: 100136.

James, K.E. (2002) 'Disentangling the "grass roots" in Tonga: "Traditional enterprise" and autonomy in the moral and market economy', *Asia Pacific Viewpoint*, 43(3): 269–292.

Kessler, M. and R. Harcourt (2010) 'Aligning tourist, industry and government expectations: A case study from the swim with whales industry in Tonga', *Marine Policy*, 34(6): 1350–1356.

Kessler, M. and R. Harcourt (2012) 'Management implications for the changing interactions between people and whales in Ha'apai, Tonga', *Marine Policy*, 36(2): 440–445.

Kingdom of Tonga (1988) *Act of Constitution of Tonga* (rev. edn). Available online at https://www.ilo.org/dyn/natlex/docs/ELECTRONIC/35656/110368/F222186159/TON35656%202.pdf (accessed 5 May, 2022).

Kirch, D.C. (1984) 'Tourism as conflict in Polynesia: Status degradation of Tongan handicraft sellers', *Unpublished PhD Thesis*, Hawaii: University of Hawai'i at Manoa.

Lee, H. (2016) 'The friendly islands? Tonga's ambivalent relationship with tourism'. In: J. Taylor and K. Alexeyeff (eds.) *Tourism Pacific Culture: Mobility, Engagement and Value*, Acton, ACT: ANU Press, pp. 393–401.

MacCannell, D. (1999) *The Tourist: A New Theory of the Leisure Class*, Berkeley: University of California Press (first published in 1976 by Schocken Books Inc).

Matangi Tonga Online (2018) 'Cyclone Gita cost Tonga $356 million', 15 May. Available online at https://matangitonga.to/2018/05/15/cyclone-gita-cost-tonga-356-million (accessed 29 December 2021).

Matangi Tonga Online (2021) 'PM dreams of big international airport for Vava'u plus two inter-island bridges', 6 April. Available online at https://matangitonga.to/2021/04/06/pm-dreams-big-intl-airport-vavau-plus-two-inter-island-bridges-0 (accessed 11 April 2021).

MEIDECC (Ministry of Meteorology, Energy, Information, Disaster Management, Environment, Climate Change and Communications) (2019) 'Tonga's mangroves project led my community members', Government of Tonga, *Reliefweb*, 2 May. Available online at https://reliefweb.int/report/tonga/tonga-s-mangroves-project-led-community-members (accessed 30 January).

MFAT (Ministry of Foreign Affairs and Trade) (2020) 'Pacific tourism: Covid-19 impact & recovery, scenario development and recovery pathways: Report', 10 June. Available online https://www.mfat.govt.nz/assets/Uploads/Pacific-Tourism-Sector-Scenerio-Development-and-Recovery-Pathways.pdf (accessed 1 February 2022).

Milne, S. (2009) *NZAID Tonga Tourism Support Programme: Economic Linkage Study, December*, Wellington: New Zealand Aid. Available online at https://nztri.org.nz/sites/default/files/Tonga%20Linkage%20Final%20report.pdf (accessed 1 February 2022).

Morton, N. (2022) 'Half of Tonga's tourism businesses "Won't return" after tsunami, lockdown', *Stuff.co.nz*, 21 February. Available online at https://www.stuff.co.nz/world/south-pacific/300522757/half-of-tongas-tourism-businesses-wont-return-after-tsunami-lockdown#:~:text=Tongan%20business

%20experts%20say%20up,tsunami%20and%20Covid%2D19%20lockdown.&text=This%20included %20a%20number%20of,the%20main%20island%20of%20Tonga (accessed 5 May 2022).

Naidu, S., A. Chand and P. Southgate (2014) 'Determinants of innovation in the handicraft industry of Fiji and Tonga: An empirical analysis from a tourism perspective', *Journal of Enterprising Communities*, 8(4): 318–330.

Nishitani, M. (2014) 'Kinship, gender, and communication technologies: Family dramas in the Tongan diaspora', *The Australian Journal of Anthropology*, 25(2): 207–222.

Ober, K. and S. Bakumenko (2020) 'A new vulnerability: COVID-19 and tropical cyclone Harold create the perfect storm in the Pacific', United Nations Office for the Coordination of Humanitarian Affairs, *Reliefweb*, 3 June. Available online at https://reliefweb.int/report/vanuatu/issue-brief-new-vulnerability-covid-19-and-tropical-cyclone-harold-create-perfect (accessed 25 January 2022).

Orams, M. (2013) 'Economic activity derived from whale-based tourism in Vava'u, Tonga', *Coastal Management*, 41(6): 481–500.

Pratt, S. (2016) 'A gross happiness index for the Solomon Islands and Tonga: An exploratory study', *Global Social Welfare*, 3(1): 11–21.

Ratuva, S. (2019) *Contested Terrain: Reconceptualising Security in the Pacific*, Canberra: ANU Press.

RNZ (Radio New Zealand) (2020) 'Tonga still without domestic air service', *RNZ*, 1 September. Available online at https://www.rnz.co.nz/international/pacific-news/424968/tonga-still-without-domestic-air-service (accessed 19 December 2021).

RNZ (Radio New Zealand) (2021) 'Tonga extends covid state of emergency', *RNZ*, 19 January. Available online at https://www.rnz.co.nz/international/pacific-news/434769/tonga-extends-covid-state-of-emergency (accessed 19 December 2021).

Scheyvens, R. and F.M. Russell (2009) *Tourism and Poverty Reduction in the South Pacific*, Massey University: Palmerston North, New Zealand. Available online at https://apo.org.au/sites/default/files/resource-files/2009-08/apo-nid110966.pdf (accessed 20 December 2021).

SPREP (Secretariat of the Pacific Regional Environment Programme) (2018) 'Regulating Plastics in Pacific Island Countries: A Guide for Policymakers and Legislative Drafters', 24 October. Available online at https://www.sprep.org/publications/regulating-plastics-in-pacific-island-countries (accessed 1 February 2022).

SPREP (Secretariat of the Pacific Regional Environment Programme) (2017) 'Tonga reaffirms commitment to protect whales', produced by I. Tora, *Ministry of Meteorology, Energy, Information, Disaster Management, Climate Change and Communications of Tonga*, 4 April. Available online at https://www.sprep.org/news/tonga-reaffirms-commitment-protect-whales (accessed 1 February 2022).

Stanley, D. (2004) *South Pacific*, 8th edition, Emeryville: Avalon Travel Publishing.

Taumoepeau, S. (2016) 'New heights of success for Tonga's aviation and tourism', *Open Journal of International Education*, 1(2): 67–89.

Tolkach, D. and S. Pratt (2019) 'Globalisation and cultural change in Pacific island countries: The role of tourism', *Tourism Geographies*, 23(3): 371–396.

Tonga Statistics Department (2020) 'Poverty in Tonga'. Available online at https://tongastats.gov.to/statistics/social-statistics/poverty-in-tonga/ (accessed 29 December 2021).

Tonga Tourism Authority (2018) 'The Kingdom of Tonga, the true South Pacific'. Available online at http://www.thekingdomoftonga.com/ (accessed 15 April 2021).

Traveller (2012) 'Trouble in paradise: The "friendly islands" are suffering', 15 May. Available online at https://www.traveller.com.au/trouble-in-paradise-the-friendly-islands-are-suffering-1x8vb (accessed 13 April 2021).

United Nations (2022) '80 per cent of Tonga population impacted by eruption and tsunami', *UN News*, 20 January. Available online at https://news.un.org/ en/story/2022/01/1110162 (accessed 1 February 2022).

Urbanowicz, C.F. (1989) 'Tourism in Tonga revisited: Continued troubled times?' In: V.L. Smith (ed.) *Hosts and Guests: The Anthropology of Tourism*, 2nd edition, Philadelphia: University of Pennsylvania Press, pp. 105–117.

Van Der Veeken, S., E. Calgaro, L. Klint, A. Law, M. Jiang, T. De Lacy and D. Dominey-Howes (2016) 'Tourism destinations' vulnerability to climate change: Nature-based tourism in Vava'u, the Kingdom of Tonga', *Tourism and Hospitality Research*, 16(1): 50–71.

Walker, K. and B. Weiler (2017) 'A new model for guide training and transformative outcomes: A case study in sustainable marine-wildlife ecotourism', *Journal of Ecotourism*, 16(3): 269–290.

Wolf, F., P. Singh, N. Scherle, D. Reiser, J. Telesford, I.B. Miljković, P.H. Havea, C. Li, D. Surroop and M. Kovaleva (2021) 'Influences of climate change on tourism development in small Pacific island states', *Sustainability*, 13(8): 1–22.

World Bank (2020a) 'Statistical innovation and capacity building in Tonga', *World Bank Group*, 21 January. Available online at http://documents1.worldbank.org/curated/en/363631581735856497/pdf/Tonga-Statistical-Innovation-and-Capacity-Building-Project.pdf (accessed 28 December 2021).

World Bank (2020b) '$30 million in support for Tonga's most vulnerable', *World Bank Group*, 16 December. Available online at https://www.worldbank.org/en/news/press-release/2020/12/16/30-million-in-support-for-tongas-most-vulnerable (accessed 29 December 2021).

World Bank (2022) 'World Bank Provides US$8 million to Support Tonga's Volcano and tsunami Response', *World Bank Group*, 29 January. Available online at https://www.worldbank.org/en/news/press-release/2022/01/20/world-bank-provides-us-8-million-to-support-tonga-s-volcano-and-tsunami-response (accessed 1 February 2022).

WorldData.info (2020) 'Tourism in Tonga'. Available online at https://www.worlddata.info/oceania/tonga/tourism.php (accessed 29 December 2021).

World Meteorological Organization (2021) 'Climate change increases threats in south west Pacific', 10 November. Available online at https://public.wmo.int/en/media/press-release/climate-change-increases-threats-south-west-pacific#:~:text=Glasgow%2C%2010%20November%202021%20(WMO,upon%20which%20the%20region%20depends.(accessed 29 December 2021).

Worldometer (2022) 'Tongan population'. Available online at https://www.worldometers.info/world-population/tonga-population/ (accessed 5 May 2022).

Young-Leslie, H.E. and S.E. Moore (2012) 'Constructions of happiness and satisfaction in the Kingdom of Tonga'. In: H. Selin and G. Davey (eds.), *Happiness Across Cultures: Views of Happiness and Quality of Life in Non-Western Cultures*, Dordrecht: Springer, pp. 181–193.

24

YOGA TOURISM AND SUSTAINABLE DEVELOPMENT IN NIUE POST-COVID

Investigating the potential for sport tourism and wellness experiences

Nazia Ali

Introduction

The objective of this chapter is to assess the possible sustainable development of sport tourism on the small island state of Niue in the Pacific. It is proposed that, given Niue's invaluable environmental resources, sport tourism development should help safeguard against a potential threat of overtourism. A form of non-competitive sport tourism noted to promote a harmonious relationship with the environment is wellness tourism, which Niue has the natural resources to plan and develop. This also builds on existing non-sport types of wellness and well-being activities currently being provided on the island such as massages, spas and hair and beauty treatments. Wellness tourism is 'travel associated with the pursuit of maintaining or enhancing one's personal wellbeing' – a multifaceted product, service and experience in search of physical, mental, social, emotional, spiritual and environmental well-being (Global Wellness Institute 2018: iii). In the Sport Tourism 2017 report produced by the United Nations World Tourism Organization (UNWTO), health and wellness activities are considered in the context of sport tourism to which health and fitness can be noted as a related dimension (Higham and Hinch 2018). In this chapter, sport tourism is considered in the wider sense to include yoga as a niche form of health and wellness, involving a special interest vacation for those motived to travel away from their everyday environments in a pursuit to nurture the well-being of their mind, body and soul. The vacation can take place in therapeutic natural environments for purposes of escape, restoration and retreat (Lea 2008). The experience can involve a search for authenticity, which can be found in such sport tourism activities as yoga which enable individuals to be involved in 'experiencing nature, and in turn experiencing ourselves during retreats' (Lea 2008: 96).

There is little known about international tourist arrivals and receipts for sport as the purpose of travel, with no distinct segment for sport or sport tourism. Nevertheless, with reference to the purpose of the visit segments recognized by UNWTO (2018: 3), sport tourism can be considered to intersect with 'health, other' and 'leisure, recreation and holidays', where sport can

DOI: 10.4324/9780429019968-28

be aligned with health in the context of well-being and can also reflect leisure and recreation as part of the sport tourism experience. This interpretation can be further extended to wellness tourism or yoga retreat tourism, as reasons for travel can be for health benefits and/or for leisure interests.

The chapter aims to contribute to the existing debate on sustainable tourism development in the Pacific region, specifically relevant to the small island state of Niue. Accordingly, the work considers sustainability in the context of its regenerative and degrowth value, particularly in the context of a post-pandemic future. The work also attempts to demonstrate the importance of yoga tourism as a niche sport tourism market which is beneficial to 'small island developing states' (SIDS) seeking to diversify their (tourism) economies. The subsequent discussion acknowledges the tourism impact of the global pandemic on Niue and the possibilities that this presents to rethinking and restarting regenerative and degrowth forms of tourism that are inherently sustainable – beyond economic growth. This leads onto a review of the principles and practices of sustainable development which integrate theoretical, conceptual, empirical and strategic perspectives associated with sustainability and sustainable sport tourism development. The discussion of the relationship between sport and tourism that fosters wellness and well-being provides the impetus to propose wellness tourism in the form of yoga retreat (tourism) – a niche market segment on the island of Niue that is closely aligned with sustainable development.

Niue: a small island state in the Pacific

Niue (Figure 24.1), which is located northeast of New Zealand, southeast of Samoa, east of Tonga and west of Rarotonga (Cook Islands), is a small, isolated island state with an estimated population of 1,642, decreasing significantly from a peak population of 5,135 in 1970 (Worldometer 2021). Niue is the 'Rock of Polynesia' with its raised coral spanning an area of 269 square kilometres and is one of the largest coral islands in the world, with fishing as the main form of livelihood (Freddie and de Sylva 2018). Tropical temperatures are found all year round, ranging from 20 to 28 degrees Celsius in the winter months (April–November) and 22 to 30 degrees Celsius in the summer months (December–March) (Niue, n.d.a.) Niue's natural and cultural environment can be considered as its major strength, with its 'natural beauty', 'chasms and sandy coves' and 'crystal clear waters' (Niue, n.d.b.) and marketed on *The Official Website of Niue Tourism* homepage (https://www.niueisland.com/) as 'Sustainable Niue'. This commitment to sustainability is further reflected with Niue being listed as one of the top 100 sustainable destinations in the world in 2020 by Green Destinations (SPTO 2020a). The 'smallness' characteristic of Niue is not only reflected in the size of the island, with a land area of 261 square kilometres, and the size of its population but also its low gross domestic product (GDP) valued at around only NZ$35,383 (2016) (Statistics for Development Division 2019). This reinforces what Ryan (2001: 43) refers to as 'marginalisation' in terms of economic and geographic marginality, which calls for forms of development that are sustainable and do not 'compromise Niue's fragile economy and its future as a sovereign nation' (Freddie and de Sylvia 2018: 1).

The first 'national development pillar' (from seven pillars) embedded in the Niue National Strategic Plan (2016–2026) is 'financial and economic development', in which it is emphasized that tourism growth should be sustainable to source financial stability (Government of Niue 2016). However, the global COVID-19 pandemic halted international travel due to restrictions to control the contact and spread of the virus from early 2020, which impacted Niue's tourism-dependent economy (ADB 2021a). Therefore, there is a need to think about a destination recovery plan in the post-pandemic future, one that is financially stable and does not compromise Niue's role as a sovereign state.

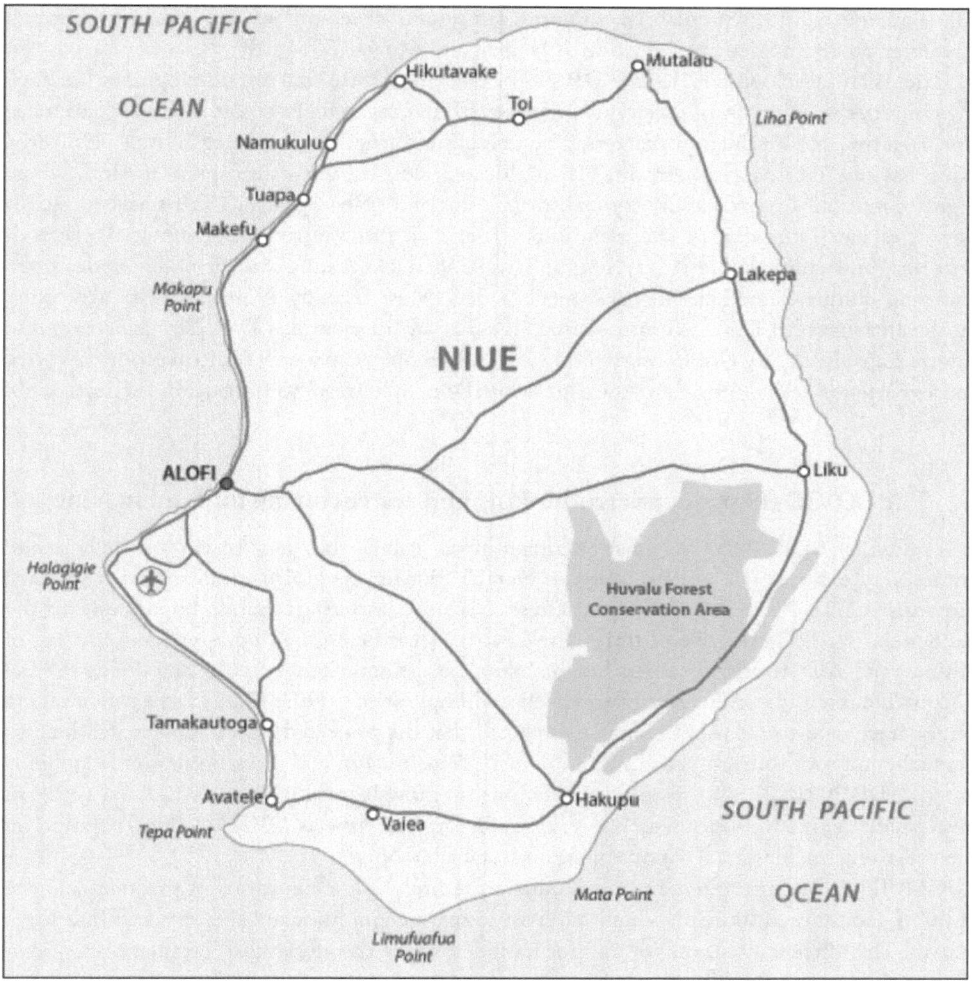

Figure 24.1 Map of Niue, indicating the country's capital Alofi. Source: Rainer Lesniewski (https://www.shutterstock.com/image-vector/niue-map-151779485).

Niue became a British protectorate at the beginning of the twentieth century and was then controlled by New Zealand, gaining independence in 1974 (Treloar and Hall 2005). Niue is a self-governing state in free association with New Zealand, meaning that Niueans, as New Zealand citizens, have the right to enter and reside in New Zealand and Australia (Department of Foreign Affairs and Trade n.d.). This close proximity and freedom of movement to New Zealand has led to out-migration to New Zealand for economic opportunities (e.g., education, employment, enterprise and employability skills) and, consequently, is a contributing factor to Niue's decreasing population (Connell 2008). New Zealand also provides economic aid and administrative assistance to Niue (Overton and Murray 2014), classifying the island as a 'migration, remittance, aid and bureaucracy' (MIRAB) economy. MIRAB economies are 'economies based in the islands but tied to the region's metropolitan nation, economically and politically, through combined effects of migration (MI), remittance (R), aid (A) and bureaucracy

(B)' (Barker 2000: 192). A MIRAB economy can sustain 'development at the margins' through a counter process of 'demarginalisation' (Overton and Murray 2014: 18).

The Asian Development Bank (ADB 2021a: 342) notes that 'tourism development has been the country's main driver of economic growth, with improved flight connectivity, accommodation facilities, and destination marketing contributing to increase in visitor arrivals'. However, Niue has one international airport, Hanan, located close to the state capital of Alofi, where flights commonly arrived twice a week from Auckland (New Zealand) and serviced by Air New Zealand (Niue n.d.a.). The island offers tourists a range of tours and activities, places of historical and cultural interest, cycling and walking tracks, eating and drinking options, art, craft and culture-based activities and retail outlets. Most recently, Niue has been accredited by the International Dark-Sky Association (IDA) as an International Dark Sky Sanctuary and International Dark Sky Community (Niue 2021). This opens up a new tourism product, service and experience concerning astro-tourism, whereby people travel to stargaze in the dark night skies.

The COVID-19 pandemic: rethinking and restructuring tourism in Niue

It is noted by ADB (2020: xvi) that countries in the Pacific that 'rely heavily on tourism and commodity exports will be particularly vulnerable' during the global pandemic with subdued growth in 2021. Around 84% of businesses in Niue reported a negative impact due to the pandemic (Pacific Trade Invest 2021), with key risk areas being the domestic economy, loss of business and skills, tourism business sustainability/closure and charter fishing and diving (SPTO 2020b). Despite the current negative impacts and implications of the COVID-19 pandemic on Niue's travel and tourism industry, it is expected that the projected travel bubbles will help to gradually improve tourism growth to 3.8% in 2022 as tourism and cross-border trade are reinstated (ADB 2021b). Proposals to accelerate tourism growth are also being considered as part of post-pandemic recovery plans such as sustainable tourism projects (SPTO 2020b), building on the pre-pandemic destination positioning of sustainable Niue.

COVID-19 has presented an opportunity to rethink and reconsider past practices and the political economy of tourism, which did little to serve economies and societies and find alternatives. This directs attention to 'transformation towards the *regenerative* paradigm and *regenerative economic systems*' which are 'based on the natural cycles of renewal and regeneration' (Ateljevic 2020: 1 (emphasis in original)). These new forms of regenerative tourism are inherently transformative in terms of making a positive contribution to economies and societies, in which 'human well-being' must concern the 'environmental, social, economic and cultural' components (Becken 2021 cited in Business World 2021). The transformation involves a shift away from neoliberalist forms of tourism that aid the political economy of tourism to being able to 'embrace alternative non-capitalist forms of ownership, non-monetary exchange and beneficial community-based development' such as regenerative tourism (Sheller 2021: 1437). Regenerative tourism echoes the principles and practices of sustainable development, going beyond traditional notions of sustainability to extend its role in cultural, economic, environmental and social regeneration by tackling the negative impacts of mass tourism (Duxbury et al. 2021).

The first point directs attention to preventing the pre-COVID state of overtourism happening when international travel restarts, especially through reconsidering growth strategies. Thus, there is a shift from 'tourism growth' to 'tourism degrowth', whereby degrowth can be seen as a social movement or a form of activism calling for alternative management and development practices (Milano et al. 2019). Accordingly, there is a need to rethink the value of tourism in terms of regenerative tourism that is connected to the degrowth debate, which has brought to the surface shifting paradigms, perspectives and practices of sustainability and positioning of sustainable tourism with the discourse of degrowth. Therefore, yoga (retreat) tourism in Niue, as proposed in this chapter, arguably provides a viable alternative, as it is a small-scale, community-based development seeking to stimulate a tourism economy which does not really exist on the island.

The Head of Niue Tourism is vigilant to the effects of increasing tourist numbers on a small island's capacity to manage growth, asserting:

> We're very mindful of managing our growth carefully. And the main reason for this is the effect on our environment. Which is the reason people come to Niue. [...] It's such a pristine, wonderful, environmentally conscious place to visit, and we don't want to ruin that.
>
> *(RNZ 2018)*

Niue received 9,834 tourists in 2019 (SPTO 2021) which is almost six times higher than the population of the island. Therefore, it is arguably imperative to look for sustainable alternatives to avoid the challenges associated with overtourism and avoid returning to the growth levels that Niue experienced prior to the pandemic.

Principles and practices of sustainable development for sport tourism

Sustainable development emphasizes the relationship between three pillars: the environment, economy and society (Jędrusik 2014). Similarly, for sustainable tourism development, the pillars are economic, social and environmental (Gibson et al. 2012). From the perspective of the government of Niue, sustainability is associated with 'development that leads to prosperity for Niue and future generations', as well as 'infinite ways for sustainable development and management and to better conserve the environment towards a prosperous Niue' (Government of Niue 2016: 20), especially to ensure 'sustainable growth in tourism' (2016: 27). The principles and practices of sustainability, regeneration and degrowth in the context of tourism can be further extended to situating sport within the UNs Sustainable Development Goals (SDGs) 2030. The United Nations (n.d.) documented how sport is recognized as an 'enabler' for sustainable development, and SDGs have been aligned with sport tourism development, for example, 'SDG 1: No Poverty – End poverty in all its forms everywhere' proposes:

> Sport is itself a productive industry with the ability to lift people out of poverty through employment and contributing to local economies. Sport and sustainable sport tourism can promote livelihoods, including in host communities of sport events.
>
> *(United Nations, n.d.: 3)*

However, conceptualizations of sustainability and sustainable development are embedded with complexities and multiple interpretations which limit its transferability to non-Western environments. Therefore, Western-centric/developed world principles and practices of sustainable development should be applied with caution to developing nations/small island states because the ideal of sustainable development in the context of tourism is rather reductionist and renders an incomplete 'narrative' for Pacific Island countries (PICs) (Panakera et al. 2011). As Panakera et al. (2011: 241) explain:

> the narrative concerning sustainable tourism development in the South Pacific is incomplete and the predominant narrative viewing the South Pacific nations as economically and environmentally vulnerable is too simplistic.

Moreover, the very conditions for sustainable development – the environment (i.e., vulnerability of a natural resources), economy (i.e., small, inflexible and little diversification) and society (i.e., under-population) can be a barrier for development; consequently, sustainable development is nothing more than an 'elusive utopia' (Jędrusik 2014: 26). Therefore, it is crucial to put aside Western-centric conceptualizations of sustainability, especially to embrace Niue's environmental vulnerabilities, socio-cultural realities and economic fragilities in the development of a niche form of sustainable sport tourism.

Sport tourism and small island destinations

Sport tourism refers to travel to participate or spectate in sport activities for competitive (e.g., professional or elite) and/or non-competitive (e.g., leisure or recreational) reasons and to visit sport attractions and sites (e.g., sport museums or sport halls of fame). To participate, to watch and to visit can involve states of being passive and/or active, and the purpose for travel can either be for sport or tourism depending on which is the primary tourist motivation (Ritchie and Aldair 2004). For Taleghani and Ghafary (2014: 296) there is an integration of the sport and tourism industry to create sport tourism, and the main areas in sport tourism are 'attractiveness of nature, fun places, sea travel, events, and tours'. Gibson (1998: 45) conceptualizes sport tourism into three categories: active sport tourism (actively participating), event sport tourism (spectating) and nostalgia sport tourism (visiting/paying homage). However, definitions, categorizations, methodologies and typologies of sport tourism are not so clear-cut, independent or static but are paradigmatically emergent, evolving and shifting (Gammon et al. 2017). Therefore, locating yoga within the ontological and epistemological domains of sport and tourism encourages recognition of the fluidity and 'terminological chaos' (Butowski 2016: 89) inherent in sport tourism.

Sport tourism has been used by small island destinations to diversify their tourism offerings to niche or specific (special) interest groups to reduce dependency on mass tourism (e.g., sun, sand and sea) markets and to enhance destination competitiveness (Azzopardi and Nash 2016; Bull and Weed 1999). Small island states may already possess natural and physical assets and international tourism (target) market strongholds that can be drawn upon in a diversification shift to sport tourism. Also, for small islands living with the economic impacts of a decline in exports due to the consequences of a resource curse and with little or no tourism industry can consider developing sport tourism products, services and experiences by harnessing the power of their 'natural capital' (Bass and Dalal-Clayton 1995: n.p.). According to Bull and Weed's (1999) early analysis of sport tourism development in the small island state of Malta, they found that the country had some key resources to host sport tourism products, services and experiences such as the Mediterranean climate for all-year sports activities and a central locale for ease of accessibility, especially for its European markets, and natural and coastal landscapes.

There are certain resource constraints that small island destinations seeking to develop sport tourism should be alert to, as these have an effect on supply and demand (e.g., locals versus tourists) and infrastructure advancements (e.g., sport facilities, hotels and transport networks). These constraints are resource shortages (e.g., water and land), government decision-making and planning controls, limited sport development initiatives in place and geographical remoteness in terms of location (Azzopardi and Nash 2016; Bull and Weed 1999). Thus, because of the limited capacity and constraints and the fragility of economies, sport tourism is likely to be small-scale and niche or a form of special interest that is sustainable in terms of its regenerative and degrowth characteristics.

Developing sport tourism in Niue: potential for wellness tourism

Wellness tourism, as established earlier in the introduction, centres on well-being for emotional, environmental, physical, psychological, social and spiritual states. Wellness travellers comprise:

> **primary wellness travelers**, who are motivated by wellness to take a trip or choose their destination based on its wellness offerings (e.g., someone visiting a wellness resort or participating in a yoga retreat); and **secondary wellness travelers**, who seek to maintain wellness or engage in wellness activities during any kind of travel (e.g., someone who visits a gym, gets a massage, or prioritizes healthy food when they take a trip).
>
> *(Global Wellness Institute 2018: iv (emphasis in original))*

Yoga tourism, a sub-segment of wellness tourism, has been linked with health and well-being as reiterated by the Prime Minister of India, Narendra Modi, who indicated that 'yoga is a symbol of universal aspiration for health and well-being' (Modi 2016 cited in Bowers and Cheer 2017: 208). According to the Global Wellness Institute (2018: iii), the Asia-Pacific region has gained the most from wellness visitors (258 million trips) and wellness tourism expenditures (US$136.7 billion) in 2017 and a projected average annual growth rate (2017–2022) of 13% – the highest in comparison to other continents across the world (Global Wellness Institute 2018: vii). The destination image or location of yoga (wellness) retreats is a contributing factor to the satisfaction levels of yoga or wellness travellers, as there is a symbolic relationship with emotions experienced (e.g., joy, love and surprise) (Sharma and Nayak 2018). Motivations for participation in yoga tourism are for wellness, well-being, spiritual guidance, healing or transformational/self-improvement, religious, medical and therapeutic purposes (Global Wellness Institute 2018; Sharma and Nayak 2018). Clearly, Niue has the capacity to be a sport tourism destination given its image (e.g., tranquil), location (e.g., Asia-Pacific) and natural environments (e.g., coastal) that are prerequisites for yoga (retreat) tourism.

Yoga retreats are not a new phenomenon to Niue, as there is a yoga, fitness and adventure retreat accommodated at the Scenic Matavai Resort (in the south), which is a combined package of yoga (e.g., at sunset), wellness (e.g., massage) and adventure (e.g., hiking) activities hosted by yoga, nutritional and fitness experts from New Zealand (Niue, n.d.a.). The location for a possible yoga retreat could be on the coast of Lakepa, a village in the northeast part of the island, about 12 miles or 20 kilometres from the capital city Alofi and some 16 miles or 26 kilometres from Hanan International Airport. In 2017, Lakepa had a population of 88 residents and ranked eighth in a list of the largest villages in Niue, compared to Alofi (South) ranked first (358 inhabitants) and Alofi (North) ranked second (256 habitants) (World Atlas 2019). There is very little information that can be found about the tourism sector in Lakepa; a tourist map of Niue Island shows the main tourism services are Lialagi accommodation with two self-contained units and also sleeping rooms for larger groups, and a guided tour of the nearby Ulupaka Cave (Niue n.d.b.). There is also a Lakepa Village Show Day that takes place annually in August, which is a celebration of local culture – food, dance and costume. Tourist arrivals have been dominated by New Zealanders, representing 79.1%, Australians representing 9.7%, Europeans 3.4%, Americans

2.4% and Pacific Islanders 2%, for whom the main purpose of visiting the country is for leisure, representing 67.6% of tourist arrivals (SPTO 2018: 31).

The coast of Lakepa is ideal for a yoga retreat because of its remoteness, located away from the city centre with its extensive tourist activities, for example: transport (air, road and sea), accommodation (guesthouses, apartments and villas, hotels and holiday homes), hospitality (restaurants, cafes and bars), visitor attractions (caves, chasm, beach and resorts) and leisure/sport facilities/activities (diving, swimming, bike track, golf and bowls). The main lodgings in Lakepa are the Lialagi accommodation or units surrounded by green open space, which can be potentially used to practice yoga in the outdoors and can house large retreat groups that can reserve the entire building with the self-contained units and sleeping quarters (see Niue Island 2013). The Lialagi accommodation and units are about 1 kilometre away from the main coast, which lends itself to water or aqua yoga because of warm sea temperatures or practicing '*yoga on the rocks*' – a concept borrowed from 'Yoga on the Rocks' that takes place in the outdoors at the Red Rock Amphitheatre in Colorado (Red Rocks 2022). Niue, named Rock of Polynesia, is inevitably the ideal location for marketing 'yoga on the rocks' in and around coastal areas in Lakepa. Therefore, such rural-based accommodation can establish links with local farmers to supply meals for yoga retreat groups involving fresh, organic and raw foods as well as authentic ('farm-to-table') culinary experiences. The Lialagi self-contained units with beds, a kitchen, tea and coffee facilities, a bathroom, linen and towels, represent a social enterprise-type project involving the once abandoned building that has been restored by the local people (Niue Island 2013). Therefore, given the social investment from local people, it is imperative that hosts benefit from the proposed yoga retreat as a sustainable sport tourism initiative and a form of regenerative tourism. Benefits to local people and places can be further extended to integrate wellness tourism into regional and community development policies, as the Global Wellness Institute (2018: 38) observes for Ibiza:

> as part of its 'Ibiza is Wellness' rebranding campaign, Ibiza, Spain, has created an online map and directory of local businesses, facilities, and services across 16 categories of wellness. This effort aims to support local residents in accessing local healthy lifestyle services, while also cultivating a healthy, holistic, and sustainable tourism product that builds on the island's natural and cultural assets and combats seasonality and low-value mass tourism.

Niue's market segment is not composed of typical sun, sand and sea tourists (de Hass 2002). Accordingly, such niche sport markets as yoga (wellness) tourism provide an alternative for those seeking unique well-being experiences and, at the same time, still wanting to enjoy the natural landscapes and seascapes of the island. The travel motivations for performing and practicing yoga in Lakepa, particularly in the outdoors, can be examined by identifying the push and pull factors associated with the motives of domestic tourists participating in wellness tourism in the United States (Tuzunkan 2018). The push motivation factors could be the need for spiritual health, stress release and life improvement, while the pull motivation factors would be core products provided by the destination itself (e.g., yoga facilities), destination accessibility and connectivity (e.g., easy access to and at the destination, climate and the gastronomy supply) and the destination reputation (e.g., environment and atmosphere) (see Tuzunkan 2018: 656–657).

As research on wellness tourism has shown, potential markets to attract yoga groups are female and male tourists from developed countries, individuals from educated backgrounds and professional occupations and those who are high-end spenders with high disposable incomes (Global Wellness Institute 2018; Tuzunkan 2018). This is feasible given that the main growth markets arriving in Niue in 2019 are from Australia and New Zealand, where the top groups of visitors in the last quarter (October–December 2019) of 2019 were 'technicians' and 'associated professionals' (Niue Statistics Office 2020). Moreover, targeting high-end tourists for yoga tourism continues to offer a viable post-pandemic commercial model for Niue, especially as many potential upmarket tourists are likely to have more saving or spending power because of non-consuming during the pandemic and, thus, can afford to pay high prices for yoga tourism. It is anticipated that holiday prices are likely to increase and tourism to economically and environmentally vulnerable nations or small island states could be expensive as a measure to rejuvenate tourism-dependent economies, which were in a state of fragility, instability and marginalization in the COVID-19 era.

It is not uncommon for small island states to rejuvenate a stagnant economy through tourism, as Pratt (2015: 148) states: 'Tourism has become the backbone of the economy, often when there is little other economic activity'. This ethos can continue to serve during the recovery period of the pandemic as an increase in prices for accommodation, visitor attractions, local transport and food and drink ensures that more local people benefit from the tourism sector, which reflects the sustainable and regenerative value of yoga tourism. Furthermore, yoga tourism, as a sustainable and regenerative type of activity, can be extended to incorporate a degrowth strategy of planning to control tourism in the post-COVID-19 era, especially when numbers could increase significantly when people feel safe to travel. Thus, more high-end, niche, specialist and less impactful forms of tourism that are wellness-led are crucial as a solution to mass tourism or overtourism to ensure sustainable forms of sport tourism development, whereby negative impacts are minimized. Accordingly:

> Because wellness travelers tend to be high-spenders and favor experiences that are authentic and unique, there is less pressure for destinations to engage in a 'race to the bottom' strategy that competes on price and quantity.
>
> *(Global Wellness Institute 2018: 17)*

By locating yoga as possible 'retreat tourism' in Lakepa may reduce high levels of tourism activity and subsequently minimizes the likelihood of 'uncontrolled tourism', which according to De Bono et al. (2005: 2) 'is putting pressure on many of the planets sensitive locations especially Small Island Developing States (SIDs)'. In terms of developing a sport tourism economy in Lakepa, prospects of a wellness tourism product addresses two main themes that have been identified as having a direct impact on tourism development in PICs: economic linkages and leakages and tourism skills and training (Cheer et al. 2018). Yoga tourism offers Lakepa an opportunity to establish economic links and stimulate flows of revenue with towns in the north, south and east part of the island, where much of the tourism activity is concentrated. As there is little economic activity in Lakepa, yoga retreat tourism could possibly have a positive impact by creating job opportunities in the community, securing accommodation with local providers and sourcing produce from local farmers, thereby creating linkages between tourism and various parts of the economy. This is particularly important given outward migration or 'brain-drain' to New Zealand in search of employment and workforce development opportunities by young Niueans. In the Asia-Pacific region, wellness tourism jobs accounted for 10.1 million in this niche market, including many operations such as lodging (e.g., destination spas, health resorts and retreats) and activities and excursions (e.g., spas, bathing and mediation) (Global Wellness Institute 2018).

The National Strategic Plan (2016–2026) proposed by Niue echoes the UN's SDGs – 'good health and well-being' (SDG 3) and 'decent work and economic growth' (SDG 8), which can be enhanced through wellness tourism such as yoga retreats. Similar to the Ibiza illustration above and given Lakepa's natural and cultural resources, the yoga (wellness) retreat as a sustainable sport tourism product, service and experience can be made accessible to locals which, as a recommendation, can potentially bridge the deficit between seasonal and non-seasonal demand and encourage staycations. As noted by the Global Wellness Institute (2018), wellness tourism can reduce seasonality of tourists throughout the year, therefore, alleviating the negative effects of dependency due to seasonality concerns. This would work well for Niue as the wet season is between January and March, and Lakepa can thus host indoor yoga retreats; and the dry season from April to September would be more suited for outdoor yoga.

Conclusions and research implications

As a small island state with natural (and cultural) assets, it is recommended that Niue pursue wellness tourism as a form of sustainable sport development by developing yoga (retreat) tourism that focuses on mind–body–spiritual enhancements. Lakepa, located in the west part of Niue, has been identified in this chapter to have the potential to develop yoga (retreat) tourism which, as a niche market, can promote the principles and practices of sustainable (sport) tourism development. The main reason Lakepa has been identified as a possible destination for yoga (retreat) tourism is because this area is not heavily populated with inhabitants, and there is very little tourism business in the area compared to the north, northwest, south and southwest regions of the island. In light of a post-pandemic tourism industry, wellness types of tourism respond to calls for regenerative and degrowth forms of development that complement community-based sustainable endeavours. The economy, environment and society in Lakepa are rather dynamic, and it is this dynamism that sustainable sport tourism development is responsible for maintaining. This is addressed by ensuring that the proposal for wellness tourism (i.e., yoga retreat) in Lakepa responds to strategic plans and policies devised by the government of Niue to help fulfil the vision and mission associated with sustainability. Therefore, for Niue, wellness and well-being should promote the values of sustainable sport tourism development that are inherent in Niue's National Strategic Plan (2016–2026) and wider UN strategic priorities and to extend its value as a regenerative and degrowth form of tourism in the 'new' world of COVID-19. The key contention of this chapter is to call for the integration of sport tourism into current and future government policies as a distinct category of tourist activity and to consider diversification of (tourism) economies to sport-led developments to facilitate niche or special interest markets such as yoga (retreat) tourism.

From the analysis of potential sustainable sport tourism development on a small island state such as Niue, this chapter has provided new theoretical and conceptual insights for future application. This inquiry has developed ontological and epistemological underpinnings of sport tourism to include broader categorizations, classifications and characteristics associated with definitional statements. The meaning of sport tourism has a transdisciplinary stretch, thus not just cutting across disciplines but also demonstrates an integrative creation of knowledge that is of use to both academia (e.g., teaching, learning and scholarship) and industry (e.g., policy, strategy and organizations). There is acknowledgement of a paradigmatic fluidity in application, which emerged when extending mainly Western-centric interpretations of sport tourism and sustainable development to non-Western settings. Importantly, the evolution of the term sport tourism and sustainable development has been the outcome of examining a non-Western and non-European small island – had this not happened then possibly conventional articulations of sport tourism and sustainable development may have remained unchallenged. These are therefore welcomed encounters that have opened a space for decolonized dialogue, definitions and debate for those exploring or seeking to explore small island states in non-Western settings and to present solutions to ongoing problems. These new theoretical and conceptual insights can be further applied to empirical research by determining ways forward, methodologically, for data retrieval.

Qualitative and quantitative studies need to inspect both the market viability of yoga (retreat) tourism for small island states and examine people's perceptions of yoga holidays based on various demographic elements. This is necessary before any decision is made to proceed with the sustainable development of yoga (retreat) tourism, especially as a sub-set and niche sport tourism market in Niue. As the potential of sustainable sport tourism development in this chapter is based upon conceptualizations rather than an empirical study, research enquiry needs to be conducted to determine the extent to which Niue has the capacity and capability, post-COVID-19, to host a wellness product and experience such as yoga tourism. Moreover, it would be useful to conduct qualitative research into community perceptions to ascertain whether local people would accept or need yoga tourism, especially in the context of community participation in the development of sustainable sport tourism projects as regenerative and degrowth strategies. Finally, this paper is concerned with sustainable sport tourism development in relation to yoga tourism, but it is important to consider and explore sport tourism development beyond yoga tourism. In light of the pandemic, there is a need to re-stimulate the economy due to lack of tourism and thus it is recommended that other forms of sport tourism that can be sustainable and regenerative are worthy of inspection, such as bicycle, walking and running tourism. These forms of sport tourism could continue to engage people who, during the pandemic, established physical activity routines to maintain their health and well-being but now have the opportunity to transport these interests and lifestyles to remote aesthetic and tranquil destinations elsewhere.

References

ADB (Asian Development Bank) (2020) *Asian Development Outlook 2020: What Drives Innovation in Asia?* Manilla, Philippines: Asian Development Bank. Available online at https://doi.org/10.22617/FLS200119-3 (accessed 1 May 2021).

ADB (Asian Development Bank) (2021a) 'Asian development bank and Niue: Fact sheet'. Available online at https://www.adb.org/publications/niue-fact-sheet (accessed 29 April 2021).

ADB (Asian Development Bank) (2021b) 'Pacific to return to positive growth in 2021'. Available online at https://www.adb.org/news/pacific-return-positive-growth-2021-ad (accessed 1 May 2021).

Ateljevic, I. (2020) 'Transforming the (tourism) world for good and (re)generating the potential "new normal"', *Tourism Geographies*, 22(3): 467–475.

Azzopardi, E. and R. Nash (2016) 'A framework for island destination competitiveness – perspectives from the island of Malta', *Current Issues in Tourism*, 19(3): 253–281.

Barker, J. C. (2000) 'Hurricanes and socio-economic development on Niue island', *Asia Pacific Viewpoint*, 41(2): 191–205.

Bass, S. and B. Dalal-Clayton (1995) 'Small island states and sustainable development: Strategic issues and experiences', *Environmental Planning Issues,* No 8. London: International Institute for Environment and Development.

Bowers, H. and J. M. Cheer (2017) 'Yoga tourism: Commodification and western enhancement of eastern practice', *Tourism Management Perspectives*, 24: 208–216.

Bull, C. and M. Weed (1999) 'Niche markets and small island tourism: The development of sports tourism in Malta', *Managing Leisure*, 4(3): 142–155.

Business World (2021) 'Reshaping tourism for post COVID-19 world'. Available online at https://search.proquest.com/magazines/reshaping-tourism-post-covid-19-world/docview/2510158508/se-2?accountid=17234 (accessed 1 May 2021).

Butowski, L. (2016) 'Sport tourism: Real or virtual?' *Tourism / Turyzm*, 26(1): 89–90.

Cheer, J. M., S. Pratt, D. Tolkach, A. Bailey, S. Taumoepeau and A. Movono (2018) 'Tourism in Pacific island countries: A status quo round up', *Asia and the Pacific Policy Studies*, 5(3): 442–561.

Connell, J. (2008) 'Niue: Embracing a culture of migration', *Journal of Ethnic and Migration Studies*, 34(6): 1021–1040.

De Bono, A., S. Kluser, G. Giuliani and P. Peduzzi (2005) 'Tourism expansion: Increasing threats of conservation opportunities', *Environment Alert Bulletin, 6*, United Nations Environment Programme, GRID Europe.

de Hass, H. (2002) 'Sustainability of small-scale ecotourism: The case of Niue, South Pacific', *Current Issues in Tourism*, 5(3–4): 319–337.

DFAT (Department of Foreign Affairs and Trade) (n.d.) 'Niue country brief', DFAT, Australian Government. Available online at https://dfat.gov.au/geo/niue/Pages/niue-country-brief.aspx (accessed 10 June 2019).

Duxbury, N., F. E. Bakas, T. Vinagre de Castro and S. Silva (2021) 'Creative tourism development models towards sustainable and regenerative tourism', *Sustainability*, 13(2): 1–17.

Freddie, A. and S. de Sylva (2018) 'Disaster resistance: A sustainable way for Niue', *IOP Conference Series: Materials Science and Engineering*, 371: 1–8.

Gammon, S., G. Ramshaw and R. Wright (2017) 'Theory in sport tourism: Some critical reflections', *Journal of Sport and Tourism*, 21(2): 69–74.

Gibson, H. J. (1998) 'Sport tourism: A critical analysis of research', *Sport Management Review*, 1(1): 45–76.

Gibson, H. J., K. Kaplanidou and S. J. Kang (2012) 'Small-scale event sport tourism: A case study in sustainable tourism', *Sport Management Review*, 15(2): 160–170.

Global Wellness Institute (2018) *Global Wellness Tourism Economy*, November 2018, Miami, FL. Available online at https://globalwellnessinstitute.org/wp-content/uploads/2018/11/GWI_GlobalWellnessTourismEconomyReport.pdf (accessed 8 June 2019).

Government of Niue (2016) *Niue National Strategic Plan 2016–2026*. Available online at http://www.gov.nu/wb/ (accessed 11 June 2019).

Highman, J. and T. Hinch (2018) *Sport Tourism Development*, Bristol: Channel View Publications.

Jędrusik, M. (2014) 'The elusive sustainable development of small tropical islands', *Miscellanea Geographies-Regional Studies on Development*, 18(3): 26–30.

Lea, J. (2008) 'Retreating to nature: Rethinking "therapeutic landscapes"', *Area*, 40(1): 90–98.

Milano, C., M. Novelli and J. M. Cheer (2019) 'Overtourism and degrowth: A social movements perspective', *Journal of Sustainable Tourism*, 27(12): 1857–1875.

Niue (n.d.a) 'Facts'. Available online at https://www.niueisland.com/facts/ (accessed 10 June 2019).

Niue (n.d.b) 'Island guide'. Available online at https://static1.squarespace.com/static/5558638ce4b01c8e63a63f0f/t/5a1e023971c10b644b39bea1/1511916111875/D0903+NIU+Arrivals+Guide+17%3A18-10Web.pdf (accessed 10 June 2019).

Niue (2021) 'Stargazing'. Available online at https://www.niueisland.com/darkskynation (accessed 31 January 2022).

Niue Island (2013) 'Lialagi units'. Available online at http://niueisland.com.au/lialagi-units/ (accessed 14 June 2019).

Niue Statistics Office (2020) 'Migration and tourism statistics'. Available online at https://niue.prism.spc.int/category/population/migration/ (accessed 26 April 2021).

Overton, J. and W. E. Murray (2014) 'Sovereignty for sale? Coping with marginality in the South Pacific – The example of Niue', *Hrvatski Geografski Glasnik*, 76(1): 5–25.

Pacific Trade Invest (2021) *PTI Pacific Business Monitor*. Available online at https://pacifictradeinvest.com/media/1726/pti-pacific-business-monitor-report-12-niue-focus.pdf (accessed 30 April 2021).

Panakera, C., G. Wilson, C. Ryan and G. Liu (2011) 'Considerations for sustainable tourism development in developing countries: Perspectives from the South Pacific', *Tourismos: An International Multidisciplinary Journal of Tourism*, 6(2): 241–262.

Pratt, S. (2015) 'The economic impact of tourism in SIDS', *Annals of Tourism Research*, 52: 148–160.

Red Rocks (2022) 'Yoga on the rocks'. Available online at https://www.redrocksonline.com/yoga/ (accessed 4 February 2022).

Ritchie, B. W. and D. Adair (2004) *Sport Tourism: Interrationships, Issues and Impacts*, Clevedon: Channel View Publications.

RNZ (Radio New Zealand) (2018) 'Niue in world top ten for tourism growth', *Radio New Zealand*, 11 September. Available online at https://www.rnz.co.nz/international/pacific-news/366200/niue-in-world-top-ten-for-tourism-growth (accessed 6 June 2019).

Ryan, C. (2001) 'Tourism in the South Pacific – a case of marginalities', *Tourism Recreation Research*, 26(3): 43–49.

Sharma, P. and J. Nayak (2018) 'Testing the role of tourists' emotional experiences in predicting destination image, satisfaction, and behavioural intentions: A case of wellness tourism', *Tourism Management Perspectives*, 28: 41–52.

Sheller, M. (2021) 'Reconstructing tourism in the Caribbean: Connecting pandemic recovery, climate resilience and sustainable tourism through mobility justice', *Journal of Sustainable Tourism*, 29(9): 1436–1449.

SPTO (South Pacific Tourism Organisation) (2018) *Annual Review of Visitor Arrivals in Pacific Island Countries 2017*, Suva, Fij. Available online at https://www.corporate.southpacificislands.travel/wp-content/uploads/2017/02/2017-AnnualTourist-Arrivals-Review-F.pdf (accessed 17 June 2019).

SPTO (Pacific Tourism Organisation) (2020a) 'Niue makes the list of the 2020 top 100 global sustainable destinations', Suva, Fiji. Available online at https://corporate.southpacificislands.travel/niue-makes-the-list-of-the-2020-top-100-global-sustainable-destinations/ (accessed 26 April 2021).

SPTO (Pacific Tourism Organisation) (2020b) *Pacific Tourism: Covid 19 Impact and Recovery,* Suva, Fiji. Available online at https://corporate.southpacificislands.travel/spto-releases-pacific-tourism-impact-report/ (accessed 30 April 2021).

SPTO (Pacific Tourism Organisation) (2021) *2020 Annual Visitor Arrivals Snapshot, 2020*, Suva, Fiji.

Statistics for Development Division (2019) 'Economic statistics'. Available online at https://sdd.spc.int/en/stats-by-topic/economic-statistics (accessed 6 June 2019).

Taleghani, G. R. and A. Ghafary (2014) 'Providing a management model for the development of sports tourism', *Procedia – Social and Behavioural Sciences*, 120: 289–298.

Treloar, P. and C. M. Hall (2005) 'Tourism in the Pacific islands'. In: C. Cooper and C.M. Hall (eds.) *Oceania: A Tourism Handbook*, Toronto: Channel View Publications, pp. 173–294.

Tuzunkan, D. (2018) 'Wellness tourism: What motivates tourists to participate?' *International Journal of Applied Engineering Research*, 13(1): 651–661.

United Nations (n.d.) *Sport and the Sustainable Development Goals: An Overview Outlining the Contribution of Sport to the SDGs*, United Nations Office on Sport for Development and Peace. Available online at https://www.un.org/sport/sites/www.un.org.sport/files/ckfiles/files/Sport_for_SDGs_finalversion9.pdf (accessed 11 June 2019).

UNWTO (United Nations World Tourism Organization) (2018) *UNWTO Tourism Highlights: 2018 Edition*, Madrid: UNWTO. Available online at https://www.e-unwto.org/doi/pdf/10.18111/9789284419876 (accessed 17 June 2019).

Worldometer (2021) 'Niue population (live)'. Available online at https://www.worldometers.info/world-population/niue-population/ (accessed 17 December 2021).

World Atlas (2019) 'The largest villages in Niue'. Available online at https://www.worldatlas.com/articles/the-largest-villages-in-niue.html (accessed 13 June 2019).

25

TOURISM DEVELOPMENT IN THE COOK ISLANDS

Deconstructing the impacts and identifying a sustainable framework

Kamelia Chaichi and Mei Kei Leong

Introduction

The Cook Islands is named after Captain James Cook, who navigated the islands in the latter part of the eighteenth century, and consists of 15 islands spread across almost 2 million square kilometres that are scattered between French Polynesia in the east and American Samoa in the west. The population of the Cook Islands was 17,575 in 2021, and the most populous island is Rarotonga with more than 10,000 inhabitants and is home to the nation's capital, Avarua, with a population of more than 5,400 people, which is the main seat of government and the country's commercial centre (World Population Review 2021).

Penrhyn, which is 1,365 kilometres in distance from Rarotonga International Airport, is the northernmost island, and Mangaia is the southernmost island, located 204 kilometres from Rarotonga (Stanley 2000). The 15 islands that comprise the Cook Islands are allocated into two groups. The 'northern group' consists of six main islands north of Palmerston: Manihiki, Nassau, Penrhyn, Pukapuka, Rakahanga and Suwarrow. The 'southern group' includes nine main islands south of Suwarrow: Aitutaki, Atiu, Mangaia, Manuae, Mauke, Mitiaro, Palmerston, Rarotonga and Takutea (Figure 25.1). All 15 islands are inhabited and were initially populated by Polynesians. Spanish explorers sighted Pukapuka in 1595 and Rakahanga in 1606; Captain Cook sighted Manuae in 1773 and then Atiu, Mangaia, Palmerston and Takutea in 1777 (Taylor 2001). Britain was the protectorate over the Cook Islands from 1888, where a single federal parliament was instituted and the scattered islands first experienced a united government (Crocombe et al. 2018). However, New Zealand governed the Cook Islands from 1900 to 1965, becoming a self-governing state in free association with New Zealand (Government of the Cook Islands 2016).

The chapter will seek to identify the impact of tourism development in relation to the country's economy, culture, human resources and physical environment. The chapter considers the challenges faced by the Cook Islands in relation to developing a sustainable development agenda. In light of the COVID-19 pandemic, the concerns of 'undertourism' and its impact on the Cook Islands are critically acknowledged, and the underlying argument becomes clear in the subsequent discussions indicating that the country needs to move ahead cautiously towards a tourism development rethink, with careful planning and policy directives in place.

DOI: 10.4324/9780429019968-29

Figure 25.1 Map of the Cook Islands, including the capital city of Avarua, Rarotonga. Source: Peter Hermes Furian (www.shutterstock.com/image-vector/cook-islands-political-map-capital-avarua-1731270859).

Tourism overview

Tourists have been attracted to the Cook Islands since the early 1900s, though admittedly in smaller numbers. Hotel Rarotonga was built in 1906 to stimulate stop-over tourism, in which tourists arrived on freighters of the Union Ship Company of New Zealand and the Matson Line of the United States (US) (Crocombe 1992). Nonetheless, as Crocombe (1992) further notes, in the 1950s, the Cook Islands was incorporated as part of the Coral Route, which was an air service to Norfolk Island, Fiji, Tonga, Samoa, Cook Islands and Tahiti. Nonetheless, the agricultural and fishing sectors largely represented the country's economy up to the 1970s, which started to change once the runway at Rarotonga Airport was significantly extended and upgraded in 1974 (Taylor 2001). Therefore, the mid- to late-1970s is commonly noted as the period when the tourism industry started to evolve in the Cook Islands, reaching 78,328 arrivals in 2003 (Hajkowicz and Okotai 2005).

Prior to the outbreak of the COVID-19 pandemic in early 2020, the tourism industry was considered the major contributor of economic growth, generating more than 67% of the country's gross domestic product (GDP) (Government of the Cook Islands 2019). The Cook Islands received a record number of 170,611 visitors in 2019 (Ministry of Finance and Economic Management 2019), which actually increased from 113, 000 in 2011 (AIIB 2021). Air New Zealand, Virgin Australia, Jetstar and Air Tahiti are the international airlines that service the Cook Islands, with flights arriving from such key locations as New Zealand, Los Angeles and Sydney (Government of the Cook Islands 2019). Nonetheless, the prime minister had hoped that the country would soon reach 200,000 tourists (Evans 2019). Despite the challenges that tourism destinations are experiencing, especially in a world of uncertainty and fractured mobilities, the *Lonely Planet* announced the Cook Islands as their 'number one travel destination' for 2022 (Campbell 2021).

The economic impact of tourism, 'undertourism' and a tourism development rethink

However, with the closure of borders in March 2020 and a ban on receiving international tourists, the Cook Islands has been experiencing 'undertourism'. This is rather ironic as there was a concern that if the Cook Islands had continued to increase their tourism visitation numbers beyond the 2019 record number of visitors, then 'overtourism' would have dominated the country at the expense of the socio-cultural fabric of Cook Island life; this would have been a particular concern for Rarotonga and Aitutaki, as they traditionally attract the most tourists (Tahana 2020).

The halting of international tourism due to the border closure meant that 65% of the country's economy also shut down instantly (Cooke et al. 2021). The government responded swiftly to the COVID-19 pandemic by preventing an economic recession through establishing an economic response package worth US$42.6 million, which included wage subsidies and grants to support workers to prevent the closure of businesses (McGarry 2021). It is worth noting that the Cook Islands was classified as a 'high income country' by the Organisation for Economic Co-operation and Development (OECD) on 1 January 2020, being the first Pacific Island to have achieved this economic classification and, thus, ending its entitlement for Overseas Development Assistance (Cooke et al. 2021).

The level of dependency that the Cook Islands had on international tourism has meant that the country needs to re-establish itself as a tourism destination, particularly within the context of the post-pandemic era. In the meantime, however, the Cook Islands is attempting to find ways to receive international tourists, especially from key markets. The country opened a 'safe travel corridor' in which nearly 7,500 tourists arrived from New Zealand for the months of May–June 2021 (Sen and Kenney 2021). However, the travel bubble was short lived and only lasted from May to August due to the outbreak of the Delta variant of COVID-19 in New Zealand (Movono and Scheyvens 2021). The country's tourism industry is, thus, economically vulnerable within the context of a pandemic that is difficult to control because of new virus mutations.

Despite the economic challenges and low annual tourism arrivals since 2020, it is arguably the right time for the Cook Islands to re-visit and rethink the type of tourism that the country should focus on in terms of its long-term development plan. In fact, in the pre-pandemic period, the government was contemplating the role of sustainability within the context of its development agenda. The National Sustainable Development Plan 2016–2020 (NSDP) aimed to promote wellbeing, foster environmentally sustainable practices and promote sustainable livelihoods and resilience as well as develop infrastructure for economic growth (Government of the Cook

Islands 2016). This planning directive understood the threat that tourism development presented to the agricultural sector, stating:

> The rise in tourism and consumerism has coincided with the decline of agriculture as an industry. One consequence of this, especially on Rarotonga, is that land once used for agricultural production has been converted to residential or commercial (usually tourist) use. The less land that is available for agriculture, the greater the limitations on our ability to produce food. We need to ensure that there is land available to increase and improve food production and security.
>
> *(2016: 39)*

The plan implicates the importance of green economy principles, especially through emphasizing the establishment of renewable energy directives and setting standards to conserve marine resources. Furthermore, the government also introduced the Sustainable Tourism Development Policy Framework, a policy recognizing the interplay of the social, economic and environmental dimensions of tourism development (Cook Islands Tourism Corporation 2017). Although the policy objectives, in many ways, complement the National Sustainable Development Plan, there is emphasis also on developing a sustainable workforce and providing career opportunities for Cook Island nationals, optimizing sustainable economic outcomes, implementing destination marketing strategies that produce sustainable outcomes and ensuring that tourists are safe and secure when they reside in the Cook Islands.

Nevertheless, whatever the post-pandemic scenario for the tourism industry of the Cook Islands, it is important to acknowledge how tourism significantly impacted the socio-cultural and environmental fabric of the country in ways which were detrimental. Recognition of the past realities of tourism development may well be helpful in terms of seriously pursuing a tourism development rethink.

Socio-cultural impacts of tourism

Prior to the COVID-19 pandemic, tourism growth in the Cook Islands caused significant cultural and social disruptions to island life. Traditionally, in the Cook Islands, hospitality was defined in terms of welcoming guests, not in terms of economic transactions, in which traditional norms commanded the generosity of hosts towards their guests Cook Islanders have thus concentrated more on receiving compensation for their goods and services that they provide to visitors, disregarding their cultural responsibilities and becoming commercialized (Berno 1999). The commercialization of hosts and local communities is often an impediment to the development of forms of tourism development that are beneficial to communities and economies (de Kadt 1979). One long-term concern in the Cook Islands relates to the way in which traditional musical performances and dances are transformed to please tourists, as well as being culturally displaced from their traditional context in terms of telling stories, legends and tribal histories. As Taylor (2001: 74) notes:

> These performances were lengthy, methodical, and slow. The unaided human performance was considered perfect. Costumes were laboriously hand made of natural fibres, seeds, leaves, shells, bark, and husks. Costumes were modest and covered much of the body. Tourism has changed much of this tradition. Tourists have short attention spans, are often impatient, and want to experience as much as possible during their vacation. Now cultural displays of music and dancing include non-Cook Island components, the performances are shorter and faster, and the costumes are less modest, baring more

> of the body. Frequent performances cause the costumes to deteriorate quickly, and thus they often are made of or repaired with non-natural materials (to increase durability). The music and entertainment is almost a Polynesian 'chop suey'.

However, contemporary dance performances and costumes can arguably manifest visitor perceptions concerning the sexualization of women and, thus, have detrimental effects, including female frontline employees suffering from inappropriate harassment, including unsolicited physical contact and sexual gestures, remarks and propositions (Auckland University of Technology 2018). Alexeyeff (2008: 289) emphasizes that the promotional images of the tourism industry denote 'pleasure' and 'sensual exhibitionism' which have been influenced by 'Western representation' of 'Polynesian paradise' in which Cook Islanders are advertised as 'naturally warm, friendly and generous, qualities promoted as being freely available to tourists who visit their shores'.

However, there has been some opposition to tourism expansion, and this has been noticeable in Aitutaki, which celebrated 200 years of Christianity in October 2021 – since the arrival of Reverend John Williams who came to the island from the London Missionary Society to convert the local inhabitants (Utanga 2021). There have been protests on the island since 2008 when there was a decision to introduce Sunday flights, which was locally seen to be at odds with the Christian belief concerning the importance of not making a profit or being involved in frivolous activities on the Sabbath (McMahon 2008; Tolkach and Pratt 2021). These protests are supported by signs placed strategically on key roads in Aitutaki (Figure 25.2).

Tourism and the need for skilled human resources

The tourism industry is highly dependent on the human factor, as it is a service industry and provides a high level of employment for less-developed nations (Nunkoo and Ramkissoon 2010).

Figure 25.2 Roadside sign constructed by a local group who oppose flights arriving on Aitutaki soil on the Sabbath. Source: M. Stephenson, May 2017.

Small islands are notorious for their lack of skilled workers (Corbett and Connell 2015). In the growth years of tourism development, until the outbreak of the pandemic in March 2020, the Cook Islands struggled to retain islanders in tourism and hospitality employment. Emigration has been compounded by the extreme geographical fragmentation of the Cook Islands, making labour allocation and all forms of service delivery particularly difficult, especially for the Outer Islands (Connell 2005). One of the facilitators of high outward migration relates to the country's free association with New Zealand, where Cook Islanders have the right to live and, thus, can seek higher standards of living and better work opportunities. The pandemic exacerbated the situation, and there was a net decrease of 690 residents between January and July 2021 (Cooke et al. 2021), thus, causing labour shortages associated with depopulation (Fotheringham 2021).

The demand for foreign workers had increased in the Cook Islands, as there was a need to fill low-paid jobs (e.g., waiters and waitresses), as these positions were deemed uninteresting for local people, and a need to fill higher tourism positions when local expertise was sparse (ADB 2008). Subsequently, it has been recognized that improved hospitality and tourism skills through specialized training provision is required for Cook Island nationals, and the promotion of career pathways in tourism-related sectors would be constructive in encouraging a sustainable workforce (Cook Islands Tourism Cooperation 2016). Nonetheless, the Cook Islands experienced a growing migrant workforce from such countries as Fiji, Indonesia and the Philippines (Cook Island Government 2015), as foreign workers represented 20% of the population of the Cook Islands in 2018 (Etches 2018). Nonetheless, the prevailing challenge is that many of these employees have not been fully working since international mobility was curtailed in early 2020. The president of the Cook Islands Tourism Industry Council urged expatriate workers not to leave the Cook Islands, as there would be insufficient skilled labour and little opportunity to transit potential employees to the Cook Islands once the border opens (Lacanivalu 2021), which it did from May to August 2021 when a travel bubble was temporarily established with New Zealand.

Tourism, environmental impact and sustainability

More than two decades ago, Taylor (2001: 80) noted that one of the main concerns of tourism development in the Cook Islands is its effect on the environment, notably the impact of sewage and solid waste disposal on such natural resources as fresh water supplies, the lagoons and coral reefs. Therefore, the country has been exposed to significant levels of pollution, especially the popular tourist island of Rarotonga, where waste products have affected the Muri Lagoon (Cyronak et al. 2013). It is one of the most significant tourist attractions in the Cook Islands, but the government labelled the pollution of Muri Lagoon as a 'national disaster', where dark patches of toxic algal bloom spread across the floor of the lagoon, clouding the water and deterring tourists from engaging in water-based activities (Evans 2019). The rapid growth of tourists, resort negligence and properties with inadequate septic tanks were seen as the main determinants of such pollution.

The establishment of an efficient waste management system in the Cook Islands is a significant challenge. In Rarotonga and Aitutaki, garbage disposal has taken place in the wetlands and foreshore (Dusevic 2001). Waste disposal had caused problems to the drainage system in such areas of Rarotonga as Avatiu, Atupa wetlands and Panama, where there are solid waste dump sites (Islands Friends Ltd 2004). There have also been concerns with such hazardous waste as batteries, fertilizers and unwanted or expired medications and hospital supplies, as tourism creates more waste management concerns given the increased importation of products (Government of the Cook Islands 2018).

One of the crucial objectives of sustainable tourism development is to protect the natural environment of the host country. The government of the Cook Islands created the National Environment Act in 2003 and the National Environment Strategic Action Framework (NESAF) in 2005, especially intending to conserve the environment and natural resources (NESAF 2004). The government is aiming for all the islands to have renewable energy by 2025, for which five islands are intending to host solar plants that are being partly funded by the Asian Development Bank, European Union, Global Environment Fund, the Japanese government and the Green Climate Fund (Bellini 2020).

Although the government understands the importance of renewable energy to the future of the Cook Islands, there are other areas of environmental stewardship that remain a concern. There have been recent moves by the government to license seabed mining operators to prospect its exclusive economic zone for manganese and cobalt nodules, despite concerns raised by environmental groups such as Ocean Foundation and MiningWatch Canada relating to the profound environmental degradation that mining can cause. These concerns relate to the effect of mining on sea species, the water column and tuna fisheries. Nevertheless, the Seabed Minerals Commissioner of the Cook Islands believes that because the country's economy suffered significantly from non-tourism activities, the seabed minerals industry could provide substantial economic diversity and encourage the state to be less reliant on tourism in the future. Nonetheless, such speculation is a concern as the Cook Islands converted its 'exclusive economic zone' in 2017, a zone which traverses nearly 2 million square kilometres, into one of the world's largest mixed-use marine protected areas in the world (Evans 2020). The creation of this marine park, known as Marae Moana, and the establishment of the MANA Tiaki Eco Certification Programme were key features that contributed to the Cook Island's 'wow factor' at the *Lonely Planet* awards (indicated earlier) (Campbell 2021). This programme provides eco-certification for businesses wishing to achieve a higher level of environmental sustainability, which is aligned to the Cook Islands Tourism Accreditation and Quality Assurance Scheme that is having the positive effect of encouraging businesses to ensure that they adhere to sustainable policies and planning directives (TRC 2020).

One key sustainable objective for contemplating future tourism development should be to work towards a re-orientation of the tourism and hospitality industry within the wider context of a post-pandemic strategy. Accordingly, there remains distinct geographical and ecological spaces and places in the Cook Islands that can be fully developed to become ecotourism destinations. There has been a long-term recognition that Atiu and Mauke in the southern group have potential for becoming such destinations (Taylor 2001). Interestingly, the pandemic is encouraging local communities to focus more on their immediate environment and island-scape. Locals launched nature-based initiatives by planting flowers and beautifying villages (RNZ 2020). Also, the government launched a subsidy to assist growers in ensuring food security amid the crisis (Clarke-Mamanu 2020). Such developments positively influence community appreciation of the importance of agriculture and land in the daily lives of Cook Islanders.

Conclusion and research implications

The Cook Islands is highly dependent upon the tourism industry for its economic growth. Nonetheless, the COVID-19 pandemic arguably encourages a re-direction in tourism development thinking. Being highly dependent on the tourism industry has been negatively impactful to the socio-cultural and environmental fabric of the Cook Islands also creating and aggravating labour force concerns. Although this chapter draws caution to the perils of over-relying on a tourism-led development approach, undertourism could also continue to impact the economy. There arguably needs to be a compromise, whereby the country aims to re-attract international tourists, once there is long-term stability and safety assured to international tourists, but within a more sustainable context. This would involve ensuring that tourism is sustainable, not only to the physical/ecological environment but also to the society, culture and economy of the Cook Islands.

Future research focusing on tourism development issues and concerns in the Cook Islands needs to consider the complexities of each or a particular island destination, especially in terms of their evolutionary stage in tourism development and particularly as the Cook Islands is ecologically, culturally and geopolitically diverse. Traditionally, tourism enquiries have drawn popular reference to the islands of Rarotonga and Aitutaki, which is not surprising, as these are popular destinations for tourists and are easier to physically access than the outer islands, notably the northern group. Nonetheless, the geographical isolation of this group of islands has often ensured that they have not been subjected to significant levels of touristification. Accordingly, it would be interesting to see how these islands interact with the visiting friends and relatives (VFR) market and domestic forms of tourism and travel, especially to understand how tourism and culture can be intertwined but in constructive ways. Beyond these forms of tourism and from a tourism management perspective, studying these distant destinations would be useful in contributing to critical insights concerning the functioning and strategic importance of niche forms of tourism.

Obviously, it would be crucial to comprehend how Cook Islanders feel about tourism and tourists now that the tourism industry has been economically, structurally and geographically destabilized and to know more about localized conceptions and perceptions of undertourism/non-tourism. Therefore, research ought to be directed to understanding the type of tourism that the Cook Islands should attract in the post-pandemic era, one that does not manifest the problems inherent in the pre-pandemic era, when tourism was considered to be a major national priority for economic growth. Moreover, the current situation offers a real opportunity for key stakeholders to come together and work productively towards the development and employment of a new strategy that will help revive tourism in the Cook Islands but for the betterment of society, culture and the ecological environment. Action research would be useful to examine ways in which tourism development can focus on a more sustainable approach in light of the lessons learnt from an era of tourism growth and the strategic measures needed for a re-positioning of the Cook Islands as a tourism destination in a post-pandemic world, particularly from the intersecting perspectives of key stakeholders.

References

ADB (Asian Development Bank) (2008) *Cook Islands Social and Economic Report*, Philippines: ADB. Available online at https://www.adb.org/sites/default/files/publication/29732/cook-islands-economic-report-2008.pdf (accessed 5 May 2020).

Alexeyeff, K. (2008) 'Are you being served? Sex, humour and globalisation in the Cook Islands', *Anthropological Forum*, 18(3): 287–293.

AIIB (Asian Infrastructure Investment Bank) (2021) 'Cook Islands: COVID-19 active response and economic support program'. Available online at https://www.aiib.org/en/projects/details/2020/approved/Cook-Islands-COVID-19-Active-Response-and-Economic-Support-Program.html (accessed 23 December 2021).

Auckland University of Technology (2018) 'Harassment in Cook Islands hospitality' (study conducted by Lisa Sadaraka), 10 April. Available online at https://news.aut.ac.nz/news/harassment-in-cook-islands-hospitality (accessed 1 March 2021).

Bellini, E. (2020) 'Solar-plus-storage for the Cook Islands', *PV Magazine*, 7 September. Available online at https://www.pv-magazine.com/2020/09/07/solar-plus-storage-for-the-cook-islands/ (accessed 7 January 2021).

Berno, T. (1999) 'When a guest is a guest: Cook Islanders view tourism', *Annals of Tourism Research*, 26(3): 656–675.

Campbell, M. (2021) 'Cook Islands: Still COVID-free after positive test is found to be a false alarm', *Euro News*, 7 December. Available online at https://www.euronews.com/travel/2021/12/07/cook-islands-plan-your-2022-trip-to-the-beach-paradise-that-never-got-covid (accessed 1 January 2022).

Clarke-Mamanu, M. (2020) 'Cook Islands offers growers assistance to ensure food security', *Te Ao: Maori News*, 12 April. Available online at https://www.teaomaori.news/cook-islands-offers-growers-assistance-ensure-food-security (accessed 23 February 2021).

Connell, J. (2005) 'A nation in decline? Migration and emigration from the Cook Islands', *Asian and Pacific Migration Journal*, 14(3): 327–350.

Cook Island Government (2015) *Economic Activity and Labour Force: Analysis of the 2011 Population and Housing Census, Ministry of Internal Affairs and the Cook Islands Statistics Office (CISO), with the support from the United Nations Population Fund (UNFPA)*, February. Available online at http://www.mfem.gov.ck/images/New_Stats_Website/12.Other_Content/Labour/UNFPA-Economic-activity-and-labour-force-of-the-Cook-Islands_Reduced.pdf (accessed 2 January 2022).

Cook Islands Tourism Cooperation (2016) *Protecting Our Future: Cook Islands Sustainable Tourism Development Policy Framework & Goals*, Cook Islands Tourism Cooperation, The Government of the Cook Islands. Available online at http://extwprlegs1.fao.org/docs/pdf/cok180226.pdf (accessed 14 December 2021).

Cooke, N., I. Hayes and S. Moncada (2021) 'Resilience building to COVID-19 in the Pacific: The case of the Cook Islands', *Small States and Territories*, 4(2): 259–278.

Corbett, J. and J. Connell (2015) 'All the world is a stage: Global governance, human resources, and the "problem" of smallness', *The Pacific Review*, 28(3): 435–459.

Crocombe, R.G. (1992) *Pacific Neighbours: New Zealand's Relations with other Pacific Islands*, Macmillan Brown Centre for Peace Studies of the University of Canterbury and the Institute of Peace Studies at the University of the South Pacific.

Crocombe, R.G., M.T. Crocombe and S. Foster (2018) 'Cook Islands', *Encyclopedia Britannica*, 26 January. Available online at https://www.britannica.com/place/Cook-Islands (accessed 6 February 2021).

Cyronak, T., I.R. Santos, D.V. Erler and B.D. Eyre (2013) 'Groundwater and pore water as major sources of alkalinity to a fringing coral reef lagoon (Muri Lagoon, Cook Islands)', *Bio Geosciences*, 10: 2467–2480.

de Kadt, E. (ed.) (1979) *Tourism: Passport to Development? Perspectives on the Social and Cultural Effects of Tourism in Developing Countries*, Oxford: Oxford University Press.

Dusevic, T. (2001) 'Modern lifestyles and tourism mean more trash, but recycling isn't enough: Old ways of treating land and sea must also be discarded', *Time*, 20 August. Available online at http://content.time.com/time/world/article/0,8599,2056143,00.html (accessed 4 May 2020).

Etches, M. (2018) 'Foreign worker numbers rise', *Cook Islands News*, 3 November. Available online at https://www.cookislandsnews.com/economy/foreign-worker-numbers-rise/ (accessed 1 January 2021).

Evans, M. (2019) 'Paradise, polluted: Cook Islands tries to clean up its tourism', *Mongabay Series: Sea Change*, 23 September. Available online at https://news.mongabay.com/2019/09/paradise-polluted-cook-islands-tries-to-clean-up-its-tourism-sector/ (accessed 6 February 2020).

Evans, M. (2020) 'Cook Islands to allow seabed mining licences despite negative environmental impact', *EcoWatch*, 18 June. Available online at https://www.ecowatch.com/cook-islands-mining-2646199832.html#toggle-gdpr (accessed 30 November 2020).

Fotheringham, C. (2021) 'People are leaving the country and not coming back', *Cook Islands News*, 20 February. Available online at https://www.cookislandsnews.com/internal/features/weekend/people-are-leaving-the-country-and-not-coming-back/ (accessed 21 February 2021).

Government of the Cook Islands (2016) *Te Kaveinga Nui: National Sustainable Development Plan 2016–2020*, Central Policy and Planning Office, Office of the Prime Minister, January. Available online at https://www.adb.org/sites/default/files/linked-documents/cobp-coo-2017-2019-ld-01.pdf (accessed 10 November 2020).

Government of the Cook Islands (2018) *Fiscal Strategy and Economic Update: September 2018*, Economics Division, Ministry of Finance and Economic Management, 25 September. Available online at http://www.mfem.gov.ck/images/documents/CEO_docs/Economics_Division/Fiscal_stratecon_update_Sep_2018.pdf (accessed 20 January 2021).

Government of the Cook Islands (2019) *Cook Islands Tourism Forecasts 2020 to 2024: Information Paper*, Economic Planning Division, Ministry of Finance and Economic Management, 7 November. Available online at http://www.mfem.gov.ck/images/WP3_CI_Tourism_model_paperV5.pdf (accessed 10 November 2020).

Hajkowicz, S. and P. Okotai (2005) 'An economic valuation of watershed pollution in Rarotonga, the Cook Islands', *IWP-Pacific Technical Report (International Waters Project) No. 18*. Apia, Samoa: SPREP – Secretariat of the Pacific Regional Environment Programme.

Island Friends Ltd (2004) 'Cook Islands priority environmental problems (PEC) report: A review and assessment of the priority environmental concerns', *IWP-Pacific Technical Report (International Waters Project) No. 11*. Apia, Samoa: SREP – Secretariat of the Pacific Regional Environmental Programme. Available online at https://www.sprep.org/attachments/000460_IWP_PTR11.pdf (accessed 8 July 2020).

Lacanivalu, L. (2021) 'Expatriate workers urged to consider all options before departing', *Cook Islands News*, 18 February. Available online at https://www.cookislandsnews.com/national/economy/expatriate-workers-urged-to-consider-all-options-before-departing/ (accessed 26 February 2021).

McGarry, D. (2021) 'Deserted islands: Pacific resorts struggle to survive a year without tourists', *The Guardian*, 2 April. Available online at https://www.theguardian.com/world /2021/apr/03/covid-coronavirus-deserted-islands-pacific-resorts-struggle-to-survive-a-year-without-tourists to keep workers on the job and businesses from shutting down their operations (accessed 1 January 2022).

McMahon, B. (2008) 'Sunday flights disturb paradise islands', *The Guardian*, 30 June. Available online at https://www.theguardian.com/world/2008/jun/30/cookislands.religion?gusrc=rss&feed=networkfront (accessed 2 January 2022).

Ministry of Finance and Economic Management (2019) *Cook Islands Tourism Forecasts from 2020 to 2024*, Economic Planning Division, 7 November. Available online at http://www.mfem.gov.ck/images/WP3_CI_Tourism_model_paperV5.pdf (accessed 17 January 2020).

Movono, A. and R. Scheyvens (2021) 'Vax and vaccination? Why that Pacific island will still mean "traveller beware"', *The Conversation*, 28 September. Available online at https://theconversation.com/vax-and-vacation-why-that-pacific-island-holiday-will-still-mean-traveller-beware-168380 (accessed 28 December 2021).

National Environment Strategic Action Framework (2004) *Cook Islands National Environment Strategic Action Framework 2005–2009*, December, Tuaere Tangianau of Upoko Solutions Ltd, Cook Islands. Commissioned by the National Environment Service, December. Available online at https://www.preventionweb.net/files/27076_nesaf.pdf (accessed 4 July 2020).

Nunkoo, R. and H. Ramkissoon (2010) 'Small island urban tourism: A residents' perspective', *Current Issues in Tourism*, 13(1): 37–60.

RNZ (Radio New Zealand) (2020) 'Covid-19: Resort owners on Cooks' Aitutaki focus on post-pandemic future', *RNZ*, 27 March. Available online at https://www.rnz.co.nz/international/pacific-news/412728/covid-19-resort-owners-on-cooks-aitutaki-focus-on-post-pandemic-future (accessed 7 November 2020).

Sen, K. and T. Kenney (2021) 'Pacific islands: Economic outlook', *Bluenotes*, 17 August. Available online at https://bluenotes.anz.com/posts/2021/08/anz-research-pacific-islands-economic-forecast-covid19 (accessed 30 December 2021).

Stanley, D. (2000) *South Pacific Handbook* (seventh edition), Emeryville, CA, USA: Avalon Travel Publishing.

Tahana, J. (2020) '"Huge crisis": Coronavirus fears knock back Pacific tourism', *Radio New Zealand*, 6 March. Available online at https://www.rnz.co.nz/international/pacific-news/411132/huge-crisis-coronavirus-fears-knock-back-pacific-tourism (accessed 3 November 2020).

Taylor, J.E. (2001) 'Tourism to the Cook Islands', *Cornell Hotel and Restaurant Administration Quarterly*, 42(2): 70–81.

Tolkach, D. and S. Pratt (2021) 'What role does tourism play in globalization of the Pacific island countries?' *Tourism Geographical Editor*, 13 May. Available online at https://medium.com/tourism-geographic/what-role-does-tourism-play-in-the-globalization-of-pacific-island-countries-2cb61d1fd94f (accessed 2 January 2022).

TRC (2020) 'Cook Islands steps up its commitment to sustainability', *Tourism Recreation Conservation*, 28 August. Available online at https://trctourism.com/cook-islands-steps-up-its-commitment-to-sustainability/ (accessed 24 January 2021).

Utanga, J. (2021) 'Cook Islands mark 200 years of Christianity on Aitutaki', *Tagata Pacific*, 20 November. Available online at https://tpplus.co.nz/pacific-region/cook-islands-mark-200-years-of-christianity-on-aitutaki/ (accessed 30 December 2021).

World Population Review (2021) 'Cook Islands population 2021 (Live)'. Available online at https://worldpopulationreview.com/countries/cook-islands-population (accessed 27 February 2021).

26
TUVALU
A paradise rarely visited

Bruce Prideaux and Alexander Trupp

Introduction

In one of the few papers that have examined aspects of Tuvalu's tourism industry, Prideaux and McNamara (2013: 585) observe that 'leisure tourism is almost non-existent relegating the country to the status of a "non-destination"'. Little substantial progress in developing a tourism industry has occurred in the years following these observations. This chapter reviews the literature that comments on the problems faced by 'small island developing states' (SIDS) to provide a platform from which to review the status of tourism in Tuvalu and the issues that the country will face if it elects to pursue a policy of tourism development. Major issues include the country's peripheral location, low level of awareness in potential tourism generating regions, lack of capital, the country's weak competitive advantage and concerns about the future impact of climate change. In stark contrast to those countries which have suffered from overtourism (Fletcher et al. 2019), Tuvalu is an example of undertourism where the destination suffers from too few tourists relative to its capacity to host tourism activity. The situation now facing Tuvalu is examined from two perspectives: a theoretical perspective applying three models often used in tourism research and a scan of social media to analyze the country's current image. The discussion raises a number of questions about the role tourism could play as a new export sector as well as outlining some of the problems that the country must confront if it decided to pursue the expansion of its current tourism sector. The chapter concludes with an analysis of the country's comparative and competitive position in relation to other Pacific Island nations that compete in the tourism space, finding that it is possible for remote islands to develop a tourism sector.

Tuvalu: contextual overview and the COVID-19 pandemic

The origins of most Tuvalu Islanders lie in Polynesia, following migration from Samoa, Tokelau and Tonga (Macdonald 1982). Small areas of the country's northern islands were settled by Micronesians from Kiribati, and the first recorded sighting of the archipelago by Europeans was by Spanish navigator Alvaro de Mendana in 1568 (Faaniu 1983). The island group was later named the Ellice Islands in 1819 by British hydrographer Alexander Findlay, becoming a British protectorate in 1892 and gaining political independence in 1978. Tuvalu's location, resource base and recent history have been significant factors in shaping the role that tourism plays in the

DOI: 10.4324/9780429019968-30

country's economy. Located midway between Australia and Hawaii, Tuvalu had a population of 11,646 in 2019 (World Bank 2021) spread over three reef islands and six atolls that are scattered over an area of 750,000 square kilometres (Borovnik 2012). The country has a total land area of 24.4 square kilometres and is the world's fourth smallest country based on land mass and the second smallest based on population. The maximum elevation of the country is 5 metres above sea level with an average elevation of 1 metre. The absence of a continental shelf leaves the country exposed to tropical cyclones and storm surges associated with cyclones. Tuvalu's remote location poses problems for both cargo and passenger transport. Prior to the outbreak of COVID-19 and the rapid slowdown in international mobility from March 2020, Fiji Airways operated three weekly flights between Fiji and Funafuti, the main island, with a return flight costing between AUD$1,000 (US$ 724.09) and AUD$2000 (US$1,448.18). Flights were disrupted by the pandemic but never ceased altogether. Air Kiribati also offers weekly flights from the capital of the Republic of Kiribati, the Tarawa atoll, to Funafuti. Tuvalu is not on the global cruise-ship circuit.

Tuvalu's agricultural sector is small and based on subsistence farming, which centres on coconuts and traditional crops such as *pulaka* (swamp taro) and limited exports of coconut and copra (Odekon 2015). Increasing salination of the soil from high tide inundation has reduced agricultural production in recent years. Nonetheless, there are abundant supplies of seafood. In the future, rising sea levels are likely to affect ground water reserves through saltwater intrusion. The country's reliance on septic sewage systems also poses a threat. As a consequence, rainwater is the only reliable source of potable water for human consumption (South Pacific Secretariat Pacific Islands Applied Geosciences Commission n.d.). Apart from the resources provided by the sea, the country has no other exploitable natural resources.

In 2020, the national gross domestic product (GDP) was estimated to be US$49 million with a per capita GDP of US$4,633 (Countryeconomy.com 2020). The main source of government revenue is the sale of fishing licenses, which contributed 56% of GDP in 2020 (IMF 2021). Nonetheless, the economy is characterized by aid dependence; earnings from a trust fund established by the United Kingdom (UK), New Zealand and Australia in 1987; and reliance on remittances from Tuvaluan nationals working or living offshore. Tuvalu's Marine Training Institute is a key element in the remittance economy as it trains a large number of Tuvaluans who work as crew on ships throughout the Pacific. One report estimated that about 20% of all households receive remittances from family members working overseas (Milan et al. 2016). The recorded rate for people formally living in poverty was estimated at 26% of the population in 2017 and the estimated average life expectancy in 2021 was 68 years (IMF 2021).

Compared to many tourism-dependant Pacific Island states, the COVID-19 pandemic had little effect on Tuvalu, which reflects the small size of the country's tourism sector. As of January 2022, Tuvalu was one of the very few countries that had successfully avoided the community transmission of COVID-19. The government's rapid response to the pandemic, declaring a state of emergency on 20 March 2020, led to the closure of all ports of entry and the imposition of quarantine on all inbound travellers. A report by the Asian Development Bank (ADB 2020) in December 2020 indicated that the country was unlikely to suffer a decline in GDP because of the pandemic. Nonetheless, in 2020 international donors did provide AU$6.8 million (US$4.67 million) of COVID-19-related support (IMF 2021). The ADB (2020) report also noted that 17 of the 40 people employed in the hospitality and tourism sector (including local restaurants) lost their jobs, though the government's social assistance programme provided financial assistance to all citizens in the form of a universal cash transfer. Reopening the country's borders post COVID-19 is unlikely to result in a significant surge in tourism, as no additional funds appear to have been invested in tourism infrastructure. A further report by the International Organization for Migration (IOM) and the International Labour Organization (ILO) emphasized that, as the principal employers in Tuvalu are largely the public sector and the government, the pandemic's impact on the workforce was rather negligible (IOM and ILO 2021).

Tourism status

It was not until the decade after gaining independence that the government identified tourism as a sector with potential for providing employment and diversifying the country's economic base (Tourism Council of the South Pacific 1988). Recent national development plans such as the 1993–2002 Tuvalu Tourism Development Plan (Tourism Council of the South Pacific 1992) and the 2005–2016 National Strategy for Sustainable Development Plan (Tuvalu Government 2005) highlighted the need to increase international tourist arrivals. However, tourism growth has been limited, rising from 399 international arrivals in 1983 (Tourism Council of the South Pacific 1988) to 2,700 arrivals in 2018, and despite government interest in tourism development, the country is one of the world's least visited international destinations, with a ranking of 203 in total arrivals in 2018 (World Data n.d.). Tuvalu's major tourist source markets are other Pacific Islands, Australia, Asia and New Zealand (SPTO 2018). Tourism infrastructure is limited, and the accommodation sector is serviced by five lodges, one hotel and one property that offer serviced apartments.

Tuvalu's official tourism website (https://www.timelesstuvalu.com/) lists the country's key tourism experiences as diving and snorkelling, yachting, cultural and traditional dance, participation in local sports, touring outer islands and historic and archaeological attractions (based on World War II relics). The main nature-based attraction is the Funafuti Conservation Area, a marine conservation area that is approximately 33 square kilometres in size and includes examples of islets and reefs on the western side of Funafuti atoll. The conservation area is also a nesting area for green sea turtles. The clear waters of the lagoon provide an ideal location for scuba diving and snorkelling. PADI Travel, an affiliated company of the international diving organization known as Professional Association of Diving Instructors (PADI), describes diving in Tuvalu as 'amazing' (PADI, n.d.). Other internet sites about diving offer similar commentary. However, there are currently no diving shops or dive operators in the country, although the country's official tourism website advises that most commercial accommodation providers have dive equipment for rental. Tuvalu is not marketed as a dive destination by dive tour operators and perhaps not seen to have the same potential as other islands in the region.

The main obstacles for tourism development include limited air access, high air fares, lack of tourism and hospitality infrastructure and challenges related to climate change (Prideaux and McNamara 2013; Walters 2017). The country's state of tourism development can be labelled as one of 'undertourism', in contrast to a number of high-volume destinations where there were growing international concerns about overtourism (Fletcher et al. 2019) prior to the temporary reduction in travel caused by the COVID-19 pandemic.

Tourism and development issues faced by 'small island developing states' (SIDS)

Small island countries have a long history of attracting the attention of research tourists and researchers. Prideaux (2009: 231) captured the essence of the attractiveness of many SIDSs in the following observation: 'Islands are intriguing tourism destinations … their location confers upon them a range of often unique attractions including landscapes, heritage, climate, flora, fauna and indigenous culture'. However, the benefits of tourism are contested. From an islander perspective, Baldacchino (2008: 42) reminds us:

> Some islanders may be just as confused by how they are seen and objectified as 'paradises' by mainlanders, while they may struggle at home against un-/under-employment, aid dependency, loss of talent, waste mountains, atrophied coasts and lagoons, sewage overflows, drug running, money laundering, HIV/AIDS, soil erosion, potable water shortage, depopulation or overpopulation.

The potential benefits of tourism development need to be balanced against the potential costs for the communities concerned. For example, while tourism can provide new economic opportunities, including jobs and government revenue, it can also instigate social, economic and environmental change that may or may not be welcome by island residents. Commenting on the literature on SIDS and tourism, Baldacchino (2000) warns that, while some writers are pessimistic about reliance on tourism as a path for development, the potential for small islands to succeed and chart their economic future should not be overlooked. Baldacchino (2000: 28), for example, stated that, while some authors equate smallness with being 'synonymous with being powerless, vulnerable and non-viable', there is another reality that views 'smallness as an inherent advantage and as characteristically associated with above-average economic growth'. Similarly, Scheyvens and Momsen (2008a) conclude that the existing tourism and development literature has often overstated dependence and vulnerability of small island states. More recently, the notion of 'resilience', which broadly focuses on coping with shocks caused by ecological, social and economic change, has been applied to tourism contexts (Lew and Cheer 2017). The vulnerability of communities in SIDS may be balanced by resilience based on strong kinship, community ties, community efforts towards adaptation (e.g., building sea walls or constructing more robust houses), seeking additional employment opportunities in industries other than tourism, remittance support from diaspora communities and strong local governance (Harrison and Pratt 2015).

Issues that affect general economic growth in SIDS include political instability (Beirman 2003), balancing the desire for environmental sustainability with the need for development (de Hass 2002), climate change (Sem and Moore 2009), lack of capital for investment, economic leakages and poverty reduction (Scheyvens and Momsen 2008b). Despite these issues, tourism development has been advocated as an avenue for economic development in many SIDSs, particularly those that have few other exploitable resources apart from tourism (Weaver 2017). Tourism-related opportunities include employment both in direct and indirect tourism-related businesses and capacity building stemming from downstream investment that can assist in the development of value chains between agriculture and tourism (UNWTO n.d.) as well as between construction and tourism and retail and tourism. Tourism can also promote the protection of natural resources by giving a monetary value to the environment. The development of so-called blue and green economies based on sustainable coastal and marine development provides further opportunities for tourism development and can also lead to improved human welfare and social equality while protecting the environment (Smith-Godfrey 2016).

However, the costs of tourism development in both economic and social terms can be high. A recent report by the United Nations World Tourism Organization (UNWTO n.d.) on tourism in SIDS listed a range of challenges, including connectivity (high costs of air transport), adverse impacts on scarce and fragile resources (including water and fragile ecosystems), climate change, economic leakage (small economic base resulting in significant imports) and community engagement (ensuring that the community is engaged and empowered). To this list can be added concerns about commodification of culture, cultural appropriation and loss of authenticity. Scheyvens and Momsen (2008b) argue for the necessity of an approach that values social sustainability as well as economic and environmental sustainability Participation in decision making and legislation to protect local land rights are examples of measures that need to be adopted by governments to enhance social sustainability.

Climate change: effects and responses

As global temperatures increase, the problems generated by climate change add another layer of complexity to the problems faced by small islands seeking to develop a competitive advantage in the international tourism industry. As the UNWTO (n.d.) observed, while small island states produce less than 1% of greenhouse emissions (GHE), they are faced with serious climate change problems including erosion, coral bleaching, ecosystem damage, storm surges, ocean acidification and rising sea levels. Climate change has been recognized as a major problem for many SIDS. For example, the problems that will confront the Maldives tourism industry as global warming accelerates have been known for some time (Agnew and Viner 2001) and discussed in a number of papers (Sovacool 2012; Shakeela and Becken 2015).

In its 2018 report, the United Nation's (UN) Intergovernmental Panel on Climate Change (IPCC) highlight the dangers of climate change, including the potential for widespread loss of coral species and rising sea levels (IPCC 2018). Beyond coral bleaching, there are concerns that low-lying island nations such as Tuvalu, Kiribati and the Maldives will suffer extensive sea water inundation. In its 2021 report, the IPCC (2021: 150) state that small islands will 'very likely' experience continued relative sea level rise which 'along with storm surges and waves, will exacerbate coastal inundation'. Major global scale coral bleaching events in 2016 and 2017 highlight the immediate problem faced by many tropical islands that promote coral reefs as a key destination pull factor.

Tuvalu is highly vulnerable to sea level rise and exposed to climatic- and geo-hazards, including cyclones, tsunamis, droughts and flooding (Siose 2017). The most obvious effect is the flooding of low-lying areas, including the airport during king and spring tides (Mason 2019). A recent report observed that Tuvalu is likely to experience rising sea levels, increasing ocean acidification, more frequent and longer lasting coral bleaching and less frequent but more intense cyclones (Australian Bureau of Meteorology and CSIRO 2014: 5). The Environmental Vulnerability Index (EVI) developed by the South Pacific Applied Geoscience Commission (SOPAC) and the United Nations Environmental Program (UNEP) is a tool for measuring economic and social vulnerability indices (SOPAC 2015). The EVI is able to provide insights into a range of factors that can have a negative influence on sustainable development. Tuvalu's EVI rating is listed as extremely vulnerable.

The government of Tuvalu has been quite vocal in international forums about the potential impact of climate change on SIDS, including Tuvalu, and has developed and adopted a National Strategic Action Plan for Climate Change and Disaster Risk Management (Government of Tuvalu 2012) to combat the impacts of climate change. In 2019, Tuvalu hosted the 50th Pacific Islands Forum where climate change was identified 'as the single greatest threat to the livelihoods, security and wellbeing of the peoples of the Pacific' (Pacific Islands Forum Secretariat 2019). Concern about the effect of rising sea levels has also led to reports on the future viability of the country. Milan et al. (2016) suggest that alternative strategies need to be considered including policies encouraging Tuvaluan nationals to permanently emigrate to other less-at-risk countries such as New Zealand and Australia. While the Tuvaluan government's stated desire is to encourage the international community to adopt mitigation strategies outlined in international climate accords, such as the Paris 2015 Agreement, the long-term sustainability of the country is doubtful if sea levels rise to such an extent that many parts of the country will become permanently submerged. The World Bank (2021) has approved of US$17.5 million to support the country's climate adaptation endeavours.

Issues affecting tourism development in Tuvalu

In a discussion on options for tourism development in Tuvalu, Prideaux and McNamara (2013) observe that government officials identified the main factors hampering tourism development were appropriate tourism-related infrastructure and access to international airlines. One senior government official stated that 'the most serious challenge to us now is money, because there is no money to build capital, infrastructure. We are trying to build local capacity, but what can you do without resources?' (Prideaux and McNamara 2013: 589). These problems were also acknowledged in Tuvalu's National Strategy for Sustainable Development which stated that 'poor urban environmental management, under-manned and poorly situated tourism office, and lack of visitor information' are key challenges of tourism development (Tuvalu Government 2005: 41). While arrivals had increased prior to the COVID-19 pandemic, remoteness and lack of investment in tourism-related infrastructure and services remain the major inhibitors to further growth. Given these problems, Prideaux and McNamara (2013) identify four types of tourism that could be offered by Tuvalu: ecotourism based on the country's reef systems, diving, 'off-the-beaten track' experiences and 'last chance tourism'. To these can be added 'bucket list' tourism.

Sea level rise and the potential disappearance of entire nation states has received increasing attention from media, air agencies and, more recently, from tourism researchers. This notion is captured by the term 'last chance tourism', a niche tourism product that focuses on experiencing or gazing upon a place or phenomenon before it disappears (Piggot-McKellar and McNamara 2017). Australia's Great Barrier Reef (GBR) is an example of a tourist icon that was identified as a 'last chance' location after extensive coral bleaching in 2016 and 2017, although in the case of the GBR the danger has receded for now (Piggott-McKellar and McNamara 2017). In the Pacific, Kiribati's official travel guide features 'climate change tours' and a homestay for 'climate change enthusiasts' (Kiribati National Tourism Office 2017).

According to Hübner (2013), Tuvalu's National Tourism Organization openly recognizes the impacts of climate change and has positioned the country as a climate change victim, which may help in attracting international tourists. While Tuvalu's official tourism website does not promote last chance tourism or climate change related activities, these issues are widely canvassed in the media. CNN Travel (2019) introduces Tuvalu as the antithesis to overtourism, describing the country as one of the most isolated and least visited nations worldwide, which should be visited before 'rising seas could swamp the low-lying country'. The 'South Pacific Specialist' (a travel website) also published a story that can be interpreted as promoting last chance tourism:

> Another recent attraction is the causeway at northern side of Funafuti Island (Capital) which also known as Fongafale. It has recently eroded due to sea level rise and therefore the pathway which connects the road to the other side is getting thinner and more tourists are being attracted to it to take pictures.
>
> *(South Specific Specialist n.d.)*

Small island states which are isolated, rarely visited and in danger of being swamped by rising seas may also match the desires of the bucket list tourist. The concept of the 'bucket list' refers to a list of experiences an individual wishes to complete before she or he dies. Bucket list destinations are perceived as either famous, authentic or extreme (Thurnell-Read 2017). It can be argued that tourism to such sites and attractions, when based on educational and sustainable principles, can lead to an increasing awareness towards climate change. However, do the ends justify the means? Promotion of long-haul flights to remote destinations will generate significant GHE, disrupting the whole notion of sustainable development. Further, empirical evidence shows that tourists' increased awareness does not necessarily lead to more sustainable travel behaviour (Antimova et al. 2012; McKercher et al. 2010).

In addition, promoting last chance or bucket tourism as the key pull factor is likely to have a negative effect on destination image and possibly hinder efforts to promote other forms of tourism such as 'off-the-beaten-track' and ecotourism. For example, promoting the recently eroded Fongafale pathway as an attraction has the potential to encourage 'voyeuristic tourism' which may fuel the victimization felt by Tuvaluans in the light of existent and future environmental changes and disasters (Borovnik 2012: 1369). Tuvalu is pinned on the world map as a country that is vulnerable and that aims to cope with the effects of climate change. Simultaneously, there is some potential for it to become a niche tourism destination juxtaposed to the neo-colonial gaze of wealthy climate change tourists. Local residents in such settings are perceived as passive victims or objects of tourism consumption rather than active agents who can pursue their own ideas.

Dilemmas for developing tourism in SIDs

Three tourism-related models (periphery, push-pull and comparative and competitive advantage) can be used to understand the dilemma faced by SIDS interested in developing a tourism industry. The location of Tuvalu on the global periphery is a major problem. Peripheral describes the location of a destination relative to its potential generating regions and can be described in terms of the time, cost and inconvenience of travel (Chaperon and Bramwell 2013). Overcoming problems related to a peripheral location are difficult and, in the case of tourism, may depend on the level of uniqueness a destination is able to offer in relation to its competitors (Prideaux 2002; Weaver 2017). From this perspective, the diving opportunities offered by Tuvalu are not unique and are available in almost all other Pacific Island countries, generally at lower cost.

The push-pull framework (Crompton 1979; Dann 1977) provides a useful tool for examining demand and supply issues. Push factors are described as the factors that encourage people to travel, including curiosity and the need for adventure and recreation. Pull factors describe the features of a destination such as nature, landscape and culture that attract potential visitors (Klenosky 2002). To attract visitors, destinations such as Tuvalu must match their destination pull factors with the push factors of their target markets. Although Tuvalu has the potential to offer quality dive experiences, the destination has faced difficulty in attracting the attention of the dive market because of remoteness, cost, lack of infrastructure and, importantly, a lack of visibility in the commercial supply chains that sell dive tours. In Tuvalu, pull factors based on potential dive experiences only partially match the push factors of the potential dive market. The country has not been able to convince either dive tour operators or individual divers that Tuvalu offers a unique dive experience that is not able to be replicated in other Pacific Ocean destinations.

Comparative and competitive advantage theory (Ritchie and Crouch 2003) also provides a useful tool for understanding the problems faced by Tuvalu and similar SIDSs. Comparative advantage refers to the resources that destinations offer tourists but which are not necessarily developed to the stage where they can be offered as a tourism product. Prideaux et al. (2014) described resources as those factors used in the production of tourism-related products and experiences. A competitive position emerges when destination resources are used to create experiences and products that are perceived by consumers to be better than their competitors, perhaps because these resources are unique, novel, low priced and, importantly, able to meet customer expectations. Competitive success is achieved when existing resources that offer a comparative advantage are transformed into products and experiences able to attract tourists.

The level of success can be measured in demand-side terms by visitor numbers, growth trends and revenue generation. From a supply-side perspective, competitiveness can be measured by factors such as hotel yield, profits and employment statistics. Although Tuvalu has yet to develop a competitive advantage, there is scope to leverage undeveloped comparative advantages to create last chance, off-the-beaten-track and bucket list tourism experiences.

The previous argument presupposes that tourism development is a desirable national objective. Baldacchino (2008) reminds us that not all communities necessarily desire tourism growth. Concerns about the impact climate change is having on all of the Earth's communities – plant, animal and human – raises questions about the desirability of growth as a key national objective (McKercher and Prideaux 2020). One of the solutions proposed to combat overtourism is degrowth in which the scale of tourism and other economic sectors are wound back to a more community and environmentally acceptable scale (Prideaux and Pabel 2021).

Media scan

In recent years, the internet has become a powerful tool for tourists undertaking searches for possible holiday destinations. A number of studies have found that factors that reduce a destination's positive image will reduce the strength of its pull factors and adversely affect its competitive advantage (Beerli and Martin 2004; Pike 2002). Information that is located through media searches has the potential to influence destination choice (Xiang and Gretzel 2010). To gain an understanding of the images that could emerge from an internet search for 'Tuvalu tourism', 23 stories that focused on an aspect of travel to Tuvalu were downloaded and analyzed. The downloaded articles included a range of sources, including news, tourism blogs, UN stories and business advice.

Analysis of the sites found that 61% expressed concerns about climate change. The underdeveloped status of tourism infrastructure was also mentioned in 61% of the sites. Many sites emphasized the country's small size, its low elevation and underdeveloped tourism infrastructure. Southerden (2016), a travel writer on a familiarization tour to Tuvalu, wrote that the country has an 'offbeat charm, and takes you back to a time when you could land in a destination without a plan or a guidebook, learn about it from the people you happened to meet and let your days unfold without an itinerary'. It was also noted that that destination lacked tour guides, tour operators, ATMs and credit card facilities, as well as being 'off the cruise ship circuit'. A post by Mar (n.d.) also noted the islands' remoteness, adding:

> The main island of Fongafale is a really really small place and … there is little development so, as expected, there are few things to do. Most visitors will stay in this island, because that is where the airport is. Going to the other atolls is hard because there are no flights and the boats depart every two weeks.

Away from the travel blogs written by travel writers, Philpott (2019), a student on a field trip to the island, painted another picture of the country that is rather less appealing:

> If traveling to Tuvalu be prepared for cold showers with no temperature gauge. Be prepared to get sick in one way or another (even if you're careful around food – the salad is washed with tap water, the door handles will have bacteria etc). Not a single member of our group of 8 had exclusively solid bowel movements. Be prepared to struggle if you are a vegetarian. I was the only strict vegetarian but the other girls struggled with the meat there as they were afraid the reef fish would contain Ciguatera, the chicken

> under cooked, and the pigs that were lined in pens alongside the runway outside our bedroom window didn't improve the appeal of the pork.

One of the more alarming stories discussed the impact of climate change, including loss of farming land, increasing incidents of inundation of the airport and the despair now being voiced by some community members about the impact of climate change and the future of the country (Mason 2019). While there were many references to the tropical beauty of the country, the overall image gained from the social media scan was negative. Two themes emerged. The first was that Tuvalu is facing an imminent threat from climate change. For investors looking for opportunities to develop tourism infrastructure, such as hotels, the takeaway message is that Tuvalu is a poor choice for investment. The second theme is that the country's tourism infrastructure is underdeveloped and that the services normally available to tourists are absent. The low quality or absence of services and high cost of airfares are likely to be a disincentive to most potential tourists.

Conclusion and research implications

This chapter has examined a number of issues that will need to be addressed if Tuvalu is to move from its current status as one of the world's least visited countries to a position where it can attract the attention of the global tourism sector as a niche destination. While location is a significant barrier, other remote SIDS, such as the Maldives and Niue, have demonstrated that it is possible to match a unique niche product to a specific market and thereby reduce the burden of being a periphery destination. Aside from its peripheral location, Tuvalu must also overcome problems associated with limited infrastructure, concerns about food supply for visitors, concerns over sea level rise and key attractions that differ only marginally from those offered by other Pacific Island countries. The opportunities for tourism development in Tuvalu are best summed up by Southerden (2016), whose following comments provide strong clues about the uniqueness of the Tuvalu tourism experience:

> Tuvalu won't be everyone's cup of coconut water. It's not the next Fiji or Vanuatu. But for those with time, curiosity, and a sense of adventure, it's the kind of place that makes you feel like a traveller again. And for the rest of us? It's a healthy reminder of a world that exists outside the tourist universe, for now.

Repackaging existing resources into a tourism experience that differs considerably from the way these resources are used by neighbouring countries thus offers an opportunity to develop a novel tourism experience. 'Off the beaten track' travel fits well with the opportunity for an authentic travel experience that has few of the markers of the mass tourism experiences offered by neighbouring Fiji. Last chance tourism also provides an opportunity to develop a unique travel experience to a country that is threatened by climate change. To capitalize on opportunities of this type, Tuvalu will need to develop an appealing story that emphasizes opportunities to engage in authentic and novel travel experiences.

The discussion in this chapter suggests at least three areas for future tourism research: (1) further investigation into how climate change impacts on key natural resources, such as coral reefs, may adversely affect opportunities to expand tourism in small island nations; (2) opportunities to develop specific forms of tourism such as 'off the beaten track travel' to mitigate the impact of a peripheral location; (3) further exploration of the concept of undertourism. Given the growing level of concern about overtourism prior to the outbreak of the COVID-19 virus and suggestions for adopting degrowth strategies as an alternative development strategy solution, further investigation of undertourism may provide ideas for rerouting tourism flows from areas of high demand to areas of low demand.

References

ADB (Asian Development Bank) (2020) *Pacific Economic Monitor*, December, ADB. Available online at https://www.adb.org/sites/default/files /publication /662406 /pem-december-2020.pdf (accessed 13 August 2020).

Agnew, M. and D.Viner (2001) 'Potential impacts of climate change on international tourism', *Tourism and Hospitality Research*, 3(1): 37–60.

Antimova, R., J. Nawijn and P. Peeters (2012) 'The awareness/attitude-gap in sustainable tourism: A theoretical perspective', *Tourism Review*, 67(3): 7–16.

Australian Bureau of Meteorology and CSIRO (2014) 'Climate variability, Extremes and Change in the Western Tropical Pacific: New Science and Updated Country Reports', *Pacific-Australia Climate Change Science and Adaptation Planning Program Technical Report*, Melbourne: Australian Bureau of Meteorology and Commonwealth Scientific and Industrial Research Organisation.

Baldacchino, G. (2000) 'An exceptional success: The case of an export-oriented, locally-owned, small-scale manufacturing firm in a small island country', *Journal of South Pacific Studies*, 23(1): 27–47.

Baldacchino, G. (2008) 'Studying islands: On whose terms? Some epistemological and methodological challenges to the pursuit of island studies', *Island Studies Journal*, 3(1): 37–56.

Beerli, A. and J.D. Martin (2004) 'Factors influencing destination image', *Annals of Tourism Research*, 31(3): 657–681.

Beirman, D. (2003) 'Restoring tourism destinations in crisis: A strategic marketing approach', [online]. In: R. Braithwaite (ed.) *CAUTHE 2003: Riding the Wave of Tourism and Hospitality Research*, Lismore: Southern Cross University, pp. 1146–1150.

Borovnik, M. (2012) 'Tuvalu'. In: G. Philander (ed.) *Encyclopedia of Global Warming & Climate Change*, Thousand Oakes: Sage, p. 1369.

Chaperon, S. and B. Bramwell (2013) 'Dependency and agency in peripheral tourism development', *Annals of Tourism Research*, 40: 132–154.

CNN Travel (2019) 'Why you should go to the world's least-visited countries . Available online at https://edition.cnn.com/travel/article/least-visited-countries-travel/index.html (accessed 20 August 2019).

Countryeconomy.com (2020) *'Tuvalu GDP – gross domestic product'*. Available online at https://countryeconomy.com/gdp/tuvalu (accessed 20 August 2019).

Crompton, J. (1979) 'Motivations for pleasure vacation', *Annals of Tourism Research*, 6(4): 409–424.

Dann, G. (1977) 'Anomie, ego-enhancement and tourism', *Annals of Tourism Research*, 4(4): 184–194.

de Haas, H. (2002) 'Sustainability of small-scale ecotourism: The case of Niue, South Pacific', *Current Issues in Tourism*, 5(3–4): 319–337.

Faaniu, S. (1983) *Tuvalu: A History*, Suva: Institute of Pacific Studies.

Fletcher, R., I. Mas, A. Blanco-Romero and M. Blázquez-Salom (2019) 'Tourism and degrowth: An emerging agenda for research and praxis', *Journal of Sustainable Tourism*, 27(12): 1745–1763.

Government of Tuvalu (2012) *Tuvalu Strategic Plan for Climate Change and Disaster Risk Management 2012–2016*, 1 January, Government of Tuvalu, Tuvalu. Available online at https://reliefweb.int/report/tuvalu/tuvalu-national-strategic-plan-climate-change-and-disaster-risk-management-2012-2016 (accessed 25 August 2019).

Harrison, D. and S. Pratt (2015) 'Tourism in Pacific islands: Current issues and future challenges'. In: S. Pratt and D. Harrison (eds.) *Tourism in Pacific islands: Current Issues and Future Challenges*, Abingdon: Routledge, pp. 3–21.

Hübner, A. (2013) 'Concepts of culture and tourism adaption to climate change in Tuvalu', *Tourism, Culture and Communication*, 13(2): 79–94.

IMF (International Monetary Fund) (2021) *Tuvalu: 2021 Article IV Consultation -press release; staff report; and statement by the executive director for Tuvalu*, August. No. 21/176, Washington, DC. Available online at file: ///C:/Users/courtsq/Downloads/1TUVEA 2021001%20(1).pdf (accessed 6 January 2022).

IOM (International Organization for Migration) and ILO (International Labour Organization) (2021) *Powering Past the Pandemic: Bolstering Tuvalu's Socioeconomic Resilience in a COVID-19 World*, Suva, Fiji:

IOM and ILO. Available online at https://publications.iom.int/system/files/pdf/powering-past-the-pandemic_1.pdf (accessed 4 January 2022).

IPCC (Intergovernmental Panel on Climate Change) (2018) *Global Warming of 1.5^{0}c.* Geneva: IPCC. Available online at https://www.ipcc.ch/sr15/ (accessed 15 September 2019).

IPCC (Intergovernmental Panel on Climate Change) (2021) *Climate Change 2021: The Physical Science Basis*, Geneva: Intergovernmental Panel on Climate Change (IPCC). Available online at https://www.ipcc.ch/report /ar6/wg1/downloads/report/IPCC_ AR6_WGI_Full_Report.pdf (accessed 11 August 2021).

Kiribati National Tourism Office (2017) *Kiribati: Tarawa and the Gilbert Outer Islands. Travel Guide 2017*, Betio: Ministry of Communications, Transport and Tourism Development.

Klenosky, D. (2002) 'The "pull" of tourism destinations: A means-end investigation', *Journal of Travel Research*, 40(4): 385–395.

Lew, A.A. and J.M. Cheer (eds.) (2017) *Tourism Resilience and Adaptation to Environmental Change: Definitions and Frameworks*, London: Taylor & Francis.

Macdonald, B. (1982) *Cinderellas of the Empire: Towards a History of Kiribati and Tuvalu*, Canberra: Australian National University Press.

Mar, (n.d.) 'Tuvalu: All you need to know'. Available online at https://www.onceinalifetimejourney.com/once-in-a-lifetime-journeys/pacific/tuvalu-travel-guide/ (accessed 18 May 2021).

Mason, M. (2019) 'Tuvalu: Flooding, global warming, and media coverage'. Available online at http://www.moyak.com/papers/tuvalu-climate-change.html (accessed 12 February 2021).

McKercher, B. and B. Prideaux (2020) *Tourism Theories, Concepts and Models*, Oxford: Goodfellow Publishers Ltd.

McKercher, B., B. Prideaux, C. Cheung and R. Law (2010) 'Achieving voluntary reductions in the carbon footprint of tourism and climate change', *Journal of Sustainable Tourism*, 18(3): 297–317.

Milan, A., R. Oakes and J. Campbell (2016) *Tuvalu: Climate Change and Migration – Relationships Between Household Vulnerability, Human Mobility and Climate Change*, Report No. 18. Bonn: United Nations University Institute for Environment and Human Security (UNU-EHS).

Odekon, M. (ed) (2015) 'Tuvalu', in *The SAGE Encyclopedia of World Poverty*, Thousand Oakes: Sage, p. 1563.

Pacific Islands Forum Secretariat (2019) *Fiftieth Pacific Islands Forum, Funafuti, Tuvalu, 13-16 August.* Available online at https://www.forumsec.org/wp-content/uploads/2019/08/50th-Pacific-Islands-Forum-Communique.pdf (accessed 16 March 2021).

Philpott, O. (2019) 'A Tuvalu travel retrospective: My experience in the least visited nation on earth', *Inspired by Maps.* Available online at https://inspiredbymaps.com/tuvalu-travel-retrospective/ (accessed 6 June 2021).

Piggott-McKellar, A. and K. McNamara (2017) 'Last chance tourism and the great barrier reef', *Journal of Sustainable Tourism*, 25(3): 397–415.

Pike, S. (2002) 'Destination image analysis-a review of 142 papers from 1973 to 2000', *Tourism Management*, 23(5): 541–549.

Prideaux, B. (2002) 'Building visitor attractions in peripheral areas', *International Journal of Tourism Research*, 4(5): 379–391.

Prideaux, B., D. Berbigier and M. Thompson (2014) 'Wellness tourism and destination competitiveness'. In: C. Pforr and C. Voigt (eds.) *Wellness Tourism – A Destination Perspective*, London: Routledge, pp. 221–248.

Prideaux, B. and K. McNamara (2013) 'Turning a global crisis into a tourism opportunity: The perspective from Tuvalu', *International Journal of Tourism Research*, 15(6): 583–594.

Prideaux, B. and A. Pabel (2021) 'Degrowth as a strategy for adjusting to the adverse impacts of climate change in a nature-based destination'. In: C.M. Hall, L. Lundmark and J. Zhang (eds.) *Degrowth and Tourism: New Perspectives on Tourism Entrepreneurship, Destinations and Policy*, London: Routledge, pp. 116–132.

PADI (Professional Association of Diving Instructors) (n.d.) *Diving in Tuvalu.* Available online at https://www.padi.com/diving-in/tuvalu/ (accessed 13 May 2022).

Ritchie, J.R. and G. Crouch (2003) *The Competitive Destination: A Sustainable Tourism Perspective*, Wallingford: CABI.

Scheyvens, R. and J. Momsen (2008a) 'Tourism in small island states: From vulnerability to strengths', *Journal of Sustainable Tourism*, 16(5): 491–510.

Scheyvens, R. and J. Momsen (2008b) 'Tourism and poverty reduction: Issues for small island states', *Tourism Geographies*, 10(1): 22–41.

Sem, G. and R. Moore (2009) *The Impact of Climate Change on the Development Prospects of the Least Developed Countries and Small Island Developing States*, New York: United Nations Office of the High

Representative for the Least Developed Countries, Landlocked Developing Countries and Small Island Developing States.

Shakeela, A. and S. Becken (2015) 'Understanding tourism leaders' perceptions of risks from climate change: An assessment of policy-making processes in the Maldives using the social amplification of risk framework (SARF)', *Journal of Sustainable Tourism*, 23(1): 65–84.

Siose, L. (2017) 'Community perception on migration as an adaptation strategy to the impact of climate change in Tuvalu', *Master Thesis*, Suva (Fiji): The University of the South Pacific.

Smith-Godfrey, S. (2016) 'Defining the blue economy, maritime affairs', *Journal of the National Maritime Foundation of India*, 12(1): 58–64.

SOPAC (South Pacific Applied Geoscience Commission) (2015) 'Environmental vulnerability index'. Available online at http://www.vulnerabilityindex.net/category/background/ (accessed 21 August 2019).

South Pacific Secretariat Pacific Islands Applied Geosciences Commission (n.d.) 'Tuvalu acts to climate proof its people'. Available online at https://web.archive.org/web/20101224205939/http://www.sopac.org/ (accessed 18 September 2019).

South Pacific Specialist (n.d.) 'Tuvalu – attractions & activities'. Available online at http://southpacificspecialist.org/tuvalu-attractions-activities/ (accessed 18 September 2019).

Southerden, L. (2016) 'Tuvalu: Visiting one of the world's tiniest countries', *Stuff*, 4 July. Available online at https://www.stuff.co.nz/travel/destinations/pacific-islands/81651362/tuvalu-visiting-one-of-the-worlds-tiniest-countries (accessed 12 September 2019).

Sovacool, B. (2012) 'Perceptions of climate change risks and resilient island planning in the Maldives', *Mitigation and Adaptation Strategies for Global Change*, 17(7): 731–752.

SPTO (South Pacific Tourism Organization) (2018) *Annual Review of Visitor Arrivals in Pacific Island Countries 2017*, Fiji: South Pacific Tourism Organisation (SPTO).

Thurnell-Read, T. (2017) '"What's on your bucket list?": Tourism, identity and imperative experiential discourse', *Annals of Tourism Research*, 67: 58–66.

Tourism Council of the South Pacific (1988) *Integrated Tourism and Tourism Linkages Development in Tuvalu with Emphasis on Accommodation Development*, Suva, Fiji: Secretariat of the Tourism Council of the South Pacific.

Tourism Council of the South Pacific (1992) *Tuvalu Tourism Development Plan 1993–2002*, Suva, Fiji: Tourism Council of the South Pacific.

Tuvalu Government (2005) *National Strategy for Sustainable Development*, Funafuti: Ministry of Finance, Economic Planning and Industries.

UNWTO (United Nations World Tourism Organization) (n.d.) *Tourism in Small Island Developing States (SIDS) Building a More Sustainable Future for the People of Islands*, Madrid: UNTO. Available online at http://cf.cdn.unwto.org/sites/all/files/docpdf/unwtotourisminsidsa4wtables.pdf (accessed 15 August 2019).

Walters, T. (2017) 'Tuvalu'. In: L. Lowry (ed.) *The SAGE international Encyclopedia of Travel and Tourism*, Thousand Oakes: SAGE, pp. 1365–1366.

Weaver, D.B. (2017) 'Core–periphery relationships and the sustainability paradox of small island tourism', *Tourism Recreation Research*, 42(1): 11–21.

World Bank (2021) 'Boost for Tuvalu's economic, social and climate resilience', 8 December. Available online at https://www.worldbank.org/en/news/press-release/2021/12/08/boost-for-tuvalu-s-economic-social-and-climate-resilience (accessed 5 January 2022).

World Data (n.d.) *Tuvalu*. Available online at https://www.worlddata.info/oceania/tuvalu/index.php (accessed 6 January 2022).

Xiang, Z. and U. Gretzel (2010) 'Role of social media in online travel information search', *Tourism Management*, 31(2): 179–188.

CONCLUSION

27

FUTURE RESEARCH TRAJECTORIES

Pacific Island tourism

Marcus L. Stephenson and Dallen J. Timothy

This volume has attempted to cover a lot of mileage in terms of dealing with the range of tourism-related matters for self-governing small island states in the Pacific Ocean. As the chapter contributions in this volume all concluded with accounts detailing the research implications derived from the respective enquiries, variously involving recommendations for pursuing future research problems, approaches and methodologies, there will be no direct attempt to replicate those propositions in this final chapter. Nonetheless, the chapter will elaborate on some of the main research themes highlighted but will, more specially, identify new thematic areas of inquiry, especially those areas of study which are believed to have efficacy in terms of advancing our critical comprehension of tourism within the wider context of Pacific Islands. There is also an undertaking to remind readers of the main issues that have emerged from those contributing chapters. Consequently, this chapter has assembled six thematic-based research trajectories: (1) tourism, sovereignty and geopolitics; (2) tourism, culture and heritage; (3) climate change, the environment and the 'blue economy'; (4) gender, frontline workers and social equality; (5) travel and tourism mobilities of Pacific Islanders; and (6) Indigenous methodologies and local diversities.

Tourism, sovereignty and geopolitics

It is clear that the supply of (and demand for) Pacific Island tourism products continue to evolve but entirely at the mercy of unpredictable events, shocks and, indeed, air-based mobility and the variables that affect it. The current battle for greater influence in the Pacific between foreign powers has implications for this situation. Australia, China, Japan, New Zealand, Taiwan and the United States are likely to continue investing greater sums of money in national airlines and transportation infrastructure as a means of exerting political and economic hegemony over some of the smaller states in the region. The increased geopolitical role of China cannot be ignored, especially in light of its recent security pact with the Solomon Islands and growing alliances with various Pacific Island states, together with its increased involvement in facilitating infrastructural development projects, including tourism infrastructure, as part of its Road and Belt initiative in the region (Gunia 2022).

DOI: 10.4324/9780429019968-32

The recovery from the current pandemic is likely to happen more quickly in the larger and more affluent states, exacerbating the differences between them and the smaller and less affluent states. This could raise the stakes in these unbalanced relationships as the smaller countries turn to their larger overlords for recovery assistance and, of course, for the supply of tourists. All of these geopolitical relations may empower and disempower the Pacific Island states simultaneously – disempowering by giving up a degree of sovereignty to their financial supporters and empowering by being able to provide more efficient mobilities and possibly experiencing fewer transportation failures. Future research must address all these issues in relation to sovereignty and power relations in the Pacific Islands and the sorts of consequences that might result from relinquishing a degree of autonomy to outside authorities.

Pacific Island states can utilize tourism-related policies to drive forward the national and economic objectives of the state, which can also reveal the intent to which a state wishes to take control of tourism development or pursue a less centralist approach. Either way, it is important that policies (and related planning directives) are implemented and that their impacts and the degree of efficacy are duly monitored, particularly for the betterment of the society, economy and the environment (see Chapters 2, 3, 5, 11, 12, 16, 17, 19, 21, 22 and 23). It would be valuable to research cross-country comparisons of tourism policies and planning directives between self-governing states and dependent territories (e.g., American Samoa, Guam, Hawai'i, New Caledonia, Tahiti and Tokelau). This would help to comprehend more clearly the functionality and effectiveness of self-governing states and degrees of autonomy concerning tourism development decisions and practices. This would also help to evaluate the impacts of tourism governance on the economy (see Armstrong and Read 2021). Moreover, there is room for another Routledge handbook concerning tourism and the dependent island territories in the Pacific, of which there are many, and these too are increasingly caught up in the geopolitical power relations of the Pacific.

Nonetheless, the Pacific Island region is not politically static, and we may well see new states being formed in the not-too-distant future, opening up new levels of academic interest in the field of tourism, development and self-determination. The independence movement in the island of Chuuk has been pressing for separation from the Federated States of Micronesia (FSM), and Bougainville witnessed a 97.7% majority vote for independence from Papua New Guinea in the 2019 referendum, with a schedule in place for full independence in 2027 (Mckenna and Ariku 2021). Such geopolitical struggles for self-governance and national sovereignty are worth monitoring, generating fundamental questions such as: how will tourism be planned by the new state? What type of policies and planning directives will be implemented? Whose interests will be prioritized?

Tourism, culture and heritage

One key concern relates to the extent to which tangible and intangible forms of heritage can be utilized more efficiently to ensure that there is another dimension to the Pacific tourism experience - beyond sun, sea and sand tourism. The educational processes of heritage tourism can help to shape and challenge the established and stereotypical perceptions of Pacific Islanders, which have been historically reinforced through travel writing and contemporary tourism media representations, including popularized imagery of Pacific peoples (see Chapter 9). The study of heritage tourism in Pacific Island states is still arguably in its embryonic stage, though, as the chapter contributions from Vanuatu (see Chapter 14) and the FSM (see Chapter 21) indicate, there is substantial scope to ensure that community-based heritage participation and conservation have important roles to play for the interests of the local economy and society. We do know that community ownership of heritage is a concern in Pacific communities because of the impact of tourism demand and commodification, as we have seen with the firewalking ceremony which originally evolved from a clan in Fiji and with the *nagol* (land diving) in Pentecost Island in Vanuatu, as these two popular localized tourism events have been challenged by cultural adoption elsewhere in Fiji and Vanuatu (see Pigliasco 2009).

The challenge of course is the extent to which Pacific heritage reaches a wider international audience. From the 13 Pacific Island states, there are only seven UNESCO World Heritage Sites in total: Levuka Historical Port Town (Fiji), Chief Roi Mata's Domain (Vanuatu), Rock Islands Southern Lagoon (Palau), Phoenix Islands Protected Area (Kiribati), Bikini Atoll Nuclear Test Sites (Marshall Islands), East Rennell (Solomon Islands) and Nan Madol (Ceremonial Centre of Eastern Micronesia) (FSM). The latter two sites are on the UNESCO World Heritage Endangered List, which speaks volumes of the challenges that heritage conservation faces in the region. The Pacific World Heritage Action Plan 2016–2020 (UNESCO 2015: 3) recognizes that the underrepresentation of heritage sites in the Pacific Islands is due to such factors as: 'limited awareness of Pacific cultural and natural heritage outside the region'; 'lack of adequate representation of the unique and special heritage of the Pacific on the World Heritage List'; 'isolation and resource limitations that restrict access to information and assistance'; 'limited financial and human resources, skills and capacities within communities, and institutions to adequately manage the region's cultural and natural heritage'; 'a need for increased awareness within communities of the great value that the World Heritage Convention contributes to the protection and vitality of cultural and biological diversity', and; 'climate change, financial instability, globalization of society and economy, technological development, commercialization, energy supply and demand, natural disasters and tourism growth'. Consequently, these challenges are useful starting points in any research agenda addressing ways in which heritage tourism in the region can be pursued in a culturally and economically sustainable manner.

There are more than 20 heritage sites in Pacific Island states that are registered on the tentative World Heritage List, which encourages state parties to submit properties that are considered to represent cultural and/or natural heritage of outstanding universal value (UNESCO n.d.). It would be useful to examine the local opinions and interpretations of these sites and the extent to which local heritage could be adequately represented by global/world heritage practices and supported (or not) by national governments. In the Solomon Islands, there are concerns that the UNESCO World Heritage Site of East Rennell (Solomon Islands) is not producing economic and cultural benefits to the local community, and local communities living around Marovo Lagoon have not been receiving full support from key stakeholders and interest groups (see Chapter 10).

Climate change, the environment and the 'blue economy'

Many of the conditions that create challenges to human mobility in the region also mean that the Pacific Island states are particularly sensitive to external forces. As Wong et al. (2013: 53) note: 'Pacific island tourism has a relatively low resistance to external shocks due to its isolation from major markets, small populations, inadequate transportation links'. Three decades ago, Milne (1992) stressed the region's susceptibility to natural disasters. And the one 'external shock' that is fast becoming an internal phenomenon is climate change itself (Christensen and Mertz 2010). This volume has given substantial treatment to climate change and its complex relationship with tourism development (see Chapters 2, 3, 4, 7, 11, 17, 19, 21, 22, 23 and 26). Engaging with the viewpoints and perceptions of Pacific Islanders towards climate change is fundamental (Weir and Pittock 2017) and especially on how ecological change and its impact on tourism livelihoods are understood. Empirically framing people's reactions and degrees of adaptation to climate change could have valuable policy implications, contributing to models and frameworks concerning community resilience in tourism destinations in the Pacific (Jiang et al. 2015; Tolkach and Pratt 2019). Social scientific research surely has a role to play in looking at how the socio-cultural elements of society influence perceptions and experiences of climate change (and of course natural disasters). Given the level of religiosity within Pacific Island communities, it would be of no surprise to suggest that faith-based narratives influence climate change perceptions and adaptation processes (Bertana 2020). Nunn (2017) goes so far as to argue that the failure of foreign governments and agencies to institute effective climate change interventions and adaptation strategies is due to the continual use of a secular approach and an inability to directly communicate with impacted communities.

Scientifically, however, it is clear that climate change is affecting many islands in the Pacific Ocean through rising sea levels, increasing numbers of cyclones, more intense storms, increased rainfall and rising temperatures (Barnett and McMichael 2018; Wong et al. 2013). However, we are still unsure how such changes will affect transportation mobility in the near future, but with the potential for increased flooding and coastal erosion, there is little doubt that getting access to particular Pacific Island states will be a significant problem. More research is necessary to understand how human mobility, including migration and tourism, is affected by climatic changes and their broader outcomes. How might geomorphological changes (i.e., insular erosion and coral deterioration) affect the cruise and yachting sector or water-based activities? To what extent will demand for tropical holiday destinations be affected, and will these be replaced by other destinations that suffer less from climate change?

Much of the social, economic and political zones of Pacific Island states are located in coastal areas along with many tourist centres and districts, and, thus, significant investments have been made in climate adaptation, protective systems and resource restoration. Investments, for instance, centre around such project initiatives as mangrove and coral restoration (Fijian Government 2021), beach nourishment (Ichikawa et al. 2017) and sustainable groundwater use (IUCN 2022). Subsequently, it would be instrumental for research to track such investments to gauge the level of effectiveness of these initiatives and the long-term value of the investments. The restoration and rehabilitation of mangroves, for instance, not only equates to ecological value but also socio-economic, tourism and recreational value (Ellison et al. 2020).

One real way forward is to approach these environmental concerns through strong levels of Pacific Island governance, including the robust support of Pacific regionalism (see Chapter 3). Therefore, the importance of 'regional cooperation', 'inter-governmental cooperation' and 'binding agreements' concerning the advancement of a sustainable tourism development framework has been emphasized (Scheveyns and Movono 2018: 137). Whilst each Pacific Island state shares the broader concerns of climate change, they have their own experiences, needs and complexities. Accordingly, research ought to initially deal with island-specific assessments, and the next logical step forward would be for action-based research to consider the mutual synergies between both regionalism and localism.

The current social movement of many people refusing to travel by air, owing to the high carbon footprint associated with airlines, could have a significant impact on Pacific Island tourism if the crusade continues to grow as projected. However, new technologies are rendering isolated places less remote. Prior to the pandemic, new niche markets grew significantly with uptakes in arrivals to many of yesterday's less-accessible destinations. As the people of the Pacific have a long history of resilience, it is likely that they will continue to adapt to the changes that are, for all intents and purposes, outside of their control. There is a chance that scattered geographies, isolation and smallness may become more of a blessing than a blight for the sustainable growth of tourism in the Pacific Islands, but considerable longitudinal and interdisciplinary research is needed to confirm this.

The Palau destination case scenario illustrates that the decisive way forward is for governments to lead by example and not to prioritize the needs of the tourists over the needs of the environment (see Chapter 18), thereby instilling a sense of responsibility on both the private sector and the tourists themselves. The real issue concerns the extent to which Pacific Island states, through the instrument of self-government, can develop tourism-related policies which apply the principle of 'environmental state sovereignty', in which the message is clear: 'the right not to be harmed environmentally by other states' (Penz 1996: 47). Here we are referring to citizens of those states that generate international tourists who impact and pollute other (more fragile and vulnerable) states.

There are certainly strong arguments laid out in this volume for Pacific Island states to seriously contemplate sustainable forms of special interest tourism, of which there are many types: from energy tourism in Kiribati (see Chapter 19) to yoga tourism in Niue (see Chapter 24), and there is certainly scope for marketing research to address the viability of particular market segments in terms of responding to the economic needs of local communities. This will be useful in terms of pursuing the potentiality of regenerative forms of tourism within the context of a post-pandemic recovery agenda (see Chapter 6). Nonetheless, future research focusing on developing solutions to the challenges that small island states face should take heed of Armstrong and Read's (2021: 231) advice to avoid 'over-specialisation' in specific markets, as this can make economies more vulnerable to external shocks. This too may involve looking beyond tourism options to diversify the economy through advancements of other niche activities or developments. Tourism scholars should not feel compelled to always look towards tourism for solutions. Tourism could well be the source of the problem. At this point, it is worth mentioning David Harvey's (2020 n.p.) critique of overconsumption, an argument that was vocalized within months of the closure of the international borders in early 2020, where he states:

> To the degree that contemporary consumerism was becoming excessive it was verging on what Marx described as 'overconsumption and insane consumption, signifying, by its turn to the monstrous and the bizarre, the downfall' of the whole system. The recklessness of this overconsumption has played a major role in environmental degradation. The cancellation of airline flights and radical curbing of transportation and movement has had positive consequence with respect to greenhouse gas emissions. … Ecotourist sites will have a time to re-cover from trampling feet … To the degree that the taste for reckless and senseless overconsumerism is curbed, there could be some long-term benefits.

We have witnessed overconsumption being played out in terms of overtourism (Milano et al. 2018), including the way in which some Pacific Island states, especially the Cook Islands, Fiji, Palau and Niue have been displaying elements (and impacts) of mass tourism prior to the pandemic (see Chapter 7, especially, and also Chapters 1[illegible], 18 and 25). In terms of tourism development, overconsumption applies not only to a high quantity of tourists visiting an island destination but also to 'high-end and low volume' forms of tourism, which can have a profound impact on the destination and natural resources (see Chapter 18). On the point of the overconsumption of resources, there is considerable scope for environmental and scientific-based research to rigorously engage in research agendas focusing on clean energy use and effective forms of energy planning within the hotel and transport sectors and encouraging the Pacific Islands to strengthen their own role in confronting climate change.

For environmentally responsible forms of tourism to be economically sustainable, the role of the 'blue economy' for Pacific Island states should be evaluated in terms of its ability to protect the environment and simultaneously produce revenue directly for island communities. Phelan et al. (2020: 1665) emphasize that coastal and marine-based tourism initiatives may be able to seek economic leverage through utilizing ocean resources for 'sustainable economic development', 'improved livelihoods' and 'ocean ecosystem health'. International organizations are also optimistic about the future of the blue economy for SIDS, where it is suggested that ocean economies can create opportunities to confront the economic challenges associated with the pandemic and the continual burden of debt (OECD 2021)(see Chapter 1). Nonetheless, the jury is still out on whether the blue economy will be the 'golden goose' or the 'lame duck' of sustainable tourism, particularly within the context of small Pacific Island states. What we do know is that coastal and marine tourism has historically been limited in terms of the 'sharing of benefits and value of tourism among local communities' (Sotiriadis and Shen 2020: 2). As these authors imply, it is important that research addresses practical outcomes and solutions for sustainable tourism initiatives which prioritize the economic and environmental needs of local communities.

Gender, frontline workers and social equality

There is considerable opportunity for tourism research to concentrate on frontline workers in the tourism and hospitality industries in Pacific Island states and especially in terms of pursuing an emancipatory research approach (Mura and Wijesinghe 2021; Noel 2016). Connell and Taulealo (2021) draw attention to the challenges that the pandemic has caused to female employees in the tourism industry, noting that women represented 80% of the formal tourism industry workforce in Samoa, and, thus, the need for tourist corridors with other tourism-generating countries is economically vital for women workers, as they have struggled financially during the pandemic. Tourism studies has, for some time, recognized the challenges that women face in the tourism industry (Kinnaird and Hall 1994), though regional case illustrations in Pacific Island states could be more forthcoming, especially in terms of deeply assessing the extent to which gender inequalities and inequities exist within the tourism and hospitality industry. In Fiji, only one-quarter of the tourism sector's managerial and professional positions are women, and women are principally employed in such minimum wage jobs as front desk work and cleaning (Hamilton 2020). For an effective emancipatory approach to research, the objective is to work towards informing social policy and labour organizations representing the rights of women in the workplace, though levels of labour representation vary in each Pacific Island state.

As tourism has been a major employer in the Pacific Island region, particularly in the Cook Islands, Fiji, Palau, Samoa and Vanuatu, it would be advantageous to conduct comparative investigations between such countries to look at organizational business culture and human resource practices in, for instance, the hotel sector. It would be valuable to draw attention to hiring and promotion processes as well as opportunities and initiatives for encouraging women to take up supervisory and managerial positions. What could be equally revealing would be to look at comparative data sets concerning salaries and equal opportunity practices (and directives) across the Pacific Island states as well as draw attention to women working in smaller businesses and accommodation units, for which monitoring of equal opportunities is more difficult to track in contrast to larger organizations. In addition to the Vanuatu study presented in this volume, which examined the challenges for women in managing small accommodation units (see Chapter 12), there is emerging tourism research identifying the catalysts for change, as research is advancing in the field of female empowerment and entrepreneurship in the tourism industry (Movono and Dahles 2017; Persson et al. 2022; Trupp et al. 2021). Whilst this development is encouraging, the social, occupational and financial disparities faced by women in the tourism and hospitality sector warrants critical treatment and potential recommendations for justice-based solutions.

Travel and tourism mobilities of Pacific Islanders

Tourism enquiries concerning the Pacific Islands have focused predominantly on the supply-side perspective of tourism, and this is certainly reflected in this volume, on which contributing chapters concern the destination, its attributes and resources, with some attention to the role of tourist patterns. This perhaps reflects the approach of Pacific tourism studies in general, where there is far more focus on the Pacific Islands as tourism-receiving destinations rather than recognizing their role as tourist-generating societies. Consequently, there are overwhelming opportunities to conduct research on Pacific Islanders' perceptions and experiences of travel and tourism – who are commonly perceived as 'tourees' rather than as 'tourists' (see van den Berghe and Keyes 1984). However, Hau'ofa, (1994: 154) reminds us that that Pacific Islanders have a deep history of travel and voyaging, stating:

> From one island to another they sailed to trade and to marry, thereby expanding social networks for greater flows of wealth. They travelled to visit relatives in a wide variety of natural and cultural surroundings, to quench their thirst for adventure, and even to fight and dominate.

The type of tourism involving Pacific Islanders that has been subjected to investigation concerns the tourism behaviour and travel patterns of migrant returnees and home-comers, i.e., the 'visiting of friends and relatives' (VFR) market, which largely, though not exclusively, concentrates on the Polynesian diaspora (Cave and Hall 2015; Cave and Koloto 2015; Hall and Duval 2004; Takahashi 2021; Taufatofua and Craig-Smith 2010). However, the VFR market has been less concerned with investigating Pacific Islander experiences of outbound travel for the purpose of visiting relatives and family members in New Zealand or Australia, for instance. From the few enquiries that are available, we do know that outbound travel by Pacific Islanders can have economic and policy implications (Hazledine and Collins 2011) as well as socio-cultural consequences (Schänzel et al. 2014). The travel preferences of Pacific Islanders are more likely to be associated with culture, religion, ethnicity and kinship than the more self-oriented objectives and motivations (Gibson et al. 2020; Trupp et al. 2022; Trupp and Stephenson 2018). Particular islands have their own socio-cultural idiosyncrasies when it comes to travel. Cook Islanders, for instance, participate in 'travelling parties', involving cultural activities or having family reunions in New Zealand (Alexeyeff 2009). Yet more empirically based enquiries are needed to understand the travel behaviour of Pacific Islanders and how outbound travel can help sustain familial, cultural and social obligations, indicating that VFR travel is not a one-way process.

It would be edifying to understand more the circular movements of Pacific Islanders, whether this involves travelling overseas for work, visiting family or friends oversees, travelling domestically to other islands and urban to rural travel. Intriguingly, the latter form of travel has recently emerged as a distinct category during the COVID-19 pandemic. Movono et al. (2022: 5) note that, in Fiji, for instance, there has been a 'return to the *vanua*' (Fijian for 'land'), particularly as people move back to their family villages and reconnect with agricultural and planting activities. Moreover, as a response to undertourism as a consequence of the global pandemic lockdown, Pacific Island states have been more proactive in promoting domestic tourism, as we have seen in Fiji (Chapter 15) and in the Solomon Islands (Chapter 10). Although focusing on the tourism habits and experiences of Pacific Islanders is crucial in representing an emancipatory vision of the role of research (and tourism), people's immobilities cannot be overlooked. The economic pressures of Pacific Island life, such as rural unemployment, underemployment and poverty, have an unfortunate effect on people's social and human right to enjoy tourism and travel (Bianchi and Stephenson 2014). These issues would contribute to the study of 'social tourism' (Diekmann and McCabe 2011) but within a non-Western and subaltern context.

Indigenous methodologies and local diversities

Chambers and Buzinde (2015: 3) infer that there is more opportunity for critical tourism studies to ensure 'engagement with indigenous and local peoples and epistemologies in the co-creation of tourism knowledge'. These authors call for decolonial perspectives and approaches which 'enable us to envisage other ways of thinking, being and knowing better about tourism' (2015: 4). Helu-Thaman (2003: 4) argues for the decolonization of knowledge through the representation of Indigenous people's interpretations of their histories, knowledge and traditions, emphasizing that understanding Indigenous knowledge represents a 'mutually beneficial collaboration between indigenous and nonindigenous peoples', especially as it 'improves their treatment of each other as equals' (2003: 11). The emphasis is for tourism studies to access the cultural and parochial knowledge of Pacific Island communities whilst respecting the integrity of the local situational context and communication norms as well as ensuring fair representation of others. This could be enabled through the increasing use of culturally driven and locally recognized approaches, such as the conversation and sharing of ideas (known as *Talanoa* in such Pacific Islands as Fiji and Samoa) or interactive talking with a purpose (known as *Talanga* in Tonga) (Ofanoa et al. 2015; Vaioleti 2006).

There are some evolving enquiries concerning Indigenous tourism, communities and enterprises (Hutchison et al. 2021; Pratt et al. 2013; Trupp et al. 2021), in which localized issues and knowledge can be sought, represented and circulated to inform tourism policy and planning. Accordingly, it is argued that research is required to examine the Sustainable Development Goals (SDGs) in a way which they positively reflect 'Indigenous people's priorities and aspirations' (Scheyvens et al. 2021: 31; see also Chapter 2). Therefore, although the SDGs are based on good intentions, there is a need to examine the extent to which the tourism industry is capable of ensuring full application of the SDGs and whether the particularistic needs, interests, perceptions, values and structural circumstances of Pacific Islanders actually align to the SDGs (see Boluk et al. 2019). As we are now halfway towards the deadline for Agenda 2030 and for nations to satisfy the SDGs, we enter a critical stage in this process. As the global pandemic aggravated steps towards advancing socio-economic forms of sustainable development (see Nhamo et al. 2020), there is an urgent need to engage with the local communities to look at the main concerns ahead. This will help identify the gaps and the strategies to move forward.

At this juncture, however, it is imperative to mention that more conceptual awareness and empirical investigation is required to decipher what we actually mean by the 'indigenous perspective' or 'indigenous tourism', for that matter. Accordingly, there is a politically sensitive need to epistemologically deconstruct who represents the Indigenous voices – who is included and who is excluded. Fundamentally, tourism research should also pay critical attention to the ethnic composition of island inhabitants, especially as island nations are by no means solely composed of a homogenous grouping. When the terminology of 'indigenous' is articulated in the Pacific Island region, it is popularly interpreted as having entrenched geographical, ethnic and socio-cultural ties to a given space or territory. This, in itself, can drive destination marketers and the tourism industry to use and misuse indigeneity as a marketing tactic, and those not considered to be Indigenous, such as Fijians of Indian descent, for instance, are marginalized from the 'paradisiacal' tourism image of the South Pacific (White 2005: 168–169).

For an emancipatory vision of tourism research to fully evolve in the region, we do need to investigate the multi-ethnic ('local') perspectives of tourism, acknowledging the significant and long-term presence of 'ethnic others' in the Pacific, paying special attention, for instance, to the Gilbertise communities who have lived in the Solomon Islands since the 1960s, the Banabans who have lived in Fiji since the 1940s, the older Chinese community who has migrated to the Solomon Islands since the 1930s and, of course, the large Fijian community of Indian descent residing in Fiji since the 1880s. Understanding these and other voices in tourism will at least help us to appreciate far more the complexities, diversities and pluralities of Pacific Island states. Moreover, it may enable us to unveil Indigenous approaches and methodologies more carefully, also taking on board the wider *local* voices.

This volume has come a long way to exposing many tourism development concerns and challenges faced by small Pacific Island states, being able to also strategically construct 'ways forward' for both the tourism industry and tourism development. Nevertheless, as this concluding chapter implies, research and critical enquiry still has a long journey ahead, especially to fully comprehend ways in which Pacific Island states can navigate favourable tourism pathways towards self-determination, sustainable development, socio-cultural advancement and economic autonomy.

References

Alexeyeff, K. (2009) 'Travelling parties: Cook Islanders transnational movement'. In: H. Lee and S.T. Francis (eds.) *Migration and Transnationalism: Pacific Perspectives*, Canberra: ANU Press, pp. 91–102.

Armstrong, H.W. and R. Read (2021) 'The non-sovereign territories: Economic and environmental challenges of sectoral and geographic over-specialisation in tourism and financial services', *European Urban and Regional Studies*, 28(3): 213–240.

Barnett, J. and C. McMichael (2018) 'The effects of climate change on the geography and timing of human mobility', *Population and Environment*, 39(4): 339–356.

Bertana, A. (2020) 'The impact of faith-based narratives on climate change adaptation in Narikoso, Fiji', *Anthropological Forum*, 30(3): 254–273.

Bianchi, R. and M.L. Stephenson (2014) *Tourism and Citizenship: Rights, Freedoms and Responsibilities in the Global Order*, London: Routledge.

Boluk, K.A., C.T. Cavaliere and F. Higgins-Desbiolles (2019) 'A critical framework for interrogating the United Nations sustainable development goals 2030 agenda in tourism', *Journal of Sustainable Tourism*, 27(7): 847–864.

Cave, J. and C.M. Hall (2015) 'Do families hold the Pacific together? VFR, voyaging, and new expressions of diasporic networks'. In: E. Backer and B. King (eds.) *VFR Travel Research: International Perspectives*, Bristol: Channel View, pp. 187–204.

Cave, J. and A.H.A.I. Koloto (2015) 'Short-term visits and Tongan livelihoods Enterprise and transnational exchange', *Population, Space and Place*, 21(7): 669–688.

Chambers, D. and C. Buzinde (2015) 'Tourism and decolonisation: Locating research and self', *Annals of Tourism Research*, 51: 1–16.

Christensen, A.E. and O. Mertz (2010) 'Researching Pacific island livelihoods: Mobility, natural resource management and nissology', *Asia Pacific Viewpoint*, 51(3): 278–287.

Connell, J. and T. Taulealo (2021) 'Island tourism and COVID-19 in Vanuatu and Samoa: An unfolding crisis', *Small States & Territories*, 4(1): 105–124.

Diekmann, A. and S. McCabe (2011) 'Systems of social tourism in the European Union: A critical review', *Current Issues in Tourism*, 14(5): 417–430.

Ellison, A.M., A.J. Felson and D.A. Friess (2020) 'Mangrove rehabilitation and restoration as experimental adaptive management', *Frontiers in Marine Science*, 15 May. Retrieved online at https://www.frontiersin.org/articles/10.3389/fmars.2020.00327/full (accessed 26 May 2020).

Fijian Government (2021) 'Environment ministry secures funding from Bezos Earth Fund to protect and restore mangroves in Fiji', 26 July. Available online at https://www.fiji.gov.fj/media-centre/news/environment-ministry-secures-funding-from-bezos-ea (accessed 8 June 2022).

Gibson, D., S. Pratt and B.L. Iaquinto (2020) 'Samoan perceptions of travel and tourism mobilities–the concept of Malaga', *Tourism Geographies*: 1–22. https://doi.org/10.1080/14616688.2020.1780632.

Gunia, A. (2022) 'An archipelago in the South Pacific is becoming the newest scene of tensions between China and the U.S', *Time*, 20 April. Available online at https://time.com/6168173/solomon-islands-china-security-pact/ (accessed 7 June 2022).

Hall, C.M. and D.T. Duval (2004) 'Linking diasporas and tourism: Transnational mobilities of Pacific islanders resident in New Zealand'. In: T. Coles and D.J. Timothy (eds.) *Tourism, Diasporas and Space*, London: Routledge, pp. 92–108.

Hamilton, W. (2020) 'Action on Covid-19 and gender: A policy review from Fiji', *ALiGN -Advancing Learning and Innovation on Gender Norms*, May. Available online at https://www.alignplatform.org/sites/default/files/2020-05/align_fiji_pacific_report_0.pdf (accessed 15 June 2022).

Harvey, D. (2020) 'Anti-capitalist politics in the time of COVID-19', *Jacobin Magazine*, Available online at https://jacobinmag.com/2020/03/david-harvey-coronavirus-political-economy-disruptions (accessed 20 January 2022).

Hau'ofa, E. (1994) 'Our sea of islands', *Contemporary Pacific*, 6(1): 148–161.

Hazledine, T. and S. Collins (2011) 'Paying the pilot? The economics of subsidising international air travel to small remote island nations with large diaspora', *Journal of Air Transport Management*, 17(3): 187–194.

Helu-Thaman, K. (2003) 'Decolonizing Pacific studies: Indigenous perspectives, knowledge, and wisdom in higher education', *The Contemporary Pacific*, 15(1): 1–17.

Hutchison, B., A. Movono and R. Scheyvens (2021) 'Resetting tourism post-COVID-19: Why Indigenous peoples must be central to the conversation', *Tourism Recreation Research*, 46(2): 261–275.

Ichikawa, S., S. Onaka, T. Uda, J. Hirano and H. Sawada (2017) 'Approaches to establish a community-based beach management in the Pacific island country'. In: *Asian and Pacific Coasts, Proceedings of the 9th International Conference on APAC 2017*, pp. 663–674.

IUCN (International Union for Conservation of Nature) (2022) 'Improving the sustainable use of groundwater in Tongatapu', 23 March. Available online at iucn.org/news/oceania/202203/improving-sustainable-use-groundwater-tongatapu (accessed 26 May 2022).

Jiang, M., E. Calgaro, L.M. Klint and D. Dominey-Howes (2015) 'Understanding climate change vulnerability and resilience of tourism destinations: An example of community-based tourism in Samoa'. In: D. Harrison (ed.) *Tourism in Pacific Islands*, London: Routledge, pp. 263–280.

Kinnaird, V. and D. Hall (eds.) (1994) *Tourism: A Gender Analysis*, Chichester: Wiley.

McKenna, K. and E. Ariku (2022) 'Bougainville independence: Recalling promises of international help', *The Interpreter*, 19 November. Available online at https://www.lowyinstitute.org/the-interpreter/bougainville-independence-recalling-promises-international-help (accessed 11 June 2022).

Milano, C., J.M. Cheer and M. Novelli (2018) 'Overtourism: A growing global problem', *The Conversation*, 18 July: 1–5.

Milne, S. (1992) 'Tourism and development in South Pacific microstates', *Annals of Tourism Research*, 19(2): 191–212.

Movono, A. and H. Dahles (2017) 'Female empowerment and tourism: A focus on businesses in a Fijian village', *Asia Pacific Journal of Tourism Research*, 22(6): 681–692.

Movono, A., R. Scheyvens and S. Auckram (2022) 'Silver linings around dark clouds: Tourism, Covid-19 and a return to traditional values, villages and the vanua', *Asia Pacific Viewpoint*: 1–16. https://doi.org/10.1111/apv.12340.

Mura, P. and S.N. Wijesinghe (2021) 'Critical theories in tourism–a systematic literature review', *Tourism Geographies*, 1–21. https://doi.org/10.1080/14616688.2021.1925733.

Nhamo, G., K. Dube and D. Chikodzi (2020) 'Global tourism value chains, sustainable development goals and COVID-19'. In: G. Nhamo, K. Dube and D. Chikodzi (eds.) *Counting the Cost of COVID-19 on the Global Tourism Industry*, Cham: Springer, pp. 27–51.

Noel, L. (2016) 'Promoting an emancipatory research paradigm in design education and practice'. In: P. Lloyd and E. Bohemia (eds.), *Future Focused Thinking - DRS International Conference 2016*, 27–30 June, Brighton, United Kingdom.

Nunn, P.D. (2017) 'Sidelining God: Why secular climate projects in the Pacific islands are failing', *The Conversation*, 16 May. Available online at https://theconversation.com/sidelining-god-why-secular-climate-projects-in-the-pacific-islands-are-failing-77623 (accessed 21 May 2022).

OECD (Organisation for Economic Co-operation and Development) (2021) 'COVID-19 Pandemic: Towards a blue recovery in small island developing states', 26 January. Available online at https://www

.oecd.org/coronavirus/policy-responses/covid-19-pandemic-towards-a-blue-recovery-in-small-island-developing-states-241271b7/ (accessed 8 June 2022).

Ofanoa, M., T. Percival, P. Huggard and S. Buetow (2015) 'Talanga: The Tongan way enquiry', *Sociology Study*, 5(4): 334–340.

Penz, P. (1996) 'Environmental victims and state sovereignty: A normative analysis', *Social Justice*, 23(4): 41–61.

Persson, K., K. Zampoukos and I. Ljunggren (2022) 'No (wo) man is an island–socio-cultural context and women's empowerment in Samoa', *Gender, Place and Culture*, 29(4): 482–501.

Phelan, A.A., L. Ruhanen and J. Mair (2020) 'Ecosystem services approach for community-based eco-tourism: Towards an equitable and sustainable blue economy', *Journal of Sustainable Tourism*, 28(10): 1665–1685.

Pigliasco, G. (2009) 'Intangible cultural property, tangible databases, visible debates: The Sawau Project', *International Journal of Cultural Property*, 16(3): 255–272.

Pratt, S., D. Gibson and A. Movono (2013) 'Tribal tourism in Fiji: An application and extension of Smith's 4Hs of indigenous tourism', *Asia Pacific Journal of Tourism Research*, 18(8): 894–912.

Schänzel, H.A., M. Brocx and L. Sadaraka (2014) '(Un)conditional hospitality: The host experience of the Polynesian community in Auckland', *Hospitality and Society*, 4(2): 135–154.

Scheyvens, R., A. Carr, A. Movono, E. Hughes, F. Higgins-Desbiolles and J.P. Mika (2021) 'Indigenous tourism and the sustainable development goals', *Annals of Tourism Research*, 90. https://doi.org/10.1016/j.annals.2021.103260.

Scheyvens, R. and A. Movono (2018) 'Development and change: Reflections on tourism in the South Pacific', *Development Bulletin*, 80: 134–139.

Sotiriadis, M. and S. Shen (2020) 'Blue economy and sustainable tourism management in coastal zones: Learning from experiences', *ADBI Working Paper 1174*. Tokyo: Asian Development Bank Institute. Available online at https://www.adb.org/publications/blue-economysustainable-tourism-management-coastal-zone (accessed 6 June 2022).

Takahashi, K. (2021) 'The impact of migrants on tourism demand: The case of Fiji', *E-Review of Tourism Research*, 18(5): 779–793.

Taufatofua, R.G. and S. Craig-Smith (2010) 'The socio-cultural impacts of visiting friends and relatives on hosts: A Samoan study', *WIT Transactions on Ecology and the Environment*, 130: 89–100.

Tolkach, D. and S. Pratt (2019) 'Globalisation and cultural change in Pacific island countries: The role of tourism', *Tourism Geographies*, *23*(3): 371–396.

Trupp, A., I. Matatolu and A. Movono (2021) 'Gender and benefit-sharing in Indigenous tourism microentrepreneurship'. In: D.B. Morais (ed.), *Tourism Microentrepreneurship*, Bingley: Emerald, pp. 51–62.

Trupp, A., S. Pratt, M.L. Stephenson, I. Matatolu and D. Gibson (2022) 'Representing and evaluating the travel motivations of Pacific islanders', *International Journal of Tourism Research*, 24: 653–666.

Trupp, A. and M.L. Stephenson (2018) 'Tourism mobilities and immobilities from a South Pacific islands perspective', *Proceedings of the Travel and Tourism Research Association Asia Pacific, Chapter 6th Annual Conference*, Ho Chi Minh City, Vietnam, pp. 108–111.

UNESCO (United Nations Educational, Scientific and Cultural Organization) (2015) *Pacific World Heritage Action Plan 2016–2020*, UNESCO. Available online at https://whc.unesco.org/document/142213 (accessed 9 September 2019).

UNESCO (United Nations Educational, Scientific and Cultural Organization) (n.d.) 'Tentative lists'. Available online at https://whc.unesco.org/en/tentativelists/ (accessed 13 June 2022).

Vaioleti, T.M. (2006) 'Talanoa research methodology: A developing position on Pacific research', *Waikato Journal of Education*, 12(1): 21–34.

van den Berghe, P. and C. Keyes (1984) 'Introduction: Tourism and re-created ethnicity', *Annals of Tourism Research*, 11(3): 343–352.

Weir, T. and J. Pittock (2017) 'Human dimensions of environmental change in small island developing states: Some common themes', *Regional Environmental Change*, 17(4): 949–958.

White, C. (2005) 'Tourism as an ethnic landscape and the landscape of ethnic tourism: The case of Fiji', *Race, Gender and Class*, 12(3–4): 155–175.

Wong, E., M. Jiang, L. Klint, T. DeLacy, D. Harrison and D. Dominey-Howes (2013) 'Policy environment for the tourism sector's adaptation to climate change in the South Pacific – The case of Samoa', *Asia Pacific Journal of Tourism Research*, 18(1–2): 52–71.

INDEX

Note: Page numbers in *italics* indicate figures, **bold** indicate tables in the text